Compendium of Organic Synthetic Methods

Compendium of Organic Synthetic Methods

Volume 8

MICHAEL B. SMITH

DEPARTMENT OF CHEMISTRY
THE UNIVERSITY OF CONNECTICUT
STORRS, CONNECTICUT

A Wiley-Interscience Publication

JOHN WILEY & SONS, INC.
New York • Chichester • Brisbane • Toronto • Singapore

Cover illustration was adapted from "Disconnect By the Numbers: A Beginner's Guide to Synthesis" by M. B. Smith. *Journal of Chemical Education*, **1990**, 67, 848–856.

This text is printed on acid-free paper.

Library of Congress Catalog Card Number: 71-162800

ISBN 0-471-57319-1

Printed in the United States of America

10 9 8 7 6 5 4 3 2 1

DEDICATION

This book is dedicated to Dr. Ted Hoffman, who recently retired as Senior Editor at Wiley-Interscience after a long and distinguished career. Ted was responsible for getting the *Compendium* started and for keeping it going for more than twenty years. The many books and series for which Ted served as editor have contributed enormously both to chemical education and to chemical research. It would be difficult to find a bookshelf in an organic chemistry research lab anywhere in the world that does not contain one or more of the books for which Ted served as editor. On a more personal level, the publication of Volume 6 of this series in 1988 began my career as an author, and many new opportunities have been opened to me because of this. There are certainly many others who can similarly thank Ted.

I want to use this dedication to thank Ted for his help to me, for his help to the previous authors of the *Compendium*, and for his contributions to the chemistry community in general (particularly the organic chemistry community). Those contributions are many and important. Thank you, and good luck in your retirement.

MICHAEL B. SMITH

PREFACE

Since the original volume in this series by Ian and Shuyen Harrison, the goal of the *Compendium of Organic Synthetic Methods* was to facilitate the search for functional group transformations in the original literature of Organic Chemistry. In Volume 2, difunctional compounds were added and this compilation was continued by Louis Hegedus and Leroy Wade for Volume 3 of the series. Wade became the author for Volume 4 and continued with Volume 5. I began editing the series with Volume 6, where I introduced an author index for the first time and added a new chapter (Chapter 15, Oxides). Volume 7 introduced Section 378 (Oxides–Alkynes) through Section 390 (Oxides–Oxides). The *Compendium* is a handy desktop reference that will remain a valuable tool to the working organic chemist, allowing a "quick check" of the literature. It also allows one to "browse" for new reactions and transformations that may be of interest. The body of organic literature is very large and the *Compendium* is a focused and highly representative review of the literature and is offered in that context.

Compendium of Organic Synthetic Methods, Volume 8, contains both functional group transformations and carbon–carbon bond forming reactions from the literature appearing in the years 1990, 1991, and 1992. The classification schemes used for Volumes 6 and 7 have been continued. Difunctional compounds appear in Chapter 16. The experienced user of the *Compendium* will require no special instructions for the use of Volume 8. Author citations and the Author Index have been continued as in Volumes 6 and 7.

Every effort has been made to keep the manuscript error free. Where there are errors, I take full responsibility. If there are questions or comments, the reader is encouraged to contact me directly at the address, phone, fax, or E-mail addresses given below. Note that the phone and fax area code for this section of Connecticut is tentatively scheduled to change in 1996 to (860).

The manuscript for Volume 8 was prepared on a Macintosh-IIci™ PC using Microsoft Word™ (version 5.1) for word processing. All structures were prepared using ChemDraw™ (version 3.1) and the manuscript was printed with a LaserWriter™ II printer. The graph presented in the introduction was prepared using Cricket-Graph.™

As I have throughout my writing career, I want to thank my wife Sarah and my son Steven, who have shown unfailing patience and devotion during this work. I also thank Ms. Carla Fjerstad, who edited this volume and was responsible for its publication.

MICHAEL B. SMITH

University of Connecticut
Department of Chemistry
215 Glenbrook Rd., Room 151
Storrs, CT USA 06269-3060

Voice Phone: 203-486-2881
Fax: 203-486-2981
E-mail: MBSMITH@UCONNVM.edu

CONTENTS

ABBREVIATIONS

Ac	Acetyl	O, CH_3
acac	Acetylacetonate	
AIBN	*azo-bis*-Isobutyronitrile	
aq.	Aqueous	
B	9-Borabicyclo[3.3.1]nonyl	
9-BBN	9-Borabicyclo[3.3.1]nonyl	
BINAP	*2R,3S*-2,2'-*bis*-(diphenylphosphino)-1,1'-binapthyl	
Bn	Benzyl	
Bz	Benzoyl	
BOC	*t*-Butoxycarbonyl	O, O*t*-Bu
Bpy (Bipy)	2,2'-Bipyridyl	
Bu	*n*-Butyl	$-CH_2CH_2CH_2CH_3$
CAM	Carboxamidomethyl	
CAN	Ceric ammonium nitrate	$(NH)_2Ce(NO_3)_6$
c-	Cyclo-	
Cat.	Catalytic	
Cbz	Carbobenzyloxy	O, OCH_2Ph
Chirald	2S,3R-(+)-4-Dimethylamino-1,2-diphenyl-3-methylbutan-2-ol	
COD	1,5-Cyclooctadienyl	
COT	1,3,5-Cyclooctatrienyl	
Cp	Cyclopentadienyl	
CSA	Camphorsulfonic acid	
CTAB	cetyltrimethylammonium bromide	$C_{16}H_{33}NMe_3^+ Br^-$
Cy (*c*-C6H11)	Cyclohexyl	
°C	Temperature in Degrees Centigrade	
DABCO	1,4-Diazabicyclo[2.2.2]octane	
Dba	Dibenzylidene acetone	
DBE	1,2-Dibromoethane	$BrCH_2CH_2Br$
DBN	1,5-Diazabicyclo[4.3.0]non-5-ene	
DBU	1,8-Diazabicyclo[5.4.0]undec-7-ene	
DCC	1,3-Dicyclohexylcarbodiimide	$c\text{-}C_6H_{13}\text{-}N{=}C{=}N\text{-}c\text{-}C_6H_{13}$
DCE	1,2-Dichloroethane	$ClCH_2CH_2Cl$

DDQ	2,3-Dichloro-5,6-dicyano-1,4-benzoquinone	
% de	% Diastereomeric excess	
DEA	Diethylamine	$HN(CH_2CH_3)_2$
DEAD	Diethylazodicarboxylate	$EtO_2C\text{-}N{=}NCO_2Et$
Dibal-H	Diisobutylaluminum hydride	$(Me_2CHCH_2)_2AlH$
Diphos (**dppe**)	1,2-*bis*-(Diphenylphosphino)ethane	$Ph_2PCH_2CH_2PPh_2$
Diphos-4 (**dppb**)	1,4-*bis*-(Diphenylphosphino)butane	$Ph_2P(CH_2)_4PPh_2$
DMAP	4-Dimethylaminopyridine	
DME	Dimethoxyethane	$MeOCH_2CH_2OMe$
DMF	*N,N'*-Dimethylformamide	$HC(=O)N(CH_3)_2$
Dmp	*bis*-[1,3-Di(*p*-methoxyphenyl)-1,3-propanedionato]	
Dpm	Dipivaloylmethanato	
Dppb	1,4-*bis*-(Diphenylphosphino)butane	$Ph_2P(CH_2)_4PPh_2$
Dppe	1,2-*bis*-(Diphenylphosphino)ethane	$Ph_2PCH_2CH_2PPh_2$
Dppf	*bis*-(Diphenylphosphino)ferrocene	
Dppp	1,3-*bis*-(Diphenylphosphino)propane	$Ph_2P(CH_2)_3PPh_2$
Dvb	Divinylbenzene	
e^-	Electrolysis	
% ee	% Enantiomeric excess	
EE	1-Ethoxyethyl	EtO(Me)CHO-
Et	Ethyl	$-CH_2CH_3$
EDA	Ethylenediamine	$H_2NCH_2CH_2NH_2$
EDTA	Ethylenediaminetetraacetic acid	
FMN	Flavin mononucleotide	
Fod	*tris*-(6,6,7,7,8,8,8)-Heptafluoro-2,2-dimethyl-3,5-octanedionate	
Fp	Cyclopentadienyl-*bis*-carbonyl iron	
FVP	Flash Vacuum Pyrolysis	
h	Hour (hours)	
hν	Irradiation with light	
1,5-HD	1,5-Hexadienyl	
HMPA	Hexamethylphosphoramide	$(Me_3N)_2P{=}O$
HMPT	Hexamethylphosphorus triamide	$(Me_3N)_3P$
iPr	Isopropyl	$-CH(CH_3)_2$
LICA (LIPCA)	Lithium cyclohexylisopropylamide	
LDA	Lithium diisopropylamide	$LiN(iPr)_2$
LHMDS	Lithium hexamethyldisilazide	$LiN(SiMe_3)_2$
LTMP	Lithium 2,2,6,6-tetramethylpiperidide	
MABR	Methylaluminum *bis*-(4-bromo-2,6-di-*tert*-butylphenoxide)	
MAD	*bis*-(2,6-Di-*t*-butyl-4-methylphenoxy)methyl aluminum	
mCPBA	*meta*-Chloroperoxybenzoic acid	
Me	Methyl	$-CH_3$
MEM	β-Methoxyethoxymethyl	$MeOCH_2CH_2OCH_2-$
Mes	Mesityl	2,4,6-tri-Me-C_6H_2

MOM	Methoxymethyl	$MeOCH_2$-
Ms	Methanesulfonyl	CH_3SO_2-
MS	Molecular Sieves (3Å or 4Å)	
MTM	Methylthiomethyl	$MeSCH_2$-
NAD	Nicotinamide adenine dinucleotide	
NADP	Sodium triphosphopyridine nucleotide	
Napth	Naphthyl ($C_{10}H_7$)	
NBD	Norbornadiene	
NBS	*N*-Bromosuccinimide	
NCS	*N*-Chlorosuccinimide	
NIS	*N*-Iodosuccinimide	
Ni(R)	Raney nickel	
Oxone	2 $KHSO_5$•$KHSO_4$•K_2SO_4	
 (P)	Polymeric backbone	
PCC	Pyridinium chlorochromate	
PDC	Pyridinium dichromate	
PEG	Polyethylene glycol	
Ph	Phenyl	
PhH	Benzene	
PhMe	Toluene	
Phth	Phthaloyl	
Pic	2-Pyridinecarboxylate	
Pip	Piperidyl	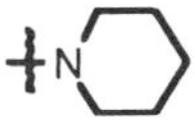 N
Pr	*n*-Propyl	$-CH_2CH_2CH_3$
Py	Pyridine	 N
quant.	Quantitative yield	
Red-Al	$[(MeOCH_2CH_2O)_2AlH_2]Na$	
sBu	*sec*-Butyl	$CH_3CH_2\underline{C}H(CH_3)$
sBuLi	*sec*-Butyllithium	$CH_3CH_2CH(Li)CH_3$
Siamyl	Diisoamyl	$[(CH_3)_2CH\underline{C}H(CH_3)-]_2$
TASF	*tris*-(Diethylamino)sulfonium difluorotrimethyl silicate	
TBAF	Tetrabutylammonium fluoride	$n\text{-}Bu_4N^+ F^-$
TBDMS	*t*-Butyldimethylsilyl	$t\text{-}BuMe_2Si$
TBHP (*t*-BuOOH)	*t*-Butylhydroperoxide	Me_3COOH
t-Bu	*tert*-Butyl	$-C(CH_3)_3$
TEBA	Triethylbenzylammonium	$Bn(Et)_3N^+$
TEMPO	Tetramethylpiperidinyloxy free radical	
TFA	Trifluoroacetic acid	CF_3COOH

TFAA	Trifluoroacetic anhydride	$(CF_3CO)_2O$
Tf (OTf)	Triflate	$-SO_2CF_3$ ($-OSO_2CF_3$)
THF	Tetrahydrofuran	
THP	Tetrahydropyran	
TMEDA	Tetramethylethylenediamine	$Me_2NCH_2CH_2NMe_2$
TMS	Trimethylsilyl	$-Si(CH_3)_3$
TMP	2,2,6,6-Tetramethylpiperidine	
TPAP	tetra-*n*-Propylammonium perruthenate	
Tol	Tolyl	$4-C_6H_4CH_3$
Tr	Trityl	$-CPh_3$
TRIS	Triisopropylphenylsulfonyl	
Ts(Tos)	Tosyl = *p*-Toluenesulfonyl	$4-MeC_6H_4SO_2-$
)))))))	Sonication	
X_c	Chiral auxiliary	

INDEX, MONOFUNCTIONAL COMPOUNDS

Sections—heavy type
Pages—light type

PREPARATION OF →
FROM ↓

	Alkynes	Carboxylic acids, acid halides, anhydrides	Alcohols, phenols	Aldehydes	Alkyls, methylenes, aryls	Amides	Amines	Esters	Ethers, epoxides	Halides, sulfonates	Hydrides (RH)	Ketones	Nitriles	Alkenes	Oxides
Alkynes	**1** 1	**16** 6			**61** 66	**76** 117	**91** 139	**106** 166	**121** 184			**166** 220		**196** 247	
Carboxylic acids acid halides, anhydrides	**2** 2	**17** 6	**32** 16	**47** 54	**62** 67	**77** 117	**92** 139	**107** 166			**152** 210	**167** 220		**197** 251	**212** 267
Alcohols, phenols		**18** 7	**33** 18	**48** 54	**63** 68	**78** 119	**93** 140	**108** 168	**123** 184	**138** 200	**153** 210	**168** 222	**183** 242	**198** 251	**213** 267
Aldehydes	**4** 2	**19** 8	**34** 18	**49** 58	**64** 69	**79** 119	**94** 140	**109** 172	**124** 186	**139** 201	**154** 212	**169** 225	**184** 242	**199** 252	**214** 267
Alkyls, methylenes, aryls		**20** 9	**35** 28	**50** 58	**65** 69							**170** 227		**200** 253	
Amides		**21** 9	**36** 29	**51** 59		**81** 120	**96** 142	**111** 173				**171** 227	**186** 243	**201** 254	
Amines	**7** 3			**52** 59	**67** 70	**82** 127	**97** 144	**112** 174	**127** 188	**142** 202	**157** 213	**172** 228	**187** 243	**202** 254	**217** 268
Esters		**23** 10	**38** 29	**53** 60	**68** 70	**83** 131	**98** 153	**113** 174	**128** 188	**143** 203	**158** 213	**173** 229		**203** 255	
Ethers, epoxides		**24** 11	**39** 30	**54** 61	**69** 71		**99** 153	**114** 176	**129** 189		**159** 214	**174** 230	**189** 244	**204** 255	**219** 269
Halides, sulfonates, sulfates	**10** 3	**25** 11	**40** 35	**55** 61	**70** 72	**85** 132	**100** 153	**115** 177	**130** 191	**145** 203	**160** 215	**175** 231	**190** 244	**205** 256	**220** 272
Hydrides (RH)			**41** 36	**56** 62	**71** 78	**86** 133	**101** 155	**116** 178		**146** 205		**176** 233	**191** 244	**206** 258	**221** 272
Ketones	**12** 4	**27** 13	**42** 39	**57** 63	**72** 79	**87** 133	**102** 155	**117** 179	**132** 193	**147** 207	**162** 218	**177** 234		**207** 259	**222** 273
Nitriles		**28** 13	**43** 45	**58** 63	**73** 80	**88** 134	**103** 157	**118** 182			**163** 219	**178** 237	**193** 245		
Alkenes		**29** 14	**44** 45	**59** 63	**74** 81	**89** 135	**104** 158	**119** 182	**134** 194	**149** 207		**179** 237		**209** 261	
Miscellaneous compounds	**15** 5	**30** 14	**45** 47	**60** 64	**75** 116	**90** 136	**105** 159	**120** 183	**135** 198		**165** 219	**180** 238	**195** 245	**210** 264	**225** 273

PROTECTION

	Sect.	Pg.
Carboxylic acids	30A	14
Alcohols, phenols	45A	48
Aldehydes	60A	64
Amides	90A	138
Amines	105A	163
Ketones	180A	239

Blanks in the table correspond to sections for which no additional examples were found in the literature.

INDEX, DIFUNCTIONAL COMPOUNDS

Sections—heavy type
Pages—light type

Blanks in the table correspond to sections for which no additional examples were found in the literature.

INTRODUCTION

Relationship between Volume 8 and Previous Volumes. *Compendium of Organic Synthetic Methods,* Volume 8, presents about 1400 examples of published reactions for the preparation of monofunctional compounds, updating the 9250 in Volumes 1-7. Volume 8 contains about 1640 examples of reactions which prepare difunctional compounds with various functional groups. Reviews have long been a feature of this series and Volume 8 adds almost 60 pertinent reviews in the various sections. It is interesting that Volume 8 contains approximately 1200 more entries than Volume 7 for an identical three-year period. One or two important new journals appear (such as *SynLett*), but the most dramatic increase was in reactions that prepared difunctional compounds. This increase in reactions led me to "compress" the format of the *Compendium* somewhat, in an attempt to keep the length of the manuscript in line with previous volumes.

Chapters 1-14 continue as in Volumes 1-7, as does Chapter 15, introduced in Volume 6. Difunctional compounds appear in Chapter 16, as in Volumes 6 and 7. The sections on oxides as part of difunctional compounds, introduced in Volume 7, continue in Chapter 16 of Volume 8 with Section 378 (Oxides-Alkynes) through Section 390 (Oxides-Oxides).

Following Chapter 16 is a complete alphabetical listing of all authors (last name, initials). The authors for each citation appear below the reaction. The principle author is indicated by underlining (i.e., Kwon, T. W.; Smith, M. B.), as in Volume 7.

Classification and Organization of Reactions Forming Monofunctional Compounds. Chemical transformations are classified according to the reacting functional group of the starting material and the functional group formed. Those reactions that give products with the same functional group form a chapter. The reactions in each chapter are further classified into sections on the basis of the functional group of the starting material. Within each section, reactions are loosely arranged in ascending order of year cited (1990-1992), although an effort has been made to put similar reactions together when possible. Review articles are collected at the end of each appropriate section.

The classification is unaffected by allylic, vinylic, or acetylenic unsaturation appearing in both starting material and product, or by increases or decreases in the length of carbon chains; for example, the reactions t-BuOH → t-BuCOOH, $PhCH_2OH$ → PhCOOH, and $PhCH{=}CHCH_2OH$ → $PhCH{=}CHCOOH$ would all be considered as preparations of carboxylic acids from alcohols. Conjugate reduction and alkylation of unsaturated ketones, aldehydes, esters, acids, and nitriles have been placed in Sections 74D and 74E (Alkyls from Alkenes), respectively.

The terms hydrides, alkyls, and aryls classify compounds containing reacting hydrogens, alkyl groups, and aryl groups, respectively; for example, $RCH_2\text{-}H \rightarrow RCH_2COOH$ (carboxylic acids from hydrides), $RMe \rightarrow RCOOH$ (carboxylic acids from alkyls), $RPh \rightarrow RCOOH$ (carboxylic acids from aryls). Note the distinction between $R_2CO \rightarrow R_2CH_2$ (methylenes from ketones) and $RCOR' \rightarrow RH$ (hydrides from ketones). Alkylations involving additions across double bonds are found in Section 74 (Alkyls, Methylenes, and Aryls from Alkenes).

The following examples illustrate the classification of some potentially confusing cases:

$RCH{=}CHCOOH \rightarrow RCH{=}CH_2$	Hydrides from carboxylic acids
$RCH{=}CH_2 \rightarrow RCH{=}CHCOOH$	Carboxylic acids from hydrides
$ArH \rightarrow ArCOOH$	Carboxylic acids from hydrides
$ArH \rightarrow ArOAc$	Esters from hydrides
$RCHO \rightarrow RH$	Hydrides from aldehydes
$RCH{=}CHCHO \rightarrow RCH{=}CH_2$	Hydrides from aldehydes
$RCHO \rightarrow RCH_3$	Alkyls from aldehydes
$R_2CH_2 \rightarrow R_2CO$	Ketones from aldehydes
$RCH_2COR \rightarrow R_2CHCOR$	Ketones from ketones
$RCH{=}CH_2 \rightarrow RCH_2CH_3$	Alkyls from alkenes (Hydrogenation of Alkenes)
$RBr + HC{\equiv}CH \rightarrow RC{\equiv}CR$	Acetylenes from halides; also acetylenes from acetylenes
$ROH + RCOOH \rightarrow RCOOR$	Esters from alcohols; also esters from carboxylic acids
$RCH{\equiv}CHCHO \rightarrow RCH_2CH_2CHO$	Alkyls from alkenes (Conjugate Reduction)
$RCH{\equiv}CHCN \rightarrow RCH_2CH_2CN$	Alkyls from alkenes (Conjugate Reduction)

How to Use the Book to Locate Examples of the Preparation or Protection of Monofunctional Compounds. Examples of the preparation of one functional group from another are found in the monofunctional index on p. xvii, which lists the corresponding section and page. Sections that contain examples of the reactions of a functional group are found in the horizontal rows of this index. Section 1 gives examples of the reactions of acetylenes that form new acetylenes; Section 16 gives reactions of acetylenes that form carboxylic acids; and Section 31 gives reactions of acetylenes that form alcohols.

Examples of alkylation, dealkylation, homologation, isomerization, and transposition are found in Sections 1, 17, 33, and so on, lying close to a diagonal of the index. These sections correspond to such topics as the preparation of acetylenes from acetylenes; carboxylic acids from carboxylic acids; and alcohols, thiols, and phenols from alcohols, thiols, and phenols. Alkylations that involve conjugate additions across a double bond are found in Section 74E (Alkyls, Methylenes, and Aryls from Alkenes).

Examples of name reactions can be found by first considering the nature of the starting material and product. The Wittig reaction, for instance, is in Section 199 (Alkenes from Aldehydes) and Section 207 (Alkenes from Ketones). The aldol condensation can be found in the chapters on difunctional compounds in Section 324 (Alcohol, Thiol-Aldehyde) and in Section 330 (Alcohol, Thiol-Ketone).

Examples of the protection of acetylenes, carboxylic acids, alcohols, phenols, aldehydes, amides, amines, esters, ketones, and alkenes are also indexed on p. xvii. Sections (designated with an A: 15A, 30A, etc.) with "protecting group" reactions are located at the end of pertinent chapters.

Some pairs of functional groups such as alcohol, ester; carboxylic acid, ester; amine, amide; and carboxylic acid, amide can be interconverted by simple reactions. When a member of these groups is the desired product or starting material, the other member should also be consulted in the text.

The original literature must be used to determine the generality of reactions, although this is occasionally stated in the citation. This is only done in cases where such generality is stated clearly in the original citation. A reaction given in this book for a primary aliphatic substrate may also be applicable to tertiary or aromatic compounds. This book provides very limited experimental conditions or precautions and the reader is referred to the original literature before attempting a reaction. **In no instance should a citation in this book be taken as a complete experimental procedure. Failure to refer to the original literature prior to beginning laboratory work could be hazardous.** The original papers usually yield a further set of references to previous work. Papers that appear after those publications can usually be found by consulting *Chemical Abstracts* and the *Science Citation Index.*

Classification and Organization of Reactions Forming Difunctional Compounds. This chapter considers all possible difunctional compounds formed from the groups acetylene, carboxylic acid, alcohol, thiol, aldehyde, amide, amine, ester, ether, epoxide, thioether, halide, ketone, nitrile, and alkene. Reactions that form difunctional compounds are classified into sections on the basis of two functional groups in the product that are pertinent to the reaction. The relative positions of the groups do not affect the classification. Thus preparations of 1,2-aminoalcohols, 1,3-aminoalcohols, and 1,4-aminoalcohols are included in a single section (Section 326, Alcohol-Amine). Difunctional compounds that have an oxide as the second group are found in the appropriate section (Sections 278–290). The nitroketone product of oxidation of a nitroalcohol is found in Section 386 (Ketone-Oxide). Conversion of an oxide to another functional group is generally found in the "Miscellaneous" section of the sections concerning monofunctional compounds. Conversion of a nitroalkane to an amine, for example, is found in Section 105 (Amines from Miscellaneous Compounds). The following examples illustrate applications of this classification system:

Difunctional Product	*Section Title*
RC≡C-C≡CR	Acetylene-Acetylene
RCH(OH)COOH	Carboxylic acid-Alcohol
RCH=CHOMe	Ether-Alkene
$RCHF_2$	Halide-Halide
$RCH(Br)CH_2F$	Halide-Halide
$RCH(OAc)CH_2OH$	Alcohol-Ester
$RCH(OH)CO_2Me$	Alcohol-Ester
$RCH{=}CHCH_2CO_2Me$	Ester-Alkene
RCH=CHOAc	Ester-Alkene
$RCH(OMe)CH_2SO_2CH_2CH_2OH$	Alcohol-Ether
$RSO_2CH_2CH_2OH$	Alcohol-Oxide

How to Use the Book to Locate Examples of the Preparation of Difunctional Compounds. The difunctional index on p. xviii gives the section and page corresponding to each difunctional product. Thus Section 327 (Alcohol, Thiol-Ester) contains examples of the preparation of hydroxyesters; Section 323 (Alcohol, Thiol-Alcohol, Thiol) contains examples of the preparation of diols.

Some preparations of alkene and acetylenic compounds from alkene and acetylenic starting materials can, in principle, be classified in either the monofunctional or difunctional sections; for example, the transformation RCH=CHBr → RCH=CHCOOH could be consider as preparing carboxylic acids from halides (Section 25, Monofunctional Compounds) or preparing a carboxylic acid-alkene (Section 322, Difunctional Compounds). The choice usually depends on the focus of the particular paper where this reaction was found. In such cases both sections should be consulted.

Reactions applicable to both aldehyde and ketone starting materials are in many cases illustrated by an example that uses only one of them. Likewise, many citations for reactions found in the Aldehyde-X sections, will include examples that can be placed in the Ketone-X section. Again, the choice is dictated by the paper where the reaction was found.

Many literature preparations of difunctional compounds are extensions of the methods applicable to monofunctional compounds. As an example, the reaction RCl → ROH might be used for the preparation of diols from an appropriate dichloro compound. Such methods are difficult to categorize and may be found in either the monofunctional or difunctional sections, depending on the focus of the original paper.

The user should bear in mind that the pairs of functional groups alcohol, ester; carboxylic acids, ester; amine, amide; and carboxylic acid, amide can be interconverted by simple reactions. Compounds of the type $RCH(OAc)CH_2OAc$ (ester-ester) would thus be of interest to anyone preparing the diol $RCH(OH)CH_2OH$ (alcohol-alcohol).

Sources of Literature Citations. I thought it would be useful for a reader of this *Compendium* to see the distribution of citations used in this book (i.e., which journals have the most new synthetic methodology). As seen in the accompanying graph, *Tetrahedron Letters* and *Journal of Organic Chemistry* account for roughly two-thirds of all the citations in Volume 8. This book was not edited to favor one category or type of article over another. Certainly, my own personal preferences undoubtedly creep into the selection of methods, but I believe that this compilation is an accurate representation of new synthetic methods that appear in the literature. Therefore, I believe the accompanying graph accurately reflects those journals where new synthetic methodology is located. I should point out that the category "23 other journals" includes: *Accts. Chem. Res.; Acta Chem. Scand.; Angew. Chem. Int. Ed. Engl.; Aust. J. Chem.; Bull. Chim. Soc. Belg.; Bull. Chim. Soc. Fr.; Can. J. Chem.; Chem. Ber.; Coll. Czech. Chem. Commun.; Gazz. Chim. Ital.; Heterocycles; Ind. J. Chem.; Isr. J. Chem.; Izv. Akad. Nauk. SSSR; J. Chem. Soc.; J. Het. Chem.; J. Indian Chem. Soc.; Liebigs Ann. Chem.; Org. Prep. Proceed Int.; Perkin Trans. I; Recl. Trav. Chim. Pays-Bas; Tetrahedron Asymmetry;* and, *Zhur. Org. Khim.* In addition, six more journals were examined but no references were recorded.

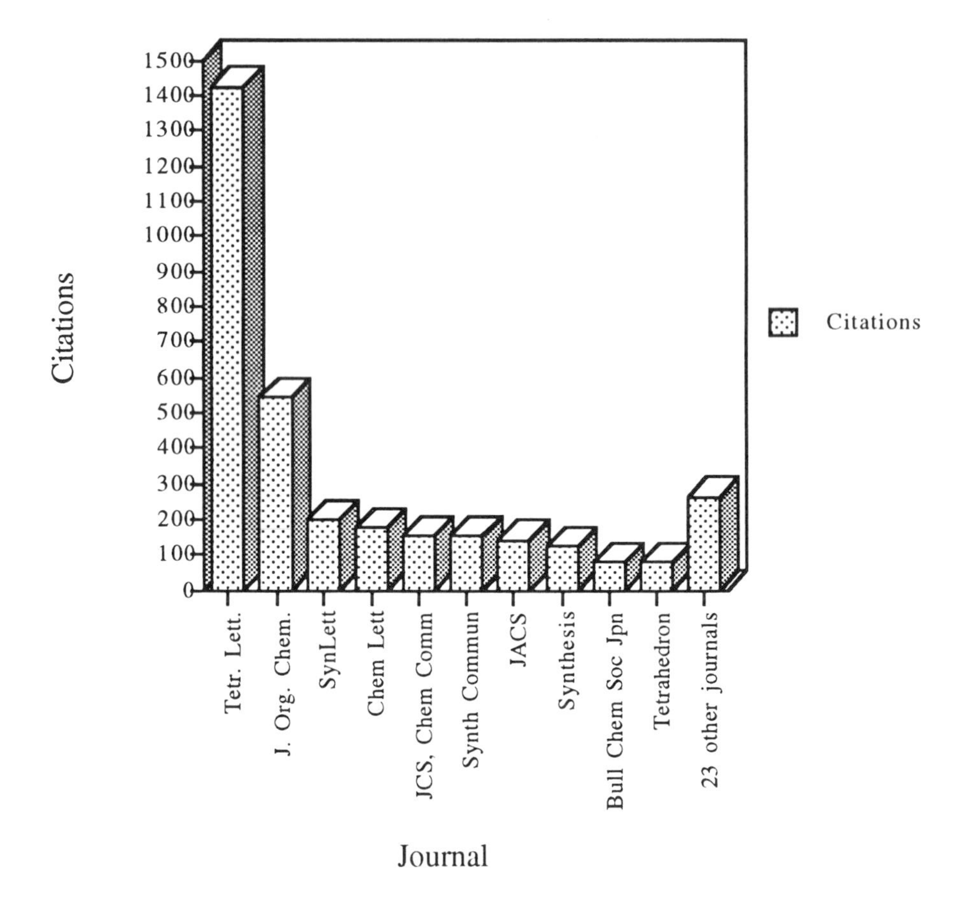
Citations
1500
1400
1300
1200
1100
1000
900
800
700
600
500
400
300
200
100
0
Citations
Tetr. Lett.
J. Org. Chem.
SynLett
Chem Lett
JCS, Chem Comm
Synth Commun
JACS
Synthesis
Bull Chem Soc Jpn
Tetrahedron
23 other journals
Journal

Compendium of Organic Synthetic Methods

CHAPTER 1

PREPARATION OF ALKYNES

SECTION 1: ALKYNES FROM ALKYNES

1. $Co_2(CO)_8$, CH_2Cl_2
2. $BF_3•OEt_2$, CH_2Cl_2
3. CAN, MEOH

MeO HO MeO MeO

(85 : 15) 59%

Grove, D.D.; Miskevich, F.; Smith, C.C.; Corte, J.R. *Tetrahedron Lett.*, ***1990***, *31*, 6277.

Br CN → PhC≡C-H, 10% Pd/C, 4% CuI / 16% PPh_3, NEt_3, MeCN reflux → C≡C-Ph CN 82%

De la Rosa, M.A.; Velarde, E.; Guzmán, A. *Synth. Commun.*, ***1990***, *20*, 2059.

Br → Me_3Si-C≡C-H, CuI, PPh_3 / $Pd(PPh_3)_4$, piperidine → C≡C-$SiMe_3$ 76%

Brandsma, L.; van den Heuvel, H.G.M.; Verkruijsse, H.D. *Synth. Commun.*, ***1990***, *20*, 1889.

Ph-C≡C-H → 1. BuLi, THF 2. $ZnCl_2$ 3. PhI, $Pd(PPh_3)_4$ → Ph-C≡C-Ph 96%

Yoneda, N.; Matsuoka, S.; Miyaura, N.; Fukuhara, T.; Suzuki, A. *Bull. Chem. Soc. Jpn.*, ***1990***, *63*, 2124.

$$O_2N\text{-}C_6H_4\text{-}I \xrightarrow[PdCl_2[PPh_2(m\text{-}C_6H_4SO_3Na)]_2]{HC{\equiv}C\text{-}CH_2OH\text{ , }NEt_3\text{; }H_2O\text{ , }20°C} O_2N\text{-}C_6H_4\text{-}C{\equiv}C\text{-}CH_2OH \quad 69\%$$

Bumagin, N.A.; Bykov, V.V.; Beletskaya,I.P. *Izv. Akad. Nauk. SSSR*, ***1990***, *39*, 2665 (Engl., p. 2418).

$$H\text{-}C{\equiv}C\text{-}n\text{-}C_5H_{11} \xrightarrow[K_2CO_3\text{ , }120°C\text{ , }26\text{ h}]{PhI\text{ , }CuI\text{-}2\ PPh_3\text{ , }DMF} Ph\text{-}C{\equiv}C\text{-}n\text{-}C_5H_{11} \quad 96\%$$

Okura, K.; Furuune, M.; Miura, M.; Nomura, M. *Tetrahedron Lett.*, ***1992***, *33*, 5363.

SECTION 2: ALKYNES FROM ACID DERIVATIVES

$$PhC(O)Cl \xrightarrow{1.\ Ph_3P{=}CHCO_2Et\text{ , }NEt_3\text{; }2.\ FVP\ (750°C)} Ph\text{-}C{\equiv}C\text{-}H \quad 42\%$$

Aitken, R.A.; Seth, S. *SynLett*, ***1990***, 211, 212.

SECTION 3: ALKYNES FROM ALCOHOLS AND THIOLS

NO ADDITIONAL EXAMPLES

SECTION 4: ALKYNES FROM ALDEHYDES

$$BocHN\text{-}CH(Me)\text{-}CHO \xrightarrow[(MeO)_2P(O)\text{-}CHN_2]{} BocHN\text{-}CH(Me)\text{-}C{\equiv}C\text{-}H \quad 67\%$$

Hauske, J.R.; Dorff, P.; Julin, S.; Martinelli, G.; Bussolari, J. *Tetrahedron Lett.*, ***1992***, *33*, 3715.

SECTION 5: ALKYNES FROM ALKYLS, METHYLENES AND ARYLS

NO ADDITIONAL EXAMPLES

SECTION 6: ALKYNES FROM AMIDES

NO ADDITIONAL EXAMPLES

SECTION 7: ALKYNES FROM AMINES

1. *t*-BuOK , DMF, 75°C
2. 3 eq. *t*-BuOK, DMF , 75°C

Ph–C≡C–Ph + $PhNH_2$

75%

Paventi, M.; Elce, E.; Jackman, R.J.; Hay, A.S. *Tetrahedron Lett.*, ***1992***, *33*, 6405.

SECTION 8: ALKYNES FROM ESTERS

NO ADDITIONAL EXAMPLES

SECTION 9: ALKYNES FROM ETHERS, EPOXIDES AND THIOETHERS

NO ADDITIONAL EXAMPLES

SECTION 10: ALKYNES FROM HALIDES AND SULFONATES

OTf → C≡C–*n*-C_6H_{13}

LiC≡C-*n*-C_6H_{13}, THF-DMPU, -20°C , 15 min

57%

Kotsuki, H.; Kadota, I.; Ochi, M. *Tetrahedron Lett.*, ***1990***, *31*, 4609.

Br, NHCbz, CO_2Me → MeO_2C, C≡C-Me, NHCbz

Bu_3Sn-C≡C-Me; $ZnCl_2$, CCl_4 , reflux

84%

Williams, R.M.; Aldous, D.J.; Aldous, S.C. *J. Org. Chem.*, ***1990***, *55*, 4657.

CO_2Et, Br → CO_2Et, C≡C-H

H·C≡C–Cu(CN)ZnI; -78°C → 0°C , 30 min

86%

Knoess, H.P.; Furlong, M.T.; Rozema, M.J.; Knochel, P. *J. Org. Chem.*, ***1991***, *56*, 5974.

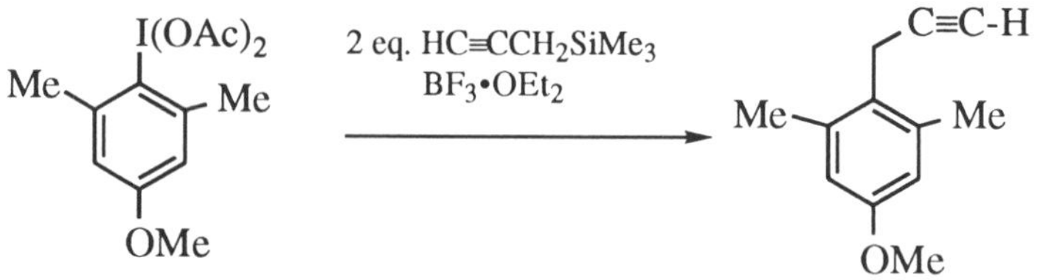

2 eq. $HC{\equiv}CCH_2SiMe_3$
$BF_3{\bullet}OEt_2$

57%

Ochiai, M.; Ito, T.; Masaki, Y. *J. Chem. Soc., Chem. Commun.*, ***1992***, 15.

SECTION 11: ALKYNES FROM HYDRIDES

For examples of the reaction $RC{\equiv}CH \rightarrow RC{\equiv}C\text{-}C{\equiv}CR^1$, see section 300 (Alkyne-Alkyne).

NO ADDITIONAL EXAMPLES

SECTION 12: ALKYNES FROM KETONES

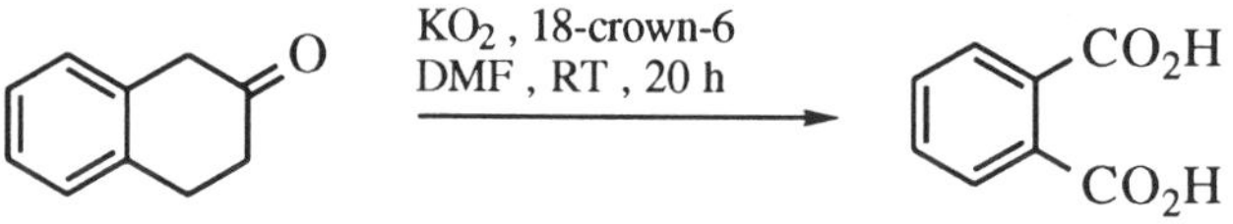

KO_2 , 18-crown-6
DMF , RT , 20 h

56%

Sotirious, C.; Lee, W.; Giese, R.W. *J. Org. Chem.*, **1990**, 55, 2159.

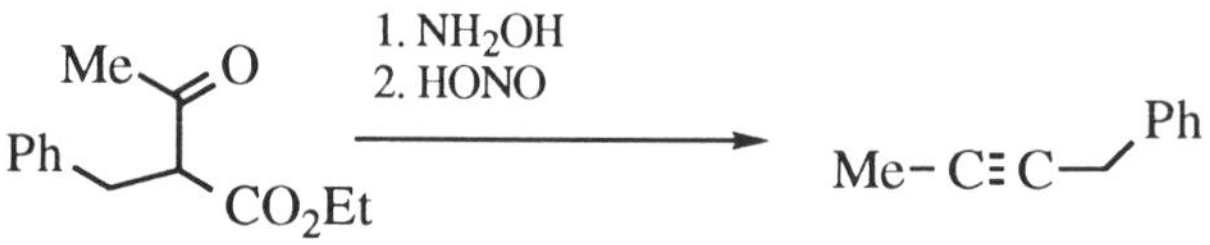

1. NH_2OH
2. HONO

59%

Boivin, J.; Elkaim, L.; Ferro, P.G.; Zard, S.Z. *Tetrahedron Lett.*, ***1991***, *32*, 5321.

SECTION 13: ALKYNES FROM NITRILES

NO ADDITIONAL EXAMPLES

SECTION 14: ALKYNES FROM ALKENES

NO ADDITIONAL EXAMPLES

SECTION 15: ALKYNES FROM MISCELLANEOUS COMPOUNDS

1-benzylbenzotriazole (N, N, N, Ph) — 1. LDA; 2. TsN=CHPh; 3. *t*-BuOK, DMF, 85°C → Ph-C≡C-Ph 70%

Katritzky, A.R.; Gordeev, M.F. *J. Chem. Soc., Perkin Trans. I*, ***1992***, 1295.

SECTION 15A: PROTECTION OF ALKYNES

NO ADDITIONAL EXAMPLES

CHAPTER 2

PREPARATION OF ACID DERIVATIVES

SECTION 16: ACID DERIVATIVES FROM ALKYNES

Ph-C≡C-H → ($CoCl_2$•6 H_2O, KCN, CO; 5N KOH, PhMe, 90°C; $Ni(CN)_2$•4 H_2O, 1 atm; PEG - 400) → CO_2H (Ph) 65% + Ph CO_2H 2%

Lee, J-T.; Alper, H. *Tetrahedron Lett.*, ***1991***, *32*, 1769.

SECTION 17: ACID DERIVATIVES FROM ACID DERIVATIVES

O, Cl → (SiH_2I_2, $CDCl_3$; 25°C . 10 min) → O, I quant.

Keinan, E.; Sahai, M. *J. Org. Chem.*, ***1990***, *55*, 3922.

O, Ph $CO_2^-K^+$ → (Ac_2O; O_2N-C6H4-CH2-CO_2H) → O_2N, O, O, O, Ph 80%

Fields, E.K.; Behrend, S.J.; Meyerson, S.; Wizenburg, M.L.; Ortega, B.R.; Hall, H.K. Jr. *J. Org. Chem.*, ***1990***, *55*, 5165.

Ph CO_2H → (1. DCC, CH_2Cl_2; 2. CH_2=$CHSO_2Ph$; 3. mCPBA, CH_2Cl_2; 4. TFAA, 22°C; 5. K_2CO_3, MeCN, 22°C) → Ph CO_2H 57% overall

Barton, D.H.R.; Chern, C-Y.; Jaszberenyi, J.Cs. *Tetrahedron Lett.*, ***1991***, *32*, 3309.

$n\text{-}C_{11}H_{23}\text{-}CO_2H$ → (1. 2-methoxy-2-propyl hydroperoxide (OMe, OOH), DCC; 2. AcOH) → $n\text{-}C_{11}H_{23}C(O)OOH$ 92%

Dussault, P.; Sahli, A. *J. Org. Chem.*, ***1992***, *57*, 1009.

PhC(O)Cl → ($MeO\text{-}C_6H_4\text{-}CO_2H$; $CoCl_2$, CH_2Cl_2, 40°C, 8 h) → PhC(O)OC(O)C$_6$H$_4$OMe 91%

Srivastava, R.R.; Kabalka, G.W. *Tetrahedron Lett.*, ***1992***, *33*, 593.

$CH_3(CH_2)_4CO_2H$ → (BBr_3, Al_2O_3, RT, 20 h) → $CH_3(CH_2)_4C(O)Br$ 67%

Bains, S.; Green, J.; Tan, L.C.; Pagni, R.M.; Kabalka, G.W. *Tetrahedron Lett.*, ***1992***, *33*, 7475.

SECTION 18: ACID DERIVATIVES FROM ALCOHOLS AND THIOLS

Cyclohexylmethanol (OH) → (1. $Ca(OCl)_2$, t-BuOH, CCl_4, HCl; 2. H_3O^+) → cyclohexyl-CO_2H 83%

Kabalka, G.W.; Chatla, N.; Wadgaonkar, P.P.; Deshpande, S.M. *Synth. Commun.*, ***1990***, *20*, 1617.

$PhCH_2OH$ → ($[(C_8H_{17})_3NMe]_3^+ \{XO_4[W(O)(O_2)_2]_4\}^{-3}$; 4% aq. H_2O_2, 90°C) → Ph-COOH 85%

Venturello, C.; Gambaro, M. *J. Org. Chem.*, ***1991***, *56*, 5924.

$(CH_3)_2CHCH_2OH$ → ($O_2^{\bullet -}$ / O_2, e^-) → $(CH_3)_2CHCO_2H$ 65%

Singh, M.; Singh, K.N.; Dwivedi, S.; Misri, R.A. *Synthesis*, ***1991***, 291.

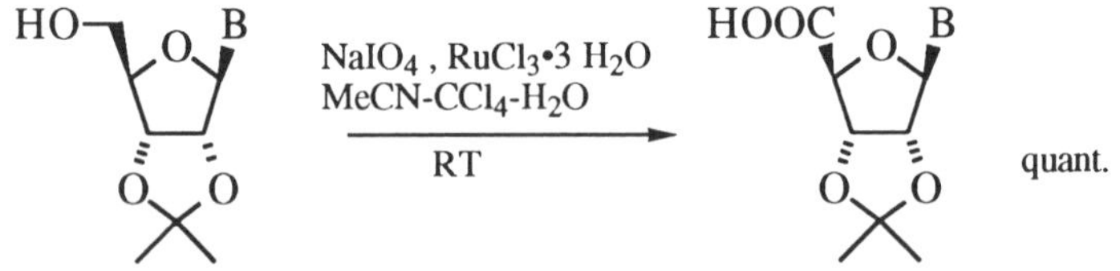

$NaIO_4$, $RuCl_3$•3 H_2O
MeCN-CCl_4-H_2O
RT

quant.

Singh, A.K.; Varma, R.S. *Tetrahedron Lett.*, ***1992***, *33*, 2307.

SECTION 19: ACID DERIVATIVES FROM ALDEHYDES

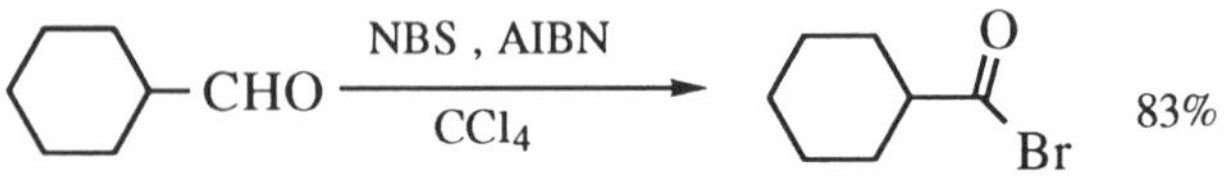

CHO → NBS, AIBN / CCl_4 → Br 83%

Markó, I.E.; Mekhalfia, A. *Tetrahedron Lett.*, ***1990***, *31*, 7237.

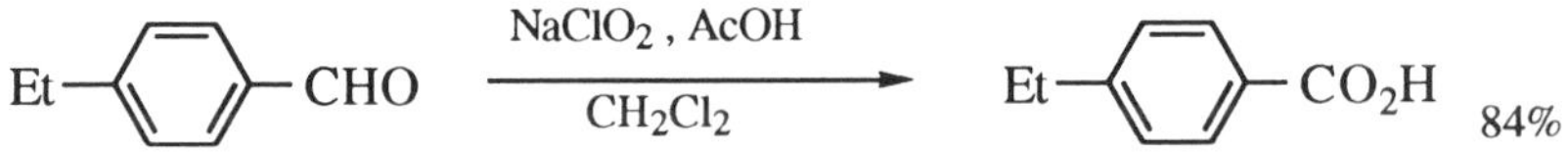

Et–C₆H₄–CHO → $NaClO_2$, AcOH / CH_2Cl_2 → Et–C₆H₄–CO_2H 84%

Bayle, J.P.; Perez, F.; Courtieu, J. *Bull. Soc. Chim. Fr.*, ***1990***, *127*, 565.

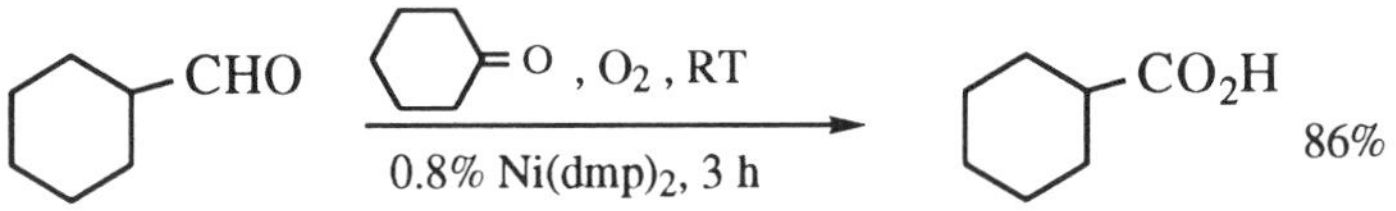

CHO → =O, O_2, RT / 0.8% Ni(dmp)$_2$, 3 h → CO_2H 86%

Ni(dmp)$_2$ = bis-[1,3-di(*p*-methoxyphenyl)-1,3-propanedionato nickel (II)

Yamada, T.; Rhode, O.; Takai, T.; Mukaiyama, T. *Chem. Lett.*, ***1991***, 5.

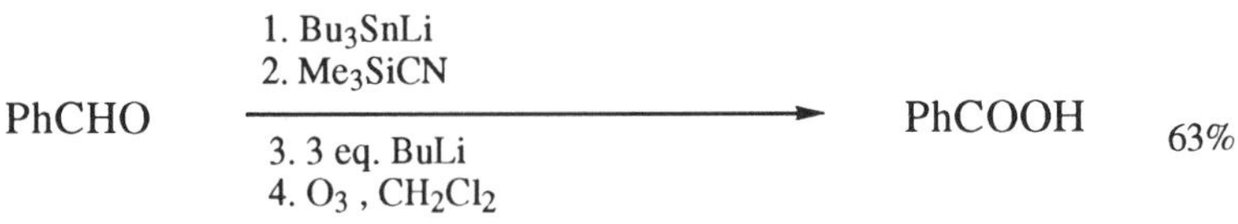

PhCHO → 1. Bu_3SnLi 2. Me_3SiCN / 3. 3 eq. BuLi 4. O_3, CH_2Cl_2 → PhCOOH 63%

Linderman, R.J.; Chen, K. *Tetrahedron Lett.*, ***1992***, *33*, 6767.

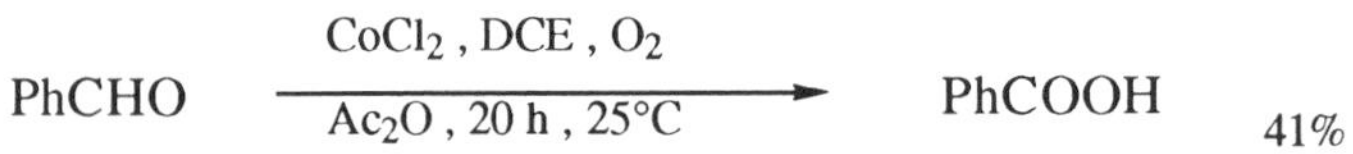

PhCHO → $CoCl_2$, DCE, O_2 / Ac_2O, 20 h, 25°C → PhCOOH 41%

Bhatia, B.; Iqbal, J. *Tetrahedron Lett.*, ***1992***, *33*, 7961.

SECTION 20: ACID DERIVATIVES FROM ALKYLS, METHYLENES AND ARYLS

Ph–CH2CH2CH(OAc)CH2CH2CH3 → ($RuCl_3$, HIO_4, CCl_4 / MeCN, H_2O, 25-40°C) → HO_2C–CH2CH(OAc)CH2CH2CH3 76%

Nuñez, M.T.; Martín, V.S. *J. Org. Chem.*, ***1990***, *55*, 1928.

Benzene → (1. $TiCl_4$, CH_2Cl_2, RT, PhSe–CH(Br)–CO_2Et; 2. NaOH, $NaBH_4$, EtOH, RT, 5 min, (2-thienyl-Te)$_2$) → Ph–CO_2H (phenylacetic acid) 70%

Silveira, C.C.; Lenardão, E.J.; Comasseto, J.V.; Babdoub, M.J. *Tetrahedron Lett.*, ***1991***, *32*, 5741.

SECTION 21: ACID DERIVATIVES FROM AMIDES

Amide (Me, Me, NH*n*-Bu), syn/anti=99.5/0.5 → (1. 3 eq. AcO–CH2–C(Me)(Me)–C(O)Cl, 3 NEt_3, 0.1 DMAP, CH_2Cl_2, RT, 11h; 2. 4 LiOH, RT, 30h) → acid (Me, Me, OH) 76%, syn/anti=99.2/0.8

Tsunoda, T.; Sasaki, O.; Itô, S. *Tetrahedron Lett.*, ***1990***, *31*, 731.

PhC(O)NH_2 → (N_2O_4, MeCN / -20°C) → Ph-CO_2H 86%

Kim, Y.H.; Kim, K.; Park, Y.J. *Tetrahedron Lett.*, ***1990***, *31*, 3893.

SECTION 22: ACID DERIVATIVES FROM AMINES

NO ADDITIONAL EXAMPLES

SECTION 23: ACID DERIVATIVES FROM ESTERS

Me_3Si ... CO_2Et / CO_2Et / Me —pig liver esterase→ Me_3Si ... CO_2H / CO_2Et / Me

60% (93% ee)

De Jeso, B.; Belair, N.; Deleuze, H.; Rascle, M.-C.; Maillard, B. *Tetrahedron Lett.*, ***1990***, *31*, 653.

Ph, OMe, O —1. AlI_3, MeCN, 82°C, 30 min; 2. H_2O→ Ph-CO_2H 88%

Mahajan, A.R.; Dutta, D.K.; Boruah, R.C.; Sandhu, J.S. *Tetrahedron Lett.*, ***1990***, *31*, 3943.

O, Me, Ph, O, OMe —DDQ, reflux, 8 h→ $PhCO_2H$ 90%

Yoo, S-E.; Kim, H.R.; Yi, K.Y. *Tetrahedron Lett.*, ***1990***, *31*, 5913.

Ph, O, Br, O —10% Se, $NaBH_4$, DMF; EtOH→ Ph, CO_2H 81%

Huang, Z-Z.; Zhou, X-J. *Synthesis*, ***1990***, 633.

OH, Ph, CO_2Me —iPr_2NEt, MECN-H_2O, 30°C, 10 kbar, 4 d→ OH, Ph, CO_2H 91%

Yamamoto, Y.; Furuta, T.; Matsuo, J.; Kurata, T. *J. Org. Chem.*, ***1991***, *56*, 5737.

n-$C_{11}H_{23}CO_2Bn$ —1. PhMe, MgI_2, reflux, 2 h; 2. aq. HCl→ n-$C_{11}H_{23}CO_2H$ 70%

Martínez, A.G.; Barcina, J.O.; del Veccio, G.H.; Hanack, M.; Subramanian, L.R. *Tetrahedron Lett.*, ***1991***,*32*, 5931.

$$\xrightarrow[\text{3 h}]{\text{HCOOH , H}_2\text{O , reflux}}$$ 84%

Schmid, C.R. *Tetrahedron Lett.*, ***1992***, *33*, 757.

$$\xrightarrow[\text{Hg cathode}]{e^- \text{, Me}_3\text{SiCl , MeCN}}$$ CO_2H CO_2H 68%

Troll, T.; Wiedemann, J. *Tetrahedron Lett.*, ***1992***, *33*, 3847.

Ph(C=O)TeBu $\xrightarrow[\text{RT}]{\text{CuCl}_2\text{•2 H}_2\text{O , acetone}}$ $PhCO_2H$ 96%

Dabdoub, M.J.; Viana, L.H. *Synth. Commun.*, ***1992***, *22*, 1619.

$PhCO_2Me$ $\xrightarrow[\text{H}_2\text{O , 80°C}]{\text{BI}_3\text{•NEt}_2\text{Ph , PhH , 2 h}}$ $PhCO_2H$ 92%

Kabalka, G.W.; Narayana, C.; Reddy, N.K. *Synth. Commun.*, ***1992***, *22*, 1793.

Other reactions useful for the hydrolysis of esters may be found in Section 30A (Protection of Carboxylic Acids).

SECTION 24: ACID DERIVATIVES FROM ETHERS, EPOXIDES AND THIOETHERS

n-C_6H_{13}(epoxide) $\xrightarrow[\text{2. KMnO}_4\text{, acetone; 3N H}_2\text{SO}_4]{\text{1. Me}_3\text{Al}}$ n-C_6H_{13}CH(Me)CO_2H 44%

Fukumasa, M.; Furuhashi, K.; Umezama, J.; Takahashi, O.; Hirai, T. *Tetrahedron Lett.*, ***1991***, *32*, 1059.

SECTION 25: ACID DERIVATIVES FROM HALIDES AND SULFONATES

$PhCH_2Cl$ $\xrightarrow[\text{K}_2\text{CO}_3\text{ , 60°C}]{e^-\text{ (Ni anode/OH}^-\text{) , }t\text{-BuOH/H}_2\text{O}}$ PhCOOH 70%

Borsotti, G.P.; Foa', M.; Gatti, N. *Synthesis*, ***1990***, 207.

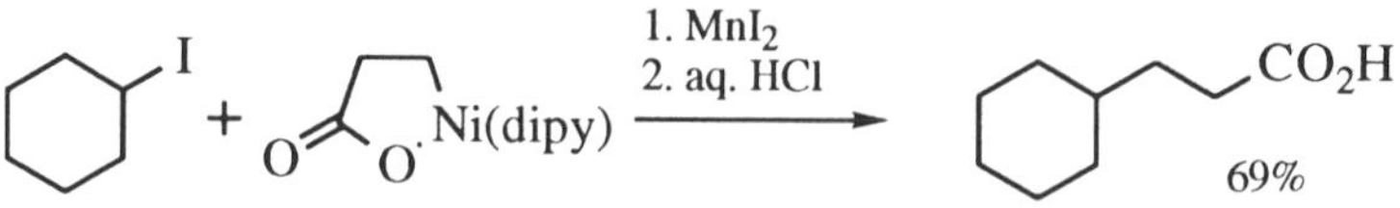

1. MnI_2
2. aq. HCl

CO_2H

Ni(dipy)

69%

Schönecker, B.; Walther, D.; Fischer, R.; Nestler, B.; Bräunlich, G.; Eibisch, H.; Droescher, P. *Tetrahedron Lett.*, ***1990***, *31*, 1257.

$PhCH_2Br$ → (20 eq. NaOCl, MeCN, RT; 2 h,))))))))) → PhCHO (22%) + PhCOOH (74%)

Khurana, J.M.; Sahoo, P.K.; Titus, S.S.; Maikap, G.C. *Synth. Commun.*, ***1990***, *20*, 1357.

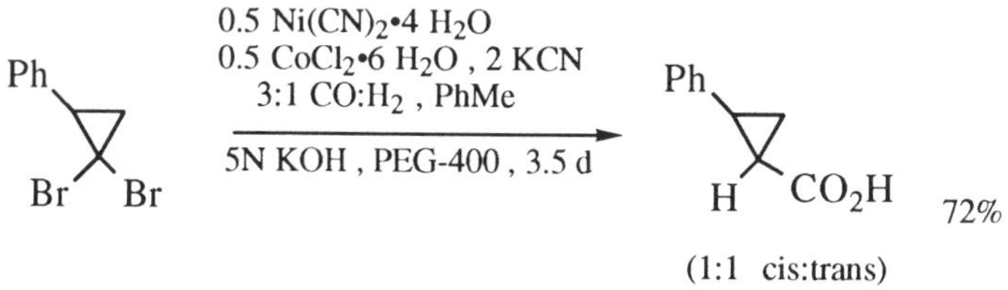

0.5 $Ni(CN)_2$•4 H_2O
0.5 $CoCl_2$•6 H_2O, 2 KCN
3:1 CO:H_2, PhMe
5N KOH, PEG-400, 3.5 d

Ph, Br, Br → Ph, H, CO_2H 72%

(1:1 cis:trans)

Grushin, V.V.; Alper, H. *Tetrahedron Lett.*, ***1991***, *32*, 3349.

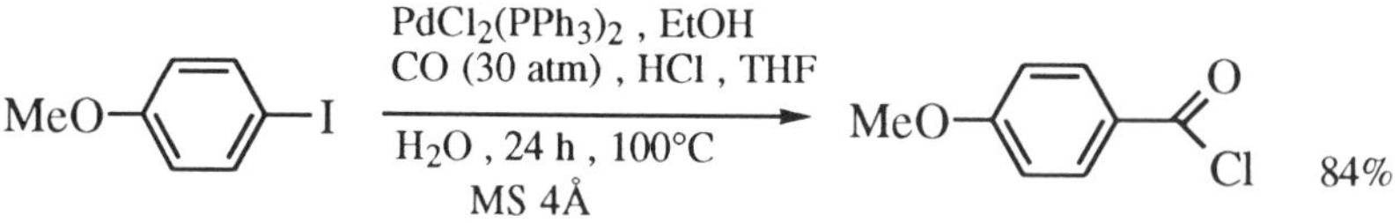

$PdCl_2(PPh_3)_2$, EtOH
CO (30 atm), HCl, THF
H_2O, 24 h, 100°C
MS 4Å

MeO–C6H4–I → MeO–C6H4–COCl 84%

Urata, H.; Hu, N-X.; Maekawa, H.; Fuchikami, T. *Tetrahedron Lett.*, ***1991***, *32*, 4733.

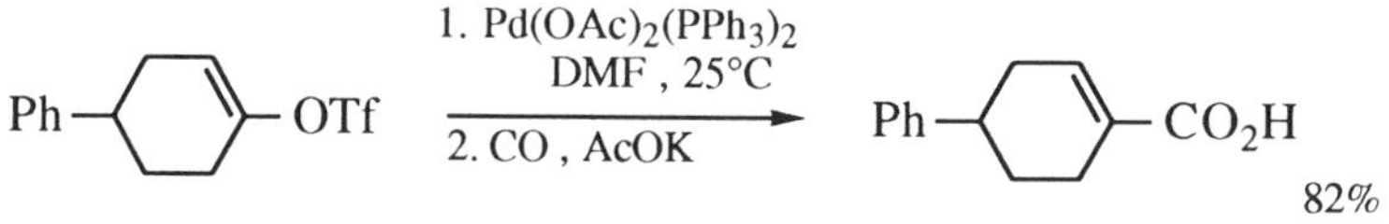

1. $Pd(OAc)_2(PPh_3)_2$
DMF, 25°C
2. CO, AcOK

Ph–(cyclohexenyl)–OTf → Ph–(cyclohexenyl)–CO_2H 82%

Cacchi, S.; Lupi, A. *Tetrahedron Lett.*, ***1992***, *33*, 3939.

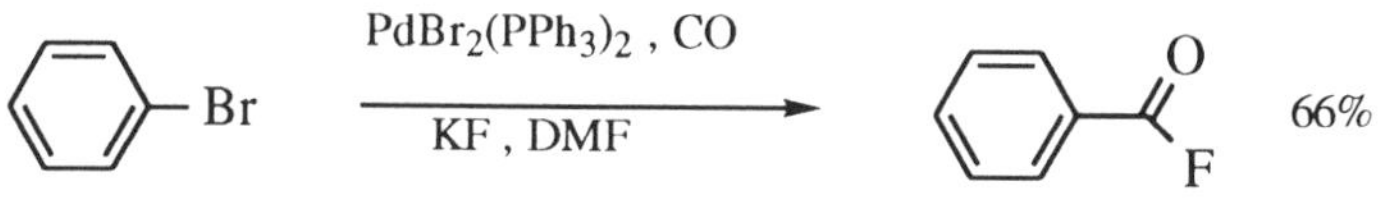

$PdBr_2(PPh_3)_2$, CO
KF, DMF

PhBr → PhCOF 66%

Okano, T.; Harada, N.; Kiji, J. *Bull. Chem. Soc. Jpn.*, ***1992***, *65*, 1741.

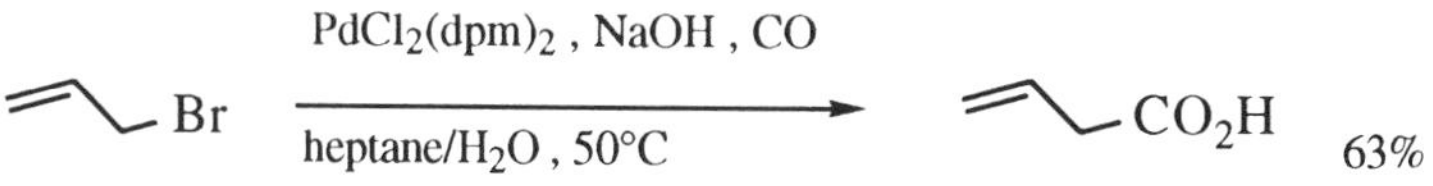

$PdCl_2(dpm)_2$, NaOH, CO
heptane/H_2O, 50°C

allyl–Br → allyl–CO_2H 63%

Okano, T., Okabe, N.; Kiji, J. *Bull. Chem. Soc. Jpn.*, ***1992***, *65*, 2589.

SECTION 26: ACID DERIVATIVES FROM HYDRIDES

NO ADDITIONAL EXAMPLES

SECTION 27: ACID DERIVATIVES FROM KETONES

hν (Pyrex), 5% aq. acetone, propylene oxide

58%

Sonawane, H.R.; Kulkarni, D.G.; Ayyangar, N.R. *Tetrahedron Lett.*, ***1990***, *31*, 7493.

$NaNO_2$, Py/(HF)$_x$

80%

Olah, G.A.; Ramos, M.T.; Wang, Q.; Surya Prakash, G.K. *SynLett*, ***1991***, 41.

1. *t*-BuOK, $CHCl_3$, hexane
2. H_2, Pd-C

65%

Llama, E.F.; del Campo, C.; Sinisterra, J.V. *Org. Prep. Proceed. Int.*, ***1992***, *24*, 165.

SECTION 28: ACID DERIVATIVES FROM NITRILES

nitrile hydralase SP 361, pH 7, 30°C, 23 h

71%

Cohen, M.A.; Sawden, J.; Turner, N.J. *Tetrahedron Lett.*, ***1990***, *31*, 7223.

Rhodococcus sp, strain CH5, pH 7 buffer, 435 h, RT

(44% conversion) 98%

Hönicke-Schmidt, P.; Schneider, M.P. *J. Chem. Soc., Chem. Commun.*, ***1990***, 648.

NC~~~COOH → (enzyme from *Rhodococcus* sp.) → HOOC~~~COOH 68%

de Raadt, A.; Klempier, N.; Faber, K.; Griengl, H. *J. Chem. Soc., Perkin Trans. I*, ***1992***, 137.

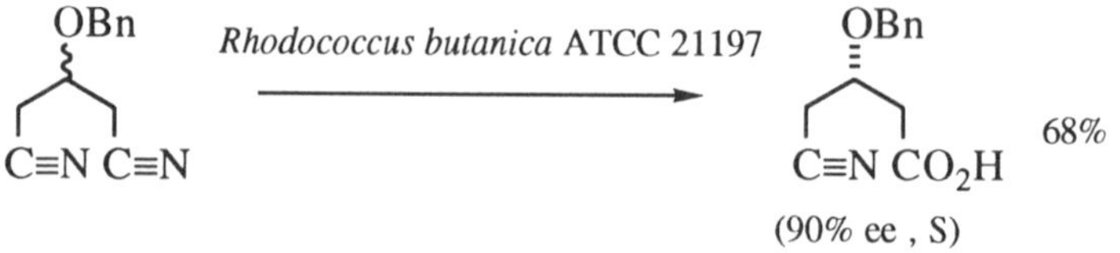

Kakeya, H.; Sakai, N.; Sano, A.; Yokoyama, M.; Sugai, T.; Ohta, H. *Chem. Lett.*, ***1991***, 1823.

SECTION 29: ACID DERIVATIVES FROM ALKENES

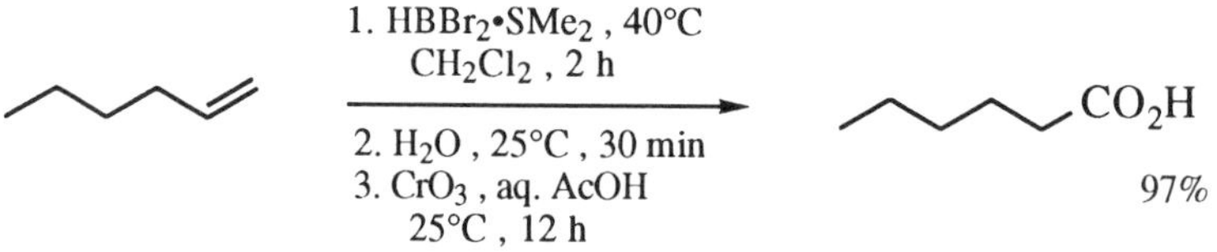

Racherla, U.S.; Khanna, V.V.; Brown, H.C. *Tetrahedron Lett.*, ***1992***, *33*, 1037.

SECTION 30: ACID DERIVATIVES FROM MISCELLANEOUS COMPOUNDS

$$PhN_2^+\ BF_4^- \xrightarrow[]{\text{1. } CuCl_2 \quad \text{2. } H_2O} PhCOOH \quad 42\%$$

Olah, G.A.; Wu, A.; Bagno, A.; Surya Prakash, G.K.S. *SynLett*, ***1990***, 596.

SECTION 30A: PROTECTION OF CARBOXYLIC ACID DERIVATIVES

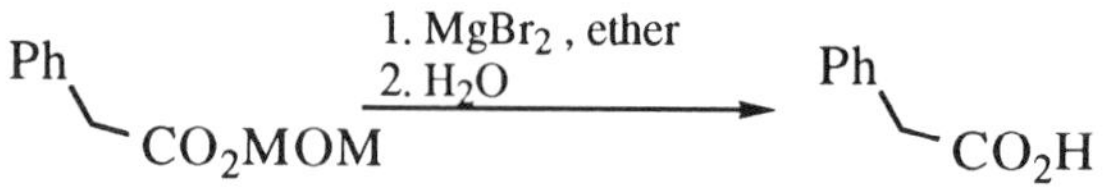

Kim, S.; Park, Y.H.; Kee, I.S. *Tetrahedron Lett.*, ***1991***, *32*, 3099.

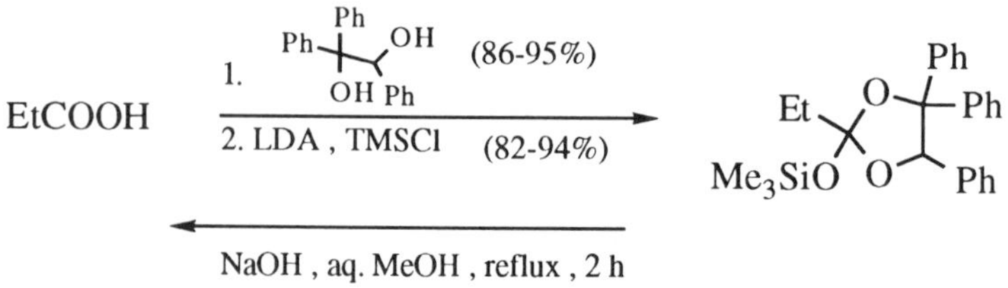

Waldmüller, D.; Braun, M.; Steigel, A. *SynLett*, ***1991***, 160.

BnO–CH$_2$–CH(OBn)–CH$_2$–CO_2Bn $\xrightarrow[\text{RT , cyclohexadiene}]{\text{EtOH , 10\% Pd-C , 23°C}}$ BnO–CH$_2$–CH(OBn)–CH$_2$–CO_2H 90%

Bajwa, J.S. *Tetrahedron Lett.*, ***1992***, *33*, 2299.

Other reactions useful for the protection of carboxylic acids are included in Section 107 (Esters from Carboxylic Acids and Acid Halides) and Section 23 (Carboxylic Acids from Esters).

CHAPTER 3

PREPARATION OF ALCOHOLS

SECTION 31: ALCOHOLS AND THIOLS FROM ALKYNES

NO ADDITIONAL EXAMPLES

SECTION 32: ALCOHOLS AND THIOLS FROM ACID DERIVATIVES

$C_5H_{11}CO_2H$ —[$Et_3N{\bullet}BH_3$, 80°C , 4 h]→ $C_6H_{13}OH$ 96%

Kříž, O.; Plzák, Z.; Plešek, J. *Coll. Czech. Chem. Commun.*, ***1990***, *55*, 2956.

CbzHN–CH(CH_2Ph)–CO_2H —[1. N-methylmorpholine N-oxide, $ClCO_2Et$, THF , -10°C; 2. $NaBH_4$, MeOH , 20 min]→ CbzHN–CH(CH_2Ph)–CH_2OH 84%

Kokotos, G. *Synthesis*, ***1990***, 299.

PhCOOH —[$BnEt_3N^+$ BH_4^- , Me_3SiCl; CH_2Cl_2]→ $PhCH_2OH$ 92%

Das, J.; Chandrasekaran, S. *Synth. Commun.*, ***1990***, *20*, 907.

$PhCO_2Me$ —[1. [5% Cp_2TiCl_2/10% BuLi] , RT, THF , -78°C, 2 eq. $HSi(OEt)_3$; 2. conc. HCl]→ $PhCH_2OH$ 93%

Berk, S.C.; Kreutzer, K.A.; Buchwald, S.L. *J. Am. Chem. Soc.*, ***1991***, *113*, 5093.

$PhCH_2CO_2H$ —[$NaBH_4$, I_2 , RT, THF]→ $PhCH_2CH_2OH$ 98%

Kanth, J.V.B.; Periasamy, M. *J. Org. Chem.*, ***1991***, *56*, 5964.

2 eq. SmI_2 , THF/H_2O, NaOH , RT, 58 sec: cyclohexyl-CO_2H → cyclohexyl-CH_2OH 78%

Kamochi, Y.; Kudo, T. *Chem. Lett.*, ***1991***, 893.

n-$C_8H_{17}COOH$ —($NaBH_4$, catechol; TFA , 4 h)→ n-$C_8H_{17}CH_2OH$ 87%

Suseela, Y.; Periasamy, M. *Tetrahedron*, ***1992***, *48*, 371.

Ph-CO_2H —(SmI_2-85% H_3PO_4; THF , 3 sec)→ $PhCH_2OH$ 91%

Kamochi, Y.; Kudo, T. *Tetrahedron*, ***1992***, *48*, 4301.

t-Bu-C_6H_4-$CH_2CH_2CO_2H$ —(1. hv , O_2 , 2 h, acridine , PhMe; 2. Me_2S , MeOH)→ t-Bu-C_6H_4-CH_2CH_2OH 60%

Okada, K.; Okubo, K.; Oda, M. *Tetrahedron Lett.*, ***1992***, *33*, 83.

PhCH(CH_3)CO_2H —(Fe (porphyrin) , RT; PhIO , CH_2Cl_2 , 20 h)→ PhCH(CH_3)OH 46% + PhC(CH_3)=O 3%

Komuro, M.; Nagatsu, Y.; Higuchi, T.; Hirobe, M. *Tetrahedron Lett.*, ***1992***, *33*, 4949.

$(CH_3)_2CHCH(NH_2)CO_2H$ —($NaBH_4$-H_2SO_4)→ $(CH_3)_2CHCH(NH_2)CH_2OH$ 89%

Abiko, A.; Masamune, S. *Tetrahedron Lett.*, ***1992***, *33*, 5517.

n-$C_7H_{15}COOH$ —(1. e^- , $NaBH_4$; 2. H_2O)→ n-$C_7H_{15}CH_2OH$ 70%

Shundo, R.; Matsubara, Y.; Nishiguchi, I.; Hirashima, T. *Bull. Chem. Soc. Jpn.*, ***1992***, *65*, 530.

PhCOOH —(4 eq. SmI_2 , 8 eq. NaOH; H_2O)→ $PhCH_2OH$ 92%

Kamochi, Y.; Kudo, T. *Bull. Chem. Soc. Jpn.*, ***1992***, *65*, 3049.

$$n\text{-}C_7H_{15}COOH \xrightarrow[\text{TFA}]{Zn(BH_4)_2\text{, DMF, 24 h}} n\text{-}C_7H_{15}CH_2OH \quad 90\%$$

Ranu, B.C.; Das, A.R. *J. Chem. Soc., Perkin Trans. I*, ***1992***, 1561.

SECTION 33: ALCOHOLS AND THIOLS FROM ALCOHOLS AND THIOLS

$$n\text{-}C_7H_{15}CH(OSiMe_3)SnBu_3 \xrightarrow{\text{BuLi, THF, -78°C}} n\text{-}C_7H_{15}CH(OSiMe_3)OH \quad 63\%$$

reverse Brook rearrangement

Linderman, R.J.; Ghannam, A. *J. Am. Chem. Soc.*, ***1990***, *112*, 2392.

$$PhCH(OH)Me \xrightarrow[\text{2. LiAlH}_4\text{, ether}]{\text{1. MeC(O)SH, (iPrOC(O)N)}_2\text{, THF, PPh}_3} PhCH(SH)Me \quad 61\%$$

(96% ee)

Corey, E.J.; Cimprich, K.A. *Tetrahedron Lett.*, ***1992***, *33*, 4099.

SECTION 34: ALCOHOLS AND THIOLS FROM ALDEHYDES

The following reaction types are included in this section:

A. Reductions of Aldehydes to Alcohols

B. Alkylation of Aldehydes, forming Alcohols.

Coupling of Aldehydes to form Diols is found in Section 323 (Alcohol-Alcohol).

SECTION 34A: REDUCTIONS OF ALDEHYDES TO ALCOHOLS

$$Me_2C(CHO)C(O)Ph \xrightarrow{Zn(BH_4)_2\text{, THF, -10°C}} Me_2C(CH_2OH)C(O)Ph \quad \text{quant.}$$

selective reduction of aldehydes in the presence of ketones

Ranu, B.C.; Chakraborty, R. *Tetrahedron Lett.*, ***1990***, *31*, 7663.

$\xrightarrow[\text{DMF - H}_2\text{O}]{\text{SbCl}_3\text{ - Al}}$ CHO → OH 98%

also with aromatic and conjugated aldehydes

Wang, W-B.; Shi, L-L.; Huang, Y-Z. *Tetrahedron Lett.*, ***1990***, *31*, 1185.

CHO $\xrightarrow{\text{Zn-ZnCl}_2\text{, 50\% aq. THF}}$ OH 81%

+ 7% of dimeric diol

Tanaka, K.; Kishigami, S.; Toda, F. *J. Org. Chem.*, ***1990***, *55*, 2981.

CHO $\xrightarrow{\text{hν , cyclooctane , 48 h; RhCl(CO)(PMe}_3)_2}$ OH 87%

Sakakura, T.; Abe, F.; Tanaka, M. *Chem. Lett.*, ***1990***, 583, 585.

PhCHO $\xrightarrow[\text{reflux , 8 h}]{\text{XP}_4\text{ - Zn(BH}_4)_2\text{ , EtOH}}$ $PhCH_2OH$ 80%

XP_4 = crosslinked poly(4-vinylpyridine)

Firouzabadi, H.; Tamami, B.; Goudarzian, N. *Synth. Commun.*, ***1991***, *21*, 2275.

PhCHO $\xrightarrow[\text{30 min}]{\text{NiCl}_2\text{•6 H}_2\text{O , aq. DMF}}$ $PhCH_2OH$ 95%

Baruah, R.N. *Tetrahedron Lett.*, ***1992***, *33*, 5417.

PhCHO $\xrightarrow[\text{H-montmorillonite , 7 h}]{\text{NaBH}_4\text{ , PhH , H}_2\text{O , RT}}$ $PhCH_2OH$ 98%

Subba Rao, Y.V.; Choudary, B.M. *Synth. Commun.*, ***1992***, *22*, 2711.

SECTION 34B: ALKYLATION OF ALDEHYDES, FORMING ALCOHOLS

ASYMMETRIC ALKYLATIONS

CHO → 1. (Ipc)$_2$B-allyl, ether , -100°C , 30 min; 2. NaOH , H_2O_2 → OH 88% (96% ee)

Racherla, U.S.; Brown, H.C. *J. Org. Chem.*, ***1991***, *56*, 401.

$SnBu_3$ → 1. R^*_2BBr , PhMe 2. PhCHO , -78°C , PhMe → Ph, OH 92% 96% ee

R^*_2BBr = CF_3 PhPh CF_3 F_3C S N B N S O_2 O_2 Br CF_3

Corey, E.J.; Kim, S.S. *Tetrahedron Lett.*, ***1990***, *31*, 3715.

Ph N O CHO → M → Ph N O HO + Ph N O HO

M = allyl Zn/LiCl , THF , 0°C	98	:	2	(70%)
M = $AlMe_3Li$, ether , -78°C	16	:	84	(78%)

Overly, K.R.; Williams, J.M.; McGarvey, G.J. *Tetrahedron Lett.*, ***1990***, *31*, 4573.

$(CO)_5Cr$ H N O → Bn N, hν → O N N Bn O 59% (70% dr)

Hegedus, L.S.; Imwinkelried, R.; Alarid-Sargent, M. *J. Am. Chem. Soc.*, ***1990***, *112*, 1109.

PhCHO → $SiMe_3$, 20% BH_3•THF / OiPr O CO_2H CO_2H O OiPr OH → Ph OH 82% (82% ee , R)

Furata, K.; Mouri, M.; Yamamoto, H. *SynLett*, ***1991***, 561.

OBn CHO → $SnMe_3$, ether, 5M $LiClO_4$, 1.7 h → OBn OH 82% (25:1 dr)

Nehry, K.J. Jr.; Grieco, P.A.; Jagoe, C.T. *Tetrahedron Lett.*, ***1992***, *33*, 1817.

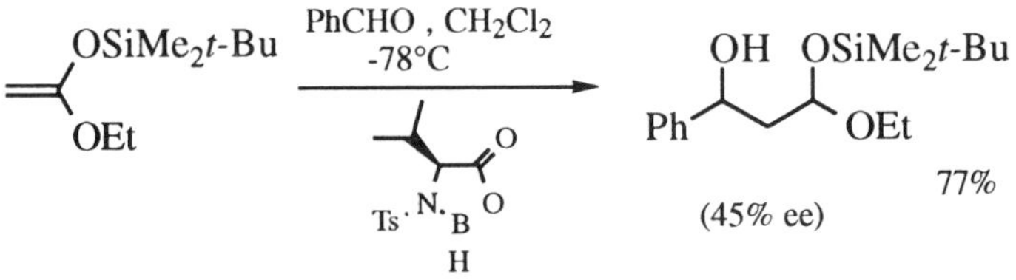

PhCHO , CH_2Cl_2 -78°C

OSiMe$_2$t-Bu, OEt → OH OSiMe$_2$t-Bu, Ph, OEt 77% (45% ee)

Ts·N·B, H, O, O

Kiyooka, S.; Kaneko, Y.; Komura, M.; Matsuo, H.; Nakano, M. *J. Org. Chem.*, ***1991***, *56*, 2276.

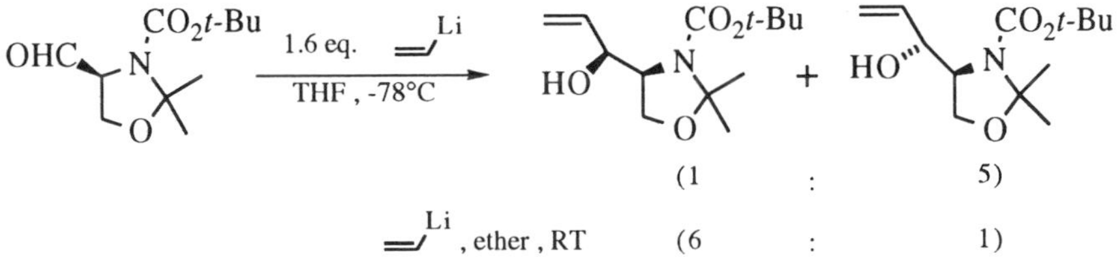

OHC, CO_2t-Bu, N, O → 1.6 eq. vinyl-Li, THF , -78°C → HO, CO_2t-Bu, N, O + HO, CO_2t-Bu, N, O

(1 : 5)

vinyl-Li , ether , RT (6 : 1)

Coleman, R.S.; Carpenter, A.J. *Tetrahedron Lett.*, ***1992***, *33*, 1697.

Reactions of Et_2Zn and PhCHO

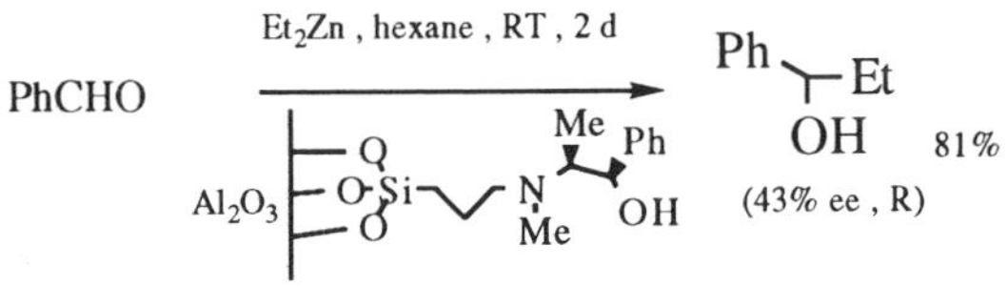

PhCHO → Et_2Zn , hexane , RT , 2 d; Al_2O_3–O–Si–(CH$_2$)$_3$–N(Me)–CH(Me)–CH(Ph)OH → Ph–CH(OH)–Et 81% (43% ee , R)

Soai, K.; Watanabe, M.; Yamamoto, A. *J. Org. Chem.*, ***1990***, *55*, 4832.

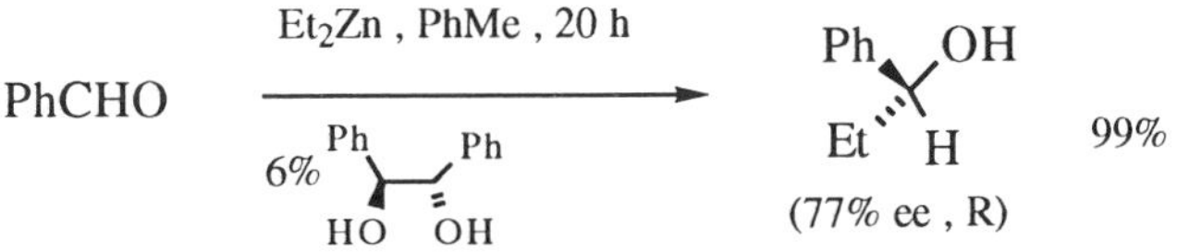

PhCHO → Et_2Zn , PhMe , 20 h; 6% Ph–CH(OH)–CH(OH)–Ph → Ph, OH, Et, H 99% (77% ee , R)

Rosini, C.; Franzini, L.; Pini, D.; Salvadori, P. *Tetrahedron Asymmetry*, ***1990***, *1*, 587.

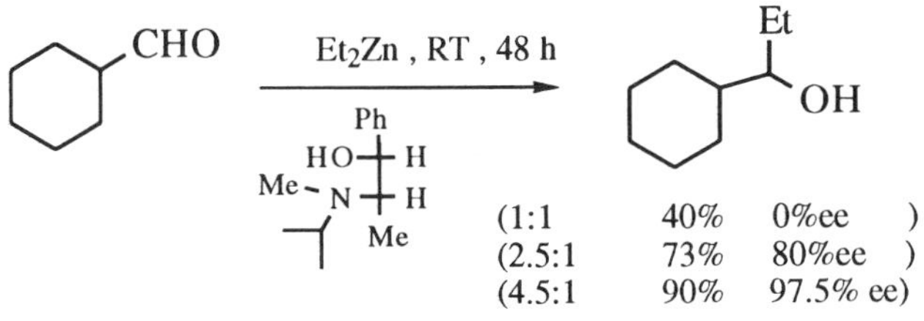

Cyclohexyl-CHO → Et_2Zn , RT , 48 h; Ph, HO–H, Me–N–H, Me → cyclohexyl–CH(Et)OH

(1:1 40% 0%ee)
(2.5:1 73% 80%ee)
(4.5:1 90% 97.5% ee)

Chaloner, P.A.; Langadianou, E. *Tetrahedron Lett.*, ***1990***, *31*, 5185.

Ph(Me)CH–CHO → (Et_2Zn, $HOCH_2CH_2NH_2$) → syn-PhCH(Me)CH(OH)Et + anti-PhCH(Me)CH(OH)Et (83% dr, syn) 55%

Soai, K.; Niwa, S.; Hatanaka, T. *Bull. Chem. Soc. Jpn.*, ***1990***, *63*, 2129.

CH_2=C(Me)CHO → (Et_2Zn, hexane, -10°C, 24 h; t-Bu–CH(OH)–CH_2–N(piperidine)) → CH_2=C(Me)CH(OH)Et 67% (95% ee, R)

Hayashi, M.; Kaneko, T.; Oguni, N. *J. Chem. Soc., Perkin Trans. I*, ***1991***, 25.

PhCHO → (Et_2Zn, 5% ephedrine) → PhCH(OH)Et 60% (66% ee, R)

Chaloner, P.A.; Langadianou, E.; Peresa, S.A.R. *J. Chem. Soc., Perkin Trans. I*, ***1991***, 2731.

PhCHO → (Et_2Zn; 5% catalyst: H, N, N Ph, H, Ph, Ph, OH) → PhCH(OH)Et 92% (96% ee)

Asami, M.; Inoue, S. *Chem. Lett.*, ***1991***, 685.

PhCHO → (Et_2Zn, PhMe, 20 h; 6% NMe_2/NMe_2 binaphthyl (S)) → PhCH(OH)Et 94% (63% ee, S)

Rosini, C.; Franzini, L.; Iuliano, A.; Pini, D.; Salvadori, P. *Tetrahedron Asymmetry*, ***1991***, *2*, 363.

PhCHO → (Et_2Zn, PhMe, 20 h; cat. Ph, HO–H, MeHN–H, Me) → PhCH(OH)Et 90% (60% ee)

Näslund, J.; Welch, C.J. *Tetrahedron Asymmetry*, ***1991***, *2*, 1123.

Et$_2$Zn , 20 h; 20% (Ph, Me, HO, NBu$_2$) — OBn, CHO → OBn OH (78% dr) 57%

Soai, K.; Hatanaka, T.; Yamashita, T. *J. Chem. Soc., Chem. Commun.*, ***1992***, 927.
Soai, K.; Watanabe, M. *J. Chem. Soc., Chem. Commun.*, ***1990***, 43.

PhCHO — ZnEt$_2$, PhMe , RT , 24 h → OH Ph Et + OH Ph Et

+ OH NMe$_2$ (91 : 2) 91%

+ OH NHSO$_2$CF$_3$ (1 : 99) 80%

Kimura, K.; Sugiyama, E.; Ishizuka, T.; Kunieda, T. *Tetrahedron Lett.*, ***1992***, *33*, 3147.

PhCHO — Et$_2$Zn , PhMe , 0°C , 3 h (N N, *t*-Bu, OH, HO, *t*-Bu) → Ph Et OH 83% (93% ee)

also - conjugate addition of Et$_2$Zn

Bolm, C.; Schlingloff, G.; Harms, K. *Chem. Ber.*, ***1992***, *125*, 1191.
Bolm, C.; Ewald, M.; Felder, M. *Chem. Ber.*, ***1992***, *125*, 1205 [conjugate addition].

PhCHO — Et$_2$Zn , -40°C; 85% (Ph, OH, N, O, N, Bn) → OH Ph Et 78% (91% ee , S)

Mori, A.; Yu, D.; Inoue, S. *SynLett*, ***1992***, 427.

PhCHO — Et$_2$Zn , Ti(OiPr)$_4$, 0°C , PhMe (Me, TosHN, OH) → Ph Et OH 81% (81% ee , S)

Ito, K.; Kimura, Y.; Okumura, H.; Katsuki, T. *SynLett*, ***1992***, 573.

PhCHO → ($ZnEt_2$; 10% $(CO)_3Cr$ complex, Me, NMe_2, OH) → Ph–CH(OH)–Et (R) 86%

Heaton, S.B.; Jones, G.B. *Tetrahedron Lett.*, ***1992***, *33*, 1693.

NON-ASYMMETRIC ALKYLATIONS

1-cyclohexenyl SPh → (1. Li *p,p'*-di-*t*-butylbiphenylide, THF, -78°C; 2. cyclohexyl-CHO) → OH product 79%

Cohen, T.; Doubleday, M.D. *J. Org. Chem.*, ***1990***, *55*, 4784.

PhCHO → (allyl I, $SbCl_3$ - Fe; 25°C, 20 h) → Ph–CH(OH)–CH$_2$CH=CH$_2$ 98%

Wang, W-B.; Shi, L-L.; Xu, R-H.; Huang, Y-Z. *J. Chem. Soc., Perkin Trans. I*, ***1990***, 424.

$Ph_2TeMe^{\oplus}$ $BPh_4^{\ominus}$ → (1. BuLi; 2. PhCHO; 3. 1 h) → Ph–CH(OH)–Ph 55%

Shi, L.L.; Zhou, J.L.; Huang, Y-Z. *J. Chem. Soc., Perkin Trans. I*, ***1990***, 2847.

4-methyl-2-pentanone → (PhCHO, $Me_2Fe{\bullet}3$ LiCl; THF, -78°C → RT; 18 h) → Ph–CH(OH)–CH$_3$ 83% + tertiary alcohol (OH) 2%

Also works with Co reagents:

Kauffmann, T.; Laarmann, B.; Menges, D. *Tetrahedron Lett.*, ***1990***, *31*, 507.

prenyl SiF_3 → (PhCHO, resorcinol; NEt_3, CH_2Cl_2, RT; 20 h) → Ph–CH(OH)–C(CH$_3$)$_2$–CH=CH$_2$ 93%

Kira, M.; Sato, K.; Sakurai, H. *J. Am. Chem. Soc.*, ***1990***, *112*, 257.

Br

e⁻ , PhCHO , DMF
Bu_4NClO_4

OH
Ph

80%

Tokuda, M.; Uchida, M.; Katoh, Y.; Suginome, H. *Chem. Lett.*, ***1990***, 461.

Ph
CHO

Bu_3SnH , AIBN

OH
Ph 80%

(1:1 trans:cis)

Yadav, V.; Fallis, A.G. *Can. J. Chem.*, ***1991***, *69*, 779.

n-C_7H_{15} Cl

1. Ba , THF
2. PhCHO

n-C_7H_{15} OH Ph 80%

(>99:1 E:Z ; 97:3 α:γ substitution)

Yanagisawa, A.; Habaue, S.; Yamamoto, H. *J. Am. Chem. Soc.*, ***1991***, *113*, 8955.

SMe
Ph $SiMe_3$

1. CsF , MeCN
2. EtCHO

SMe
Ph Et
OH
26%

+

SMe
Ph
27%

Bonini, B.F.; Masiero, S.; Mazzanti, G.; Zani, P. *Tetrahedron Lett.*, ***1991***, *32*, 815.

Br
O
O
O

1. BuLi , -78°C
2. $TiCl(OiPr)_3$
3. -60°C , 21 h
MeO_2C- C≡ C- CHO

MeO_2C- C≡ C- OH
O
O
O
70%

Haarmann, H.; Eberbach, W. *Tetrahedron Lett.*, ***1991***, *32*, 903.

ClF_2C $SiMe_3$

iPrCHO , Zn-CuCl
DMA , 100°C

F_2
C
$SiMe_3$
OH

DMA = N,N-dimethylacetamide

75%

Ishihara, T.; Miwatashi, S.; Kuroboshi, M.; Utimoto, K. *Tetrahedron Lett.*, ***1991***, *32*, 1069.

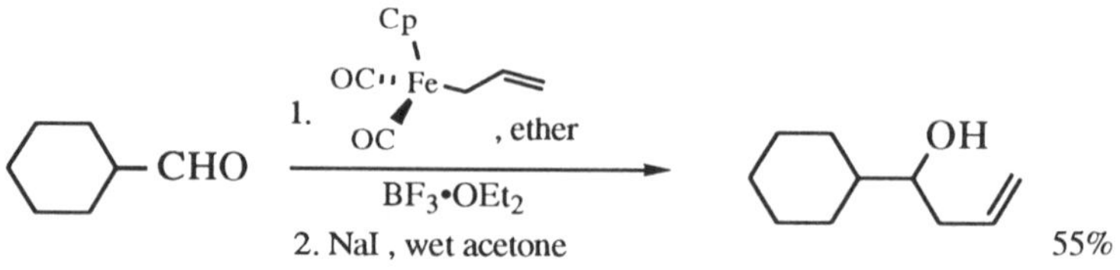

Cp, OC, Fe, 1. OC, ether
CHO
BF_3•OEt_2
2. NaI, wet acetone
OH
55%

Agoston, G.E.; Cabal, M.P.; Turos, E. *Tetrahedron Lett.*, ***1991***, *32*, 3001.

1. $C_{10}H_8Li$
2. PhCHO
Cl O O
OH O O
Ph
51%
"masked" Li bis-homoenolates

Ramón, D.J.; Yus, M. *J. Org. Chem.*, ***1991***, *56*, 3825.

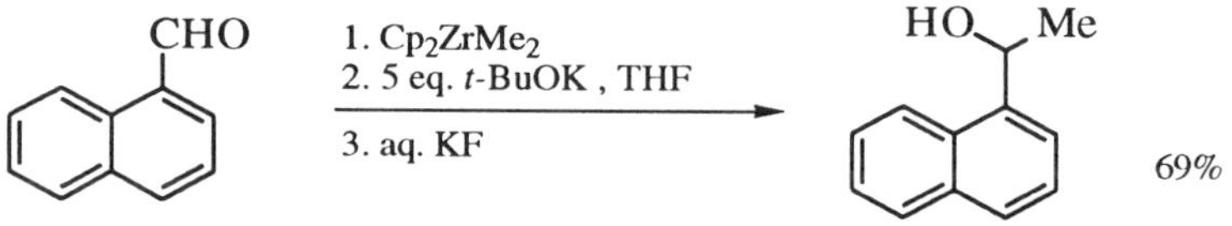

CHO
1. Cp_2ZrMe_2
2. 5 eq. *t*-BuOK, THF
3. aq. KF
HO Me
69%

Larson, A.L.; Baker, D.L.; Towne, R.W.; Straus, D.A. *Tetrahedron Lett.*, ***1991***, *32*, 5893.

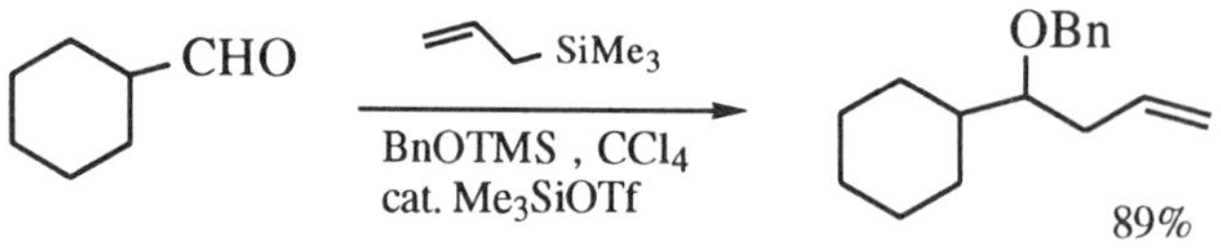

CHO
$SiMe_3$
BnOTMS, CCl_4
cat. Me_3SiOTf
OBn
89%

Mekhalfia, A.; Markó, I.E. *Tetrahedron Lett.*, ***1991***, *32*, 4779.

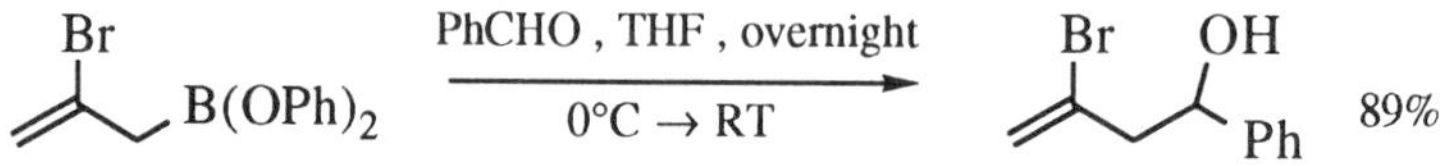

Br
$B(OPh)_2$
PhCHO, THF, overnight
0°C → RT
Br OH
Ph
89%

Hara, S.; Suzuki, A. *Tetrahedron Lett.*, ***1991***, *32*, 6749.

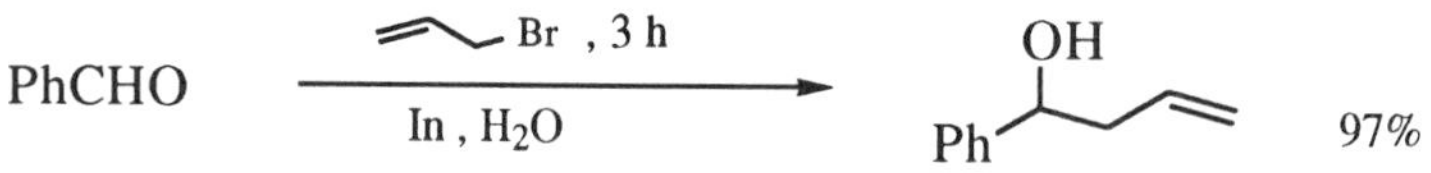

PhCHO
Br, 3 h
In, H_2O
OH
Ph
97%

Li, C.J.; Chan, T.H. *Tetrahedron Lett.*, ***1991***, *32*, 7017.

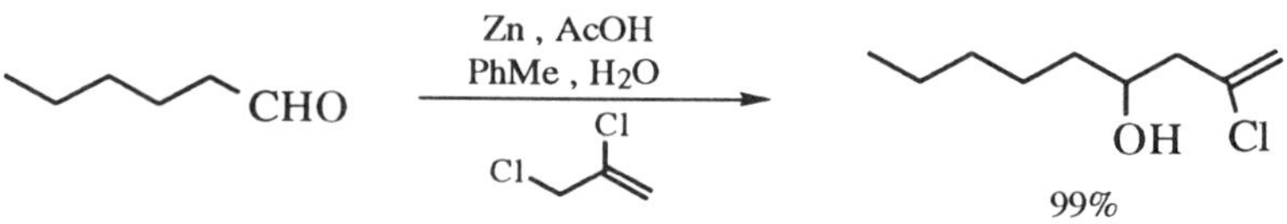

CHO
Zn, AcOH
PhMe, H_2O
Cl
Cl
OH Cl
99%

Oda, Y.; Matsuo, S.; Saito, K. *Tetrahedron Lett.*, ***1992***, *33*, 97.

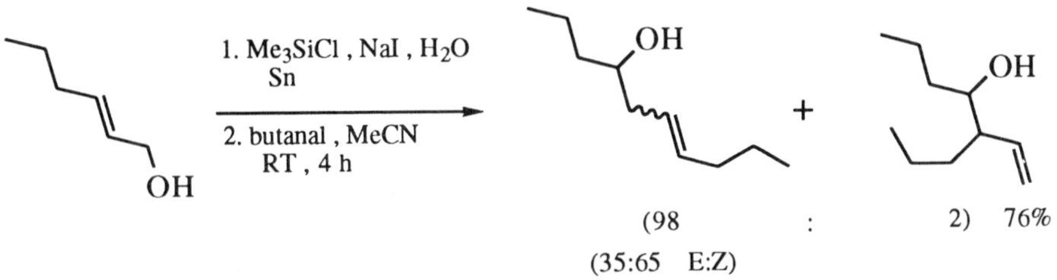

1. Me_3SiCl , NaI , H_2O
Sn
2. butanal , MeCN
RT , 4 h

OH OH + OH

(98 : 2) 76%
(35:65 E:Z)

Kanagawa, Y.; Nishiyama, Y.; Ishii, Y. *J. Org. Chem.*, ***1992***, *57*, 6988.

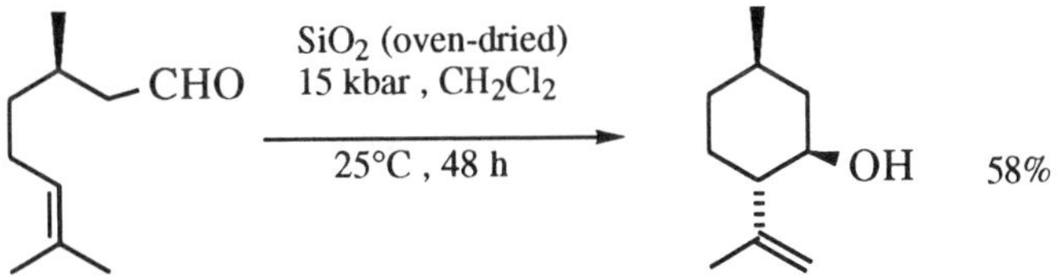

CHO SiO_2 (oven-dried)
15 kbar , CH_2Cl_2
25°C , 48 h
OH 58%

Dauben, W.G.; Hendricks, R.T. *Tetrahedron Lett.*, ***1992***, *33*, 603.

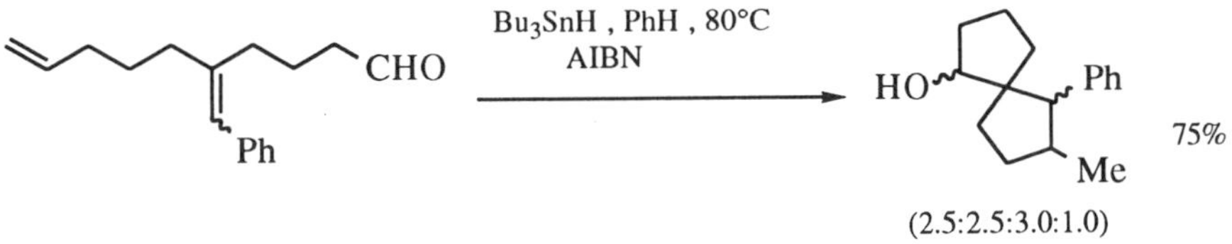

CHO Ph
Bu_3SnH , PhH , 80°C
AIBN
HO Ph Me 75%
(2.5:2.5:3.0:1.0)

Enholm, E.J.; Burroff, J.A. *Tetrahedron Lett.*, ***1992***, *33*, 1835.

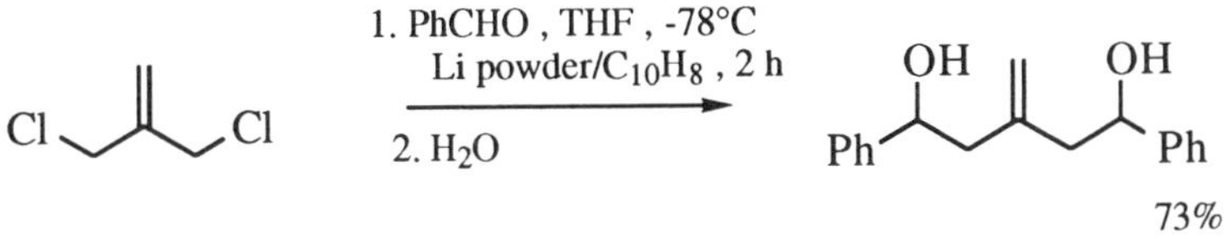

Cl Cl
1. PhCHO , THF , -78°C
Li powder/$C_{10}H_8$, 2 h
2. H_2O
OH OH Ph Ph
73%

Ramón, D.J.; Yus, M. *Tetrahedron Lett.*, ***1992***, *33*, 2217.

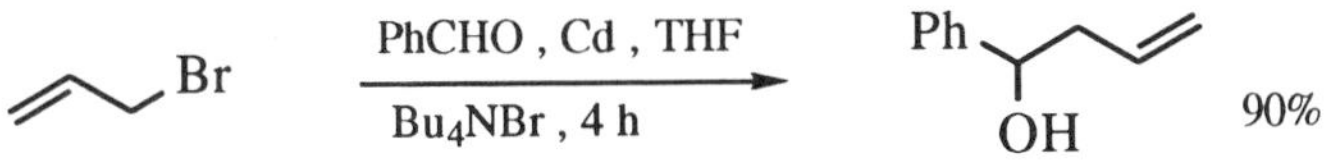

Br
PhCHO , Cd , THF
Bu_4NBr , 4 h
Ph OH 90%

Sain, B.; Prajapti, D.; Sandhu, J.S. *Tetrahedron Lett.*, ***1992***, *33*, 4795.

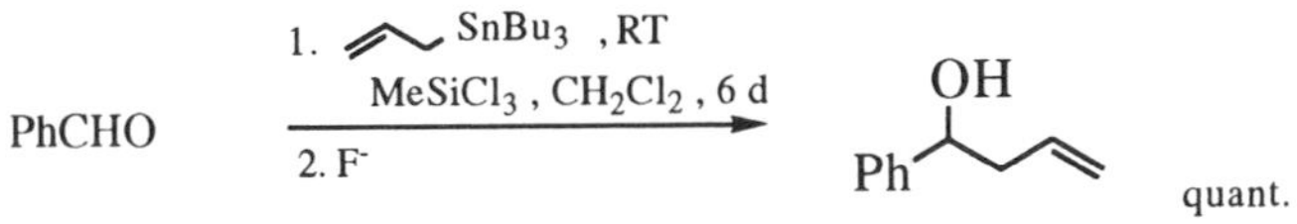

PhCHO
1. $SnBu_3$, RT
$MeSiCl_3$, CH_2Cl_2 , 6 d
2. F^-
OH Ph quant.

Marshall, R.L.; Young, D.J. *Tetrahedron Lett.*, ***1992***, *33*, 1365.

CHO — allyl bromide, THF, RT; cat. Cp_2TiCl_2, Zn, RT, 3 min → OH product 80%

Ding, Y.; Zhao, G. *Tetrahedron Lett.*, ***1992***, *33*, 8117.

PhCHO — 1. allyl $SiMe_3$, 21 h, Me_3SiOTf; 2. H_2O → Ph(OH) homoallyl alcohol 61%

Davis, A.P.; Jaspars, M. *Angew. Chem. Int. Ed. Engl.*, ***1992***, *31*, 420.

OH — PhCHO, $SnCl_2$, DMF, 2% $PdCl_2(PhCN)_2$, 25°C, 63 h → OH, Ph product 63%

(30:70 dr)

Takahara, J.P.; Masuyama, Y.; Kurusu, Y. *J. Am. Chem. Soc.*, ***1992***, *114*, 2577.

PhCHO — $MeCrCl_2(THF)_3$, THF, -60°C → 20°C, 18 h → Ph–CH(Me)OH 90%

Kauffmann, T.; Beirich, C.; Hamsen, A.; Möller, T.; Philipp, C.; Wingbermühle, D. *Chem. Ber.*, ***1992***, *125*, 157.

REVIEWS:

"Addition of Organochromium Compounds to Aldehydes: the Nozaki-Hiyama Reaction" Cintas, P. *Synthesis*, ***1992***, 248.

SECTION 35: ALCOHOLS AND THIOLS FROM ALKYLS, METHYLENES AND ARYLS

No examples of the reaction $RR^1 \rightarrow ROH$ (R^1 = alkyl, aryl, etc.) occur in the literature. For reactions of the type $RH \rightarrow ROH$ (R = alkyl or aryl) see Section 41 (Alcohols and Phenols from Hydrides).

AcO, Me, Me_2Si, AcO — 1. Br_2, CH_2Cl_2; 2. HF•Py, 0°C; 3. 30% H_2O_2, KF, $KHCO_3$, THF, MeOH → AcO, Me, OH, AcO 91%

Magar, S.S.; Fuchs, P.L. *Tetrahedron Lett.*, ***1991***, *32*, 7513.

SECTION 36: ALCOHOLS AND THIOLS FROM AMIDES

PhC(O)NH2 → (SmI_2, $LiNH_2$, THF, MeOH, RT, 3 secn) → Ph–CH2OH 72%

Kamochi, Y.; Kudo, T. *Tetrahedron Lett.*, ***1991***, *32*, 3511.

SECTION 37: ALCOHOLS AND THIOLS FROM AMINES

NO ADDITIONAL EXAMPLES

SECTION 38: ALCOHOLS AND THIOLS FROM ESTERS

MeO–C6H4–CH2OAc → ($HSiMe_3$, CO, PhH, cat. $Co_2(CO)_8$, 25°C, 1 atm.) → MeO–C6H4–CH2CH2–$OSiMe_3$ 76%

Chatani, N.; Sano, T.; Ohe, K.; Kawasaki, Y.; Murai, S. *J. Org. Chem.*, ***1990***, *55*, 5923.

PhC(O)OMe → (SmI_2, $LiNH_2$, THF, MeOH, RT, 14 min) → Ph–CH2OH 64%

Kamochi, Y.; Kudo, T. *Tetrahedron Lett.*, ***1991***, *32*, 3511.

AcO–(cyclohexene)–OAc → (*Pseudomonas cepacia* lipase P-30, 4 h) → HO–(cyclohexene)–OAc 64% (79% ee)

Harris, K.J.; Gu, Q.M.; Shih, Y-E.; Girdaukas, G.; Sih, C.J. *Tetrahedron Lett.*, ***1991***, *32*, 3941.

O_2N–C6H4–OAC → (1. Bu_3Sn_2O, PhMe, 80°C, 2 h; 2. pH 4) → O_2N–C6H4–OH 96%

Salomon, C.J.; Mata, E.G.; Mascaretti, O.A. *Tetrahedron Lett.*, ***1991***, *32*, 4239.

PhCH2CO_2Et → ($NaBH_4$, $ZnCl_2$, 2 h, $PhNMe_2$, THF, reflux) → PhCH2CH2OH 85%

Yamakawa, T.; Masaki, M.; Nohira, H. *Bull. Chem. Soc. Jpn.*, ***1991***, *64*, 2730.

n-$C_5H_{11}CHO$, DMF , 40°C
$NiCl_2(PPh_3)$-Zn , 7 h

n-C_5H_{11}
99%

Nokami, J.; Konishi, H.; Matsuura, H. *Chem. Lett.*, ***1991***, 2023.

5% $Ti(OiPr)_4$, 55°C
3 eq. $SiH(OEt)_3$, 4 h

Ph–CO_2Et → Ph–OH 89%

Berk, S.C.; Buchwald, S.L. *J. Org. Chem.*, ***1992***, *57*, 3751.

SECTION 39: ALCOHOLS AND THIOLS FROM ETHERS, EPOXIDES AND THIOETHERS

DMF , 60°C

n-Bu_3SnN_3 , 5h	15	:	1	(69%)
n-$Bu_3Sn(N_3)_2$, 4h	>99	:	1	(79%)

Saito, S.; Nishikawa, T.; Yokoyama, Y.; Moriwake, T. *Tetrahedron Lett.*, ***1990***, *31*, 221.

1. (Al reagent, t-Bu, Br, Me) CH_2Cl_2 , -20°C , 3h
2. 0°C , 12h
3. RT , 1h

1 : 9) 92%

Maruoka, K.; Sato, J.; Banno, H.; Yamamoto, H. *Tetrahedron Lett.*, ***1990***, *31*, 377.

MeLi , THF , -78°C

87%

Arjona, O.; de la Pradilla, R.F.; Mallo, A.; Plumet, J.; Viso, A. *Tetrahedron Lett.*, ***1990***, *31*, 1475.

Ph-OMe $\xrightarrow[3\ h]{BI_3\bullet PhNEt_2\ ,\ PhH}$ Ph-OH 85%

Narayana, C.; Padmanabhan, S.; Kabalka, G.W. *Tetrahedron Lett.*, ***1990***, *31*, 6977.

n-C_3H_7 (epoxide) $OSiMe_3$ $\xrightarrow[hexane]{LiBH_4\ ,\ 2\ h}$ n-C_3H_7 (OH) $OSiMe_3$ + n-C_3H_7 (OH) $OSiMe_3$

(96 : 4) 90%

Sugita, K.; Onaka, M.; Izumi, Y. *Tetrahedron Lett.*, ***1990***, *31*, 7467.

OTBDMS OTBDMS $\xrightarrow[THF\ ,\ 0°C\ \rightarrow\ RT]{secBuLi\text{-}CuCN}$ *sec*Bu OH OTBDMS OTBDMS 73%

no reaction with MeLi/CuCN or *n*-BuLi/CuCN; 78% with *t*-BuLi/CuCN

Lautens, M.; Smith, A.C.; Abd-El-Aziz, A.S.; Huboux, A.H. *Tetrahedron Lett.*, ***1990***, *31*, 3253.

Me O Cl $\xrightarrow{1\ eq.\ LDA\ ,\ THF\ \ -78°C}$ Me OH Cl + Me OH C≡C-H

82%

with 3 eq. LDA - 77%

Yadav, J.S.; Deshpande, P.K.; Sharma, G.V.M. *Tetrahedron Lett.*, ***1990***, *31*, 4495.

n-C_8H_{17} O H $\xrightarrow[2.\ MeOH\ ,\ -78°C]{1.\ LDBB\ ,\ THF\ \ -78°C\ ,\ 2\ min}$ n-C_8H_{17} HO + n-C_8H_{17} H

65% 35%

LDBB = lithium 4,4'-di-*t*-butylbiphenylide

Cohen, T.; Jeong, I-H.; Mudryk, B.; Bhupathy, M.; Awad, M.M.A. *J. Org. Chem.*, ***1990***, *55*, 1528.

Ph(OMe)(OMe) $\xrightarrow[2.\ H_2O]{1.\ Li\ ,\ THF,\ 20°C}$ Ph(Me)(OH) 70%

Azzena, U.; Denurra, T.; Melloni, G.; Piroddi, A.M. *J. Org. Chem.*, ***1990***, *55*, 5532.

Bu_3SnH , AIBN
PhH , reflux

H Me

n-C_4H_9

OH

(2.7:1)

68%

Rawal, V.H.; Newton, R.C.; Krishnamurthy, V. *J. Org. Chem.*, ***1990***, *55*, 5181.

1. MeMgCl
2. MeMgI , CuBr
3. aq. NH_4Cl

Ts

Me
Ts OH
Me

56%

Nájera, C.; Sansano, J.M. *Tetrahedron*, ***1990***, *46*, 3993.

MsO

1. $Bu_2Cu(CN)Li_2$, THF -20°C
2. $Bu_2Cu(CN)Li_2$, THF -20°C

Bu Bu OH + OH Bu Bu

(1.2 : 1) 75%

Kurth, M.J.; Abreo, M.A. *Tetrahedron*, ***1990***, *46*, 5085.

MeMgI , CuBr , PBu_3 0°C

OH

73%

Mitani, M.; Matsumoto, H.; Gouda, N.; Koyama, K. *J. Am. Chem. Soc.*, ***1990***, *112*, 1286.

1. (Li amide) N Li N
2. H_3O^+

OH

(90% ee , S)

78%

Asami, M. *Bull Chem. Soc. Jpn.*, ***1990***, *63*, 721.

Cp_2TiCl , C_6H_8

OH

91%

Rajan Babu, T.V.; Nugent, W.A.; Beattie, M.S. *J. Am. Chem. Soc.*, ***1990***, *112*, 6408.

Ph-epoxide → ($Zn(BH_4)_2/SiO_2$, THF; RT , 24 h) → Ph~~OH 90%

Ranu, B.C.; Das, A.R. *J. Chem. Soc., Chem. Commun.*, ***1990***, 1334.

2-methoxynaphthalene (OMe) → (KF•alumina, ethylene glycol; 215°C , 5 h) → 2-naphthol (OH) 80%

Radhakrishna, A.S.; Prasad Rao, K.R.K.; Suri, S.K.; Sivaprakash, K.; Singh, B.B. *Synth. Commun.*, ***1991***, *21*, 379.

Ph-epoxide → ($NaBH_4$, H_2O , RT , 48 h; β-cyclodextrin) → Ph(OH) 66% (25% ee , S) + Ph~~OH 17% 15% ee , S)

Hu, Y.; Uno, M.; Harada, A.; Takahashi, S. *Bull. Chem. Soc. Jpn.*, ***1991***, *64*, 1884.

Me, O, Ph → (Bu_3SnH , AIBN , PhH (0.05 M); syringe pump addition) → Me, OH, Ph, H 82%

Kim, S.; Lee, S.; Koh, J.S. *J. Am. Chem. Soc.*, ***1991***, *113*, 5106.

Ph~~OBn → (3 $AlCl_3$, 4 $PhNMe_2$; CH_2Cl_2 , RT; 30 min) → Ph~~OH 93%

Akiyama, T.; Hirofuji, H.; Ozaki, S. *Tetrahedron Lett.*, ***1991***, *32*, 1321.

Cl~epoxide → (Ph_3SnH , hν , 70°C) → allyl-OH

Krosley, K.W.; Gleicher, G.J.; Clapp, G.E. *J. Org. Chem.*, ***1992***, *57*, 840.

Ph-epoxide-OTs → (Dibal , CH_2Cl_2; -15°C → RT) → Ph~(OH) + Ph(OH)~ (>99 : <1) 73%

Chong, J.M. *Tetrahedron Lett.*, ***1992***, *33*, 33.

e^- , $SmCl_3$, Bu_4NBF_4
KI , DMF , RT
Mg anode , Ni foam cathode

t-Bu O → t-Bu OH 63%

Espanet, B.; Duñach, E.; Périchon, J. *Tetrahedron Lett.*, ***1992***, *33*, 2485.

hν , $MeCN-H_2O$, NEt_3
$NaBH_4$

Ph O → Ph OH 95%

Epling, G.A.; Wang, Q. *J. Chem. Soc., Chem. Commun.*, ***1992***, 1133.

O Me Me OH

BuLi , ether , 0°C

Me OH Me OH Bu 92%

Lautens, M.; Belter, R.K. *Tetrahedron Lett.*, ***1992***, *33*, 2617.

O H $OSiMe_2t$-Bu

BuLi

Br HO H $OSiMe_2t$-Bu 93%

Woo, S.; Keay, B.A. *Tetrahedron Lett.*, ***1992***, *33*, 2661.

$OSiMe_2t$-Bu O Ph

Bu_3SnH , AIBN
PhH , reflux , 3 h

t-$BuMe_2SiO$ Ph OH 68%

Kim, S.; Koh, J.S. *Tetrahedron Lett.*, ***1992***, *33*, 7391.

O

SiO_2 - $Zn(BH_4)_2$, THF

OH + OH

(98 : 2) 95%

Ranu, B.C.; Das, A.R. *J. Chem. Soc., Perkin Trans. I*, ***1992***, 1881.

O

MeCN , LiBr , RT
Amberlyst-15 resin

Br OH (>99:1)

Bonini, C.; Giuliano, C.; Righi, G.; Rossi, L. *Synth. Commun.*, ***1992***, *22*, 1863.

REVIEWS:

"Biosynthetic, Biomimetic, and Related Epoxide Cyclizations"
Taylor, S.K. *Org. Prep. Proceed. Int.*, ***1992***, *24*, 247.

Additional examples of ether cleavages may be found in Section 45A (Protection of Alcohols and Thiols).

SECTION 40: ALCOHOLS AND THIOLS FROM HALIDES AND SULFONATES

Br → SH 71%
1. $(Bu_3Sn)_2O$, 20°C, 20 h
2. TBAF•3 H_2O

Gingras, M.; Harpp, D.N. *Tetrahedron Lett.*, ***1990***, *31*, 1397.

I → OH 91%
1. *t*-BuLi, pentane/ether, -78°C, 5 min
2. pentanal

general method - preparation of 1° organolithium reagents

Bailey, W.F.; Punzalan, E.R. *J. Org. Chem.*, ***1990***, *55*, 5404.

I → OH 64%
1. *t*-BuLi, pentane-ether, -78°C
2. TMEDA
3. warm
4. MeCHO

Bailey, W.F.; Khanolkar, A.D. *J. Org. Chem.*, ***1990***, *55*, 6058.

BuMgCl → $BuCH_2OH$ 95%
1. $ClCH_2OCH_2CH_2Cl$, ether, RT
2. Na-K

Ogle, C.A.; Wilson, T.E.; Stowe, J.A. *Synthesis*, ***1990***, 495.

Ph Br → Ph SH <75%
1. TBS-SLi, ether, 0°C - 25°C, 48 h
2. HF, MeCN

Kraus, G.A.; Andersh, B. *Tetrahedron Lett.*, ***1991***, *32*, 2189.

1. BuLi, ether, -78°C
2. ethylene oxide, CuCN
-65°C → 25°C
3. 10% H_2SO_4

56%

Wang, W.; D'Andrea, S.V.; Freeman, J.P.; Szmuszkovicz, J. *J. Org. Chem.*, ***1991***, *56*, 2914.

1. *n*-PrSNaTol, NMP
reflux (Dean-Stark)
2. HCl

96%

NMP = N-methyl-2-pyrrolidinone

Shaw, J.E. *J. Org. Chem.*, ***1991***, *56*, 3728.

1. Bu_3SnH, air
PhMe, 20°C
2. $NaBH_4$, EtOH

80% 14%

Nakamura, E.; Inubushi, T.; Aoki, S.; Machii, D. *J. Am. Chem. Soc.*, ***1991***, *113*, 8980.

SECTION 41: ALCOHOLS AND THIOLS FROM HYDRIDES

6 (F3C)(Me)C(O–O), 2 h
CH_2Cl_2/TFP

90%

TFP = 1,1,1-trifluoropropanone

Mello, R.; Cassidei, L.; Fiorentino, M.; Fusco, C.; Curci, R. *Tetrahedron Lett.*, ***1990***. *31*, 3067.

$[Ru_3O(pfb)_6(OEt_2)_3]^+$
MeCN, air (3 atm), 75°C
10^{-5} mol. catalyst

(5 : 1)
15 turnovers

pfb = perfluorobutyrate

Davis, S.; Drago, R.S. *J. Chem. Soc., Chem. Commun.*, ***1990***, 250.

1. Cu°, O_2, Me_3N-O, MeCN, 75°C, 6h
2. 0.5M HCl

quant.

Reinaud, O.; Capdevielle, P.; Maumy, M. *J. Chem. Soc., Chem. Commun.*, ***1990***, 566.

OTs → Co^{+2} - montmorillonite, 180°C, 2 h → OH, Ts 96% + OH, Ts 4%

Fries rearrangement

Pitchumani, K.; Pandian, A. *J. Chem. Soc., Chem. Commun.*, ***1990***, 1613.

Me → $NaBO_3$•4 H_2O, TfOH, -10°C → Me, OH + Me, OH + Me, OH

(65 : 26 : 9) 66%

Surya Prakash, G.K.; Krass, N.; Wang, Q.; Olah, G.A. *SynLett*, ***1991***, 39.

Rhizopus arrhizus, 28°C, 18 h

42% 27%

Ouazzani,J.; Arseniyadis, S.; Alvarez-Manzaneda, R.; Cabrera, E.; Ouvisson, G. *Tetrahedron Lett.*, ***1991***, *32*, 647.

Mucor plumbeus, CBS 110-6, 2 d

66% 22%

Hammoum., A.; Revial, G.; D'Angelo, J.; Girault, J.P.; Azerad, R. *Tetrahedron Lett.*, ***1991***, *32*, 651.

Oxone, TBA, 21 h
10 Bu_4NHSO_5/CH_2Cl_2
aq. acetone, RT (dark)
$NaHCO_3$

quant.

Kumarathasan, R.; Hunter, N.R. *Org. Prep. Proceed. Int.*, ***1991***, *23*, 651.

1. HF ; SO_2ClF
2. 90% H_2O_2 , HF
-10°C → 0°C

OH

>98%

Olah, G.A.; Keumi, T.; Lecoq, J.C.; Fung, A.P.; Olah, J.A. *J. Org. Chem.*, ***1991***, *56*, 6148.

O_2 , CH_2Cl_2 , 30°C
Hhfacac-Zn
[{Fe(HBpz$_3$)(hfacac)}O]

OH

21 turnovers

Hhfacac = hexafluoroacetyl acetone
HBpz$_3$ = hydrotris-1-pyrazolylborate

Kitajima, N.; Ito, M.; Fukui, H.; Moro-oka, Y. *J. Chem. Soc., Chem. Commun.*, ***1991***, 102.

P. putida
O_2

OH 66% (45% ee) + O 34%

Boyd, D.R.; Sharma, N.D.; Stevenson, P.J.; Chima, J.; Gray, D.J.; Dalton, H. *Tetrahedron Lett.*, ***1991***, *32*, 3887.

HO, H, O, N, NO_2

1. Cu , O_2 , TMAO
MeCN , 75°C
2. H_3O^+

HO, H, O, N, NO_2, OH 80%

TMAO = Me_3N-O

Reinaud, O.; Capdevielle, P.; Maumy, M. *J. Chem. Soc., Perkin Trans. I*, ***1991***, 2129.

Me

Ru(TPP)(CO) , HBr
Cl, N⊕, O⊖, Cl , PhH , RT , 6 h

Me, OH 94%

Ohtake, H.; Highuchi, T.; Hirobe, M. *J. Am. Chem. Soc.*, ***1992***, *114*, 10660.

SECTION 42: ALCOHOLS AND THIOLS FROM KETONES

The following reaction types are included in this section:
A. Reductions of Ketones to Alcohols **B.** Alkylations of Ketones, forming Alcohols

Coupling of ketones to give diols is found in Section 323 (Alcohol → Alcohol).

1. $Rh_2(OAc)_4$, CH_2Cl_2, RT
2. $NaBH_4$, $CeCl_3$

O, n-C_5H_{11}, N_2, n-C_3H_7 → OH, n-C_5H_{11} + OH, n-C_5H_{11}, n-C_3H_7

(70 : 30) 86%

Taber, D.F.; Hennessy, M.J.; Louey, J.P. *J. Org. Chem.*, ***1992***, *57*, 436.

HO–C6H4–CHO → HO–C6H4–OH

1. $[Na_2CO_3•1.5\ H_2O_2]$ aq. THF ,)))))))
2. AcOH

86%

Kabalka, G.W.; Reddy, N.K.; Narayama, C. *Tetrahedron Lett.*, ***1992***, *33*, 865.

SECTION 42A: REDUCTION OF KETONES TO ALCOHOLS

ASYMMETRIC REDUCTION

O, CO_2Me → OH, CO_2Me

Bakers yeast, $ClCH_2CO_2Et$

51% (91% ee)

Nakamura, K.; Kawai, Y.; Ohno, A. *Tetrahedron Lett.*, ***1990***, *31*, 267.

O, $SiMe_3$ → H, OH, $SiMe_3$

1. (-)IPC_2BCl , 3 h -35°C
2. $HN(CH_2CH_2OH)_2$

67% (95% ee)

Soderquist, J.A.; Anderson, C.L.; Miranda, E.I.; Rivera, I.; Kabalka, G.W. *Tetrahedron Lett.*, ***1990***, *31*, 4677.

O, CO_2Et, Me → OH, CO_2Et, Me + OH, CO_2Et, Me

bakers yeast, methyl vinyl ketone

(96 : 4) 96%

Nakamura, K.; Kawai, Y.; Miyai, T.; Ohno, A. *Tetrahedron Lett.*, ***1990***, *31*, 3631.

O Ph Me → OH Ph Me

H B- N O H Ph Ph

0.1

90%

(87% ee , R)

Rama Rao, A.V.; Gurjar, M.K.; Sharma, P.A.; Kaiwar, V. *Tetrahedron Lett.*, ***1990***, *31*, 2341.

O Ph + O B- H O → OH Ph

C_7H_8 , -78°C , 15h

H Ph Ph N. B O Bu

0.1 equiv.

>95%

92%ee , R

Corey, E.J.; Bakshi, R.K. *Tetrahedron Lett.*, ***1990***, *31*, 611.

S S Ph O → S S Ph OH

B_2H_6

Ph Ph N. B O Ph

94% ee

DeNinno, M.P.; Perner, R.J.; Lijewski, L. *Tetrahedron Lett.*, ***1990***, *31*, 7415.

O → OH

Thermoanaerobium brockii alcohol dehydrogenase

79% ee , S

Pseudomonas alcohol dehydrogenase

97% ee , R

Shen, G-J.; Wang, Y-F.; Bradshaw, C.; Wong, C-H. *J. Chem. Soc., Chem. Commun.*, ***1990***, 677.

O → OH

BF_3•OEt_2 , PhH , 0°C

N Ph Me • BH_3

81%

(41% ee , R)

Kanth, J.V.B.; Periasamy, M. *J. Chem. Soc., Chem. Commun.*, ***1990***, 1145.

O, CO_2H, S — bakers yeast, 28 h → OH, CO_2H, S + OH, CO_2H, S

(70 : 30) 67%

(>99% ee) (98% ee)

Fujisawa, T.; Mobele, B.I.; Shimizu, M. *Tetrahedron Lett.*, ***1991***, *32*, 7055.

O, Ph — Zn(cra)*, THF, RT → OH, Ph 88%

(84% ee)

Zn(cra) = zinc/sodium chiraldate

Lalloz, L.; Vanderesse, R.; Mayesky, B.; Caubere, P. *Chem. Lett.*, ***1991***, 1961.

O, Ph — [$LiAlH_4$- (OH HO, MeO, OMe, MeO OMe OMe OMe) /EtOH], THF, -78°C → OH, Ph 84%

(94% ee)

Rawson, D.; Meyers, A.I. *J. Chem. Soc., Chem. Commun.*, ***1992***, 494.

OMe, Ph, Me, O — Bu_3SnH, TBAF, 0°C → OMe, Ph, Me, OH + OMe, Ph, Me, OH

(100 : 0) 81%

with Bu_2SnClH, 0°C (10 : 90) 72%

Shibata, I.; Yoshida, T.; Kawakami, T.; Baba, A.; Matsuda, H. *J. Org. Chem.*, ***1992***, *57* 4049.

O, n-C_5H_{11}, CCl_3 — (catecholborane) O, B-H, O, PhMe; 0.1 (H, Ph, Ph, O, N, B, Bu), -60°C, 12 h → H, OH, n-C_5H_{11}, CCl_3 96%

(95% ee)

Corey, E.J.; Link, J.O. *Tetrahedron Lett.*, ***1992***, *33*, 3431.

Corey, E.J.; Cheng, X-M.; Cimprich, K.A.; Sarshar, S. *Tetrahedron Lett.*, ***1991***, *32*, 6835.

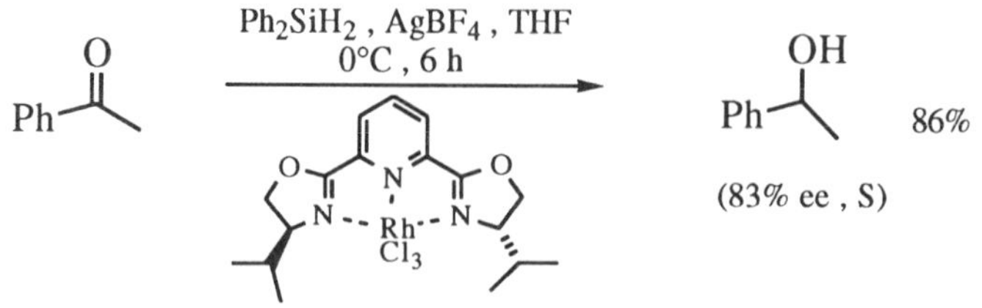

Ph_2SiH_2, $AgBF_4$, THF, 0°C, 6 h; RhCl_3 pybox catalyst → 86% (83% ee, S)

Nishiyama, H.; Yamaguchi, S.; Kondo, M.; Itoh, K. *J. Org. Chem.*, ***1992***, *57*, 4306.

NON-ASYMMETRIC REDUCTION

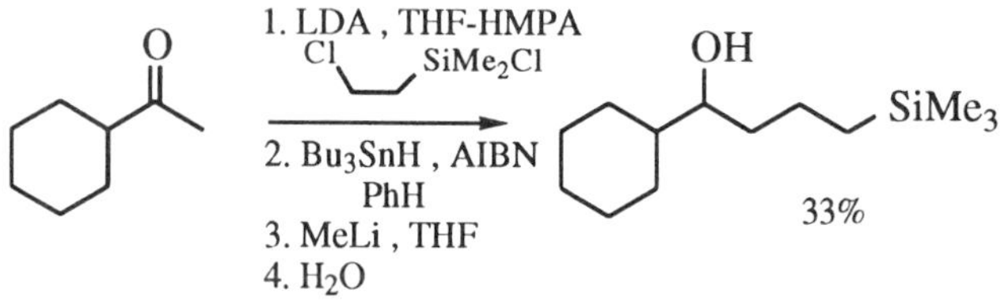

1. LDA, THF-HMPA, Cl$\sim$$SiMe_2Cl$; 2. Bu_3SnH, AIBN, PhH; 3. MeLi, THF; 4. H_2O → OH, $SiMe_3$ 33%

Walkup, R.D.; Kane, R.R.; Obeyesekene, N.U. *Tetrahedron Lett.*, ***1990***, *31*, 1531.

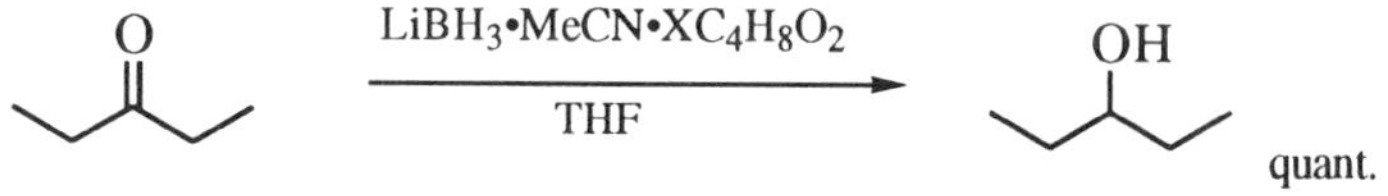

$LiBH_3$•MeCN•$XC_4H_8O_2$, THF → OH quant.

Mittakanti, M.; Peters, J.L.; Morse, K.W. *J. Org. Chem.*, ***1990***, *55*, 4464.

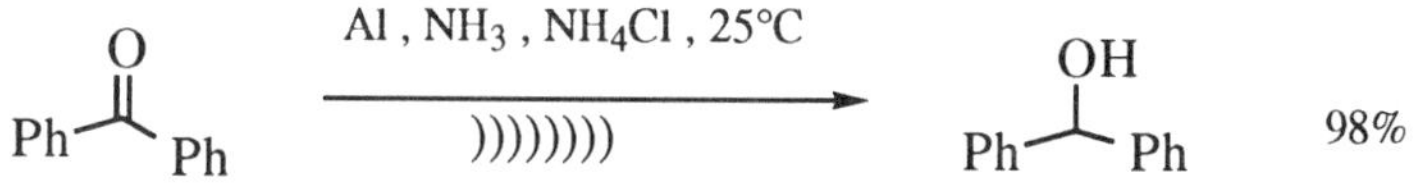

Al, NH_3, NH_4Cl, 25°C,))))))) → Ph, OH, Ph 98%

Sato, R.; Nagaoka, T.; Goto, T.; Saito, M. *Bull. Chem. Soc. Jpn.*, ***1990***, *63*, 290.

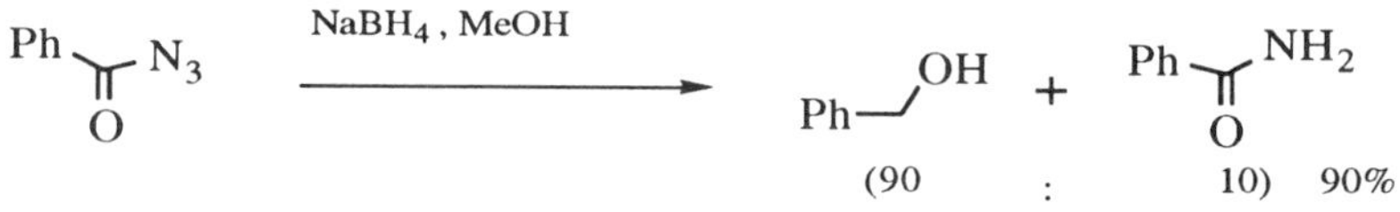

Ph, N_3, O; $NaBH_4$, MeOH → Ph, OH + Ph, NH_2, O (90 : 10) 90%

Rao, H.S.P.; *Synth. Commun.*, ***1990***, *20*, 45.

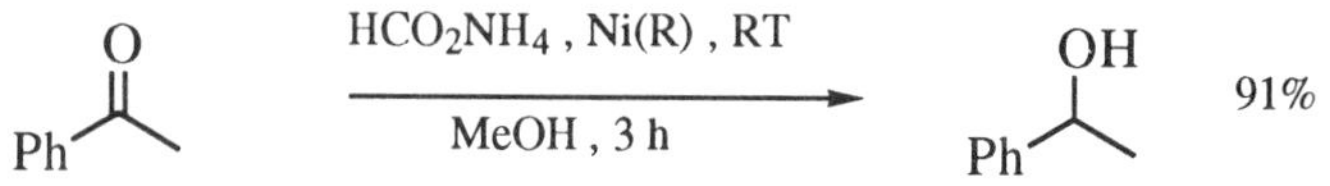

HCO_2NH_4, Ni(R), RT, MeOH, 3 h → Ph, OH 91%

Chen, F-E.; Zhang, H.; Yuan, W.; Zhang, W-W. *Synth. Commun.*, ***1991***, *21*, 107.

PhCHO → 1. TBSO–CH=C(Br)CO$_2$Et, PhLi, THF, -90°C; 2. HNTMS$_2$, TMSI, -78°C → (TBSO, Ph-substituted dihydrofuran, CO$_2$Et) 70%

Ranu, B.C.; Das, A.R. *J. Org. Chem.*, ***1991***, *56*, 4796.

Cyclohexanone → 0.1% RuCl$_2$(PPh$_3$)$_3$, 2.4 eq. NaOH, iPrOH, reflux, 1h → cyclohexanol (OH) 89%

Chowdhury, R.L.; Bäckvall, J-E. *J. Chem. Soc., Chem. Commun.*, ***1991***, 1063.

Cyclohexanone → Bu$_3$SnH, TBAF, THF, RT, 5 h → cyclohexanol (OH) 99%

Shibata, I.; Yoshida, T.; Baba, A.; Matsuda, H. *Chem. Lett.*, ***1991***, 307.

Cyclohexenone → NaBH$_4$, CaCl$_2$, MeOH → cyclohexenol (OH) + cyclohexanol (OH) (97 : 3) 92%

Fujii, H.; Oshima, K.; Utimoto, K. *Chem. Lett.*, ***1991***, 1847.

2-Methylcyclohexanone (O, Me) → pyrrolidine N·BH$_3$ Li, THF, 25°C, 30 min → 2-methylcyclohexanol (OH, Me) quant.

Fisher, G.B.; Harrison, J.; Fuller, J.C.; Goralski, C.T.; Singaram, B. *Tetrahedron Lett.*, ***1992***, *22*, 4533.

REVIEWS:

"Bakers Yeast as a Reagent in Organic Synthesis"
Servi, S. *Synthesis*, ***1990***, 1.

"Practical and Useful Methods for the Enantioselective Reduction of Unsymmetrical Ketones"
Singh, V.K. *Synthesis*, ***1992***, 605.

SECTION 42B: ALKYLATION OF KETONES, FORMING ALCOHOLS

Aldol reactions are listed in Section 330 (Ketone-Alcohol)

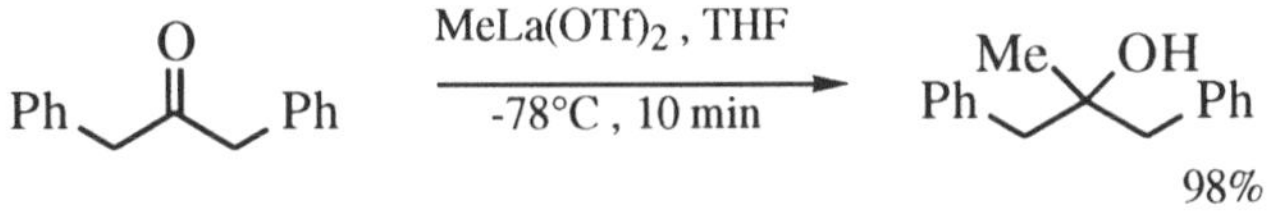

MeLa(OTf)$_2$, THF

-78°C , 10 min

98%

Collins, S.; Hong, Y.; Hoover, G.J.; Veit, J.R. *J. Org. Chem.*, ***1990***, *55*, 3565.

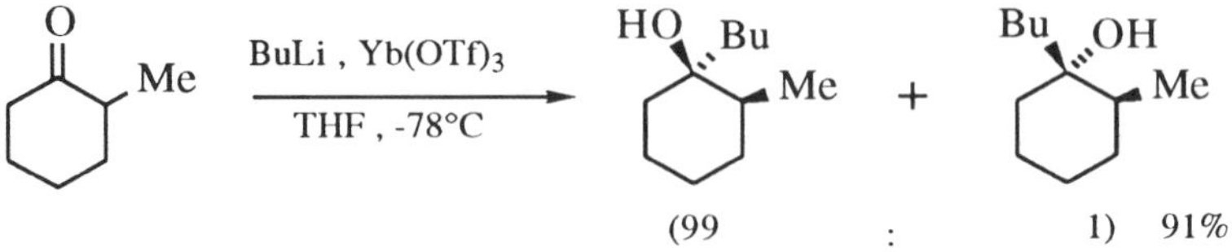

BuLi , Yb(OTf)$_3$

THF , -78°C

(99 : 1) 91%

Molander, G.A.; Burkhardt, E.R.; Weining, P. *J. Org. Chem.*, ***1990***, *55*, 4990.

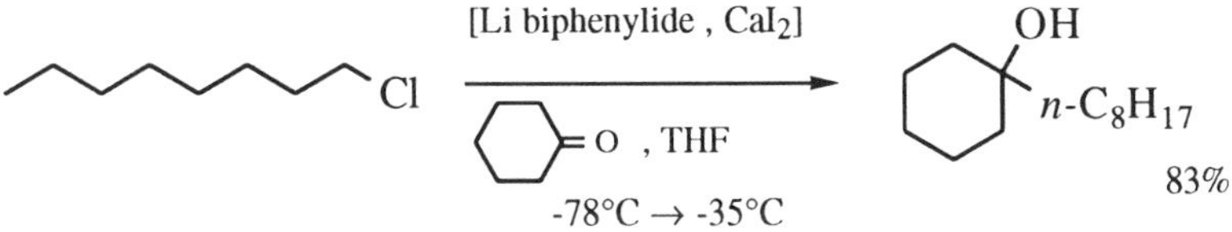

[Li biphenylide , CaI$_2$]

, THF

-78°C → -35°C

83%

Wu, T-C.; Xiong, H.; Rieke, R.D. *J. Org. Chem.*, ***1990***, *55*, 5045.

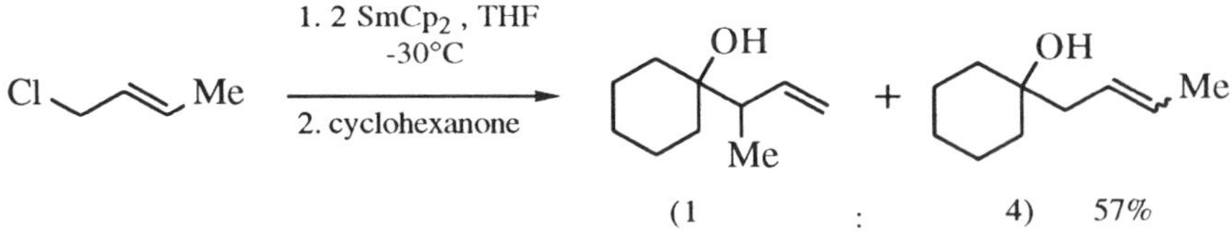

1. 2 SmCp$_2$, THF -30°C

2. cyclohexanone

(1 : 4) 57%

Collin, J.; Bied, C.; Kagan, H.B. *Tetrahedron Lett.*, ***1991***, *32*, 629.

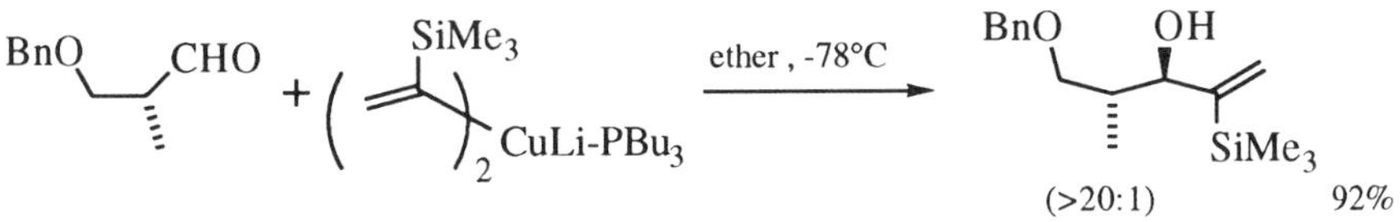

ether , -78°C

(>20:1) 92%

Burke, S.D.; Piscopio, A.D.; Marron, B.E.; Matulenko, M.A.; Pan, G. *Tetrahedron Lett.*, ***1991***, *32*, 857.

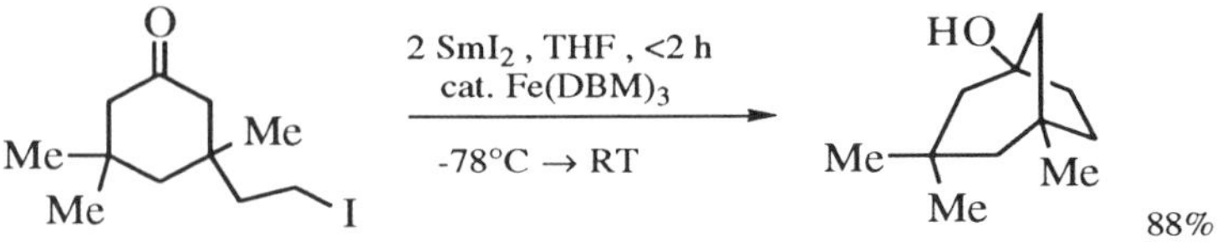

2 SmI$_2$, THF , <2 h cat. Fe(DBM)$_3$

-78°C → RT

88%

DBM = dibenzoyl methide

Molander, G.A.; McKie, J.A. *J. Org. Chem.*, ***1991***, *56*, 4112.

1. $\diagup\!\!\diagdown$ $Fe(CO)_2Cp$ BF_3, ether
2. NaI, wet acetone

77%

Jiang, S.; Turos, E. *Tetrahedron Lett.*, ***1991***, *32*, 4619.

$BuCeCl_2$

77%

ultrasound used with $CeCl_3 \cdot 7\ H_2O$ to prepare cerium reagent

Greeves, N.; Lyford, L. *Tetrahedron Lett.*, ***1992***, *33*, 4759.

SECTION 43: ALCOHOLS AND THIOLS FROM NITRILES

300°C, iPrOH
hydrous zirconium oxide

73%

Takahashi, K.; Shibagaki, M.; Matsushita, H. *Chem. Lett.*, ***1990***, 311.

SECTION 44: ALCOHOLS AND THIOLS FROM ALKENES

1. [(+)-BINAP]RhCl
PhMe, -5°C, 72h
2. NaOH, H_2O_2

73% (38%ee, S)

Sato, M.; Miyaura, N.; Suzuki, A. *Tetrahedron Lett.*, ***1990***, *31*, 231.

$Co(acac)_2$, O_2
iPrOH

46% 17% 8%

Kato, K.; Yamada, T.; Takai, T.; Inoki, S.; Isayama, S. *Bull. Chem. Soc., Jpn.*, ***1990***, *63*, 179.

$Ru_3(CO)_{12}$, 180°C
$(c\text{-}C_6H_{11})_3P$, H_2O
HCO_2Me, 10 h
'shaking' autoclave

9% 60%

Jenner, G. *Tetrahedron Lett.*, ***1991***, *32*, 505.

1. 9-BBN
2. NaOOH

$OSiMe_2t$-Bu

HO HO HO

+ +

$OSiMe_2t$-Bu $OSiMe_2t$-Bu $OSiMe_2t$-Bu

HO HO HO

(13 : 1 : -) 93%

with thexylborane (1 : 15 : 1) 91%

Harada, T.; Matsuda, Y.; Wada, I.; Uchimura, J. *J. Chem. Soc., Chem. Commun.*, ***1990***, 21.

$[RhH(PEt_3)_3]$, 120°C , EtOH

OH

94%

(*n:iso* = 2.22)

MacDougall, J.K.; Cole-Hamilton, D.J. *J. Chem. Soc., Chem. Commun.*, ***1990***, 165.

hν (Vycor) , H_2O_2

MeCN , 15 h

CN + OH

34% 66%

Sonawane, H.R.; Bellur, N.S.; Shah, V.G. *J. Chem. Soc., Chem. Commun.*, ***1990***, 1603.

1. (catecholborane: O, B H, O)

$LiBH_4$, THF

2. H_2O_2 , NaOH

OH + OH

94% 6%

Arase, A.; Nunokawa, Y.; Masuda, Y.; Hoshi, M. *J. Chem. Soc., Chem. Commun.*, ***1991***, 205.

Ph

$NaBH_4$, iPrOH , 100°C

CH_2Cl_2 , CO (500 psi)

$B^{\ominus}Ph_3$, 22 h

$\oplus$Rh (COD)

Ph OH + Ph OH

(95 : 5) 83%

Zhou, J-Q.; Alper, H. *J. Chem. Soc., Chem. Commun.*, ***1991***, 233.

n-C_9H_{19}

OH

1. 4 eq. EtMgCl , 12 h
5% Cp_2ZrCl_2

2. $B(OMe)_3$; H_2O_2

n-C_9H_{19} OH OH + n-C_9H_{19} OH OH

(>99 : 1) 75%

Hoveyda, A.H.; Xu, Z.; Morken, J.P.; Houri, A.F. *J. Am. Chem. Soc.*, ***1991***, *113*, 8950.

1. 0.01% $[PdCl(\pi\text{-}C_3H_5)]_2$
$HSiCl_3$, 0.02% R-MOP
2. KF , $KHCO_3$, H_2O_2

OH

89%

R-MOP = (R)-2-methoxy-2-diphenylphosphino-1,1'-binaphthyl

Uozumi, Y.; Lee, S-Y.; Hayashi, T. *Tetrahedron Lett.*, ***1992***, *33*, 7185.

1. 2 eq. (catecholborane) , 20°C
2% $Rh(PPh_3)Cl$, THF
2. NaOH , aq. H_2O_2
EtOH , THF

HO + OH

(99 : 1)

Evans, D.A.; Fu, G.C.; Hoveyda, A.H. *J. Am. Chem Soc.*, ***1992***, *114*, 6671.

1. (catecholborane) , 25°C
$Cp_2LaCH(SiMe_3)_2$
2. H_2O_2 , NaOH

OH

78%

Harrison, K.N.; Marks, T.J. *J. Am. Chem. Soc.*, ***1992***, *114*, 9220.

Ph

1. (catecholborane) , PPh_3
2% $RhCl(PPh_3)_3$
2. H_2O_2 , NaOH

OH Ph + Ph OH

(>99 : 1)

Burgess, K.; van der Donk, W.A.; Westcott, S.A.; Marder, T.B.; Baker, R.T.; Calabrese, J.C. *J. Am. Chem. Soc.*, ***1992***, *114*, 9350.

5% $LiBEt_3H$, THF , 5 h
20°C , $(BuO)_2BH$

OH + OH

(95 : 5) 72%

Arase, A.; Nunokawa, Y.; Masuda, Y.; Hoshi, M. *J. Chem. Soc., Chem. Commun.*, ***1992***, 51.

SECTION 45: ALCOHOLS AND THIOLS FROM MISCELLANEOUS COMPOUNDS

NO ADDITIONAL EXAMPLES

SECTION 45A: PROTECTION OF ALCOHOLS AND THIOLS

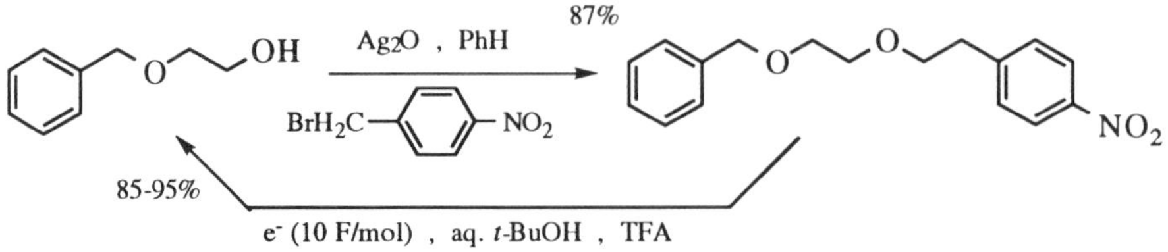

87%
Ag_2O , PhH
$BrH_2C-C_6H_4-NO_2$
NO_2
85-95%
e⁻ (10 F/mol) , aq. *t*-BuOH , TFA

Fukase, K.; Hata, N.; Oishi, T.; Kusumoto, S. *Tetrahedron Lett.*, ***1990***, *31*, 381.

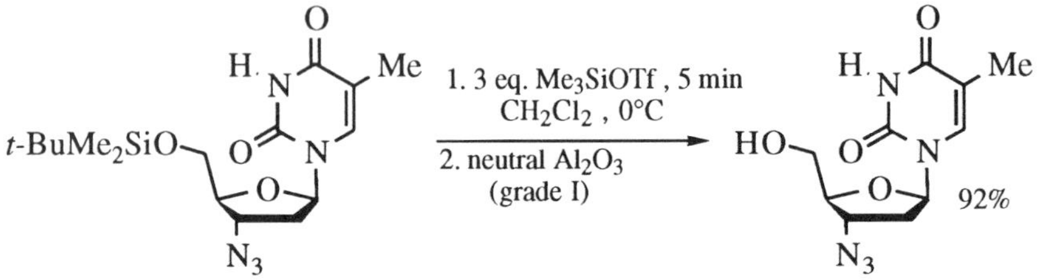

O H N Me *t*-$BuMe_2SiO$ O N O N_3
1. 3 eq. Me_3SiOTf , 5 min
CH_2Cl_2 , 0°C
2. neutral Al_2O_3
(grade I)
O H N Me HO O N O N_3 92%

Bou, V.; Vilarrasa, J. *Tetrahedron Lett.*, ***1990***, *31*, 567.

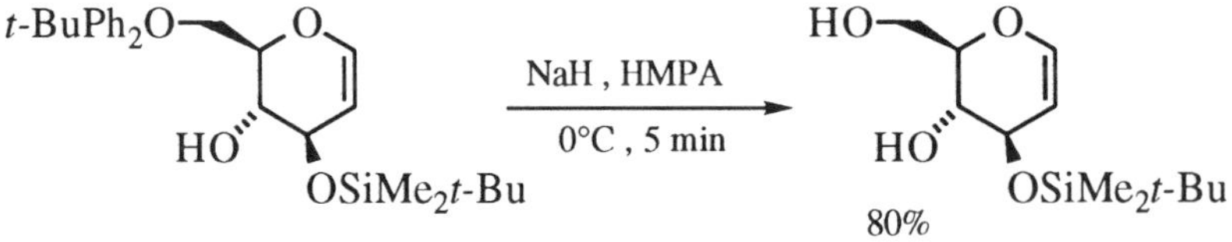

t-$BuPh_2O$ O HO $OSiMe_2t$-Bu
NaH , HMPA
0°C , 5 min
HO O HO $OSiMe_2t$-Bu
80%

Shekhani, M.S.; Khan, K.M.; Mahmood, K.; Shah, P.M.; Malik, S. *Tetrahedron Lett.*, ***1990***, *31*, 1669.

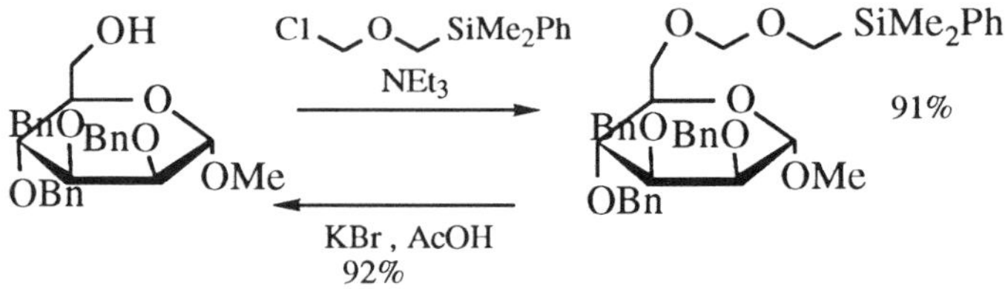

OH O BnO BnO OMe OBn
Cl O $SiMe_2Ph$
NEt_3
O O $SiMe_2Ph$ O BnO BnO OMe OBn
91%
KBr , AcOH
92%

Boons, G.J.P.H.; Elie, C.J.J.; van der Marel, G.A.; van Boom, J.H. *Tetrahedron Lett.*, ***1990***, *31*, 2197.

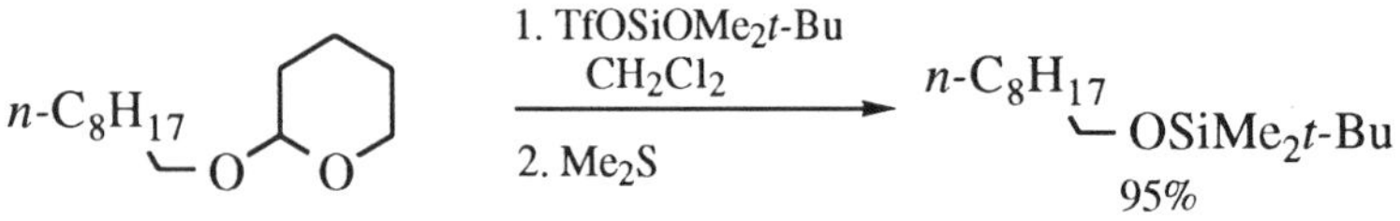

n-C_8H_{17} O O
1. $TfOSiOMe_2t$-Bu
CH_2Cl_2
2. Me_2S
n-C_8H_{17} $OSiMe_2t$-Bu
95%

Kim, S.; Kee, I.S. *Tetrahedron Lett.*, ***1990***, *31*, 2899.

H_2 , $Pd(OH)_2$
MeOH

BnO~~~~O-Si(i-Pr)(Et)... → HO~~~~OH 93%

in dioxane, Bn cleaved but not $OSiR_3$

Toshima, K.; Yanagawa, K.; Mukaiyama, S.; Tatsuta, K. *Tetrahedron Lett.*, ***1990***, *31*, 6697.

BnOH , PPh_3
DEAD

74%

Bajwa, J.S.; Anderson, R.C. *Tetrahedron Lett.*, ***1990***, *31*, 6973.

n-$C_{10}H_{21}OTMS$ —$FeCl_3$, MeCN , RT; 1 min→ n-$C_{10}H_{21}OH$

other Lewis acids were also used

Cort, A.D. *Synth. Commun.*, ***1990***, *20*, 757.

n-$C_8H_{17}OH$ —1. $(MeO)_2CH_2$, pTsOH, CH_2Cl_2 , 10 h; 2. PhSH , pTsOH, CH_2Cl_2 , 7 h→ n-$C_8H_{17}OCH_2SPh$ 56%

Dieter, R.K.; Datar, R. *Org. Prep. Proceed. Int.*, ***1990***, *22*, 63.

hν , NBS , $CaCO_3$
aq. CCl_4

TsO, BnO, Me, O, OMe → TsO, HO, Me, O, OMe 86%

Binkley, R.W.; Hehemann, D.G. *J. Org. Chem.*, ***1990***, *55*, 378.

HO, O, Adenine, OH OH —DMAP , NEt_3; t-$BuMe_2SiCl$→ t-$BuMe_2SiO$, O, Adenine, OH OH 85%

Nair, V.; Buenger, G.S. *Org. Prep. Proceed. Int.*, ***1990***, *22*, 3.

Ph~~OTHP —1. 0.9 $Me_2Sn(SMe)_2$, $BF_3 \bullet OEt_2$, 0°C , 16 h; 2. H_3O^+→ Ph~~OH 90%

Sato, T.; Otera, J.; Nozaki, H. *J. Org. Chem.*, ***1990***, *55*, 4770.

$(PPh_3)_4RhH$, TFA
EtOH , reflux

C_8H_{17} C_8H_{17} HO 98%

Ziegler, F.E.; Brown, E.G.; Sobolov, S.B. *J. Org. Chem.*, ***1990***, *55*, 3691.

Et_3SiH , CH_2Cl_2
1% $Rh_2(pfb)_4$, 2 h
pfb = perfluorobutyrate

OH 88% $OSiEt_3$

Doyle, M.P.; High, K.G.; Bagheri, V.; Pieters, R.J.; Lewis, P.J.; Pearson, M.M. *J. Org. Chem.*, ***1990***, *55*, 6082.

bentonite earth (TONSIL)
acetone, RT , 30 min

OTHP Me OH Me 77%

Cruz-Almanza, R.; Pérez-Flores, F.J.; Avila, M. *Synth. Commun.*, ***1990***, *20*, 1125.

ether , 20°C , 30 min

OH O N CF$_3$ O O CF_3 99%

Keumi, T.; Shimada, M.; Morita, T.; Kitajima, H. *Bull. Chem. Soc. Jpn.*, ***1990***, *63*, 2252.

1. Bu_2SnO , PhMe
reflux , 4 h
2. $Pd(PPh_3)_4$, THF
RT , 1 h
OAc OAc

HO HO HO O O O HO O + O O HO O (8 : 2) 89%

Holzapfel, C.W.; Huyser, J.J.; van der Merwe, T.L.; van Heerden, F.R. *Heterocycles*, ***1991***, *32*, 1445.

3 eq. $MgBr_2$, ether , RT
2.5 eq. BuSH

n-$C_9H_{19}OMOM$ → n-$C_9H_{19}OH$ 97%

Kim, S.; Kee, I.S.; Park, Y.H.; Park, J.H. *SynLett*, ***1991***, 183.

OSiMe$_2$t-Bu, OSiMe$_2$t-Bu → ($KF-Al_2O_3$, MeCN, RT) → OSiMe$_2$t-Bu, OH 78%

ultrasound used in cases when deprotection is slow

Schmittling, E.A.; Sawyer, J.S. *Tetrahedron Lett.*, ***1991***, *32*, 7207.

OH → (dihydropyran, CH_2Cl_2; sulfonated charcoal, MS 3Å, 5 h) → OTHP 97%

Patney, H.K. *Synth. Commun.*, ***1991***, *21*, 2329.

Ph O OH → (Br-CH$_2$-C$_6$H$_4$-N$_3$, NaH, DMF) → Ph O O N$_3$ 92%

← 1. H_2, Pd-C 2. PPh_3 3. DDQ, -5°C

Fukase, K.; Hashida, M.; Kusumoto, S. *Tetrahedron Lett.*, ***1991***, *32*, 3557.

MPMO OH → (Br-CH$_2$-C$_6$H$_4$-NHBoc; $Ba(OH)_2$•OBu, DMF, 32 h) → MPMO O NHBoc 58%

← DDQ

Fukase, K.; Yoshimura, T.; Hashida, M.; Kusumoto, S. *Tetrahedron Lett.*, ***1991***,*32*, 4019.

OSiMe$_2$t-Bu, TIPSO → (aq. H_2SiF_6, MeCN; NEt_3, 1 h) → OH, TIPSO 75% + OH, OH 17%

Pilcher, A.S.; Hill, D.K.; Shimshock, S.J.; Waltermire, R.E.; DeShong, P. *J. Org. Chem.*, ***1992***, *57*, 2492.

OH → (Ph_3CCl, DBU) → $OCPh_3$ 85%

Colin-Mesdsager, S.; Girard, J-P.; Ross, J-C. *Tetrahedron Lett.*, ***1992***, *33*, 2689.

1. $SiF_{4(gas)}$, MeCN, 23°C, 20 min; 2. H_2O

t-Bu, OSiMe$_2$t-Bu → t-Bu, OH 94%

Corey, E.J.; Yi, K.Y. *Tetrahedron Lett.*, ***1992***, *33*, 2289.

e^- (4 F/mol), MeOH

O, $SnBu_3$ → O, OMe 90%

Yoshida, J.; Ishichi, Y.; Nishiwaki, K.; Shiozawa, S.; Isoe, S. *Tetrahedron Lett.*, ***1992***, *33*, 2599.

BH_3•THF, 20°C, 24 h

n-$C_{12}H_{25}$-OTHP → n-$C_{12}H_{25}$-O-$(CH_2)_5$OH 84%

Cossy, J.; Bellosta, V.; Müller, M.C. *Tetrahedron Lett.*, ***1992***, *33*, 5045.

$TiCl_3$-Li-THF, reflux 14 h

n-$C_{16}H_{33}$-O-CH_2Ph → n-$C_{16}H_{33}$-OH 85%

Kadam, S.M.; Nayak, S.K.; Banerji,A. *Tetrahedron Lett.*, ***1992***, *33*, 5129.

$AlCl_3$ - $PhNMe_2$, anisole, CH_2Cl_2

Ph, OBn → Ph, OH 93%

Akiyama, T.; Hirofuji, H.; Ozaki, S. *Bull. Chem. Soc. Jpn.*, ***1992***, *65*, 1932.

dihydropyran, CH_2Cl_2, H_2SO_4, silica gel, 10 min

Ph, OH → Ph, OTHP 96%

Chávez, F.; Godinez, R. *Synth. Commun.*, ***1992***, *22*, 159.

$Al(PO_4)_3$, dihydropyran, reflux, 15 min

OH → OTHP 97%

Campelo, J.M.; Garcia, A.; Lafont, F.; Luna, D.; Marinas, J.M. *Synth. Commun.*, ***1992***, *22*, 2335.

ROH → (Ph·O-C(=CH2)CH3, cat. $POCl_3$) RO–C(CH3)2–OPh ; reverse: H_3O^+

Zandbergen, P.; Willems, H.M.G.; van der Marel, G.A.; Brussee, J.; van der Gen, A. *Synth. Commun.*, ***1992***, *22*, 2781.

OH → ($LaCl_3$, dihydropyran, H_2Cl_2, RT, 4 h) OTHP 90%

Bhuma, V.; Kantam, M.L. *Synth. Commun.*, ***1992***, *22*, 2941.

Ph–(CH2)3–O–CH2–C6H4–OMe → (3 eq. TMSCl, RT, 1.5 eq. anisole, 10% $SnCl_2$, CH_2Cl_2, 30 min) Ph–(CH2)3–OH 90%

Akiyama, T.; Shima, H.; Ozaki, S. *SynLett*, ***1992***, 415.

OH → (MeOCH2OMe, MS 3Å, anhydrous $FeCl_3$, RT, CH_2Cl_2, 1 h) O–CH2–OMe 99%

Patney, H.K. *SynLett*, ***1992***, 567.

BnO–C6H4–C(O)Et → (2.5% dimethyldioxirane, acetone, 20°C, 45 h) HO–C6H4–C(O)Et 69%

Marples, B.A.; Muxworthy, J.P.; Baggaley, K.H. *SynLett*, ***1992***, 646.

CHAPTER 4

PREPARATION OF ALDEHYDES

SECTION 46: ALDEHYDES FROM ALKYNES

NO ADDITIONAL EXAMPLES

SECTION 47: ALDEHYDES FROM ACID DERIVATIVES

6.5% $(Cy_3P)_2RhCHCl_2$, PhH/NEt_3, H_2, 50°C, 1 atm, 65 h; 2-naphthoyl chloride → 2-naphthaldehyde (CHO) 65%

Grushin, V.V.; Alper, H. *J. Org. Chem.*, ***1991***, *56*, 5159.

n-$C_9H_{21}COOH$ → 1. 2 eq. Py; 2. $LiAlH(NEt_2)_3$; 3. H_3O^+ → n-C_9H_{21}-CHO 90%

Cha, J.S.; Lee, J.C.; Lee, H.S.; Lee. S.E. *Org. Prep. Proceed. Int.*, ***1992***, *24*, 327.

PhCHO → 1. Bu_3SnLi; 2. Me_3SiCN; 3. 3 eq. BuLi; 4. O_3, CH_2Cl_2 → PhCOOH 63%

Yu, H.K.B.; Schwartz, J. *Tetrahedron Lett.*, ***1992***, *33*, 6787, 6791.

SECTION 48: ALDEHYDES FROM ALCOHOLS AND THIOLS

Cinnamyl alcohol (OH) → p-benzoquinone (O=⟨⟩=O), diglyme, 120°C, 12 h → cinnamaldehyde (CHO) 80%

Kulkarni, M.G.; Mathew, T.S. *Tetrahedron Lett.*, ***1990***, *31*, 4497.

O-Mo peroxo picolinate complex; CH_2Cl_2, 18 h → epoxy aldehyde (CHO) 60%

also with 2° alcohols to give ketones

Campestrini, S.; DiFuria, F.; Modena, G.; Bortolini, O. *J. Org. Chem.*, ***1990***, *55*, 3658.

'electrolytic' MnO_2, hexane, RT, 13 h; OH → CHO 78%

Tsuboi, S.; Ishii, N.; Sakai, T.; Tari, I.; Utaka, M. *Bull. Chem. Soc. Jpn*, ***1990***, *63*, 1888.

$PhCH_2OH$ —(CrO_3 - "wet alumina")→ PhCHO 90%

also for the preparation of aryl ketone from aryl alcohols

Hirano, M.; Kuroda, H.; Morimoto, T. *Bull. Chem. Soc. Jpn.*, ***1990***, *63*, 2433.

$PhCH_2OSiMe_3$ —(NBS, AIBN, CCl_4; 95°C, 8 min)→ PhCHO 87%

Markó, I.E.; Mekhalfia, A.; Ollis, W.D. *SynLett*, ***1990***, 345.

OH —($Me_3NHCl\bullet CrO_3$, MS 4Å; CH_2Cl_2, 20 h)→ CHO 75%

Acharya, S.P.; Rane, R.A. *Synthesis*, ***1990***, 127.

OH —(e^-, Pt electrodes, pH 6.8; $Ru(trpy)(bpy)O]^{+2}$)→ CHO 41%

trpy = 2,2',6',2"-terpyridine

Navarro, M.; De Giovani, W.F.; Romero, J.R. *Synth. Commun.*, ***1990***, *20*, 399.

Ph OH —(MnO_2, Na_2CO_3, 0°C; hexane)→ Ph CHO >95%

(13:1 retention:isomerization)

Xiao, X.; Prestwich, G.D. *Synth. Commun.*, ***1990***, *20*, 3125.

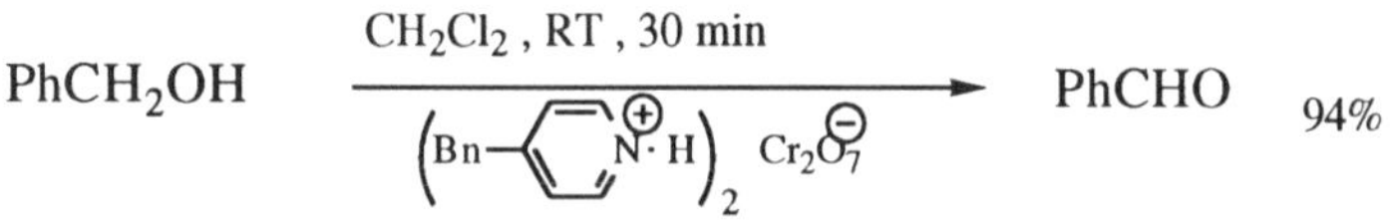

Akamanchi, K.G.; Iyer, L.G.; Meenakshi, R. *Synth. Commun.*, ***1991***, *21*, 419.

$$PhCH_2OH \xrightarrow[\text{reflux}]{\text{hydrous zirconium oxide, benzophenone, xylene}} PhCHO \quad 81\%$$

Kuno, H.; Shibagaki, M.; Takahashi, K.; Matsushita, H. *Bull. Chem. Soc. Jpn.*, ***1991***, *64*, 312.

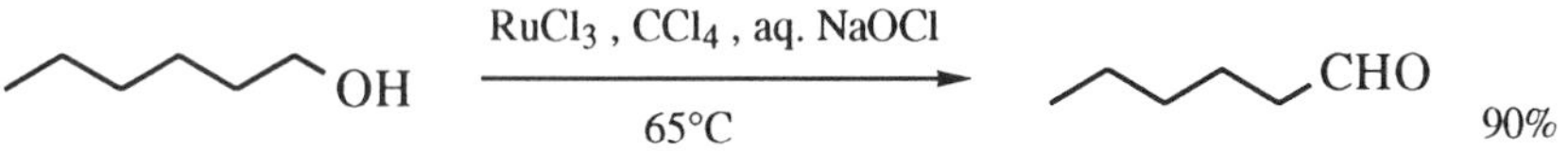

Ogibin, Yu.N.; Ilovaiskii, A.I.; Nikishin, G.I. *Izv. Akad. Nauk. SSSR*, ***1991***, *40*, 115 (Engl., p.99).

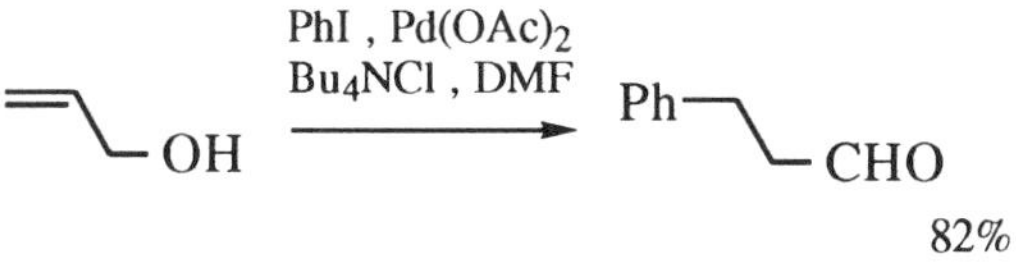

Jeffery, T. *Tetrahedron Lett.*, ***1991***, *32*, 2121.

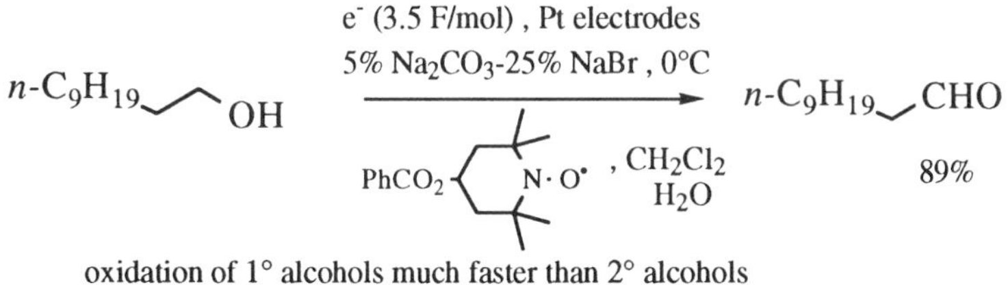

oxidation of 1° alcohols much faster than 2° alcohols

Inokuchi, T.; Matsumoto, S.; Torii, S. *J. Org. Chem.*, ***1991***, *56*, 2416.
Inokuchi, T.; Matsumoto, S.; Fukushima, M.; Torii, S. *Bull. Chem. Soc. Jpn.* ***1991***, *64*, 796.
Inokuchi, T.; Matsumoto, S.; Nishiyama, T.; Torii, S. *J. Org. Chem.*, ***1990***, *55*, 462.
Inokuchi, T.; Matsumoto, S.; Nisiyama, T.; Torii, S. *SynLett*, ***1990***, 57.

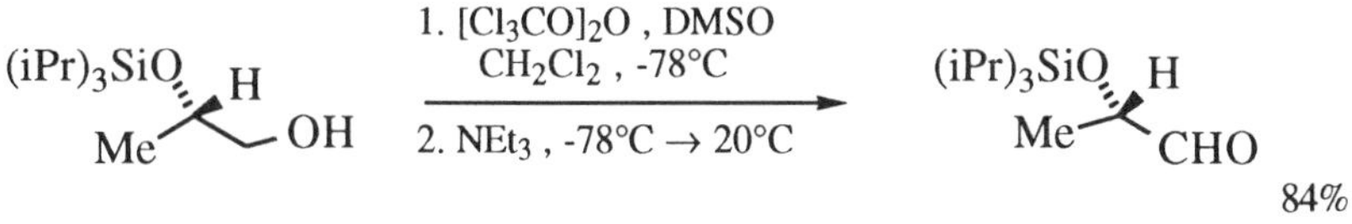

Palomo, C.; Cossio, F.P.; Ontoria, J.M.; Odriozola, J.M. *J. Org. Chem.*, ***1991***, *56*, 5948.

CH_2Cl_2 , 2 h

AcHN N.OH H ⊕ ⊖OTs

OH → CHO 94%

Ma, Z.; Bobbitt, J.M. *J. Org. Chem.*, ***1991***, *56*, 6110.

N - O , RT , overnight

RuTMP(O)2

Ph OH → Ph CHO 81%

TMP = tetramethylpyrazole

Higuchi, T.; Ohtake, H.; Hirobe, M. *Tetrahedron Lett.*, ***1991***, *32*, 7435.

1. DMSO , -78°C triphosgene

2. NEt_3 , -78°C → RT

OH → CHO 82%

Rivero, I.A.; Somanathan, R.; Hellberg, L.H. *Org. Prep. Proceed. Int*, ***1992***, *24*, 363.

2 $CuBr_2$, 2 LiBr , 13 h

Bu_3SnO*t*-Bu

Ph $OSnBu_3$ → Ph CHO 78%

Yamaguchi, J.; Takeda, T. *Chem. Lett.*, ***1992***, 423.

CrO_3 , DMSO , RT , 4 h

OH → CHO 68%

Lou, W-X.; Lou, J-D. *Synth. Commun.*, ***1992***, *22*, 767.

TPCD , DMF , 80°C

1 h

n-$C_6H_{13}CH_2OH$ → *n*-$C_6H_{13}CHO$ 80%

TPCD = tetrakis-pyridino-cobalt dichromate = $[Py_4Co(HCrO_4)_2]$

Hu, Y.; Hu, H. *Synth. Commun.*, ***1992***, *22*, 1491.

SeO_2 , SiO_2 , RT

t-BuOOH , CH_2Cl_2 15 h

HO OH → OHC OH 90%

Kalsi, P.S.; Chhabra, B.R.; Singh,. J.; Vig, R. *SynLett*, ***1992***, 425.

REVIEWS:

"Oxidation of Alcohols by Activated DMSO and Related Reactions: An Update: Tidwell, T.T. *Synthesis*, ***1990***, 857.

SECTION 49: ALDEHYDES FROM ALDEHYDES

Conjugate reductions and Michael Alkylations of conjugated aldehydes are listed in Section 74 (Alkyls from Alkenes).

$PhCH(OEt)_2 \xrightarrow{ISiCl_3\text{ , RT , 20 min}} PhCHO$ 95%

Elmorsy, S.S.; Bhatt, M.V.; Pelter, A. *Tetrahedron Lett.*, ***1992***, *33*, 1657.

OAc, OAc, AcO, OMe — CAN , SiO_2 / CH_2Cl_2 → CHO, AcO, OMe 90%

Cotelle, P.; Catteau, J-P. *Tetrahedron Lett.*, ***1992***, *33*, 3855.

Related Methods: Aldehydes from Ketones (Section 57)
Ketones from Ketones (Section 177)
Also via: Alkenyl aldehydes (Section 341)

SECTION 50: ALDEHYDES FROM ALKYLS, METHYLENES AND ARYLS

1. $CF_3SO_2CH{=}NMe_2^+$ 130°C , 48 h
2. 5% aq. NaOH

CHO 50%

Martinez, A.G.; Alvarez, R.M.; Barcina, J.O.; de la Moya Cerero, S.; Vilar, E.T.; Fraile, A.G.; Hanack, M.; Subramanian, L.R. *J. Chem. Soc., Chem. Commun.*, ***1990***, 1571.

CH_3, CH_3 — $Ce(OTf)_4$, aq. MeCN / 1.5 h → CH_3, CHO 70%

also, EtPh converted to acetophenone

Imamoto, T.; Koide, Y.; Hiyama, S. *Chem. Lett.*, ***1990***, 1445.

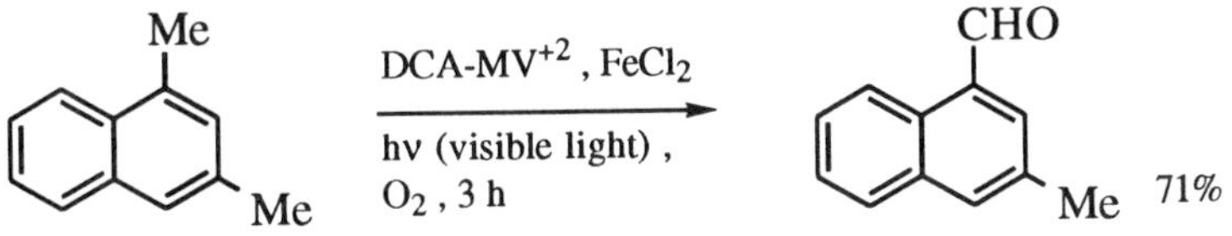
Me
Me
DCA-MV+2 , FeCl2
hν (visible light) ,
O2 , 3 h
CHO
Me
71%

DCA = 9,10-dicyanoanthracene; MV^{+2} = methyl viologen

ketones are fromed form $ArCH_2R$

Santamaria, J.; Jroundi, R. *Tetrahedron Lett.*, ***1991***, *32*, 4291.

SECTION 51: ALDEHYDES FROM AMIDES

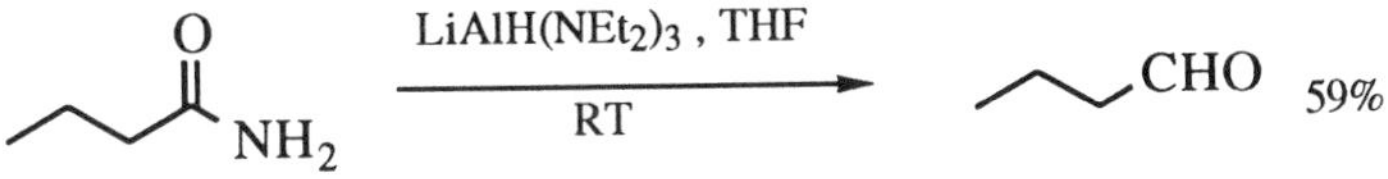
O
NH2
LiAlH(NEt2)3 , THF
RT
CHO
59%

Cha, J.S.; Lee, J.C.; Lee, H.S.; Lee, S.E.; Kim, J.M.; Kwon, O.O.; Min, S.J. *Tetrahedron Lett.*, ***1991***, *32*, 6903.

SECTION 52: ALDEHYDES FROM AMINES

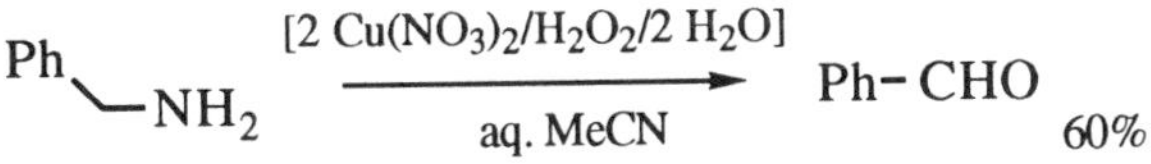
Ph
NH2
[2 Cu(NO3)2/H2O2/2 H2O]
aq. MeCN
Ph- CHO
60%

Capdevielle, P.; Maumy, M. *Tetrahedron Lett.*, ***1990***, *31*, 3891.

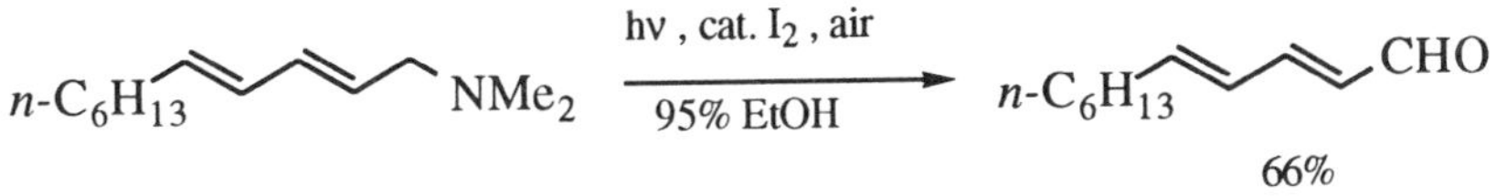
n-C6H13
NMe2
hν , cat. I2 , air
95% EtOH
n-C6H13
CHO
66%

Gangloff, A.R.; Judge, T.M.; Helquist, P. *J. Org. Chem.*, ***1990***, *55*, 3679.

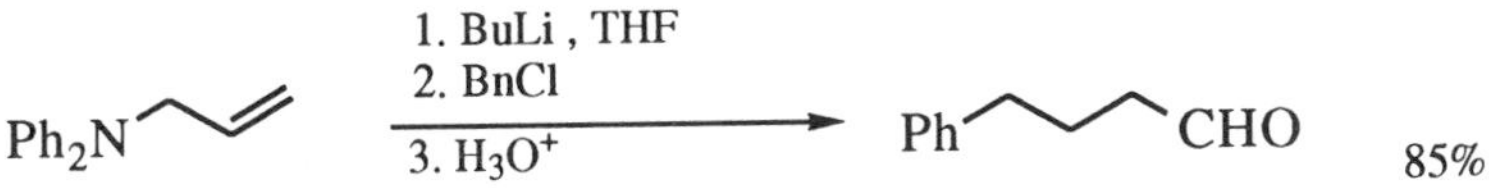
Ph2N
1. BuLi , THF
2. BnCl
3. H3O+
Ph
CHO
85%

Eisch, J.J.; Shah, J.H. *J. Org. Chem.*, ***1991***, *56*, 2955.

PhCH=N-OH —(bakers yeast)→ PhCHO 68%

(with ultrasound 96%)

Kamal, A.; Rao, M.V.; Meshram, H.M. *J. Chem. Soc., Perkin Trans. I*, ***1991***, 2056.

N-OH, n-C_7H_{15}, H → 1. [$TiCl_4$, NaI, MeCN] MeCN, 20 min; 2. aq. NaOH → O, n-C_7H_{15}, H 63%

Balicki, R.; Kaczmarek, L. *Synth. Commun.*, ***1991***, *21*, 1777.

$PhCH_2NH_2$ → 1. $PhSO_2Cl$; 2. KO_2, THF, 24 h → PhCHO 90%

Park, K.H.; Lee, J.B. *Synth. Commun.*, ***1992***, *22*, 1061.

Related Methods: Ketones from Amines (Section 172)

SECTION 53: ALDEHYDES FROM ESTERS

Ph, Me, N, PPh_2 → 1. OAc, $Pd(PPh_3)_4$, THF, RT, 45 h; 2. aq. HCl → Ph, Me, CHO 80% (48% ee)

Hiroi, K.; Abe, J. *Tetrahedron Lett.*, ***1990***, *31*, 3623.

MeO, O, O, SEt → Et_3SiH, Pd-C, acetone → MeO, O, O, H 94%

Furuyama, T.; Lin, S-C.; Li, L. *J. Am. Chem. Soc.*, ***1990***, *112*, 7050.

n-$C_3H_7CO_2Et$ → 1. $Li(Et_2N)_3AlH$ THF, -78°C; 2. H_3O^+ → n-C_3H_7-CHO 82%

Cha, J.S.; Min, S.J.; Lee, H.S. *Org. Prep. Proceed. Int.*, ***1992***, *24*, 335.

O, Ph, SPh → 1. Li, THF, -78°C; 2. MeOH → PhCHO 98%

Penn, J.H.; Owens, W.H. *Tetrahedron Lett.*, ***1992***, *33*, 3737.

SECTION 54: ALDEHYDES FROM ETHERS, EPOXIDES AND THIOETHERS

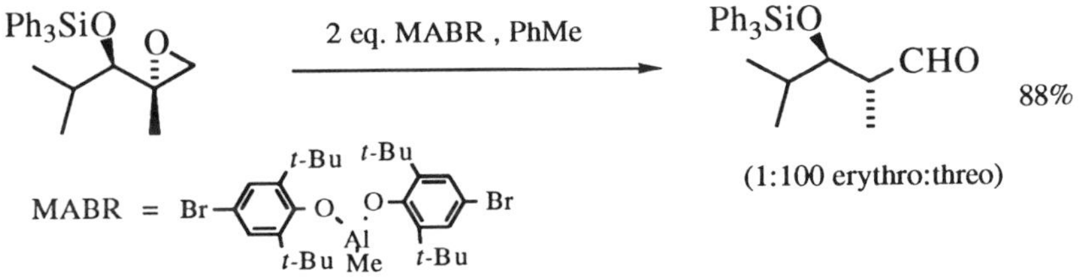

Ph_3SiO ... 2 eq. MABR, PhMe → Ph_3SiO ... CHO 88%

MABR = (t-Bu, t-Bu, Br, O, O, Al, Me, t-Bu, t-Bu)

(1:100 erythro:threo)

Maruoka, K.; Sato, J.; Yamamoto, H. *J. Am. Chem. Soc.*, ***1991***, *113*, 5449.

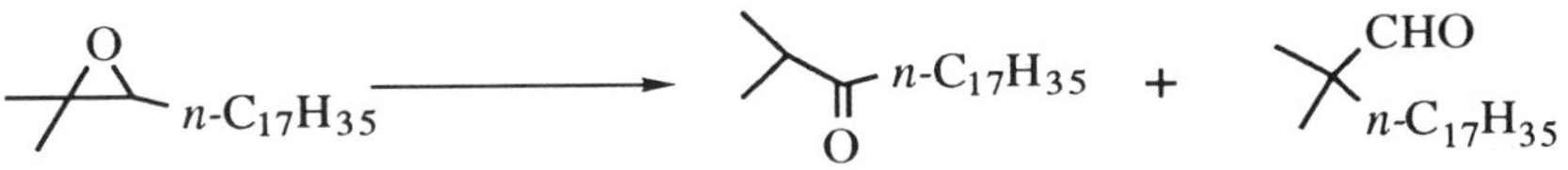

O, n-$C_{17}H_{35}$ → n-$C_{17}H_{35}$, O + CHO, n-$C_{17}H_{35}$

MABR, CH_2Cl_2, -20°C	(0 :	100)	99%
SbF_5, PhMe, -78°C	(83 :	17)	47%

MABR = methyl aluminum bis-(4-bromo-2,6-di-*t*-butylphenoxide)

Maruoka, K.; Bureau, R.; Ooi, T.; Yamamoto, H. *SynLett*, ***1991***, 491.

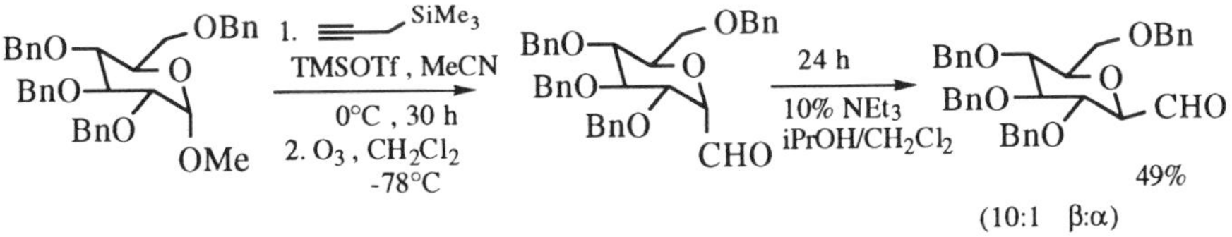

BnO, BnO, BnO, OBn, O, OMe — 1. $SiMe_3$ alkyne, TMSOTf, MeCN, 0°C, 30 h; 2. O_3, CH_2Cl_2, -78°C → BnO, BnO, BnO, OBn, O, CHO — 24 h, 10% NEt_3, iPrOH/CH_2Cl_2 → BnO, BnO, BnO, OBn, O, CHO 49%

(10:1 β:α)

Kobertz, W.R.; Bertozzi, C.R.; Bednarski, M.D. *Tetrahedron Lett.*, ***1992***, *33*, 737.

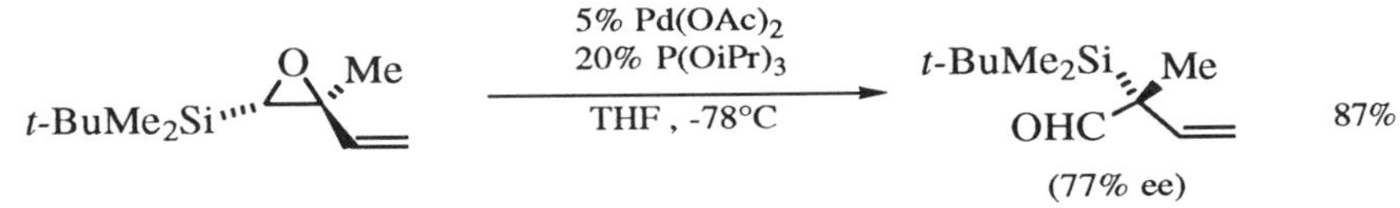

t-$BuMe_2Si$, O, Me — 5% $Pd(OAc)_2$, 20% $P(OiPr)_3$, THF, -78°C → t-$BuMe_2Si$, Me, OHC 87%

(77% ee)

Malacria, M.; Cilloir, F. *Tetrahedron Lett.*, ***1992***, *33*, 3859.

Related Methods: Ketones from Ethers and Epoxides (Section 174)

SECTION 55: ALDEHYDES FROM HALIDES AND SULFONATES

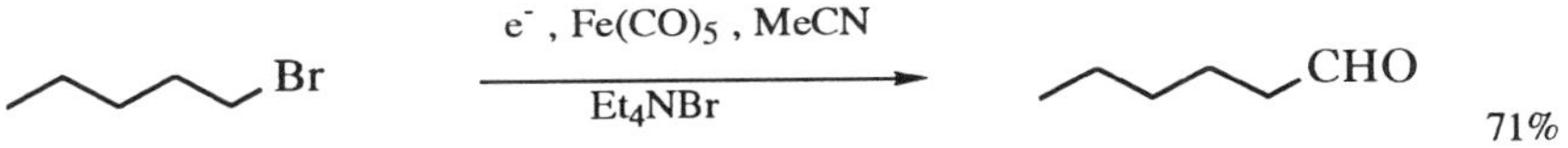

Br — e^-, $Fe(CO)_5$, MeCN, Et_4NBr → CHO 71%

Yoshida, K.; Kobayashi, M.; Amano, S. *J. Chem. Soc., Perkin Trans. I*, ***1992***, 1127.

Ph~~Br —[CO (80 atm), Bu_3SnH; AIBN, PhH, 80°C]→ Ph~~CHO 65%

Ryu, I.; Kusano, K.; Ogawa, A.; Kambe, N.; Sonoda, N. *J. Am. Chem. Soc.*, ***1990***, *112*, 1295.
Ryu, I.; Kusano, K.; Masumi, N.; Yamazaki, H.; Ogawa, A.; Sonoda, N. *Tetrahedron Lett.*, ***1990***, *31*, 6887.

n-C_8H_{17}-I —[1. MeO–C_6H_4–SeNa, vinyl isopropyl ether, EtOH/ether, 6 h; 2. NBS, H_2O]→ n-$C_8H_{17}CH_2CHO$ 77%

Uneyama, K.; Kitagawa, K. *Tetrahedron Lett.*, ***1991***, *32*, 3385.

$PhCH_2Cl$ —[TPCD, DMF, 100°C; 1 h]→ PhCHO 54%

TPCD = tetrakis-pyridino-cobalt dichromate = $[Py_4Co(HCrO_4)_2]$

Hu, Y.; Hu, H. *Synth. Commun.*, ***1992***, *22*, 1491.

SECTION 56: ALDEHYDES FROM HYDRIDES

4-Me–C_6H_4–C(O)Me —[iPrONO, HCl; iPrOH]→ 4-Me–C_6H_4–C(O)CH(OiPr)(OiPr) 60%

Gordeeva, G.N.; Kalashnikov, S.M.; Gordeev, A.N.; Popov, Yu.N.; Kruglov, E.A.; Imashev, U.B. *Zhur. Org. Khim.*, ***1990***, *26*, 2199 (Engl., p. 1896).

Biphenyl —[HF-SbF_5, CO (20 atm)]→ OHC–C_6H_4–C_6H_4–CHO 81%

(93:7 4,4':2,4')

Tanaka, M.; Souma, Y. *J. Chem. Soc., Chem. Commun.*, ***1991***, 1551.

2-Bromoaniline (NH_2, Br) —[DMSO - HCl, $CuCl_2$]→ OHC–C_6H_3(NH_2)(Br) 98%

Liedholm, B. *J. Chem. Soc., Perkin Trans. I*, ***1992***, 2235.

hν (λ = 366 nm), CO (1200 psi), Ph(C=O)-C6H4-2-CF3; cyclohexane → cyclohexane-CHO 91%

Boese, W.T.; Goldman, A.S. *Tetrahedron Lett.*, ***1992***, *33*, 2119.

Me (toluene); SbF_5-HSO_3F , CO; 168 min → Me/CHO 2% + Me/SO_2F/CHO + Me/SO_2F/CHO (76 : 24) 90%

Tanaka, M.; Iyoda, J.; Souma, Y. *J. Org. Chem.*, ***1992***, *57*, 2677.

SECTION 57: ALDEHYDES FROM KETONES

$PhMe_2Si$, O, OBn; 0.1 M , 1 atm H_2, Pd/C , RT , EtOH; 20 h → H, O, OBn 80%

Cirillo, P.F.; Panek, J.S. *Tetrahedron Lett.*, ***1991***, *32*, 457.

SECTION 58: ALDEHYDES FROM NITRILES

PhCN; 1. Li(hexyl$_2$N)$_3$AlH, THF , 0°C; 2. H_3O^+ → Ph-CHO 99%

Cha, J.S.; Lee, S.E.; Lee, H.S. *Org. Prep. Proceed. Int.*, ***1992***, *24*, 331.

SECTION 59: ALDEHYDES FROM ALKENES

O, O; e^- , $NaIO_4$, DCE, H_2O; 2% $(Bu_4N)_5(H_2O)RuSiW_{11}O_{39}$; 60°C → CHO, O, O 68%

Steckhan, E.; Kandzia, C. *SynLett*, ***1992***, 139.

Related Methods: Ketones from Alkenes (Section 179)

SECTION 60: ALDEHYDES FROM MISCELLANEOUS COMPOUNDS

n-C_9H_{19}N(NO)C(=O)CH_2Cl —NMO , DMSO→ n-C_9H_{19}CHO 50%

Nikolaides, N.; Godfrey, A.G.; Ganem, B. *Tetrahedron Lett.*, ***1990***, *31*, 6009.

n-$C_{11}H_{23}$$CH_2$$NO_2$ —1. Sn(SPh)$_4$, EtOH; 2. $SnCl_2$•2 H_2O , 3 h, L-tartaric acid , 40°C, $NaHCO_3$, $NaHSO_3$; 3. 2M HCl→ n-$C_{11}H_{23}$–CHO 81%

Urpí, F.; Vilarrasa, J. *Tetrahedron Lett.*, ***1990***, *31*, 7499.

t-BuO$CH_2$$SO_2$Ph —1. BuLi , THF , -78°C; 2. n-C_5H_{11}I , 4 eq. HMPA, -78°C - RT; 3. LiBr , K_2CO_3 , 48 h, H_2O/ether , reflux→ n-C_5H_{11}CHO 83%

Chemla, F.; Julia, M.; Uguen, D.; Zhang, D. *SynLett*, ***1991***, 501.
Julia, M.; Uguen, D.; Zhang, D. *SynLett*, ***1991***, 503.

SECTION 60A: PROTECTION OF ALDEHYDES

n-C_8H_{17}CHO —ethanedithiol , $LaCl_3$, CH_2Cl_2 , RT , 4 h→ 2-octyl-1,3-dithiolane 66%

Garlaschelli, L.; Vidari, G. *Tetrahedron Lett.*, ***1990***, *31*, 5815.

PhCHO —allyl bromide , Sb(OEt)$_3$→ PhCH(OEt)$_2$ 97%

Liao, Y.; Huang, Y-Z.; Zhu, F-H. *J. Chem. Soc., Chem. Commun.*, ***1990***, 493.

Me-epoxide (propylene oxide) —PhCHO , RT , 30 min; 1-(4-MeO-benzyl)-2-cyanopyridinium $SbF_6^{\ominus}$→ 4-Me-2-Ph-1,3-dioxolane quant.

Lee, S-B.; Takata, T.; Endo, T. *Chem. Lett.*, ***1990***, 2019.

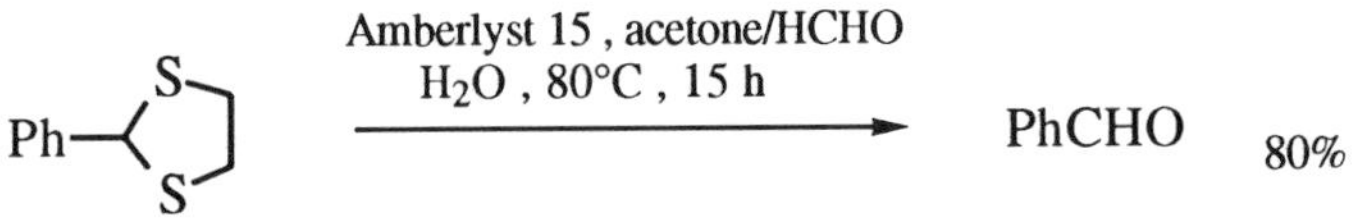
Amberlyst 15 , acetone/HCHO
H2O , 80°C , 15 h
Ph
S
S
PhCHO
80%

Ballini, R.; Petrini, M. *Synthesis, **1990***, 336.

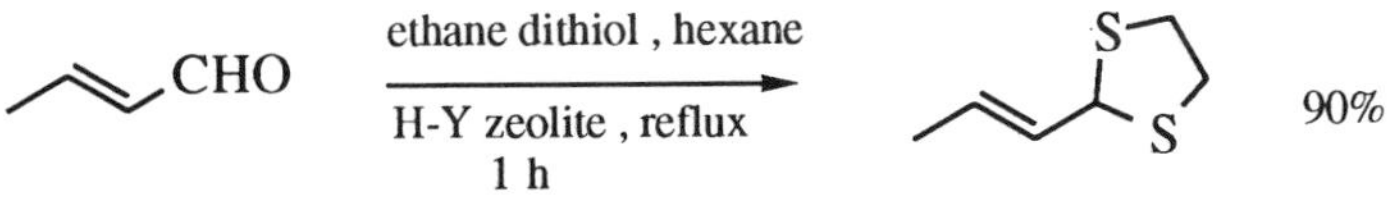
CHO
ethane dithiol , hexane
H-Y zeolite , reflux
1 h
S
S
90%

Kumar, P.; Reddy, R.S.; Singh, A.P.; Pandey, B. *Tetrahedron Lett.*, ***1992***, *33*, 825.

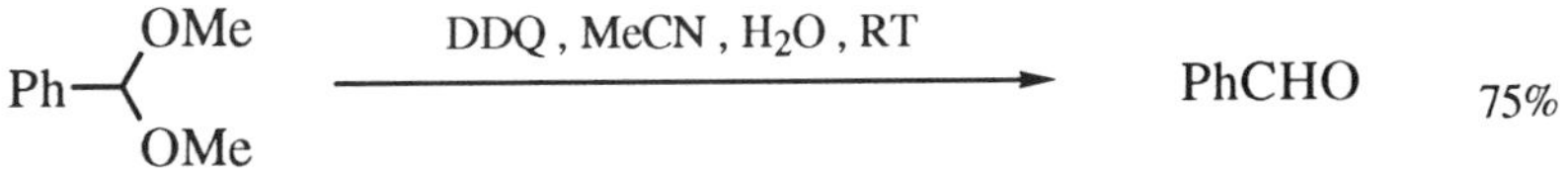
OMe
Ph
OMe
DDQ , MeCN , H2O , RT
PhCHO
75%

Tanemura, K.; Suzuki, T.; Horaguchi, T. *J. Chem. Soc., Chem. Commun.*, ***1992***, 979.

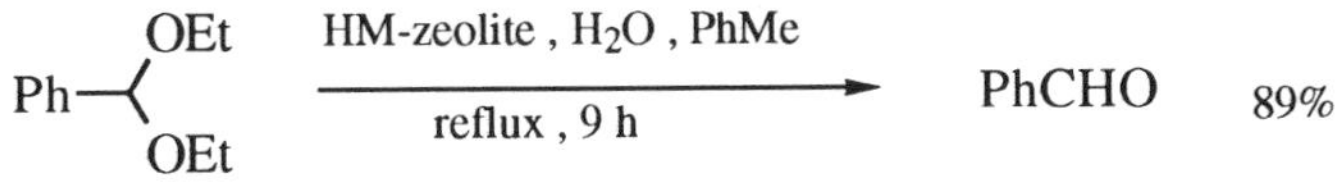
OEt
Ph
OEt
HM-zeolite , H2O , PhMe
reflux , 9 h
PhCHO
89%

Rao, M.N.; Kumar, P.; Singh, A.P.; Reddy, R.S. *Synth. Commun.*, ***1992***, *22*, 1299.

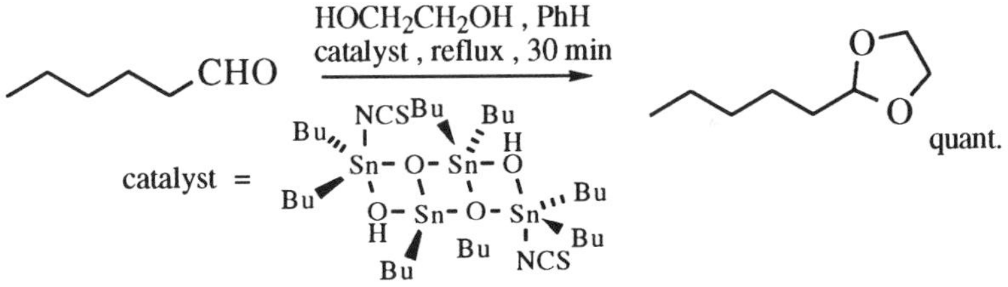
CHO
HOCH2CH2OH , PhH
catalyst , reflux , 30 min
O
O
quant.
catalyst =

Otera, J.; Dan-oh, N.; Nozaki, H. *Tetrahedron*, ***1992***, *48*, 1449.

CHAPTER 5

PREPARATION OF ALKYLS, METHYLENES AND ARYLS

This chapter lists the conversion of functional groups into methyl, ethyl, propyl, etc. as well as methylene (CH_2), phenyl, etc.

SECTION 61: ALKYLS, METHYLENES AND ARYLS FROM ALKYNES

$(iPr)_3SiO-C\equiv C(CH_2)_9$–CH=CH–Bu ; DCE , hν — 54%

Danheiser, R.L.; Cha, D.D. *Tetrahedron Lett.*, ***1990***, *31*, 1527.

1.5 eq. KOH , 37°C ; DMSO , EtOH , 24 h — 58%

Toshima, K.; Ohta, K.; Ohtake, T.; Tatsuta, K. *Tetrahedron Lett.*, ***1991***, *32*, 391.

CH_2Cl_2 , 130°C — 60%

Padwa, A.; Austin, D.J.; Chiacchio, U.; Kassir, J.M.; Rescifina, A.; Xu, S.L. *Tetrahedron Lett.*, ***1991***, *32*, 5923.

Et-C≡C-Et , 2 eq. NEt_3 ; 5% $Pd(PPh_3)_4$, DMF ; 65°C , 12 h — 40%

Negishi, E.; Harring, L.S.; Owczarczyk, Z.; Mohamud, M.M.; Ay, M. *Tetrahedron Lett.*, ***1992***, *33*, 3253.

1. MeO–CH=CH–C(=CH2)–OSiMe2t-Bu, MeCN, 25°C, 16 h, CO
2. 110°C

62%

(R = TMS:R = OH = 5.2)

Bao, J.; Draagisich, V.; Wenglowsky, S.; Wulff, W.D. *J. Am. Chem. Soc.*, ***1991***, *113*, 9873.

1. SmI_2-THF, $Pd(PPh_3)_4$, -78°C → 0°C
2. CO_2Me

57%

Inanaga, J.; Sugimoto, Y.; Hanamoto, J. *Tetrahedron Lett.*, ***1992***, *33*, 7035.

MeO_2C , PhCl, 210°C, 24 h

72%

Grissom, J.W.; Calkins, T.L. *Tetrahedron Lett.*, ***1992***, *33*, 2315.

SECTION 62: ALKYLS, METHYLENES AND ARYLS FROM ACID DERIVATIVES

PPh_3, $Pd(PhCN)_2Cl_2$

mesitylene, 165°C	72%	21%
nitrobenzene, 125°C	0%	100%

Krafft, T.E.; Rich, J.D.; McDermott, P.J. *J. Org. Chem.*, ***1990***, *55*, 5430.

SECTION 63: ALKYLS, METHYLENES AND ARYLS FROM ALCOHOLS AND THIOLS

OH

SiMe3

BF_3•OEt_2 , CH_2Cl_2
-78°C → RT

95%

(98:2 trans:cis)

Schmitt, A.; Reissig, H.U. *SynLett*, ***1990***, *40*.

OMe
Me OH

1. NBS , hν , CCl_4
2. Bu_2CuLi , ether
3. TMSI , MeCN , heat

OH
Me

74%

Kotnis, A.S. *Tetrahedron Lett.*, ***1991***, *32*, 3417.

$$Ph\text{-}CH_2OH \xrightarrow{BR_3\text{ , TfOH}} Ph\text{-}CH_2\text{-}R + Ph\text{-}CH_3$$

	Ph-CH_2-R	Ph-CH_3
R = Me	57%	43%
R = Et	11%	89%
R = iPr	0.1%	99%

Olah, G.A.; Wu, A.; Farooq, O. *J. Org. Chem.*, ***1991***, *56*, 2759.

OH

Nafion-H , PhH
95°C , 1 h

81%

Yamamoto, T.; Hideshima, C.; Surya Prakash, G.K.; Olah, G.A. *J. Org. Chem.*, ***1991***, *56*, 2089.

n-$C_{13}H_{27}$ OH

1. Cl, O, S, F, F, F, F, F
PhMe , 5 min
2. Bu_3SnH , PhMe
5 min

n-$C_{13}H_{27}$ CH_3

95%

Barton, D.H.R.; Blundell, P.; Dorchak, J.; Jang, D.O.; Jaszberenyi, J.Cs. *Tetrahedron*, ***1991***, *47*, 8969.

1. 5.7 eq. $SnCl_4$, -78°C → 25°C
2. Ph-CH=CH-Me, 1 h

77%

Angle, S.R.; Arnaiz, D.O. *J. Org. Chem.*, ***1992***, *57*, 5937.

SECTION 64: ALKYLS, METHYLENES AND ARYLS FROM ALDEHYDES

$n\text{-}C_5H_{11}-C\equiv C-n\text{-}C_5H_{11}$

1. $NbCl_5$, Zn, DME/PhH, 25°C, 10h
2. 2,6-lutidine
3. (o-phthalaldehyde: CHO, CHO), 25°C, 30 min

93%

Kataoka, Y.; Miyai, J.; Tezuka, M.; Takai, K.; Oshima, K.; Utimoto, K. *Tetrahedron Lett.*, ***1990***, *31*, 369.

MeMgI, $NiCl_2$(dppe)

93%

Tzeng, Y-L.; Yang, P-F.; Mei, N-W.; Yuan, T-M.; Yu, C-C.; Luh, T-Y. *J. Org. Chem.*, ***1991***, *56*, 5289.

REVIEWS:

"Chiral Acetals in Asymmetric Synthesis"
Alexakis, A.; Mangeney, P. *Tetrahedron Asymmetry*, ***1990***, *1*, 477.

Related Methods: Alkyls, Methylenes and Aryls from Ketones (Section 72)

SECTION 65: ALKYLS, METHYLENES AND ARYLS FROM ALKYLS, METHYLENES AND ARYLS

1. BuLi, THF, -78°C
2. $ClSiMe_3$
3. BuLi, THF, -78°C
4. Et-I

Et, $SiMe_3$

84%

Rabideau, P.W.; Dhar, R.K.; Clawson, D.K.; Zhan, Z. *Tetrahedron Lett.*, ***1991***, *32*, 3969.

1. NaH
2. NMP , $ZnCl_2$, RT
3. $PdCl_2(PPh_3)_2$, PhBr
RT , PPh_3, 2 h
4. 140°C , 2 h

Ph + Ph + Ph-Ph

40% 10% 50%

Filippini, L.; Gusmerol, M.; Riva, R. *Tetrahedron Lett.*, ***1992***, *33*, 1755.

2.4 eq. $Fe(ClO_4)_3$
TFA , CH_2Cl_2 , RT

MeO MeO MeO H O O

74%

Tanaka, M.; Mitsuhashi, H.; Wakamatsu, T. *Tetrahedron Lett.*, ***1992***, *33*, 4161.

SECTION 66: ALKYLS, METHYLENES AND ARYLS FROM AMIDES

NO ADDITIONAL EXAMPLES

SECTION 67: ALKYLS, METHYLENES AND ARYLS FROM AMINES

NMe_2 NMe_2

1. 2.2 eq. Li naphthalenide
THF , -78°C
2. Me_2SO_4

NMe_2 Me

60%

Kawase, T.; Fujino, S.; Oda, M. *Tetrahedron Lett.*, ***1990***, *31*, 545.

SECTION 68: ALKYLS, METHYLENES AND ARYLS FROM ESTERS

OAc

$NaBPh_4$, THF
$Pd(dba)_2$, PPh_3
60°C

Ph

80%

Legros, J-Y.; Fiaud, J-C. *Tetrahedron Lett.*, ***1990***, *31*, 7453.

Ph–CH=CH–CH$_2$–OAc $\xrightarrow[\text{DMF , LiCl , 23°C, 69 h}]{\text{PhSnMe}_3\text{ , 5\% Pd(dba)}_2}$ Ph–CH=CH–CH$_2$–Ph 57%

Del Valle, L.; Stille, J.K.; Hegedus, L.S. *J. Org. Chem.*, ***1990***, *55*, 3019.

PhS–CH$_2$–CH=CH–CH$_2$–OAc $\xrightarrow{\text{X eq. Li}_2\text{CuClO}_4\text{, Z min addition of BuMgBr}}$ PhS–CH$_2$–CH=CH–CH$_2$–Bu + PhS–CH$_2$–CH(Bu)–CH=CH$_2$

X = 2; Z = 2 min 75% 25%

X = 5; Z = 40 min 13% 87%

Bäckvall, J-E.; Sellén, M.; Grant, B. *J. Am. Chem. Soc.*, ***1990***, *112*, 6615.

CO_2Me CO_2Me

1. O, CO_2Me
NaH/BuLi , 0°C
2. CA(OAc)$_2$•H$_2$O , reflux
3. 2N HCl

O OH CO_2Me CO_2Me

77%

Yamaguchi, M.; Hasebe, K.; Higashi, H.; Uchida, M.; Irie, A.; Minami, T. *J. Org. Chem.*, ***1990***, *55*, 1611.

CO_2Me O O N H O

N_2^+ CO_2^-

heat

CO_2Me N H O 43%

via benzyne cycloaddition-retro Diels-Alder

Meirás, D.P.; Guitián, E.; Castedo, L. *Tetrahedron Lett.*, ***1990***, *31*, 143.

OAc

cat. Pd(dba)$_2$, PhZnCl

Ph Ph P Ph O

Ph 60%

(9.4% ee)

Fiaud, J-C.; Legros, J-Y. *Tetrahedron Lett.*, ***1991***, *32*, 5089.

O O Ph

(PhMe$_2$Si)$_2$Cu(CN)Li$_2$

THF - ether

SiMe$_2$Ph 83%

Fleming, I.; Higgins, D.; Lawrence, N.J.; Thomas, A.P. *J. Chem. Soc., Perkin Trans. I*, ***1992***, 3331.

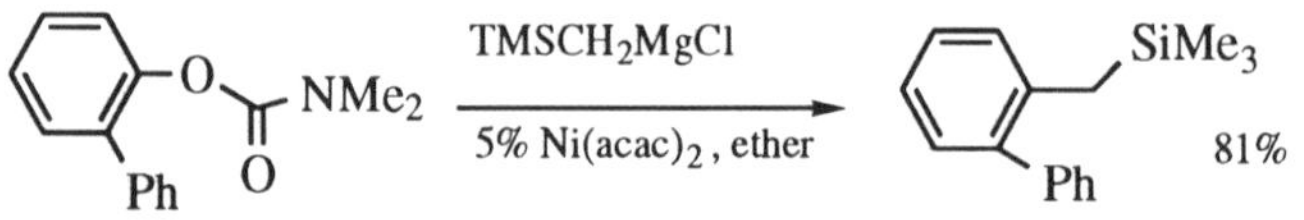

TMSCH$_2$MgCl; 5% Ni(acac)$_2$, ether — 81%

Sengupta, S.; Leite, M.; Raslan, D.S.; Quesnelle, C.; Snieckus, V. *J. Org. Chem.*, ***1992***, *57*, 4066.

SECTION 69: ALKYLS, METHYLENES AND ARYLS FROM ETHERS, EPOXIDES AND THIOETHERS

The conversion ROR → RR' (R' = alkyl, aryl) is included in this section.

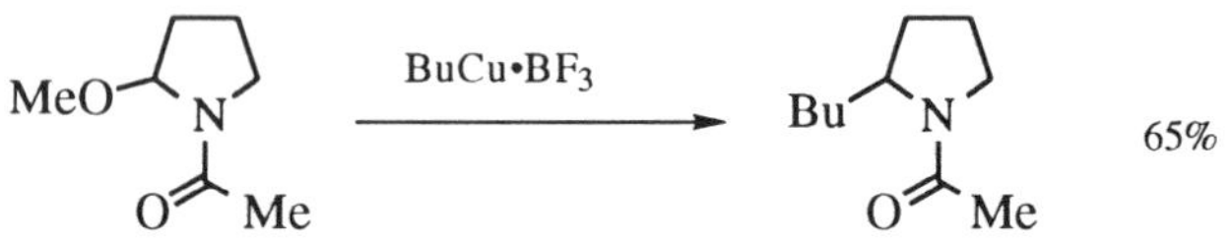

BuCu•BF$_3$ — 65%

Ludwig, C.; Wistrand, L-G. *Acta Chem. Scand.*, ***1990***, *44*, 707.

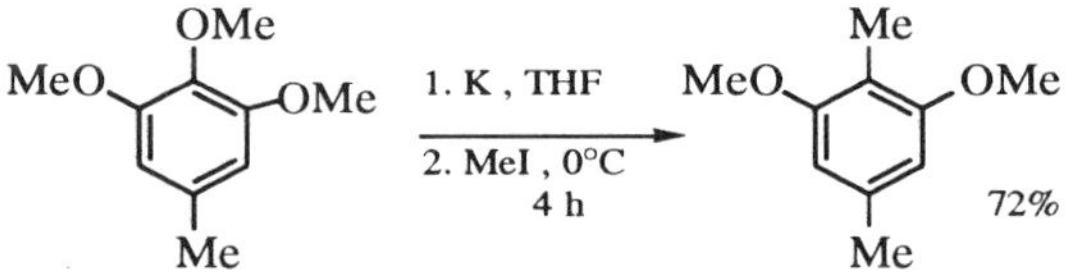

1. K, THF; 2. MeI, 0°C, 4 h — 72%

Azzena, U.; Denurra, T.; Melloni, G.; Piroddi, A.M. *J. Org. Chem.*, ***1990***, *55*, 5386.

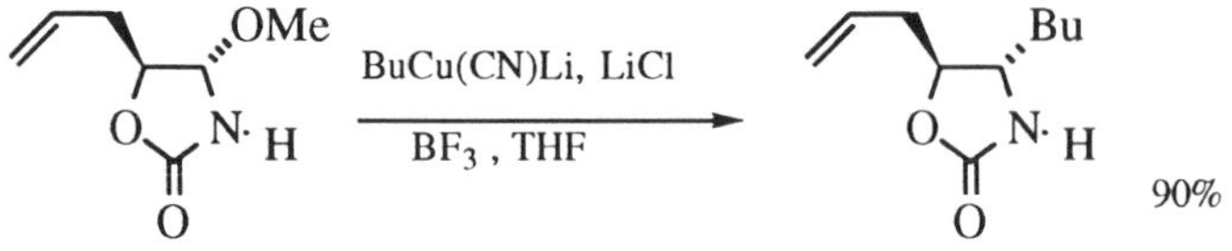

BuCu(CN)Li, LiCl; BF$_3$, THF — 90%

Ishibuchi, S.; Ikematsu, Y.; Ishizuka, T.; Kunieda, T. *Tetrahedron Lett.*, ***1991***, *32*, 3523.

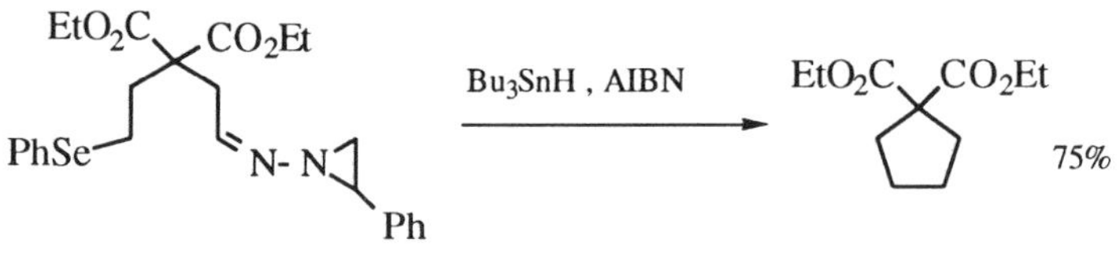

Bu$_3$SnH, AIBN — 75%

Kim, S.; Kee, I.S.; Lee, S. *J. Am. Chem. Soc.*, ***1991***, *113*, 9882.

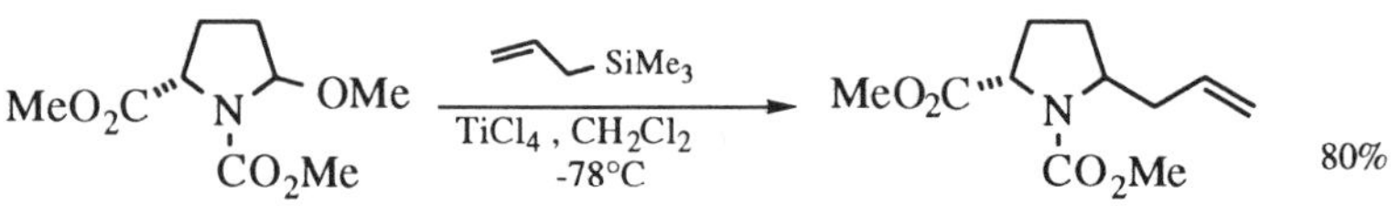

allyl-SiMe$_3$; TiCl$_4$, CH$_2$Cl$_2$, -78°C — 80% (73:27 cis:trans)

Shono, T.; Fujita, T.; Matsumura, Y. *Chem. Lett.*, ***1991***, 81.

SECTION 70: ALKYLS, METHYLENES AND ARYLS FROM HALIDES AND SULFONATES

The replacement of halogen by alkyl or aryl groups is included in this section. For the conversion of RX → RH (X = halogen) see Section 160 (Hydrides from Halides and Sulfonates).

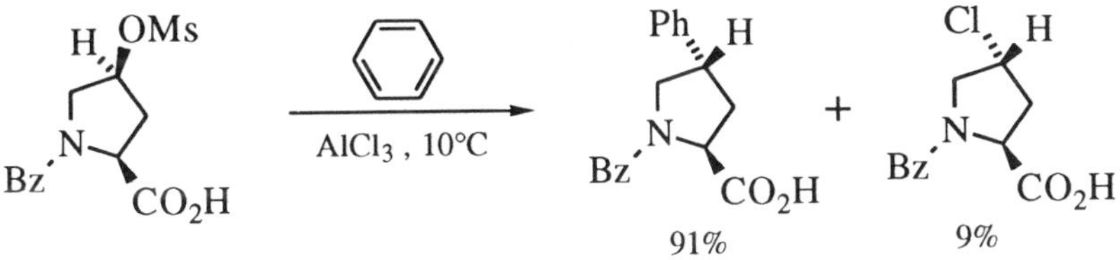
H
OMs
N
Bz
CO2H
AlCl3 , 10°C
Ph
H
N
Bz
CO2H
91%
+
Cl
H
N
Bz
CO2H
9%

Kronenthal, D.R.; Mueller, R.H.; Kuester, P.L.; Kissick, T.P.; Johnson, E.J. *Tetrahedron Lett.*, ***1990***, *31*, 1241.

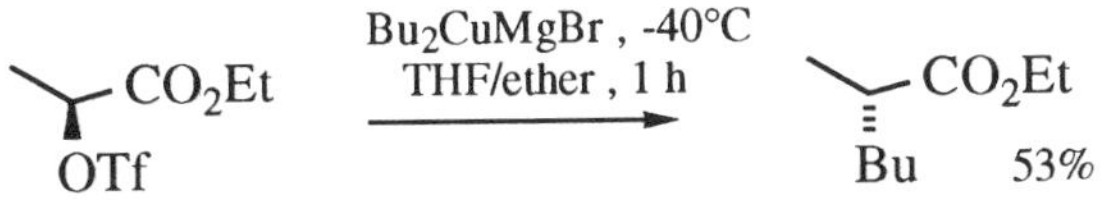
CO2Et
OTf
Bu2CuMgBr , -40°C
THF/ether , 1 h
CO2Et
Bu
53%

Petit, Y.; Sanner, C.; Larchevêque, M. *Tetrahedron Lett.*, ***1990***, *31*, 2149.

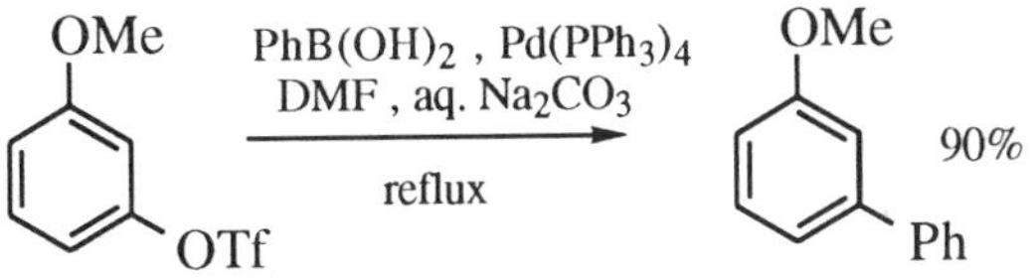
OMe
OTf
PhB(OH)2 , Pd(PPh3)4
DMF , aq. Na2CO3
reflux
OMe
Ph
90%

Fu, J-M.; Snieckus, V. *Tetrahedron Lett.*, ***1990***, *31*, 1665.

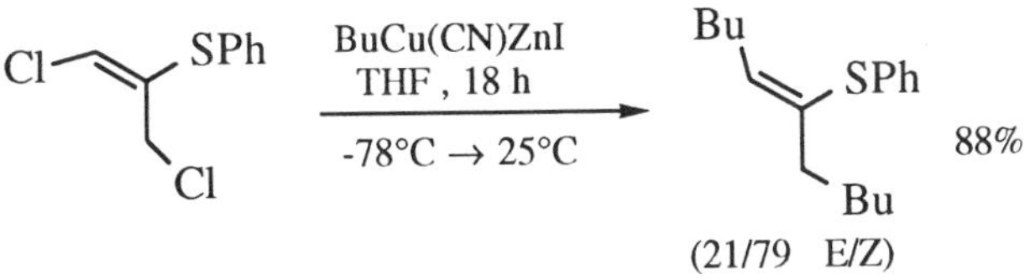
Cl
SPh
Cl
BuCu(CN)ZnI
THF , 18 h
-78°C → 25°C
Bu
SPh
Bu
88%
(21/79 E/Z)

Chen, H.G.; Gage, J.L.; Barrett, S.D.; Knochel, P. *Tetrahedron Lett.*, ***1990***, *31*, 1829.

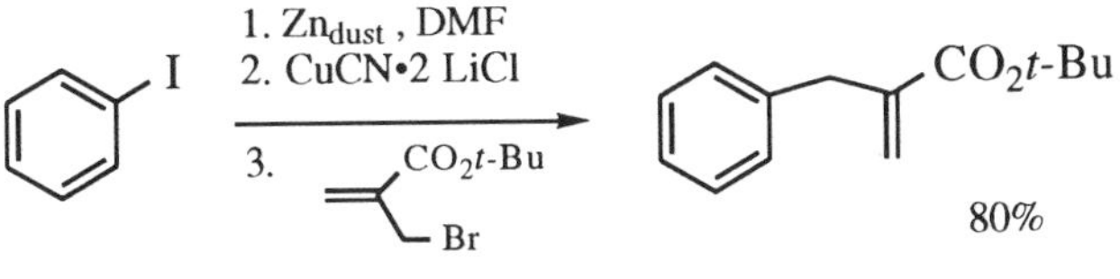
I
1. Zndust , DMF
2. CuCN•2 LiCl
3.
CO2t-Bu
Br
CO2t-Bu
80%

also undergoes conjugate addition reactions

Majid, T.N.; Knochel, P. *Tetrahedron Lett.*, ***1990***, *31*, 4413.

Br / Ph + 9-MeO-BBN (Me) — 1. $Pd(PPh_3)_4$, NaOH THF , reflux , 6 h; 2. NaOH , H_2O_2 → Ph / Me

Soderquist, J.A.; Santiago, B. *Tetrahedron Lett.*, ***1990***, *31*, 5541.
Soderquist, J.A.; Santiago, B.; Rivera, I. *Tetrahedron Lett.*, ***1990***, *31*, 4981.

O OH OTf O OH — Bu_4Sn , $PdCl_2(PPh_3)_2$ DMF , 3 eq. LiCl; 95°C , 2.5 h → O OH Bu O OH 74%

Tamayo, N.; Echavarren, A.M.; Paredes, M.C.; Fariña, F.; Noheda, P. *Tetrahedron Lett.*, ***1990***, *31*, 5189.

O B O Cl — 1. BuLi , THF , -78°C 2. PhMe , reflux , 36 h; 3. H^+ , AcOH → Bu 55%

Brown, H.C.; Rangaishenvi, M.V. *Tetrahedron Lett.*, ***1990***, *31*, 7113, 7115.

Br Br OMOM — Bu_3SnH , AIBN PhH , 80°C → OMOM + OMOM (80 : 20) 64%

Nagai, M.; Lazor, J.; Wilcox, C.S. *J. Org. Chem.*, ***1990***, *55*, 3440.

PhBr — 10% $NiBr_2(PPh_3)_2$, Zn , THF; 10% Et_4NI → Ph-Ph 99%

Iyoda, M.; Otsuka, H.; Sato, K.; Nisato, N.; Oda, M. *Bull. Chem. Soc. Jpn.*, ***1990***, *63*, 80.

OTf — 1. Sn_2Me_6 , LiCl , THF $Pd(PPh_3)_4$, reflux 2. PhBr , iPr_2NEt , PhMe; cat. 2,6-di-*t*-butyl-4-methylphenol $Pd(PPh_3)_4$, reflux → Ph + Ph (91 : 9) 58%

Stork, G.; Isaacs, R.C.A. *J. Am. Chem. Soc.*, ***1990***, *112*, 7399.

MeO–C6H4–OTf, Pd(PPh3)4, THF

ZnCl 41% OMe

Arcadi, A.; Burini, A.; Cacchi, S.; Delmastro, M.; Marinelli, F.; Pietroni, B. *SynLett*, ***1990***, 47.

SmI_2, -20°C, 30 min

Br Br 95% Br Br

Takahashi, S.; Mori, N. *J. Chem. Soc., Perkin Trans. I*, ***1991***, 2029.

$Li[Bu_4In]$, RT, 3 h

Br Bu 80%

Araki, S.; Shimizu, T.; Jin, S-J.; Butsugan, Y. *J. Chem. Soc., Chem. Commun.*, ***1991***, 824.

PhMgBr, ether, 20 h; 10% $NiCl_2$(dppf); reflux

I Ph 77%

Yuan, K.; Scott, W.J. *Tetrahedron Lett.*, ***1991***, *31*, 189.

Me_2CuLi, ether, -78°C

OMe $SiMe_3$ O O $SiMe_3$ + O $SiMe_3$

(2.59 : 1) 89%

Burke, S.D.; Piscopio, A.D.; Kort, M.E. *Tetrahedron Lett.*, ***1991***, *32*, 855.

1. Bu_3ZnLi, THF; 2. aq. AcOH

$n-C_8H_{17}CHBr_2$ → $n-C_8H_{17}CH_2Bu$ 77%

Harada, T.; Kotani, Y.; Katsuhira, T.; Oku, A. *Tetrahedron Lett.*, ***1991***, *32*, 1573.

$Ph_3C\text{-}Cl$ —Ph_2TlBr, PhLi, ether→ $Ph_3C\text{-}Ph$ 74%

Markó, I.E.; Kantam, M.L. *Tetrahedron Lett.*, ***1991***, *32*, 2255.
Markó, I.E.; Southern, J.M.; Kantam, M.L. *SynLett*, ***1991***, 235.

Cl, N (3-chloropyridine) → $Ph\text{-}B(OH)_2$ / $PdCl_2(dppb)$ → Ph, N (3-phenylpyridine) 71%

Mitchell, M.B.; Wallbank, P.J. *Tetrahedron Lett.*, ***1991***, *32*, 2273.

$SiMe_3$ → 1. BuLi, *t*-BuOK, THF, -78°C; 2. *n*-C_4H_9Br → $SiMe_3$, Bu + $SiMe_3$, Bu

(3.3 : 1) 95%

Li, L-H.; Wang, D.; Chan, T.H. *Tetrahedron Lett.*, ***1991***, *32*, 2879.

MgBr → CH_2Br_2, CuBr, THF-HMPA, reflux →

83%

Yamato, T.; Sakaue, N.; Suehiro, K.; Tashiro, M. *Org. Prep. Proceed. Int.*, ***1991***, *23*, 617.

OSO_2Me, Me, * CO_2Me, * (S) → PhH, $AlCl_3$, 6 h, 80°C → Ph, Me, CO_2Me 80%

97% ee (S)

Piccolo, O.; Azzena, U.; Melloni, G.; Delogu, G.; Valoti, E. *J. Org. Chem.*, ***1991***, *56*, 183.

OH → 1. $(FSO_2)_2O$, $EtN(iPr)_2$, 2 h, CH_2Cl_2, -78°C; 2. PhZnCl, THF, LiCl, 5% $Pd(PPh_3)_4$, 50°C → 90%

Roth, G.P.; Fuller, C.E. *J. Org. Chem.*, ***1991***, *56*, 3493.

Ph, Cl → $BuTi(OiPr)_3$, THF, -70°C, 7.5% CuI•2 LiCl, 6 h → Bu, Ph 66%

>99.8% S_N2'

Arai, M.; Nakamura, E. *J. Org. Chem.*, ***1991***, *56*, 5490.

$PhICl_2$ → BuLi, THF, -80°C → Bu-Bu 85%

Barton, D.H.R.; Jaszberenyi, J.Cs.; Leßmann, K.; Timár, T. *Tetrahedron*, ***1992***, *48*, 8881.

t-Bu, OH, t-Bu; F–C6H4–Br; hν (UV), $NH_{3(liq)}$, 2 h; pyrex; HO, t-Bu, F, t-Bu; 48%

Beugelmans, R.; Chastanet, J. *Tetrahedron Lett.*, ***1991***, *32*, 3487.

OH, OH; 1. MeO, MeO, I, Cl, O; 2. I, Cl, O; 3. 10 eq. Cu powder DMF, reflux; OMe, MeO, O, O, O, O; 71%

Takahashi, M.; Kuroda, T.; Orgiku, T.; Ohmizu, H.; Kondo, K.; Iwasaki, T. *Tetrahedron Lett.*, ***1991***, *32*, 6919.

MeO, OMe, OTf; $SnBu_3$, DMF; $PdCl_2(PPh_3)_2$, LiCl PPh_3, 120°C; MeO, OMe; 84%

Saá, J.M.; Martorell, G.; García-Raso, A. *J. Org. Chem.*, ***1992***, *57*, 678.
Martorell, G.; García-Raso, A.; Saá, J.M. *Tetrahedron Lett.*, ***1990***, *31*, 2357.

I, Br; 1. $Pd(PPh_3)_4$, DME 2. aq. Na_2CO_3, reflux 24 h; Me, $B(OH)_2$; 3. Ph-$B(OH)_2$, reflux; Ph, Me; 77%

Unrau, C.M.; Campbell, M.G.; Snieckus, V. *Tetrahedron Lett.*, ***1992***, *33*, 2773.

O, Ph, N, H, Cl; 1. BuLi 2. $LiC_{10}H_8$, -78°C; 3. PhI, -78°C → RT; O, Ph, N, H, Ph; 75%

Barluenga, J.; Montserrat, J.M.; Flórez, J. *Tetrahedron Lett.*, ***1992***, *33*, 6183.

1. Zn* , THF , RT
2. PhI , $Pd(PPh_3)_4$, THF

76%

Sakamoto, T.; Kondo, Y.; Murata, N.; Yamanaka, H. *Tetrahedron Lett.*, ***1992***, *33*, 5373.

HO Si Ph

1. BuLi
2. BnBr , HMPA

Ph Ph

63%

Hudrlik, P.F.; Abdallah, Y.M.; Hudrlik, A.M. *Tetrahedron Lett.*, ***1992***, *33*, 6747, 6743.

1. *t*-BuLi , pentane-ether , -78°C
2. TMEDA
3. warm
4. MeOH

89%

Bailey, W.F.; Khanolkar, A.D.; Gavaskar, K.V. *J. Am. Chem. Soc.*, ***1992***, *114*, 8053.
Bailey, W.F.; Khanolkar, A.D. *Tetrahedron Lett.*, ***1990***, *31*, 5993.

2 PhOTf

$2\ e^-$. 10% $PdCl_2(PPh_3)_2$
DMF , 90°C

Ph-Ph 76%

Jutand, A.; Négri, S.; Mosleh, A. *J. Chem. Soc., Chem. Commun.*, ***1992***, 1729.

B- n-C_8H_{17}

I , K_3PO_4
$Pd(PPh_3)_4$

n-$C_{14}H_{30}$ 64%

Ishiyama, T.; Abe, S.; Miyaura, N.; Suzuki, A. *Chem. Lett.*, ***1992***, 691.

OMe
I

$B(OH)_2$
2% $Pd(PPh_3)_4$, DME-H_2O
$Ba(OH)_2$, 80°C

OMe

80%

Watanabe, T.; Miyaura, N.; Suzuki, A. *SynLett*, ***1992***, 207.

SECTION 71: ALKYLS, METHYLENES AND ARYLS FROM HYDRIDES

This section lists examples of the reaction of RH → RR' (R,R' = alkyl or aryl). For the reaction C=CH → C=C-R (R = alkyl or aryl), see Section 209 (Alkenes from Alkenes). For alkylations of ketones and esters, see Section 177 (Ketones from Ketones) and Section 113 (Esters from Esters).

MeO— 0.6 $SnCl_4$, RT , 24 h; Ph⁄⁄SnBu$_3$ → MeO— —Ph 78%

(9:91 *o:p*)

Yamaguchi, J.; Takagi, Y.; Nakayama, A.; Fujiwara, T.; Takeda, T. *Chem. Lett.*, ***1991***, 133.

O, Ph, Ph 1. *t*-BuLi , THF -100°C → -78°C 2. $SOCl_2$, -78°C → *t*-Bu, O, Ph + O, Ph, *t*-Bu

(9 : 1) 52%

Olah, G.A.; Wu, A.; Farooq, O. *Synthesis*, ***1991***, 1179.

CHO, OH 1. NHMe, NHMe, EtOH , reflux 2. BuLi , TMEDA , ether RT 3. EtI 4. 2M HCl , H_2O → CHO, OH 54%

Gray, M.; Parsons, P.J. *SynLett*, ***1991***, 729.

O, S, *t*-Bu, OMOM 1. MeLi 2. MeI → Me, O, S, *t*-Bu, OMOM 58%

Quesnelle, C.; Iihama, T.; Aubert, T.; Perrier, H.; Snieckus, V. *Tetrahedron Lett.*, ***1992***, *33*, 2625.

SECTION 72: ALKYLS, METHYLENES AND ARYLS FROM KETONES

The conversions $R_2C{=}O \rightarrow$ R-R, R_2CH_2, R_2CHR', etc. are listed in this section.

O, Me $VO(OEt)Cl_2$, O_2 EtOH , 30 min → OEt, Me 90%

Hirao, T.; Mori, M.; Ohshiro, Y. *J. Org. Chem.*, ***1990***, *55*, 358.

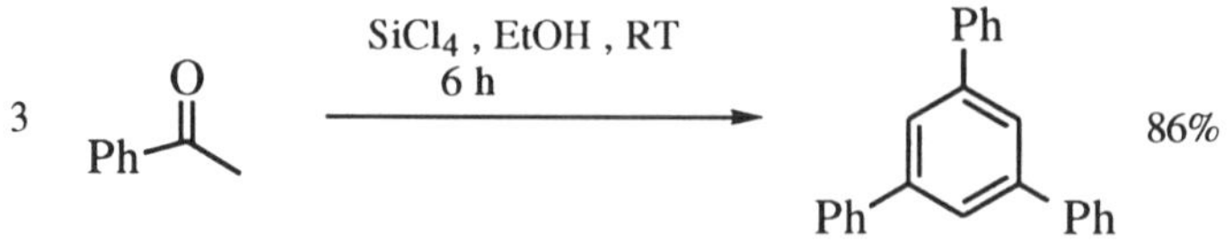

3 PhC(O)Me — $SiCl_4$, EtOH, RT, 6 h → 1,3,5-triphenylbenzene 86%

Elmorsy, S.S.; Pelter, A.; Smith, K. *Tetrahedron Lett.*, ***1991***, *32*, 4175.

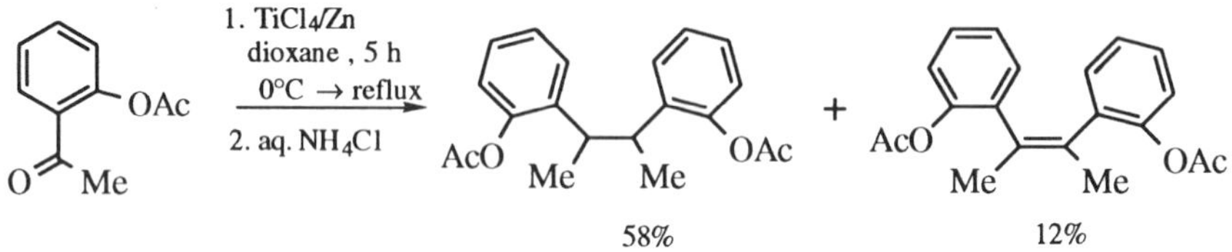

1. $TiCl_4$/Zn, dioxane, 5 h, 0°C → reflux; 2. aq. NH_4Cl

58% 12%

Nayak, S.K.; Banerji, A. *Ind. J. Chem.*, ***1991***, *30B*, 286.

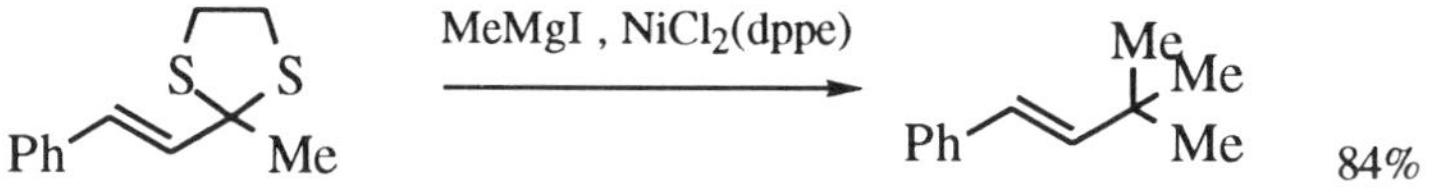

MeMgI, $NiCl_2$(dppe)

84%

Yuan, T-M.; Luh, T-Y. *J. Org. Chem.*, ***1992***, *57*, 4550.

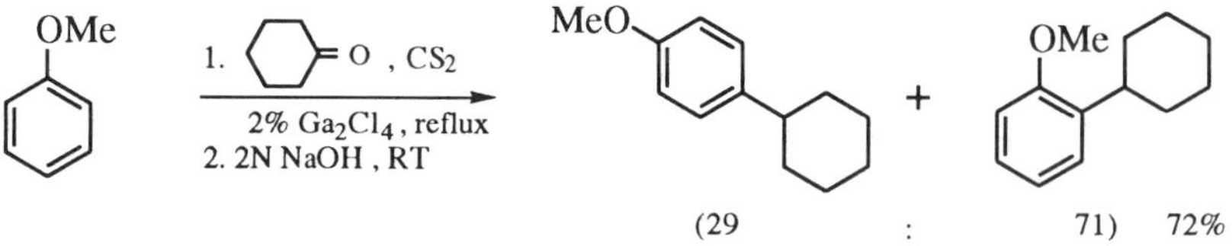

1. cyclohexanone, CS_2, 2% Ga_2Cl_4, reflux; 2. 2N NaOH, RT

(29 : 71) 72%

Hashimoto, Y.; Hirata, K.; Kihara, N.; Hasegawa, M.; Saigo, K. *Tetrahedron Lett.*, ***1992***, *33*, 6351.

SECTION 73: ALKYLS, METHYLENES AND ARYLS FROM NITRILES

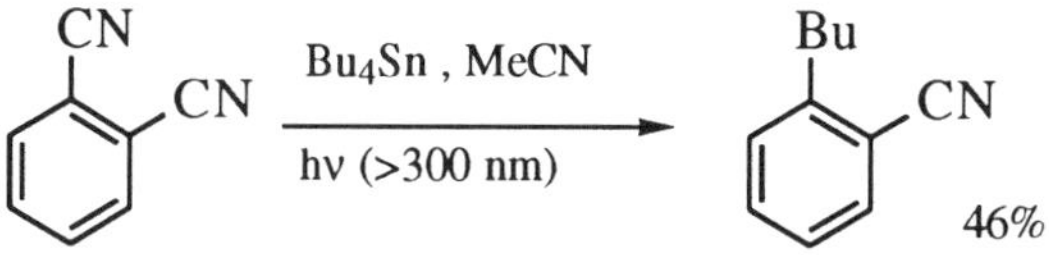

Bu_4Sn, MeCN, hν (>300 nm)

46%

Kyushin, S.; Masuda, Y.; Matsushita, K.; Nakadaira, Y.; Ohashi, M. *Tetrahedron Lett.*, ***1990***, *31*, 6395.

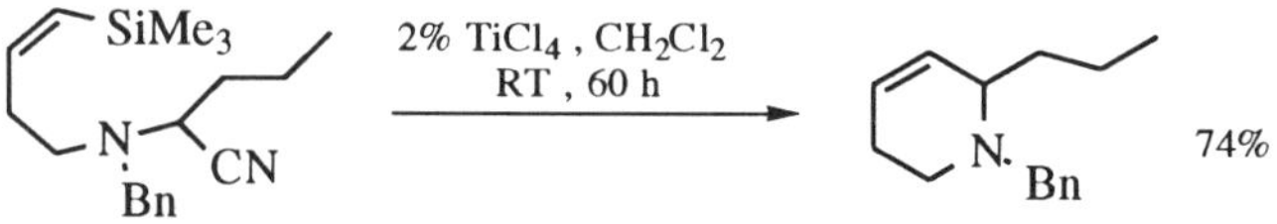

2% $TiCl_4$, CH_2Cl_2, RT, 60 h

74%

Teng, T-F.; Lin, J-H.; Yang, T-K. *Heterocycles*, ***1990***, *31*, 1201.

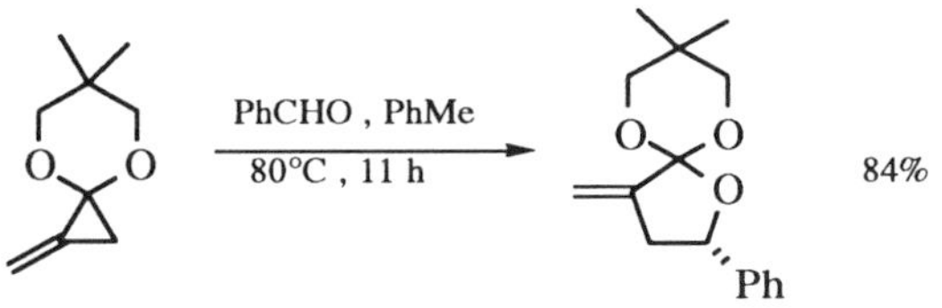
PhCHO , PhMe
80°C , 11 h
O
O
O
O
O
Ph
84%

Rychnovsky, S.D.; Zeller, S.; Skalitzky, D.J.; Griesgraber, G. *J. Org. Chem.*, ***1990***, *55*, 5550.

SECTION 74: ALKYLS, METHYLENES AND ARYLS FROM ALKENES

The following reaction types are included in this section:

A. Hydrogenation of Alkenes (and Aryls); **B.** Formation of Aryls; **C.** Alkylations and Arylations of Alkenes; **D.** Conjugate Reduction of Conjugated Carbonyl Compounds and Nitriles; **E.** Conjugate Alkylations; **F.** Cyclopropanations, including halocyclopropanations

SECTION 74A: Hydrogenation of Alkenes (and Aryls)

Reduction of aryls to dienes are listed in Section 377 (Alkene-Alkene).

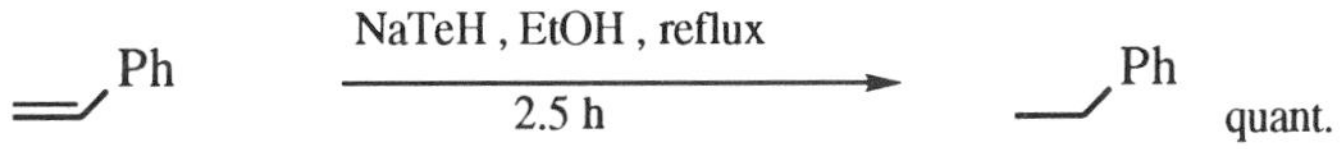
Ph
NaTeH , EtOH , reflux
2.5 h
Ph
quant.

Barton, D.H.R.; Bohé, L.; Lusinchi, X. *Tetrahedron*, ***1990***, *46*, 5273.

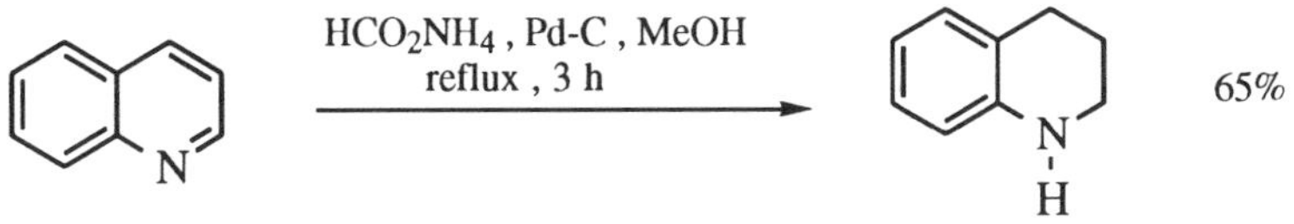
N
HCO2NH4 , Pd-C , MeOH
reflux , 3 h
N
H
65%

Bałczewski, P.; Joule, J.A. *Synth. Commun.*, ***1990***, *20*, 2815.

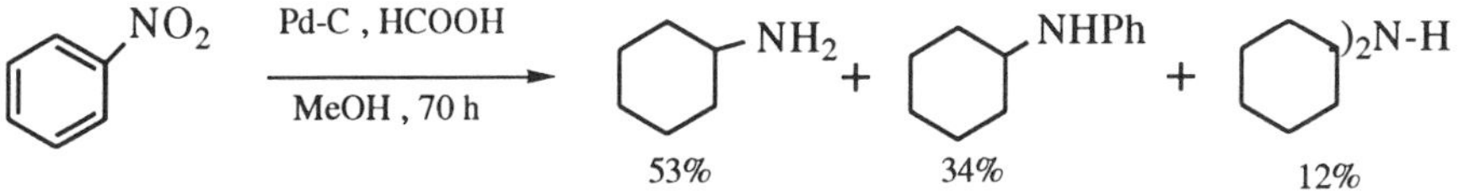
NO2
Pd-C , HCOOH
MeOH , 70 h
NH2
+
NHPh
+
)2N-H
53%
34%
12%

Ar also reduced with other Ar-X groups

Alper, H.; Vasapollo, G. *Tetrahedron Lett.*, ***1992***, *33*, 7477.

REVIEWS:

"Electrocatalytic Hydrogenation Hydrogen-Activity Electrodes"
Moutet, J-C. *Org. Prep. Proceed. Int.*, ***1992***, *24*, 309.

"Design Concepts for Developing Highly Efficient Chiral Bisphosphine Ligands in Rhodium-Catalyzed Asymmetric Hydrogenations"
Inoguchi, K.; Sakuraba, S.; Achiwa, K. *SynLett*, ***1992***, 169.

SECTION 74B: Formation of Aryls

1. Et-C≡C-OMe, DCE
hν (253.7 nm)
2. DCE, reflux, 4 h

N_2 OMe OH O

50%

Danheiser, R.L.; Brisbois, R.G.; Kowalczyk, J.J.; Miller, R.F. *J. Am. Chem. Soc.*, ***1990***, *112*, 3093.

C≡C-CO_2Me
C≡C-CO_2Me

H-C≡C-CH_2OH
$NiCl_2(PPh_3)_2$
2 eq. PPh_3
2 eq. BuLi

O O CO_2Me

78%

Bhatarah, P.; Smith, E.H. *J. Chem. Soc., Chem. Commun.*, ***1991***, 277.

PhH, hν (Hg vapor)
6 h

Ph

57%

Olsen, R.J.; Minniear, J.C.; Overton, W.M.; Sherrick, J.M. *J. Org. Chem.*, ***1991***, *56*, 989.

hν (Hg lamp/pyrex), I_2
propylene oxide, PhH

95%

Liu, L.; Yang, B.; Katz, T.J.; Poindexter, M.K. *J. Org. Chem.*, ***1991***, *56*, 3769.

H-C≡C-C(Me)(Me)OH
5% $Pd(OAc)_2$, MeCN
20% PPh_3, 5% CuI
NET_3, reflux, 14 h

MeO_2C MeO_2C C≡C-Et Br

Et MeO_2C MeO_2C Me Me OH

79%

Torii, S.; Okumoto, H.; Nishimura, A. *Tetrahedron Lett.*, ***1991***, *32*, 4167.

3 Et−C≡C−Et → ($MeO-C_6H_4-I$; NiI_2, Py, 60°C, 24 h) 39%

Kong, K-C.; Cheng, C-H. *J. Chem. Soc., Chem. Commun.*, ***1991***, 423.

$VO(OEt)Cl_2$, Me_3SiOTf; EtOH, 0°C → RT 82%

Hirao, T.; Mori, M.; Ohshiro, Y. *Chem. Lett.*, ***1991***, 783.

I_2, CAN, MeOH/MeCN; reflux, 8.5 h 82%

Horiuchi, C.A.; Fukunishi, H.; Kajita, M.; Yamaguchi, A.; Kiyomiya, H.; Kiji, S. *Chem. Lett.*, ***1991***, 1921.

$Cl^{\ominus}$ $Me_2HN^{\oplus}$ ... Ph → ($Cl^{\ominus}$; EtOH, NEt_3, heat, 2 h) OH, Ph 83%

Eichinger, K.; Nussbaumer, P. *Synthesis*, ***1991***, 663.

Me → (SeO_2-PPSE, CCl_4; reflux, 26 h) Me 96%

PPSE = trimethylsilyl polyphosphate = $P_2O_5/Me_3SiOSiMe_3$

Lee, J.G.; Kim, K.C. *Tetrahedron Lett.*, ***1992***, *33*, 6363.

O, N, C≡C-H, Br → ($Me_3SnSnMe_3$, PhNC, hν; 80°C) N, N, O 40%

Curran, D.P.; Liu, H. *J. Am. Chem. Soc.*, ***1992***, *114*, 5863.

1. $[Rh(COD)Cl]_2$, Bu-CC-H; 2. H^+, 40°C — 70%

Bz Ph Bu Bu OH

Padwa, A.; Xu, S.L. *J. Am. Chem. Soc.*, ***1992***, *114*, 5881.

$Cr(CO)_5$ OMe Ph — 2 eq. *t*-BuN≡C, dark; ether, 0°C → reflux — OMe NH*t*-Bu 90%

Merlic, C.A.; Burns, E.E.; Xu, D.; Chen, S.Y. *J. Am. Chem. Soc.*, ***1992***, *114*, 8722.

SECTION 74C: Alkylations and Arylations of Alkenes

$SiMe_3$ MgBr + OAc — $Pd(PPh_3)_4$, THF; RT — $SiMe_3$ 93%

Fotiadu, F.; Cros, P.; Faure, B.; Buono, G. *Tetrahedron Lett.*, ***1990***, *31*, 77.

Ph I OH — 1. *t*-$BuMe_2SiOTf$; 2. 2 *t*-BuLi, THF, -78°C → +23°C — Ph *t*-$BuMe_2Si$ OH 84%

Kim, K.D.; Magriotis, P.A. *Tetrahedron Lett.*, ***1990***, *31*, 6137.

O NMe_2 — PhI, K_2CO_3, DMF; Bu_4NCl, 3% $Pd(OAc)_2$; 80°C, 16 h — Ph O NMe_2 80%

Andersson, C-M.; Larsson, J.; Hallberg, A. *J. Org. Chem.*, ***1990***, *55*, 5757.

Ph ⊕ N= N– SO_2Ph ⊖ O — CO_2Et, PhH, 80°C; $Pd(PPh_3)_4$, sealed tube — Ph CO_2Et 89%

Kamigata, N.; Satoh, M.; Fukushima, T. *Bull. Chem. Soc. Jpn.*, ***1990***, *63*, 2118.

1. Mg , THF
2. Cp_2TiCl_2 , CH_2Cl_2
3. $AlCl_2Et$, PhMe , -78°C
4. HCl

Br … → Me–(cyclopentyl) 78%

Rigollier, P.; Young, J.R.; Fowley, L.A.; Stille, J.R. *J. Am. Chem. Soc.*, ***1990***, *112*, 9441.

3 eq. PhCl , MeCN , Py; Zn , $NiCl_2(PPh_3)_2$

Ph … Ph 54% + Ph … Ph 7%

Lebedev, S.A.; Pedchenko, V.V.; Lopatina, V.S.; Berestova, S.S.; Petrov, Ê.S. *Zhur. Org. Khim.*, ***1990***, *26*, 1520 (Engl., p. 1312).

hν , Me_3SiH , 20°C

Me … $Si(SiMe_3)_3$ + Me … $Si(SiMe_3)_3$ (68 : 32) 65%

Kulicke, K.J.; Giese, B. *SynLett*, ***1990***, 91.

B-$(CH_2)_{10}CO_2Me$; $PdCl_2$(dppf) , THF reflux

t-Bu … OTf → *t*-Bu … $(CH_2)_{10}CO_2Me$ 86%

Oh-e, T.; Miyaura, N.; Suzuki, A. *SynLett*, ***1990***, 221.

Bu_3SnCl , AIBN
$NaBH_3CN$, 3 d
t-BuOH , reflux

Br … OH → OH 20% + HO 43%

Bothwick, A.D.; Caddick, S.; Parsons, P.J. *Tetrahedron Lett.*, ***1990***, *31*, 6911.

1. *sec*-BuLi , THF , -70°C
2. MeI

Ph Me O I → Ph Me O Me 79%

Barluenga, J.; Rodríguez, M.A.; Campos, P.J. *SynLett*, ***1990***, 270.

$SiMeCl_2$, NEt_3 , 120°C; cat. $PdCl_2(PPh_3)_2$

PhCHO → Ph … $SiMeCl_2$ 77%

Yamashita, H.; Roan, B.L.; Tanaka, M. *Chem. Lett.*, ***1990***, 2175.

Br, Br — Ph (styrene), $Pd(OAc)_2$, DMF / TBAB, K_2CO_3, 100°C, 2 h → Ph, Ph 92%

Lansky, A.; Reiser, O.; de Meijere, A. *SynLett*, ***1990***, 405.

Ph — $HSiCl_3$, RT / $PdCl_2$ (NHSO$_2$Me, PPh$_2$) → Ph, $SiCl_3$, Me (64.8% ee, S) 57%

Okada, T.; Morimoto, T.; Achiwa, K. *Chem. Lett.*, ***1990***, 999.

S, S, O, O, N, O — Ph, Ph, OH, Ph, Me, OH, Ph, Ph •$TiCl_2(OiPr)_2$ / mesitylene, MS 4Å, 0°C, 4 d → S, S, O, N, O, O 41% (84% ee) + S, S, O, O 40% (>98% ee)

Narasaka, K.; Hayashi, Y.; Shimada, S.; Yamada, J. *Isr. J. Chem.*, ***1991***, *31*, 261.

Cl_2MeSiH, Ni*, THF, 25°C → $SiCl_2Me$ 92%

Ni* = activated Ni, from Li reduction of NiI in THF

Boudjouk, P.; Han, B-H.; Jacobsen, J.R.; Hauck, B.J. *J. Chem. Soc., Chem. Commun.*, ***1991***, 1424.

1. BuMgCl, 2.5% Cp_2ZrCl_2 2. H^+ → 94% (82:18 cis:trans)

Knight, K.S.; Waymouth, R.M. *J. Am. Chem. Soc.*, ***1991***, *113*, 6268.

SiF_3 — I, Br, TBAF / 5% $Pd(PPh_3)_4$, 37 h, 100°C → Br 78%

Hatanaka, Y.; Ebina, Y.; Hiyama, T. *J. Am. Chem, Soc.*, ***1991***, *113*, 7075.

CO_2Bn NHBoc

$Pd(OAc)_2$, Bu_4NCl
$NaHCO_3$, DMF

I, Br → CO_2Bn, NHBoc, Br 63%

Carlström, A-S.; Frejd, T. *J. Org. Chem*, ***1991***, *56*, 1289.

MeO_2C, Br, $OSiMe_2t$-Bu, OH

BEt_3 , Bu_3SnH
PhMe

→ H, $OSiMe_2t$-Bu, OH, MeO_2C, H 97%

Satoh, S.; Sodeoka, M.; Sasai, H.; Shibasaki, M. *J. Org. Chem.*, ***1991***, *56*, 2278.

EtO_2C, TfO, Br

$Bu_3SnSiMe_3$, PhMe
0.1 M , Bu_4NBr
3% $PdCl_2(PPh_3)_2$
Li_2CO_3 , 110°C

→ CO_2Et 70%

Mori, M.; Kaneta, N.; Shibasaki, M. *J. Org. Chem.*, ***1991***, *56*, 3486.

5% $Pd_2(dba)_3$, PhI
2 eq. $CH_2(CO_2Et)_2$
PhI , DMSO , 80°C
Bu_4NCl , $NaHCO_3$

→ Ph, CO_2Et, CO_2Et 82%

Larock, R.C.; Lu, Y.; Bain, A.C.; Russell, C.E. *J. Org. Chem.*, ***1991***, *56*, 4589.

1. 2 eq. Mg , MeOH
2. 3N HCl , 0°C

O, O → O, O 98%

Domínguez, C.; Csákÿ, A.G.; Plumet, J. *Tetrahedron Lett.*, ***1991***, *32*, 4183.

OSO_2CF_3, *t*-Bu

MeO–C_6H_4–$SnBu_3$
$Pd_2(dba)_3$, $AsPh_3$
RT

→ OMe, *t*-Bu 89%

tailors catlyst to substrate

Farina, V.; Roth, G.P. *Tetrahedron Lett.*, ***1991***, *32*, 4243.

1. $MeCl_2SiH$, CuCl
Parr microwave bomb
TMEDA, (750 W)
microwave oven
2. $HC(OEt)_3$

$SiMe(OEt)_3$

75%

Abramovitch, R.A.; Abramovitch, D.A.; Iyanar, K.; Tamareselvy, K. *Tetrahedron Lett.*, ***1991***, *32*, 5251.

O Me Me O

$(PhMe_2Si)Cu•LiCN$
THF, 30 min

$PhMe_2Si$ O Me Me O

quant.

Lautens, M.; Belter, R.K.; Lough, A.J. *J. Org. Chem.*, ***1992***, *57*, 422.

$SiH(SiMe_3)_3$, PhMe
AIBN, 90°C

$Si(SiMe_3)_3$

81%

Kopping, B.; Chatgilialoglu, C.; Zehnder, M.; Giese, B. *J. Org. Chem.*, ***1992***, *57*, 3994.

$Cp_2Y[CH(TMS)_2]$
14 h

SiH_2Ph

95%

Molander, G.A.; Julius, M. *J. Org. Chem.*, ***1992***, *57*, 6347.

1. hν, O_2, DCA
TMSCN
2. H^+

60%

DCA = 9,10-dicyanoanthracene

Santamaria, J.; Kaddachi, M.T.; Ferround, C. *Tetrahedron Lett.*, ***1992***, *33*, 781.

n-C_6H_{13}

1. n-$C_6H_{13}CH_2CH_2MgCl$
3% Cp_2ZrCl_2, THF
2. H_3O^+

n-C_6H_{13} n-C_5H_{11}

(4:1 E:Z) 92%

Rousset, C.J.; Negishi, E.; Suzuki, N.; Takahashi, T. *Tetrahedron Lett.*, ***1992***, *33*, 1965.

$ICHFCO_2Et$ —[=/Bu, Zn, Py, THF; $NiCl_2$•6 H_2O, 60°C]→ $BuCH_2CH_2CHFCO_2Et$ 73%

Wang, Y.; Yang, Z-Y.; Burton, D.J. *Tetrahedron Lett.*, ***1992***, *33*, 2137.

OH; $Pd(OAc)_2$, NEt_3; P(*o*-tolyl)$_3$, DMF; 60°C, 3 h

TfO Ph → HO Ph 75%

Bernocchi, E.; Carrchi, S.; Ciattini, P.G.; Morera, E.; Ortar, G. *Tetrahedron Lett.*, ***1992***, *33*, 3073.

Ph BuTe C C Ph —[1. BuLi, -78°C; 2. Me_2SO_4, -78°C → RT]→ Ph Me C C Ph 93%

Dabdoub, M.J.; Dabdoub, V.B.; Comasseto, J.V. *Tetrahedron Lett.*, ***1992***, *33*, 2261.

n-C_5H_{11} CO_2Me I —[1. $Pd(PPh_3)_4$, THF, PhZnCl; 2. aq. NH_4Cl]→ *n*-C_5H_{11} CO_2Me Ph 76%

Rossi, R.; Carpita, A.; Cossi, P. *Tetrahedron Lett.*, ***1992***, *33*, 4495.

Ph TfO —[$NaBPh_4$, $Pd(PPh_3)_4$; DMF, RT, 2 h]→ Ph Ph 73%

Ciattini, P.G.; Morena, E.; Ortar, G. *Tetrahedron Lett.*, ***1992***, *33*, 4815.

Cl O *n*-C_8H_{17} —[$SnBu_3$, DCE; $Pd(PPh_3)_4$]→ O *n*-C_8H_{17} 82%

Bhatt, R.K.; Shin, D-S.; Falck, J.R.; Mioskowski, C. *Tetrahedron Lett.*, ***1992***, *33*, 4885.

O O —[PhI, Ag_2CO_3, DMF]→ Ph O O 62% (43% ee)

Sakamoto, T.; Kondo, Y.; Yamanaka, H. *Tetrahedron Lett.*, ***1992***, *33*, 6845.

BuMgCl , 1% Cp_2ZrCl_2

1 d

95%

(86:14 trans:cis)

Wischmeyer, U.; Knight, K.S.; Waymouth, R.M. *Tetrahedron Lett.*, ***1992***, *33*, 7735.

Ph

MeO Br

B-n-C_8H_{17}

$Pd(PPh_3)_4$, dioxane

60°C , 3 eq. K_2CO_3 , 16 h

Ph

MeO n-C_8H_{17}

83%

Abe, S.; Miyaura, N.; Suzuki, A. *Bull. Chem. Soc. Jpn.*, ***1992***, *65*, 2863.

Me

OTf

OSiMe$_2$$t$-Bu

H Me Me

$Pd(PPh_3)_4$, THF

Li_2CO_3 , LiCl , reflux

Me OSiMe$_2$$t$-Bu

91%

H Me

Liang, S.; Paquette, L.A. *Acta Chem. Scand.*, ***1992***, *46*, 597.

Cp_2YMe•THF , PhH

H_2 (2 atm) , RT , 45 min

S S + S S

97% 2%

Molander, G.A.; Hoberg, J.O. *J. Am. Chem. Soc.*, ***1992***, *114*, 3123.

O Ph

5% Mo(CHCMe$_2$Ph)

(N[2,6-diisopropylphenyl])

(OCMe[CF_3]$_2$)$_2$

PhH , 20°C , 15 min

O Ph

92%

Fu, G.C.; Grubbs, R.H. *J. Am. Chem. Soc.*, ***1992***, *114*, 5426.

Ph

1. B-H , THF , 25°C

2% Rh(DiPPE)

2. NaOH , H_2O_2

Ph OH

85%

DiPPE = 1,2-bis(diisopropylphosphino)ethane

Wescott, S.A.; Blom, H.P.; Marder, T.B.; Baker, R.T. *J. Am. Chem. Soc.*, ***1992***, *114*, 8863.

Ph–CH=CH–SiMe$_2$Ph → [I–C$_6$H$_4$–OMe, 65°C; (η^3C$_3$H$_5$PdCl)$_2$, THF, 48 h] → Ph–CH=CH–C$_6$H$_4$–OMe 67%

Rossi, R.; Carpita, A.; Messeri, T. *Gazz. Chim. Ital.*, ***1992***, *122*, 65.

CH$_2$=CH–Bu → [1. F$_3$CCH$_2$I, AIBN, 85°C; 2. NaBH$_4$, DMF, 0°C → RT] → F$_3$C–CH$_2$CH$_2$CH$_2$–Bu 40%

Cloux, R.; Kováts, Ez. *Synthesis*, ***1992***, 409.

Ph–CH$_2$Cl → [CH$_2$=CH–Ph, Pd(OAc)$_2$; Nu$_3$N, 110°C, 15 h] → Ph–CH=CH–Ph 57%

Yi, P.; Zhuangyu, Z.; Hongwen, H. *Synth. Commun.*, ***1992***, *22*, 2019.

3-I–C$_6$H$_4$–NH$_2$ → [CH$_2$=CH–CO$_2$Et, 37°C; Pd(OAc)$_2$, Ph$_3$POTs, 10 h] → H$_2$N–C$_6$H$_4$–CH=CH–CO$_2$Et

Genet, J.P.; Blart, E.; Savignac, M. *SynLett*, ***1992***, 715.

SECTION 74D: Conjugate Reduction of α,β-Unsaturated Carbonyl Compounds and Nitriles

Ph–CH=CH–CO$_2$H → [1. aq. NaOH, 5 min, Pd(OAc)$_2$; 2. (EtO)$_3$SiH, RT, 10 min] → Ph–CH$_2$CH$_2$–CO$_2$H 93%

Tour, J.M.; Pendalwar, S.L. *Tetrahedron Lett.*, ***1990***, *31*, 4719.

4-OAc-3-Me-cyclohex-2-enone → [0.24 [(Ph$_3$PCuH]$_6$, 30 H$_2$O, RT, 1 h; THF] → 4-OAc-3-Me-cyclohexanone 94%

Koenig, T.M.; Daeuble, J.F.; Brestensky, D.M.; Stryker, J.M. *Tetrahedron Lett.*, ***1990***, *31*, 3237.

$NaHFe(CO)_4$

60°C , 16 h

85%

Künzer, H.; Stahnke, M.; Sauer, G.; Wiechert, R. *Tetrahedron Lett.*, ***1990***, *31*, 3859.

1. MAD , -78°C
2. L-Selectride
3. H_2O

MeO OMe Me

MAD = bis(2,6-di-*t*-butyl-4-methylphenoxy)methyl aluminum

Doty, B.J.; Morrow, G.W. *Tetrahedron Lett.*, ***1990***, *31*, 6125.

$(EtO)_3SiH$, $Pd(OAc)_2$, aq. THF

OBu

add ester prior to silane

99%

Tour, J.M.; Cooper, J.P.; Pendalwar, S.L. *J. Org. Chem.*, ***1990***, *55*, 3452.

OMOM

(*n*-C_5H_{11})$_2$ Cu(CN)Li_2

Me_3SiCl , THF
-78°C

n-C_5H_{11}

OMOM >90%

(70:30 diastereomers)

Linderman, R.J.; Griedel, B.D. *J. Org. Chem.*, ***1990***, *55*, 5428.

$RhCl(PPh_3)$, THF
catecholborane
-20°C , 12 h

CO_2Bn → CO_2Bn

82%

Evans, D.A.; Fu, G.C. *J. Org. Chem.*, ***1990***, *55*, 5678.

C≡C-CO_2Et

OH

e^- , $PhCO_2H$, DMF

Et_4NClO_4 , (0.1 M)

HO CO_2Et 87%

Arcadi, A.; Cacchi, S.; Carelli, I.; Curulli, A.; Inesi, A.; Marinelli, F. *SynLett*, ***1990***, 408.

e^- ($2e^-$/mol) , aq. MeOH

aq. KCl

C/poly PPV^{+2}-Pd electrodes

quant.

PPV = poly(pyrrole-viologen)

Coche, L.; Ehui, B.; Limosin, D.; Moutet, J-C. *J. Org. Chem.*, ***1990***, *55*, 5905.

CHO $\xrightarrow{H_2\text{, 1 h}}$ CHO quant.

Ar, Me_2 N $PdCl_2$ S Ar cat. Fe S- Ar

Ar = *p*-Me-phenyl

Ali, H.M.; Naiini, A.A.; Brubaker, C.J. Jr. *Tetrahedron Lett.*, ***1991***, *32*, 5489.

1. NaSeH , 50°C , 3 h , EtOH
2. 2N HCl

Ph Me O → Ph Me O 84%

Nishiyama, Y.; Yoshida, M.; Ohkawa, S.; Hamanaka, S. *J. Org. Chem.*, ***1991***, *56*, 6720.

CO_2H CO_2H $\xrightarrow{H_2\text{, EtOH}}$ CO_2H CO_2H

2% Me Me $P(c\text{-}C_6H_{11})_2$ Rh(NBD)Cl $P(c\text{-}C_6H_{11})_2$

(96% ee) quant.

NBD = norbornadiene

Chiba, T.; Miyashita, A.; Nohira, H.; Takaya, H. *Tetrahedron Lett.*, ***1991***, *32*, 4745.

O O OAc OAc OAc OAc

H_2 , PtO_2 , EtOAc
NEt_3 , RT , 2 d

O O OAc OAc OAc 88%

Katsuki, J.; Inanaga, J. *Tetrahedron Lett.*, ***1991***, *32*, 4963.

Ph_3SnH , BEt_3 , PhH
25°C , 6 h

O → O 98%

Nozaki, K.; Oshima, K.; Utimoto, K. *Bull. Chem. Soc. Jpn.*, ***1991***, *64*, 2585.

$n\text{-}C_5H_{11}$ (Me ketone) — bakers yeast, 64 h → $n\text{-}C_5H_{11}$ (Me ketone) (>99% ee) 24%

Sakai, T.; Matsumoto, S.; Hidaka, S.; Imajo, N.; Tsuboi, S.; Utaka, M. *Bull. Chem. Soc. Jpn.*, ***1991***, *64*, 3473.

Ph(Me)CH–CH=CH–CO_2Et — SmI_2, $AcNMe_2$, THF / EtOH, 12 h → Ph(Me)CH–CH_2CH_2–CO_2Et 96%

Inanaga, J.; Sakai, S.; Handa, Y.; Yamaguchi, M.; Yokoyama, Y. *Chem. Lett.*, ***1991***, 2117.

CO_2Et / CO_2Et (cis alkene) — Zn, AcOH, 2 h /))))))) → CO_2Et / CO_2Et 98%

Marchand, A.P.; Reddy, G.M. *Synthesis*, ***1991***, 198.

OMe — $Pd(OAc)_2$, HCO_2K / DMF, 60°C → OMe 94%

Arcadi, A.; Bernocchi, E.; Cacchi, S.; Marinelli, F. *SynLett*, ***1991***, 27.

CO_2Me — 1. SmI_2, THF 2. HMPA, RT, 30 min → CO_2Me 67%

Cabrera, A.; Alper, H. *Tetrahedron Lett.*, ***1992***, *33*, 5007.

CO_2H — 4 atm H_2, MeOH, 25°C, 24 h; catalyst: MeO, Ph, P, P, Ph, OMe, Ru, Me, Me → CO_2H (40% ee) quant.

Genet, J-P.; Pinel, C.; Mallart, S.; Juge, S.; Cailhil, N.; Laffitte, J.A. *Tetrahedron Lett.*, ***1992***, *33*, 5343.

CO_2H [RuH(S-BINAP)$_2$] PF_6 / iPrOH → CO_2H H 18%

(52% ee , R)

Saburi, M.; Ohnuki, M.; Ogawawara, M.; Takahashi, T.; Uchida, Y. *Tetrahedron Lett.*, ***1992***, *33*, 5783.

O ; CO_2Et , DBU , DMEU / 185°C , 24 h → O CO_2Et 80%

DMEU = 1,3-dimethyl-2-imidazolidinone

Hwu, J.R.; Kakimelahi, G.H.; Chou, C-T. *Tetrahedron Lett.*, ***1992***, *33*, 6469.

SECTION 74E: Conjugate Alkylations

n-Bu, Ph, $B(OiPr)_2$; Ph, Ph, O / 0.1 eq. BF_3•OEt_2 , PhMe → *n*-Bu, Ph, Ph, Ph, O 78%

Hara, S.; Hyuga, S.; Aoyama, M.; Sato, M.; Suzuki, A. *Tetrahedron Lett.*, ***1990***, *31*, 247.

O, Me ; 1. $Me_2Cu(CN)Li_2$ 2. $(PhMe_2Si)Cu(CN)Li_2$ / 3. aq. NH_4Cl 60% → O, Me, Me, $SiMe_2Ph$ (major) + O, Me, Me, $SiMe_2Ph$

Lee, S.-H.; Hulce, M. *Tetrahedron Lett.*, ***1990***, *31*, 311.

O + O, Cu, Me, $(CN)Li_2$; THF/ether/pentane / CO , -110°C <5 min → O, O

73% (12:1 trans:cis)

Lipshutz, B.H.; Elworthy, T.R. *Tetrahedron Lett.*, ***1990***, *31*, 477.

Et_2CuLi , THF
-78°C → RT , 135 min
80%

Charonnat, J.A.; Mitchell, A.L.; Keogh, B.P. *Tetrahedron Lett.*, ***1990***, *31*, 315.

1. 1.1 eq.MeLi
2. 3 eq. MeI
64%

Eid, C.N. Jr.; Konopelski, J.P. *Tetrahedron Lett.*, ***1990***, *31*, 305.

e^- , (Hg/SCE electrodes)
dimethyl malonate
n-Bu_4NBr , MeCN
(89%)

Sowell, C.G.; Wolin, R.L.; Little, R.D. *Tetrahedron Lett.*, ***1990***, *31*, 485.

1. $Me_2Cu(CN)Li_2$ ether , -78°C , 30 min
2. NaOMe , MeOH 14h , 25°C
no yield 60 : 40

Jung, M.E.; Lew, W. *Tetrahedron Lett.*, ***1990***, *31*, 623.

1.5 MeMgBr , [Cu]
3.0 TMSCl , -46°C
ether:THF , 1 h
99%

[Cu] =

Sakata, H.; Aoki, Y.; Kuwajima, I. *Tetrahedron Lett.*, ***1990***, *31*, 1161.

$Me_3SnCu{\bullet}SMe_2$, THF
$Li(Bu_3SnCuSPh)$
-78°C
$SnBu_3$
85%

Laborde, E.; Lesheski, L.E.; Kiely, J.S. *Tetrahedron Lett.*, ***1990***, *31*, 1837.

1. LDA, THF, -78°C
2. Et_2AlCl, -40°C
3.
SMe
quant.

Kpegba, K.; Metzner, P. *Tetrahedron Lett.*, ***1990***, *31*, 1853.

$Me_2AgMgBr$
CH_2Cl_2 , THF
78%
Me_3Rh , THF
61%

Kauffmann, T.; Hülsdünker, A.; Menges, D.; Nienaber, H.; Rethmeier, L.; Robber, S.; Scherler, D.; Schrickel, J.; Wingbermühle, D. *Tetrahedron Lett.*, ***1990***, *31*, 1553.

$SnBu_3$
1. MeLi
2. 0.5 $CuCN{\bullet}LiCl$
3.
83%

also reacts with halides, epoxides, etc.

Lipshutz, B.H.; Ung, C.; Elworthy, T.R.; Reuter, D.C. *Tetrahedron Lett.*, ***1990***, *31*, 4539.

4 eq. Et_3SiH , sealed tube
t-BuOO*t*-Bu , 140°C
PhH
Me_3Si
71%

also cyclization with unconjugated dienes

Kraus, G.A.; Vivas, S. *Tetrahedron Lett.*, ***1990***, *31*, 5265.

benzophenone, PhH
hν (>300 nm), 12 mM
20 min

CO_2Et → CO_2Et

57%

Nishida, A.; Nishinda, M.; Yonemitsu, O. *Tetrahedron Lett.*, ***1990***, *31*, 7035.

CO_2Me → $CHCl_3$, NEt_4; e^- (1.8 F/mol), DMF; Et_4NBr, Pt cathode → CO_2Me, CCl_3 78%

Shono, T.; Ishifune, M.; Ishige, O.; Uyama, H.; Kashimura, S. *Tetrahedron Lett.*, ***1990***, *31*, 7181.

O, Me → $Me_3Sn(Bu)Cu(CN)Li_2$ → O, $SnMe_3$, Me 89%

Lipshutz, B.H.; Sharma, S.; Reuter, D.C. *Tetrahedron Lett.*, ***1990***, *31*, 7253.

1. [Cp_2ZrCl_2, $AlMe_3$ DCE, PhMe, 0°C]
2. CuCN, BuC≡CLi THF
3. -23°C, cyclopentenone

Ireland, R.E.; Wipf, P. *J. Org. Chem.*, ***1990***, *55*, 1425.

CO_2BHA
1. BuLi, THF
2. $LiBEt_3H$, THF
3. MeI, HMPA
4. $NaBH_4$, MeOH
5. 10% aq. HCl

HO, Me, Bu 75%

BHA = 2,6-di-*t*-butyl-4-(methoxyphenyl)-1-naphthalenecarboxaldehyde

Tomioka, K.; Shindo, M.; Koga, K. *J. Org. Chem.*, ***1990***, *55*, 2276.
Tomioka, K.; Shindo, M.; Koga, A. *Tetrahedron Lett.*, ***1990***, *31*, 1739.

BuMgBr , $MnCl_2$
LiCl , THF , 0°C → 20°C
CuCl , Me_3SiCl

CO_2Et → Bu, CO_2Et 97%

Cahiez, G.; Alami, M. *Tetrahedron Lett., **1990**, 31*, 7423, 7425.

PhCH(R)Cu(CN)ZnCl
THF

R - n-C_3H_7

O → O, n-C_3H_7, SPh 78%

Achyutha Rao, S.; Tucker, C.E.; Knochel, P. *Tetrahedron Lett., **1990**, 31*, 7575.

Ph-C≡C-ZnBr , -40°C
t-$BuMe_2SiOTf$
ether-THF

O → $OSiMe_2t$-Bu, C≡C-Ph 96%

Kim, S.; Lee, J.M. *Tetrahedron Lett., **1990**, 31*, 7627.

Bu_3SnH , AIBN
PhH (0.04 M)
reflux

O, Me, O, SePh → O, Me, O 65%

Middleton, D.S.; Simpkins, N.S.; Terrett, N.K. *Tetrahedron, **1990**, 46*, 545.

hν , DMSO , 4 eq. iPrHgCl
4 eq. KI

CO_2Me, CO_2Me → CO_2Me, CO_2Me 82%

Russell, G.A.; Kim, B.H. *SynLett, **1990***, 87.

$Cu\bullet Me_3SiCl$
Li I , THF , 30 min

O → O 87%

Lipshutz, B.H.; Ellsworth, E.L.; Dimock, S.H.; Smith, R.A.J. *J. Am. Chem. Soc., **1990**, 112*, 4404.

$LiAlH_4$ - $SbCl_3$, THF

0°C

84%

Sayama, S.; Inamura, Y. *Bull. Chem. Soc. Jpn.*, ***1991***, *64*, 306.

$SnMe_3$, hν

MeCN , 15 h

51%

Takuwa, A.; Nishigaichi, Y.; Iwamoto, H. *Chem. Lett.*, ***1991***, 1013.

CO_2Et

t-BuBr , EtOH

))))))))

CO_2Et 95%

Dupuy, C.; Petrier, C.; Sarandeses, L.A.; Luche, J.L. *Synth. Commun.*, ***1991***, *21*, 643.

1. SmI_2 , THF , DMPU

2. CHO / OTBS

TBSO 60% + TBSO 20%

Curran, D.P.; Wolin, R.L. *SynLett*, ***1991***, 317.

H- C≡C- CO_2Me

iPr-I , $Bu_3SnSnBu_3$

hν (sunlamp) , 85°C
30 min

I CO_2Me 70%

(1:2.4 E:Z)

Curran, D.P.; Kim, D. *Tetrahedron*, ***1991***, *47*, 6171.
Curran, D.P.; Kim, D.; Ziegler, C. *Tetrahedron*, ***1991***, *47*, 6189.

Me

[cat. Cp_2ZrCl_2 , CH_2Cl_2 , 0°C
HC≡CCH_2CH_2OH]

CuCN•2 LiCl , THF-ether

Me Me OH 82%

(2.8:1 trans:cis)

Lipshutz, B.H.; Dimock, S.H. *J. Org. Chem.*, ***1991***, *56*, 5761.

Reagents: Ph–CH(OMOM)(SiMe$_3$); CsF, DMF — 68%

Linderman, R.J.; Ghannam, A.; Badejo, I. *J. Org. Chem.*, ***1991***, *56*, 5213.

1. $Cp_2Zr(H)Cl$, THF, RT, 20 min,))))))
2. cyclohexenone
3. $CuBr{\bullet}SMe_2$, RT, 1 h

70%

Wipf, P.; Smitrovich, J.H. *J. Org. Chem.*, ***1991***, *56*, 6494.

CO_2Et → EtO_2C … $(CF_2)_4Cl$

$Cl(CF_2)_4I$, Zn, EtOH, bromo(Py) cobaloxime (III); 20°C, 1 h

Hu, C-M.; Qiu, Y-L. *Tetrahedron Lett.*, ***1991***, *32*, 4001.

n-C_6H_{13} … OMe → n-C_6H_{13} … OMe, Bu

$Bu_2Cu(CN)Li_2$, THF, -78°C — 90%

Soderquist, J.A.; Rosado, I. *Tetrahedron Lett.*, ***1991***, *32*, 4451.

$OSiMe_2t$-Bu, OMe, ether; 2.5 M $LiClO_4$, 1.2 h → $OSiMe_2t$-Bu, CO_2Me — 95%

Grieco, P.A.; Cooke, R.J.; Henry, K.J.; Vander Roest, J.M. *Tetrahedron Lett.*, ***1991***, *32*, 4665.

Me, CO_2Et, CO_2Et + Me, $GeMe_3$, CO_2Et —TBAF→ Me, CO_2Et, CO_2Et, CO_2Et, Me

(7:3 dr) 61%

Yamamoto, Y.; Okano, H.; Yamada, J. *Tetrahedron Lett.*, ***1991***, *32*, 4749.

Me_3Sn ... $SnMe_3$ — 1. $Me_2Cu(CN)Li_2$, -78°C, 3 h; 2. (cyclohexenone) =O; 3. $Me_2Cu(CN)Li_2$; 4. (cyclohexenone) =O → 82% (91% E,E)

Lipshutz, B.H.; Lee, J.I. *Tetrahedron Lett.*, ***1991***, *32*, 7211.

CO_2t-Bu, Et, I — 1. BuLi , THF , -78°C; 2. H_3O^+ → CO_2t-Bu, Et 88% (trans/cis = 3.0)

Cooke, M.P. Jr. *J. Org. Chem.*, ***1992***, *57*, 1495.

1. MeLi; 2. aq. H_2SO_4 → Me, Me, OH 70%

Alonso,F.; Yus, M. *Tetrahedron*, ***1992***, *48*, 2709.

Bu_3Sn $Cu(CN)Li$; Et_3SiCl , -78°C → $OSiEt_3$, $SnBu_3$ 96%

Marino, J.P.; Emonds, M.V.M.; Stengel, P.J; Oliveira, A.R.M.; Simonelli, F.; Ferreira, J.T.B. *Tetrahedron Lett.*, ***1992***, *33*, 49.

$PbPh_3$, (cyclohexyl)-Br; hν , 5% $(Ph_3Pb)_2$, 2.5h → 84% (91:9 trans:cis)

Toru, T.; Watanabe, Y.; Tsusaka, M.; Gautam, R.K.; Tazawa, K.; Bakouetila, M.; Yoneda, T.; Ueno, Y. *Tetrahedron Lett.*, ***1992***, *33*, 4037.

$Me_3To{\bullet}MeLi$, ether
-40°C, 1 h

O

Me

73%

Markó, I.E.; Rebière, R. *Tetrahedron Lett.*, ***1992***, *33*, 1763.

1. 4 eq. SmI_2, THF, HMPA
2. $CuI{\bullet}P(OEt)_3$, -78°C → -20°C
3. (cyclohexenone)

Ph I

O

Ph

67%

Tottleben, M.J.; Curran, P.P.; Wipf, P. *J. Org. Chem.*, ***1992***, *57*, 1740.

Me O CO_2Me

1. $BnNH_2$, cyclohexane reflux
2. $NaBH_3CN$, MeOH

Me N Bn CO_2Me

57%

>6:1 E:Z)

Dumas, F.; d'Angelo, J. *Tetrahedron Lett.*, ***1992***, *33*, 2005.

O B·(ZnICu[CN]C=CHBu O

(cyclohexenone), Me_3SiCl
-78°C → 25°C, 12 h

O

$n\text{-}C_5H_{11}$

O

76%

Waas, J.R.; Sidduri, A.R.; Knochel, P. *Tetrahedron Lett.*, ***1992***, *33*, 3717.

MeO_2C MeO_2C CO_2Me CO_2Me O

Cs_2CO_3, MeCN

add yn-one by syringe pump

CO_2Me CO_2Me MeO_2C MeO_2C O

90%

(88:12 cis:trans)

Girard, S.; Deslongchamps, P. *Can. J. Chem.*, ***1992***, *70*, 1265.

Ph Te Ph

1. $Bu_2Cu(CN)Li_2$, THF RT
2. -78°C, (methyl vinyl ketone) O

O Ph

77%

Tucci, F.C.; Chieffi, A.; Comasseto, J.V. *Tetrahedron Lett.*, ***1992***, *33*, 5721.

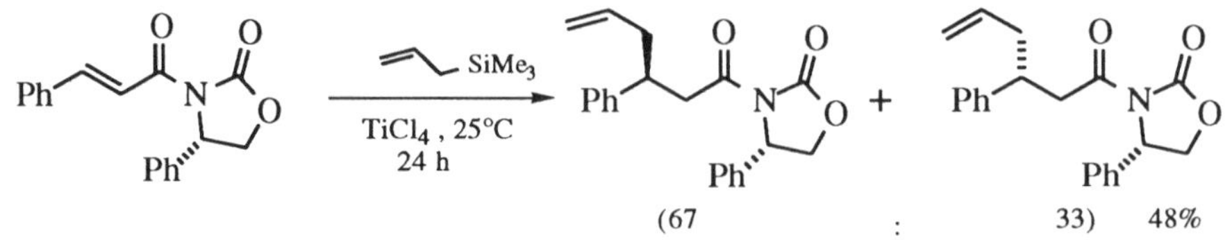

Ph, O, O, N, O, Ph — SiMe$_3$, $TiCl_4$, 25°C, 24 h → Ph, O, O, N, O, Ph + Ph, O, O, N, O, Ph

(67 : 33) 48%

Wu, M-J.; Wu, C-C.; Lee, P-C. *Tetrahedron Lett.*, ***1992***, *33*, 2547.

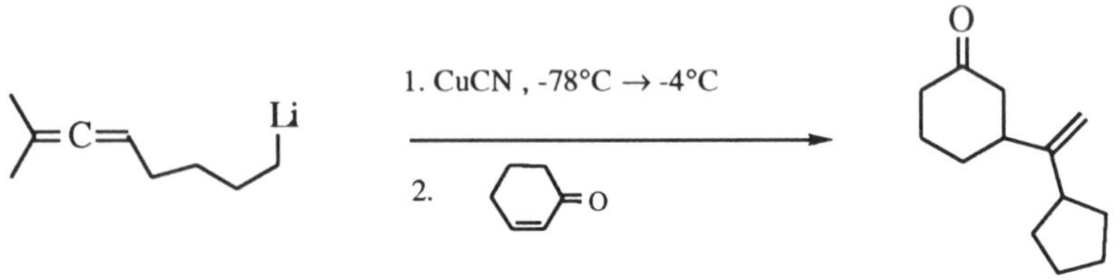

C=, Li — 1. CuCN , -78°C → -4°C; 2. (cyclohexenone) O → O

Crandall, J.K.; Ayers, T.A. *Tetrahedron Lett.*, ***1992***, *33*, 5311.

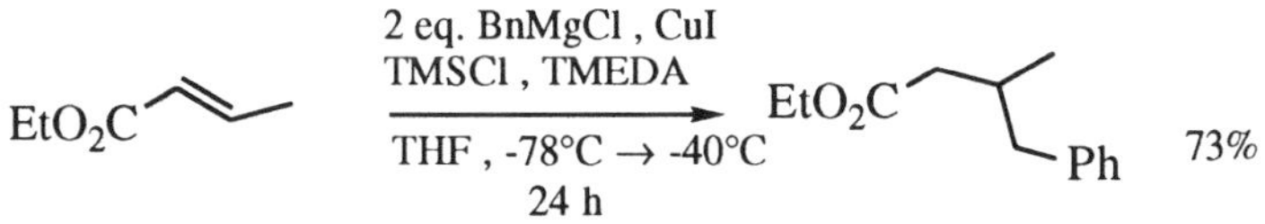

EtO_2C — 2 eq. BnMgCl , CuI; TMSCl , TMEDA; THF , -78°C → -40°C; 24 h → EtO_2C, Ph 73%

van Heerden, P.S.; Bezuidenhoudt, B.C.B.; Steenkamp, J.A.; Ferreira, D. *Tetrahedron Lett.*, ***1992***, *33*, 2383.

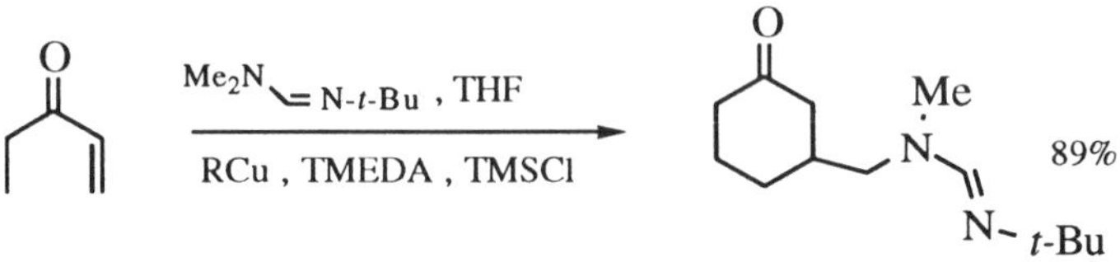

O — Me_2N=N-*t*-Bu , THF; RCu , TMEDA , TMSCl → O, Me, N, N-*t*-Bu 89%

Dieter, R.K.; Alexander, C.W. *Tetrahedron Lett.*, ***1992***, *33*, 5693.

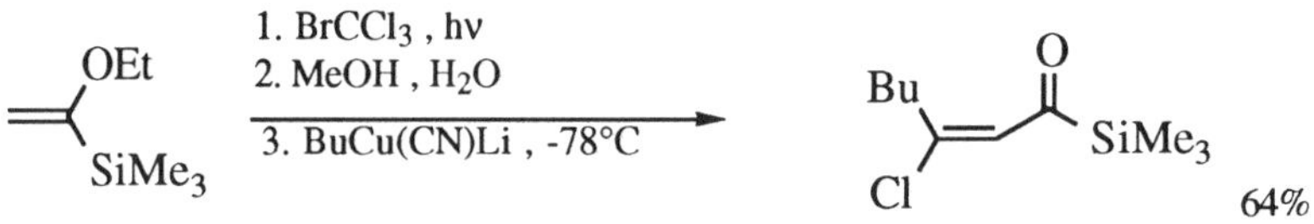

OEt, SiMe$_3$ — 1. $BrCCl_3$, hν; 2. MeOH , H_2O; 3. BuCu(CN)Li , -78°C → Bu, O, SiMe$_3$, Cl 64%

Cunico, R.F.; Zhang, C. *Tetrahedron Lett.*, ***1992***, *33*, 6751.

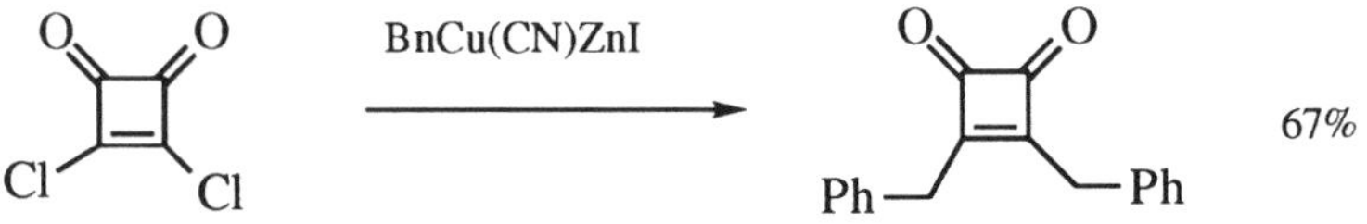

O, O, Cl, Cl — BnCu(CN)ZnI → O, O, Ph, Ph 67%

Sidduri, A.R.; Budries, N.; Laine, R.M.; Knochel, P. *Tetrahedron Lett.*, ***1992***, *33*, 7515.

$Cl(CF_2)_8I$ → ($=\!\!\diagup CO_2Et$, Zn, THF; cat. $SmCl_3$, reflux) → $Cl(CF_2)_8CH_2CH_2CO_2Et$ 76%

Ding, Y.; Zhao, G.; Huang, W. *Tetrahedron Lett.*, ***1992***, *33*, 8119.

PhBr, $Ni(bipy)_3(BF_3)_2$
DMF, $ZnBr_2$, Bu_4NBr
e^- (Zn anode, C cathode)

O, Ph 52%

Sibille, S.; Ratovelomanana, V.; Périchon, J. *J. Chem. Soc., Chem. Commun.*, ***1992***, 283.

O, I → ($=\!\!\diagup CO_2Me$, Zn, CuI; aq. EtOH, RT,))))))))) → O, CO_2Me 74%

Sarandeses, L.A.; Mouriño, A.; Luche, J-L. *J. Chem. Soc., Chem. Commun.*, ***1992***, 798.

CO_2Me, NHAc → (4 eq. furan (O), BF_3•OEt_2; 25°C, 40 h) → O, CO_2Me, NHAc quant.

Cativiela, C.; López, M.P.; Mayoral, J.A. *SynLett*, ***1992***, 121.

O, H-C≡C, $SiPh_3$ → (Bu_2CuLi, ether; -78°C) → Bu, $SiPh_3$, O 89%

Degl'Innocenti, A.; Stucchi, E.; Capperucci, A.; Mordini, A.; Reginato, G.; Ricci, A. *SynLett*, ***1992***, 329, 332.

ASYMMETRIC CONJUGATE ADDITION REACTIONS

O → (DMS, LiCu(*n*-Bu)R; R = Ph, N, H, N) → O, Bu

92% (81%ee,S)

Rossiter, B.E.; Eguchi, M. *Tetrahedron Lett.*, ***1990***, *31*, 965.

O

[BuLi/CuBr], ether

NH_2 Me

Ph N NMe_2

Me

O

Bu

63%

(68% de, R)

Dieter, R.K.; Lagu, B.; Deo, N.; Dieter, J.W. *Tetrahedron Lett.*, ***1990***, *31*, 4105.

O

Ph Ph

$ZnEt_2$, catalyst

1% $Ni(acac)_2$

O Ph

Ph *

55%

(72% ee, R)

catalyst = N N OH HO

Bolm, C.; Ewald, M. *Tetrahedron Lett.*, ***1990***, *31*, 5011.

OH O O Me

Ph_2CuLi, -50°C

30 min

OH O O Me H Ph

\+

OH O O Me H Ph

(94 : 6) 61%

Fang, C.; Suemune, H.; Sakai, K. *Tetrahedron Lett.*, ***1990***, *31*, 4751.

O

iPrMgBr, 1% $ZnCl_2$, THF

-90°C, N Me N OH

O

91%

(18% ee)

Jansen, J.F.G.A.; Feringa, B.L. *J. Org. Chem.*, ***1990***, *55*, 4168.

O

NO_2

1. MeCN 25°C H N OMe

2. $ZnBr_2$, THF

3. Me_2CuLi, -78°C → 0°C

O

Me

84%, S

Tamura, R.; Watabe, K.; Katayama, H.; Suzuki, H.; Yamamoto, Y. *J. Org. Chem.*, ***1990***, *55*, 408.

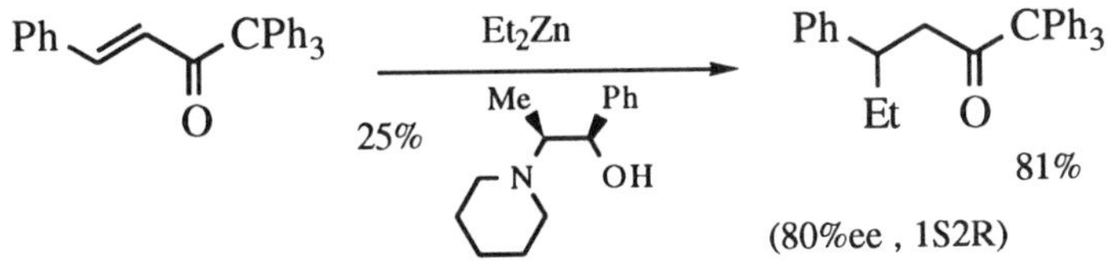

Ph CPh$_3$ O → Et_2Zn 25% Me Ph N OH → Ph CPh$_3$ Et O 81%

(80%ee , 1S2R)

Soai, K.; Okudo, M.; Okamoto, M. *Tetrahedron Lett.*, ***1991***, *32*, 95.

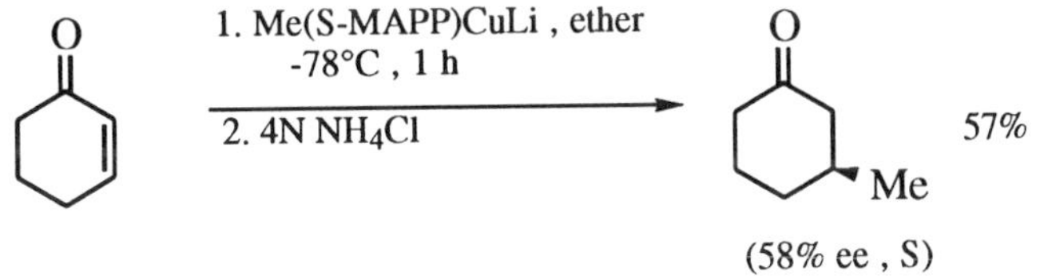

O → 1. Me(S-MAPP)CuLi , ether -78°C , 1 h; 2. 4N NH_4Cl → O Me 57%

(58% ee , S)

Rossiter, B.E.; Eguchi, M.; Hernández, A.E.; Vickers, D.; Medich, J.; Marr, J.; Heinis, D. *Tetrahedron Lett.*, ***1991***, *32*, 3973.

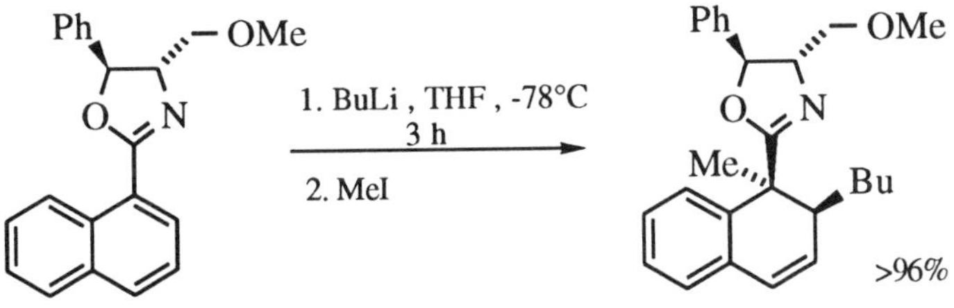

Ph — OMe O N → 1. BuLi , THF , -78°C 3 h; 2. MeI → Ph — OMe O N Me Bu >96%

Rawson, D.J.; Meyers, A.I. *Tetrahedron Lett.*, ***1991***, *32*, 2095.

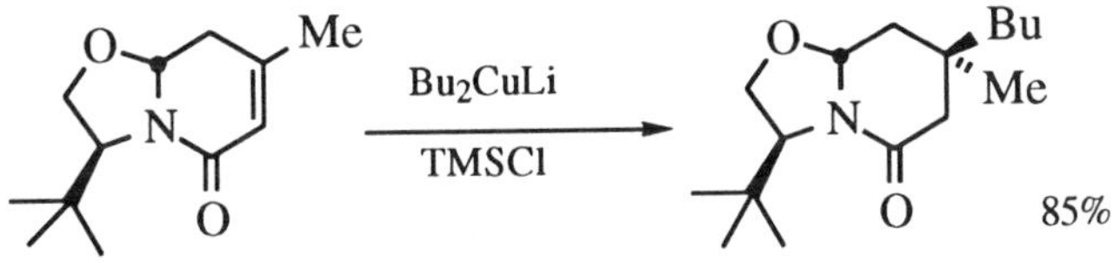

O Me N O → Bu_2CuLi; TMSCl → O Bu N Me O 85%

Meyers, A.I.; Leonard, W.R. Jr.; Romine, J.L. *Tetrahedron Lett.*, ***1991***, *32*, 597.

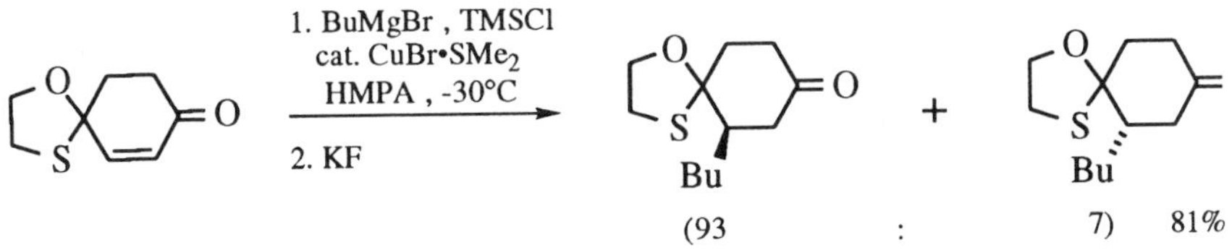

O S O → 1. BuMgBr , TMSCl cat. CuBr•SMe_2 HMPA , -30°C; 2. KF → O S O Bu + O S O Bu

(93 : 7) 81%

Sonoda, S.; Houchigai, H.; Asaoka, M.; Takei, H. *Tetrahedron Lett.*, ***1992***, *33*, 3145.

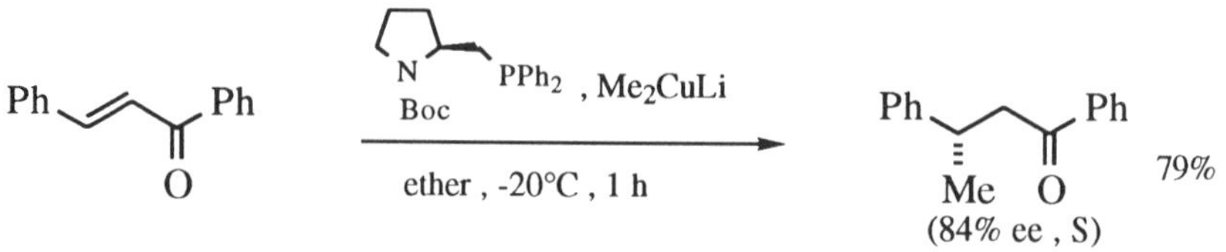

Kanai, M.; Koga, K.; Tomioka, K. *Tetrahedron Lett.*, ***1992***, *33*, 7193.

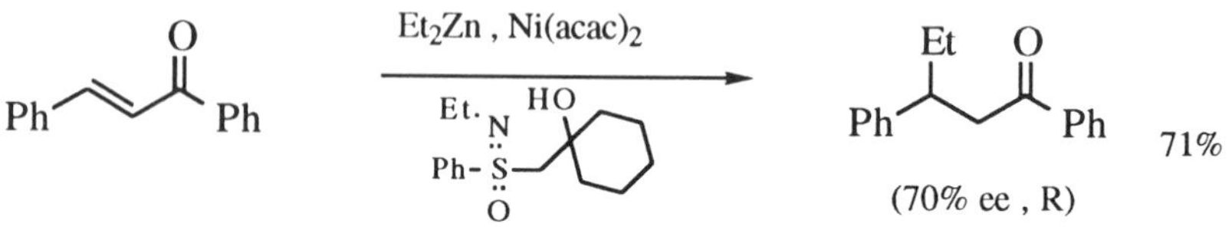

Bolm, C.; Felder, M.; Müller, J. *SynLett*, ***1992***, 439.

REVIEWS:

"The Evolution of Higher Order Cuprates"
Lipshutz, B.H. *SynLett*, ***1990***, 119.

"New Aspects of Organocopper Reagents: 1,2- and 1,2- Chiral Induction and Reaction Mechanism"
Ibuka, T.; Yamamoto, Y. *SynLett*, ***1992***, 769.

SECTION 74F: Cyclopropanations, including Halocyclopropanations

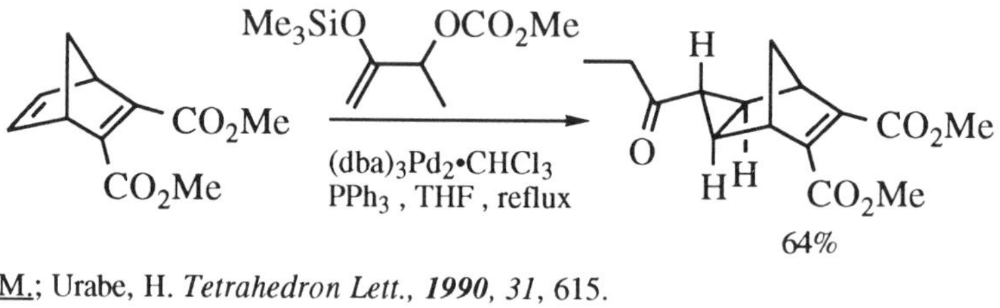

Trost, B.M.; Urabe, H. *Tetrahedron Lett.*, ***1990***, *31*, 615.

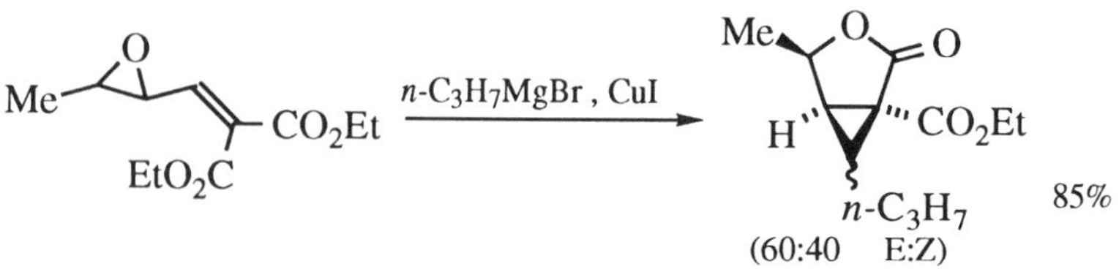

Kasatkin, A.N.; Kulak, A.N.; Biktimirov, R.Kh.; Tolstikov, G.A. *Tetrahedron Lett.*, ***1990***, *31*, 4915.

Ph (styrene) → [$N_2CHC(O)O$-*l*-menthyl; Cu salen derivative catalyst] → Ph-cyclopropane-CO_2-*l*-menthyl + Ph-cyclopropane-CO_2-*l*-menthyl

(86 : 14) 72%

98% ee 96% ee

Lowenthal, R.E.; Abiko, A.; Masamune, S. *Tetrahedron Lett.*, ***1990***, *31*, 6005.

menthyl diazoacetate (CHN_2) → [=/Ph , CH_2Cl_2; cat. dirhodium carboxamidate (CO_2Me)] → Ph-cyclopropyl ester

69% (67:33 trans:cis)

trans (48% ee, 1S,2S) cis (85% ee , 1S,2R)

Doyle, M.P.; Brandes, B.D.; Kazala, A.P.; Pieters, R.J.; Jarstfer, M.B.; Watkins, L.M.; Eagle, C.T. *Tetrahedron Lett.*, ***1990***, *31*, 6613.

Bu-CH=CH$_2$ → [$CH_2(CO_2Et)_2$, I_2 , 2 h; K_2CO_3 , reflux, tricapylMeNCl] → 2-Bu-cyclopropane-1,1-(CO_2Et)$_2$ 39%

Tóke, L.; Szabó, G.T.; Hell, Z.; Tóth, G. *Tetrahedron Lett.*, ***1990***, *31*, 7501.

diazo ketone (N_2, H) → [0.75 Dibal , PhH, RT , 1 h; 3% Cu complex (Br, N, Cu, O)$_2$] → bicyclic ketone 54%

77% ee

Dauben, W.G.; Hendricks, R.T.; Luzzio, M.J.; Ng, H.P. *Tetrahedron Lett.*, ***1990***, *31*, 6969.

0.33 Ph(Me)C=CH$_2$ → [$CFCl_3$, $TiCl_4$, 0°C; $LiAlH_4$, 30 min] → 1-Ph-1-Me-2-Cl-2-F-cyclopropane 85%

Dolbier, W.R. Jr.; Burkholder, C.R. *J. Org. Chem.*, ***1990***, *55*, 589.

Br, OMe, OMe → TCNE, THF, -10°C, 1 h → CO_2Me, NC, CN, NC, CN 88%

Lee, J-Y.; Hall, H.K. Jr. *J. Org. Chem.*, ***1990***, *55*, 4963.

Bu, $B(OH)_2$ → 1. (+)-TMTA, MS 4Å; 2. CH_2I_2, Zn-Cu, ether, reflux, 24 h; 3. H_2O_2, 2N $KHCO_3$, THF → Bu, OH 91% ee 67%

TMTA = N,N,N',N'-tetramethyl tartaramide

Imai, T.; Mineta, H.; Nishida, S. *J. Org. Chem.*, ***1990***, *55*, 4986.

OH → [Me, Me, Cl C≡CH, PhH, 5% KOH, dibenzo-18-crown-6] RT, 5 d → OH, C 43%

Sheu, J-H.; Yen, C-F.; Chan, Y-L.; Chung, J-F. *J. Org. Chem.*, ***1990***, *55*, 5232.

Ph → Zn, THF, CF_2Br_2, RT, overnight → Ph, F, F 96%

yields range from 4-96% with various alkenes

Dolbier, W.R. Jr.; Wojtowicz, H.; Burkholder, C.R. *J. Org. Chem.*, ***1990***, *55*, 5420.

OH → CH_2N_2, $PdCl_2(PhCN)_2$, CH_2Cl_2 - ether, 10°C → RT → OH 74%

also used to convert allylamines to aminomethyl cyclopropanes

Tomilov, Yu.V.; Kostitsyn, A.B.; Shulishov, E.V.; Nefedov, O.M. *Synthesis*, ***1990***, 246.

$SiMe_3$, Li → 1. CO, THF, 15°C, 2 h; 2. t-$BuMe_2SiCl$, HMPA, -78°C → Me_3Si, $OSiMe_2t$-Bu 59% + =C=, $SiMe_3$, $OSiMe_2t$-Bu 15%

Ryu, I.; Hayama. Y.; Hirai, A.; Sonoda, N.; Orita, A.; Ohe, K.; Murai, S. *J. Am. Chem. Soc.*, ***1990***, *112*, 7061.

Me$_3$P Rh ⊕ $PF_6^{\ominus}$ — 1. Me(Ph)C=CH O$^-$ K$^+$, THF, -60°C, 60 h; 2. I_2, THF, -78°C → Me, Ph, O 70%

Tjaden, E.B.; Stryker, J.M. *J. Am. Chem. Soc.*, ***1990***, *112*, 6420.

Cl, O — 1. Mg, THF; 2. aq. NH_4Cl → OH + OH (3 : 2) 75%

Akhmedov, M.A.; Akhmedov, I.M.; Mustafaeva, Z.G.; Sardarov, I.K.; Kostikov, R.R.; Musaeva, Kh.E. *Zhur. Org. Khim.*, ***1990***, *26*, 1257 (Engl., p. 1087).

Me, Me — $Mo(CO)_5$ Bu OMe; THF, 100°C → Bu, Me, MeO, Me 71%

Harvey, D.F.; Lund, K.P. *J. Am. Chem. Soc.*, ***1991***, *113*, 8916.
Harvey, D.F.; Lund, K.P.; Neil, D.A. *Tetrahedron Lett.*, ***1991***, *32*, 6311.

O, O, N_2 — Rh_2(5S-MEPY)$_4$, 25°C, CH_2Cl_2 → O, O 74% (88% ee)

MEPY = MeO_2C N O

Doyle, M.P.; Pieters, R.J.; Martin, S.F.; Austin, R.E.; Oalmann, C.J.; Müller, P. *J. Am. Chem. Soc.*, ***1991***, *113*, 1423.

CO_2Me — 1. EtMgBr, Ti(OiPr)$_4$, ether, -78°C → 0°C; 2. aq. H_2SO_4 → OH 90%

Kulinkovich, O.G.; Sviridov, S.V.; Vasilevskii, D.A.; Savchenko, A.I.; Pritytskaya, T.S. *Zhur. Org. Khim.*, ***1991***, *27*, 294 (Engl., p. 250).

$EtCO_2Me$ — 1. 2 eq. EtMgBr, 10% Ti(OiPr)$_4$, ether; 2. 5% aq. H_2SO_4, 5°C → Et, OH 76%

Kulinkovich, O.G.; Sviridov, S.V.; Vasilevskii, D.A. *Synthesis*, ***1991***, 234.

2 CBr$_2$F$_2$, CH$_2$Br$_2$, 27 h
2 KOH (60% aq)
Bu$_4$NHSO$_4$

F F 70%

Balcerzak, P.; Fedoryński, M.; Jończyk, A. *J. Chem. Soc., Chem. Commun.*, ***1991***, 826.

PhS PhS CHO

1. Me N N·Li , 2 BuLi
2. H$_2$O

PhS CHO 73%

Tanaka, K.; Matsuura, H.; Funaki, I.; Suzuki, H. *J. Chem. Soc., Chem. Commun.*, ***1991***, 1145.

Bu B-O O

1. 7 eq. CH$_2$N$_2$, cat. Pd(OAc)$_2$
ether
2. NaBO$_3$•4 H$_2$O , reflux , 50 min
dioxane - H$_2$O

Bu OH 76-82%

Fontani, P.; Carboni, B.; Vaultier, M.; Maas, G. *Synthesis*, ***1991***, 605.

n-C$_8$H$_{17}$

t-BuSO$_2$CH$_2$Li , THF
2% Ni(acac)$_2$, reflux

n-C$_8$H$_{17}$ 65%

Gai, Y.; Julia, M.; Verpeaux, J-N. *SynLett*, ***1991***, 56, 269.

Ph$_2$S=CH$_2$, Cu(acac)$_2$
NaH , THF

86%

Cimetière, B.; Julia, M. *SynLett*, ***1991***, 271.

Ph SMe SMe

1. BuLi , THF
-78°C → 0°C
2. MeOH

Ph 82%

Krief, A.; Batbeaux, P *Tetrahedron Lett.*, ***1991***, *32*, 417.

O N-OMe Me BnO

1. Me$_3$SO$^+$ I$^-$, NaH
2. *t*-BuOK , aq. ether

CO$_2$H BnO 88%

Rodriques, K.E. *Tetrahedron Lett.*, ***1991***, *32*, 1275.

$Cr(CO)_5$ OMe → (CO_2Me alkene; THF, 65°C) MeO CO_2Me + MeO CO_2Me

(46 : 54) 64%

+ PPh_3 (26 : 74) 39%

Herndon, J.W.; Tumer, S.U. *J. Org. Chem.*, ***1991***, *56*, 286.

MeO_2C MeO_2C C≡C-H C≡C-H → (Me, MeO =$Cr(CO)_5$, SiO_2; decane, 63°C, 3 h) MeO_2C MeO_2C O 94%

Katz, T.J.; Yang, G.X-Q. *Tetrahedron Lett.*, ***1991***, *32*, 5895.

Me CO_2Me Br (75:25 Z:E) → (1. *t*-BuOK, THF -70°C, 100 min; 2. 1N HCl, -70°C) CO_2Me Br 12% + MeO_2C CO_2Me Br H Me 88%

Duhamel, L.; Peschard, O.; Plé, G. *Tetrahedron Lett.*, ***1991***, *32*, 4695.

O O H O O N_2 Ph → (Ph alkene; $Rh_2(OAc)_4$) O O O O Ph Ph major + O O O O Ph Ph (89% dr) 91%

Davies, H.M.L.; Cantrell, W.R. Jr. *Tetrahedron Lett.*, ***1991***, *32*, 6509.
Davies, H.M.L.; Hu, B. *Tetrahedron Lett.*, ***1992***, *33*, 453.

Br CO_2Et → (PhMgBr, 2 SmI_2; THF - HMPA) OH Ph 90%

Fukuzawa, S.; Niimoto, Y.; Sakai, S. *Tetrahedron Lett.*, ***1991***, *32*, 7691.

n-C_4H_9 NO_2 OH → (N_2CHCO_2Et, PPh_3; PhH, RT) *n*-C_4H_9 NO_2 (10:1 trans:cis) 82%

Yu, J.; Falck, J.R.; Mioskowski, C. *J. Org. Chem.*, ***1992***, *57*, 3757.

Me_3Si ... OH — Sm(Hg), CH_2I_2; THF, -78°C → RT — HO (1 : 1.3) 84%

Lautens, M.; Delanghe, P.H.M. *J. Org. Chem.*, ***1992***, *57*, 798.

NHTs, O, H, H, Me — aq. NaOH, heat — H, N-Ts, H, Me, HO 81%

Moulines, J.; Charpentier, P.; Bats, J-P.; Nuhrich, A.; Lamidey, A-M. *Tetrahedron Lett.*, ***1992***, *33*, 487.

I O — 1. Zn, THF, reflux 2. $ClSiMe_2t$-Bu — O, $OSiMe_2t$-Bu 56%

Yasui, K.; Fugami, K.; Tanaka, S.; Tamaru, Y.; Ii, A.; Yoshida, Z.; Saidi, M.R. *Tetrahedron Lett.*, ***1992***, *33*, 785, 789.

Ph — 0.03% $Cl(PPh_3)_2Ru[H_2C_2B_9H_{10}]$; N_2CHCO_2Et, 60°C, 4 h — Ph, CO_2Et 86%

(cis:trans = 0.69)

Demonceau, A.; Saive, E.; de Froidmont, Y.; Noels, A.F.; Hubert, A.J. *Tetrahedron Lett.*, ***1992***, *33*, 2009.

Me, Ph, H, Me, MeS, OH — 1. BuLi 2. $PhSO_2Cl$ 3. BuLi, -78°C — Ph, H, Me, Me 79%

(88% dr)

Krief, A.; Hobe, M.; Dumont, W.; Badaou, E.; Guittet, E.; Evrard, G. *Tetrahedron Lett.*, ***1992***, *33*, 3381.

Krief, A.; Hobe, M. *SynLett*, ***1992***, 317.

EtO_2C, CO_2Et — 1. NaH, THF 2. I_2 3. THF, reflux — EtO_2C, EtO_2C, I 81%

Beckwith, A.L.J.; Tozer, M.J. *Tetrahedron Lett.*, ***1992***, *33*, 4975.

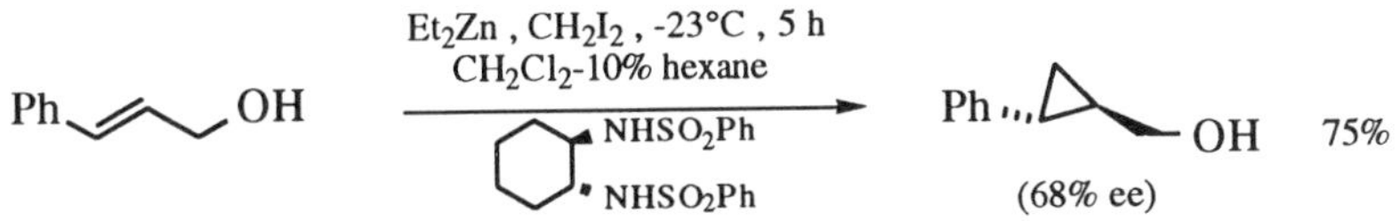

Et_2Zn, CH_2I_2, -23°C, 5 h
CH_2Cl_2-10% hexane
$NHSO_2Ph$
$NHSO_2Ph$

Ph OH → Ph OH 75%
(68% ee)

Takahashi, H.; Yoshioka, M.; Ohno, M.; Kobayashi, S. *Tetrahedron Lett.*, ***1992***, *33*, 2575.

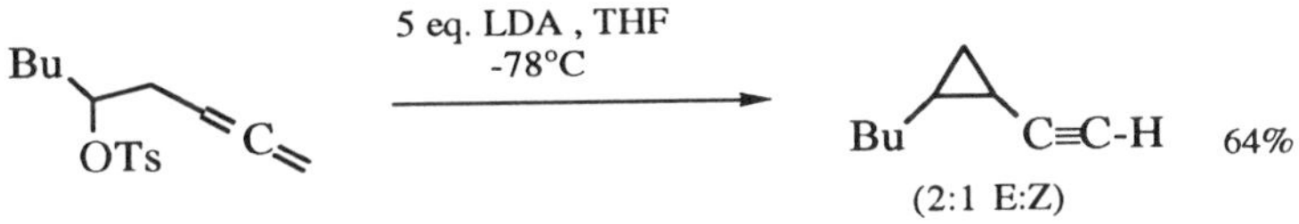

5 eq. LDA, THF
-78°C

Bu OTs C → Bu C≡C-H 64%
(2:1 E:Z)

Pyo, S.; Skowron, J.F. III; Cha, J.K. *Tetrahedron Lett.*, ***1992***, *33*, 4703.

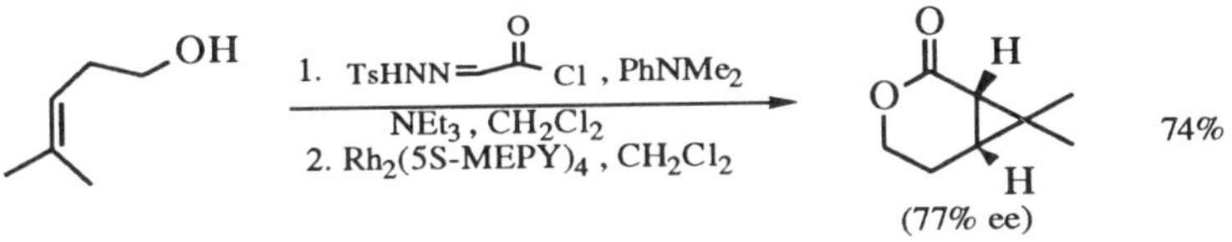

OH
1. TsHNN= Cl, $PhNMe_2$
NEt_3, CH_2Cl_2
2. $Rh_2(5S\text{-MEPY})_4$, CH_2Cl_2

O O H H 74%
(77% ee)

MEPY = tetrakis[methyl-2-pyrrolidinone-5S-carboxylate]

Martin, S.F.; Oalmann, C.J.; Liras, S. *Tetrahedron Lett.*, ***1992***, *33*, 6727.

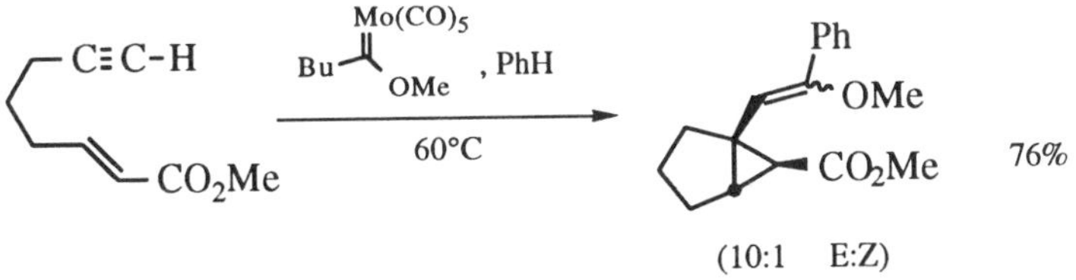

C≡C-H
CO_2Me
$Mo(CO)_5$
Bu OMe, PhH
60°C

Ph OMe CO_2Me 76%
(10:1 E:Z)

Harvey, D.F.; Lund, K.P.; Neil, D.A. *J. Am. Chem. Soc.*, ***1992***, *114*, 8424.
Harvey, D.F.; Brown, M.F. *Tetrahedron Lett.*, ***1990***, *31*, 2529.

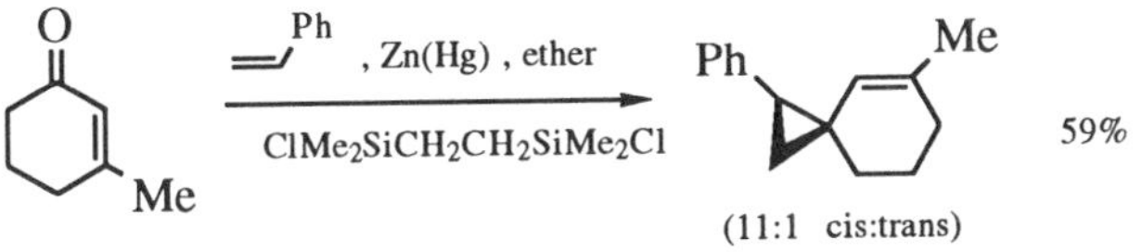

O
Me
Ph, Zn(Hg), ether
$ClMe_2SiCH_2CH_2SiMe_2Cl$

Ph Me 59%
(11:1 cis:trans)

Motherwell, W.B.; Roberts, L.R. *J. Chem. Soc., Chem. Commun.*, ***1992***, 1582.

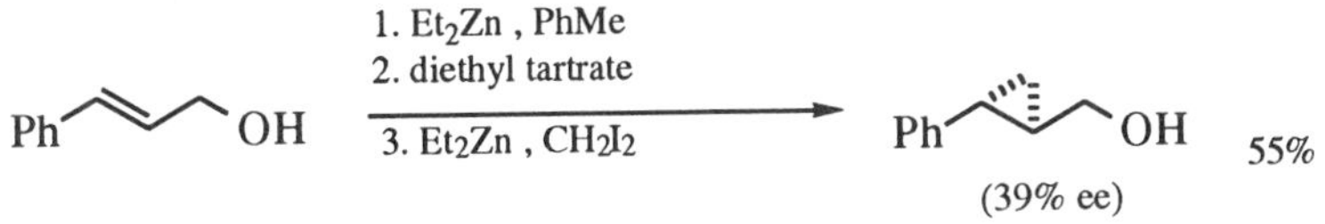

Ph OH
1. Et_2Zn, PhMe
2. diethyl tartrate
3. Et_2Zn, CH_2I_2

Ph OH 55%
(39% ee)

Ukaji, Y.; Nishimura, M.; Fujisawa, T. *Chem. Lett.*, ***1992***, 61.

Me, Ph; Me_2N, OH → 1. Et_2Zn , PhMe; 2. Ph/\/OH; 3. $(ICH_2)_2Zn$ → Ph-cyclopropyl-OH 82%

(18% ee , RR)

Denmark, S.E.; Edwards, J.P. *SynLett*, ***1992***, 229.

SECTION 75: ALKYLS, METHYLENES AND ARYLS FROM MISCELLANEOUS COMPOUNDS

Cl-C6H4-SO_2Cl → $PdCl_2(PhCN)_2$, xylene; $Ti(OiPr)_4$ → Cl-C6H4-C6H4-Cl

76%

Miura, M.; Hashimoto, H.; Itoh, K.; Nomura, M. *Chem. Lett.*, ***1990***, 449.

N-CHO piperidine-SO_2Ph → PhMgBr , $ZnBr_2$ → N-CHO piperidine-Ph 87%

Brown, D.S.; Hansson, T.; Ley, S.V. *SynLett*, ***1990***, 48.
Brown, D.S.; Charreau, P.; Ley, S.V. *SynLett*, ***1990***, 749 [with $ZnCl_2$].

$(PhO)_2PO_2$/\/Ph → BuMgCl , THF , 2 h; -70°C → Bu/\/Ph 78%

Yanagisawa, A.; Nomura, N.; Yamamoto, H. *SynLett*, ***1991***, 513.

CHAPTER 6

PREPARATION OF AMIDES

SECTION 76: AMIDES FROM ALKYNES

H- C≡ C- CMe_3 → (1. BuLi, THF, -60°C; 2. Me_3SiCl, -20°C → RT; 3. 55°C, Ph_2NH, 11 h) Me_3Si–CH_2–C(=O)–NPh_2 quant.

heating after 2. leads to ketenes

Valentí, E.; Pericàs, M.A.; Serratosa, F. *J. Org. Chem.*, ***1990***, *55*, 395.

SECTION 77: AMIDES FROM ACID DERIVATIVES

CH_2=CH–C(=O)Cl → (Ph, H, Si, N(Ph)CHO, N·Me, Me) CH_2=CH–C(=O)–N(Ph)–CHO 72%

Corriu, R.J.P.; Lanneau, G.F.; Perrot-Petta, M.; Mehta, V.D. *Tetrahedron Lett.*, ***1990***, *31*, 2585.

n-C_6H_{11}, NH_2, HO_2C → (NEt_3, MeCN; EtO·P(=O)·Cl, Cl) n-C_6H_{11}, N, H, O 81%

Kim, C.W.; Chung, B.Y.; Nam Kung, J-Y.; Lee, J.M.; Kim, S. *Tetrahedron Lett.*, ***1990***, *31*, 2905.

O, O, O → ($HN(SiMe_3)_2$, MeOH; DMF, RT, 16 h) O, N-H, O 93%

Davis, P.D.; Bit, R.A. *Tetrahedron Lett.*, ***1990***, *31*, 5201.

NEt$_3$, CH$_2$Cl$_2$, reflux

70%

Manhas, M.S.; Ghosh, M.; Bose, A.K. *J. Org. Chem.*, ***1990***, *55*, 575.

PhO, Cl, O, NEt$_3$, CH$_2$Cl$_2$

good yield

Sharma, S.D.; Pandhi, S.B. *J. Org. Chem.*, ***1990***, *55*, 2196.

pyrrolidine, CBr$_4$, NEt$_3$, PPh$_3$, RT, 15 min

89%

Einhorn, J.; Einhorn, C.; Luche, J-L. *Synth. Commun.*, ***1990***, *20*, 1105.

PhO⌒CO$_2$H → 1. N(*n*-Pr)$_3$, CH$_2$Cl$_2$; 2. Ph–CH=N–C$_6$H$_4$–OMe, reflux

85%

(15:1 cis:trans)

Georg, G.I.; Mashava, P.M.; Guan, X. *Tetrahedron Lett.*, ***1991***, *32*, 581.

CH_3COOH → *n*-C$_6$H$_{13}$NH$_2$, PhH, Ph$_3$SbO, P$_4$S$_{10}$, 40°C, 5 h → AcNH-*n*-C$_6$H$_{13}$ 90%

Nomura, R.; Nakano, T.; Yamada, Y.; Matsuda, H. *J. Org. Chem.*, ***1991***, *56*, 4076.

CO$_2$H → Ru$_3$(CO)$_{12}$, Ph-NO$_2$, 180°C, 8 h → NHPh 51%

Nahmed, E.M.; Jenner, G. *Tetrahedron Lett.*, ***1991***, *32*, 4917.

NH_2 — $(PyS)_2$-PPh_3, MeCN, reflux — 58%

Bai, D.; Shi, Y. *Tetrahedron Lett.*, ***1992***, *33*, 943.

1. H-N(Boc)OSiMe$_2$*t*-Bu, DMAP, MeCN, 5 h
2. TFA, CH_2Cl_2, 0°C, 6 h

77%

Altenburger, J.M.; Mioskowski, C.; d'Orchymont, H.; Schirlin, D.; Schalk, C.; Tarnus, C. *Tetrahedron Lett.*, ***1992***, *33*, 5055.

SECTION 78: AMIDES FROM ALCOHOLS AND THIOLS

OH — $CoCl_2$, MeCN, Ac_2O, 80°C, 15 h — NHAc 70%

Nayyar, N.K.; Reddy, M.M.; Iqbal, J. *Tetrahedron Lett.*, ***1991***, *32*, 6965.

SECTION 79: AMIDES FROM ALDEHYDES

CHO — NBS, AIBN, CCl_4, $BuNH_2$ — NHBu 80%

Markó, I.E.; Mekhalfia, A. *Tetrahedron Lett.*, ***1990***, *31*, 7237.

CHO — 1. BnO_2CNSO, -25°C, BF_3•OEt_2, CH_2Cl_2; 2. Et_3SiH — NHCbz 64%

Alexander, M.D.; Anderson, R.E.; Sisko, J.; Weinreb, S.M. *J. Org. Chem.*, ***1990***, *55*, 2563.

MsO, Br, N, H + B(OH)$_2$, CHO, O, O; $Pd(PPh_3)_4$, Na_2CO_3, DME, reflux → MsO, N, O, O, O 49%

Siddiqui, M.A.; Snieckus, V. *Tetrahedron Lett.*, ***1990***, *31*, 1523.

PhCHO —1. $PhNH_2$, HCl, KCN; 2. *t*-BuOK, O_2, THF; RT→ PhC(O)NHPh 76%

Chuang, T-H.; Yang, C-C.; Chiang, C-J.; Fang, J-M. *SynLett*, ***1990***, 733.

Ph⌒CHO —2 eq. $SiICl_3$, MeCN; 2 H_2O, CH_2Cl_2, 10 h→ Ph⌒⌒$NHCO_2Me$ 59%

Elmorsy, S.S.; Nour, M.A.; Kandeel, E.M.; Pelter, A. *Tetrahedron Lett.*, ***1991***, *32*, 1825.

2-furyl-CHO —1. $N(SnMe_3)_3$; 2. *t*-$BuPh_2SiCl$; 3. $Me_2C=C(OMe)OLi$→ β-lactam, N-$SiPh_2t$-Bu 65%

Busato, S.; Cainelli, G.; Panunzio, M.; Bandini, E.; Martelli, G.; Spunta, G. *SynLett*, ***1991***, 243.

SECTION 80: AMIDES FROM ALKYLS, METHYLENES AND ARYLS

NO ADDITIONAL EXAMPLES

SECTION 81: AMIDES FROM AMIDES

Conjugate reductions of unsaturated amides are listed in Section 74D (Alkyls from Alkenes).

2-ethyl-NHBoc-benzene —1. *sec*-BuLi; 2.DMF; 3. HCl→ 3-methyl-N-Boc-indole 60%

Clark, R.D.; Muchowski, J.M.; Souchet, M.; Repke, D.B. *SynLett*, ***1990***, 207.

1. Na , THF
2. H_2O

95%

coupling occurs in some cases

Karaman, R.; Fry, J.L. *Tetrahedron Lett.*, ***1990***, *31*, 941.

Bn Me Me N O 30%

$Ph_4B^- Na^+$

I N O

10% $Pd(OAc)_2$
20% PPh_3 , DMF
Et_4NCl , 80°C , 6 h

Me Me Me N O 70%

Grigg, R.; Dorrity, M.J.; Malone, J.F.; Sridharan, V.; Sukirthalingam, S. *Tetrahedron Lett.*, ***1990***, *31*, 1343.

NH_2 O

1. $Hg(OAc)_2$, DMF MeOH
2. NBS , DMF 12 h

H N OMe O 91%

new Hofmann rearrangement

Jew, S.; Park, H.G.; Park, H-J.; Park, M.; Cho, Y. *Tetrahedron Lett.*, ***1990***, *31*, 1559.

EtO(C=O)NH_2

1. *t*-BuOH
2. NaOH
3. BEt_3 , 65°C , 6 h
4. aq. NaOH

EtO(C=O)NHEt 80-88%

Wachter-Jurcsak, N.; Scully, F.R. Jr. *Tetrahedron Lett.*, ***1990***, *31*, 5261.

$Na_2(P_4S_{10}O)$, THF , 25°C
2 h

88%

Brillon, D. *Synth. Commun.*, ***1990***, *20*, 3085.

H(C=O)NMePh

$Se(AlMe_2)_2$, PhMe-THF
40°C , 4 h

H(C=Se)NMePh 91%

Segi, M.; Kojima, A.; Nakajima, T.; Suga, S. *SynLett*, ***1991***, 105.

$SiMe_3$ allyl; $SnCl_4$, -22°C

(82 : 18) 90%

Polniaszek, R.P.; Belmont, S.E.; Alvarez, R. *J. Org. Chem.*, ***1990***, *55*, 215.

$SnCl_4$, -23°C, 1.5 h

80%

Fisher, M.J.; Overman, L.E. *J. Org. Chem.*, ***1990***, *55*, 1447.

$(CF_3CO)_2O$, RT; CH_2Cl_2, 5 h

87%

Masuda, R.; Hojo, M.; Ichi, T.; Sasano, S.; Kobayashi, T.; Kuroda, C. *Tetrahedron Lett.*, ***1991***, *32*, 1195.

$Me_2S(=O)CH^-$, DMSO; 25°C

51%

Nadir, U.K.; Sharma, R.l.; Koul, V.K. *J. Chem. Soc., Perkin Trans. I*, ***1991***, 2015.

Bu_3SnH, AIBN; PhH

73%

(3αMe:3βMe 6:1)

Ishibashi, H.; Nakamura, N.; Sato, T.; Takeuchi, M.; Ikeda, M. *Tetrahedron Lett.*, ***1991***, *32*, 1725.

RO2CHN — 1. (MeO)(MeO)C6H3CH=CHOMe; 2. POCl3, CH2Cl2 — R = (-)-8-phenylmenthyl — (83 : 17) 68%

Comins, D.L.; Badawi, M.M. *Tetrahedron Lett.*, ***1991***, *32*, 2995.

1. LDA
2. BnBr

63% (1.8:1 trans:cis)

Baldwin, J.E.; Moloney, M.G.; Shim, S.B. *Tetrahedron Lett.*, ***1991***, *32*, 1379.

Bu_3SnCl, $NaBH_3CN$, *t*-BuOH

(4 : 1) 90%

Clark, A.J.; Jones, K.; McCarthy, C.; Storey, J.M.D. *Tetrahedron Lett.*, ***1991***, *32*, 2829.

PhCHO, PPA, 35°C, 72 h

74%

Marson, C.M.; Grabowska, U; Walgrove, T.; Eggleston, D.S.; Baures, P.W. *J. Org. Chem.*, ***1991***, *56*, 2603.

$AlCl_3$

quant.

Baba, A.; Seki, K.; Matsuda, H. *J. Org. Chem.*, ***1991***, *56*, 2684.

1. *sec*-BuLi, TMEDA THF, -78°C
2. $B(OMe)_3$
3. H_3O^+
4. Br, OMe, $Pd(PPh_3)_4$, DME, aq. Na_2CO_3

$N(iPr)_2$ — MeO — 81%

Alo, B.I.; Kandil, A.; Patil, P.A.; Sharp, M.J.; Siddiqui, M.A.; Snieckus, V.; Josephy, P.D. *J. Org. Chem.*, ***1991***, *56*, 3763.

Bu_3SnH, AIBN
PhH, 80°C

81% (1.1/1 cis/trans)

Curran, D.P.; Abraham, A.C.; Liu, H. *J. Org. Chem.*, ***1991***, *56*, 4335.

$PdCl_2(PPh_3)_2$, DMAC
DBU, 115°C, 5.5 h
95 psi CO

DMAC = N,N-dimethyl acetamide

63%

Perry, R.J.; Turner, S.R. *J. Org. Chem.*, ***1991***, *56*, 6573.

H_2 (300 psi)
CO (900 psi)
$HRh(CO)(PPh_3)_3$
2 eq. dppb, 80°C
40 h

(98 : 2) 88%

Ojima, I.; Korda, A.; Shay, W.R. *J. Org. Chem.*, ***1991***, *56*, 2024.

Bu_3SnH, AIBN
PhH, reflux, 0.02 M

95%

(1:1 α:β)

Hart, D.J.; Wu, S.C. *Tetrahedron Lett.*, ***1991***, *32*, 4099.

1. 5% $RuCl_2(PPh_3)_3$, 2 h, 140°C
2. Bu_3SnH , PhMe , 1 h, reflux

47%

Nagashima, H.; Wakamatsu, H.; Ozaki, N.; Ishii, T.; Watanabe, M.; Tajima, T.; Itoh, K. *J. Org. Chem.*, ***1992***, *57*, 1682.

1. BuLi , DABCO
2. 3 eq. $Cu(OAc)_2$

40%

Kawabata, T.; Minami, T.; Hiyama, T. *J. Org. Chem.*, ***1992***, *57*, 1864.

Bu_3SnH , AIBN
PhH , reflux

70%

Keusenkothen, P.F.; Smith, M.B. *Tetrahedron*, ***1992***, *48*, 2977.

1. LDA
2. BnBr

(>500 : 1) 76%

Kimura, K.; Murata, K.; Otsuka, K.; Ishizuka, T.; Haratake, M.; Kunieda, T. *Tetrahedron Lett.*, ***1992***, *33*, 4461.

$Rh_2(4S\text{-}MEOX)_4$
CH_2Cl_2 , reflux

(91 : 9) 82%

(71% ee) (80% ee)

MEOX = tetrakis[methyl 2-oxazolinone-4(S)-carboxylate]

Doyle, M.P.; Protopopova, M.N.; Winchester, W.R.; Daniel, K.L. *Tetrahedron Lett.*, ***1992***, *33*, 7819.

Me, O, N, NH_2 —[Ph_2NNO , PhH , 3 h; reflux]→ Me, O, N, H 51%

White, J.D.; Perri, S.T.; Toske, S.G. *Tetrahedron Lett.*, ***1992***, *33*, 433.

N, O, Boc —[$BnNH_2$, MeCN; 10 kbar , 80°C; 21 h]→ BocHN, O, NHBn 97%

Kotsuki, H.; Iwaski, M.; Nishizawa, H. *Tetrahedron Lett.*, ***1992***, *33*, 4945.

O, N-H, O —[Me–C_6H_4–$Pb(OAc)_3$; NaH , Cl_2CH_2-DMF; 70°C , 7 h]→ O, N, Me, O 86%

López-Alvarado, P.; Avendaño, C.; Menéndez, J.C. *Tetrahedron Lett.*, ***1992***, *33*, 6875.

O, H, N-N, H, O-CH_2CCl_3, O —[1. $PhI(COCF_3)_2$, CH_2Cl_2, 0°C , 2.5 h; 2. 3 eq. Zn dust , 12 h, AcOH/acetone , RT]→ O, N, H 70%

Scartozzi, M.; Grondin, R.; Leblanc, Y. *Tetrahedron Lett.*, ***1992***, *33*, 5717.

t-BuO, N, O, Ac —[1. PhLi , $CuBr{\bullet}SMe_2$, THF, Me_3SiCl , -78°C → RT; 2. Me_2NH , DMF , RT , 28 h; 3. 3 eq. allyl-$SiMe_3$, 3 eq. $BF_3{\bullet}OEt_2$, 18 h , RT, CH_2Cl_2]→ Ph, N, O, H 47%

Koot, W-J.; Hiemstra, H.; Speckamp, W.N. *Tetrahedron Lett.*, ***1992***, *33*, 7969.

Cl, N, O, Me —[Bu_3SnH , AIBN; PhMe , reflux]→ H, O, N, H, Me 63% + Me, N, O, Me 8%

Sato, T.; Nakamura, N.; Ikeda, K.; Okada, M.; Ishibashi, H.; Ikeda, M. *J. Chem. Soc., Perkin Trans. I*, ***1992***, 2399.

Ph–C(=O)–NH$_2$ —[BuBr , TBAB , KOH; 80°C]→ Ph–C(=O)–NHBu 71%

Loupy, A.; Sansoulet, J.; Díez-Barra, E.; Carrillo, J.R. *Synth. Commun.*, ***1992***, *22*, 1661.

PhS, SPh, Ph, N, Me, O —[3.3 eq. Bu_3SnH, PhMe , AIBN; 14.5 h]→ Ph, N, Me, O 75%

Sato, T.; Machigashira, N.; Ishibashi, H.; Ikeda, M. *Heterocycles*, ***1992***, *33*, 139.

O, N- H, O —[Cl⌒⌒C≡C-H , PhMe; reflux , 80 h]→ O, N⌒⌒C≡C-H, O 90%

Gesson, J.P.; Jacquesy, J.C.; Rambaud, D. *Bull. Soc. Chim. Fr.*, ***1992***, *129*, 227.

O, N, H, CH_2OSiMe_2t-Bu —[1. PhMe , PhLi; 2. n-$C_5H_{11}Br$, reflux]→ O, N, CH_2OSiMe_2t-Bu, n-C_5H_{11} 75%

Keusenkothen, P.F.; Smith, M.B. *Synth. Commun.*, ***1992***, *22*, 2935.

SECTION 82: AMIDES FROM AMINES

5 $BuNH_2$ —[1. PCl_3; 2. 2 ⌒CO_2H]→ 2 ⌒C(=O)NHBu 77%

Cabral, J.; Laszlo, P.; Montaufier, M-T.; Randriamahefa, S.L. *Tetrahedron Lett.*, ***1990***, *31*, 1705.

NH_2, CO_2H, $(CH_2)_8$ —[Bu_3N , CH_2Cl_2; 10 h addition; 35°C; Br, N· Me, ⊕, I⊖]→ H, N, O, $(CH_2)_8$ 87%

Bai, D.; Bo, Y.; Zhou, Q. *Tetrahedron Lett.*, ***1990***, *31*, 2161.

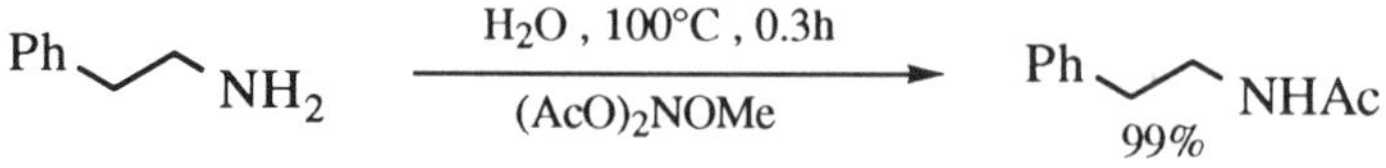
Ph
NH2
H2O , 100°C , 0.3h
(AcO)2NOMe
Ph
NHAc
99%

N-methoxy diacetamide - a new acetylating reagent
many amines are acetylated at RT

Kikugawa, Y.; Mitsui, K.; Sakamoto, T. *Tetrahedron Lett.*, ***1990***, *31*, 243.

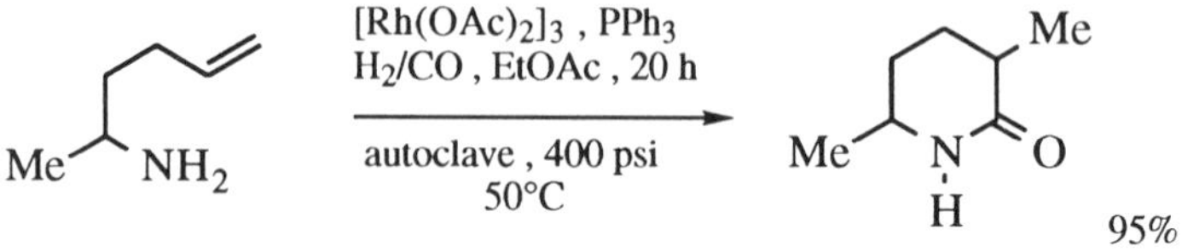
Me
NH2
[Rh(OAc)2]3 , PPh3
H2/CO , EtOAc , 20 h
autoclave , 400 psi
50°C
Me
Me
N
H
O
95%

Anastasiou, D.; Jackson, W.R. *Tetrahedron Lett.*, ***1990***, *31*, 4795.

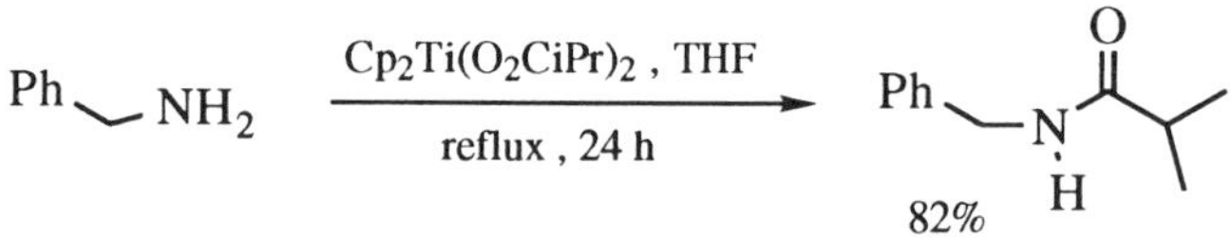
Ph
NH2
Cp2Ti(O2CiPr)2 , THF
reflux , 24 h
Ph
N
H
O
82%

Recht, J.; Cohen, B.I.; Goldman, A.S.; Kohn, J. *Tetrahedron Lett.*, ***1990***, *31*, 7281.

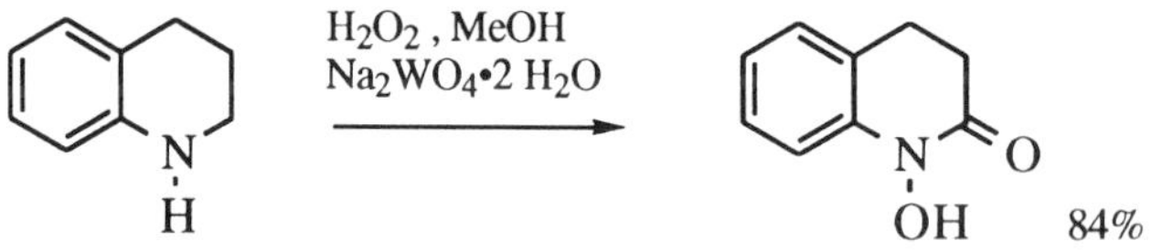
N
H
H2O2 , MeOH
Na2WO4•2 H2O
N
OH
O
84%

Murahashi, S.; Oda, T.; Sugahara, T.; Masui, Y. *J. Org. Chem.*, ***1990***, *55*, 1744.

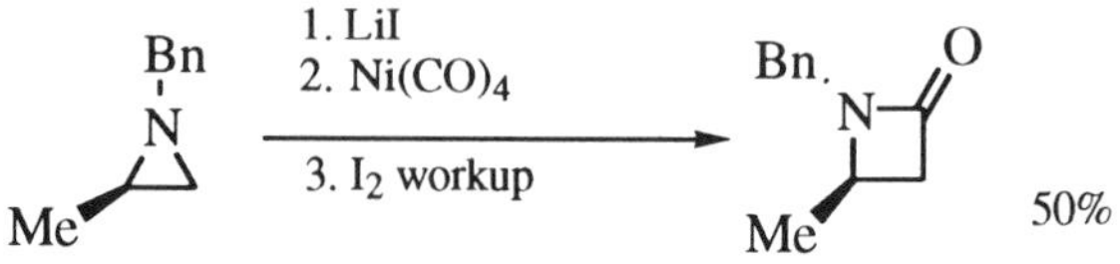
Bn
N
Me
1. LiI
2. Ni(CO)4
3. I2 workup
Bn
N
O
Me
50%

Chamchaang, W.; Pinhas, A.R. *J. Org. Chem.*, ***1990***, *55*, 2943.

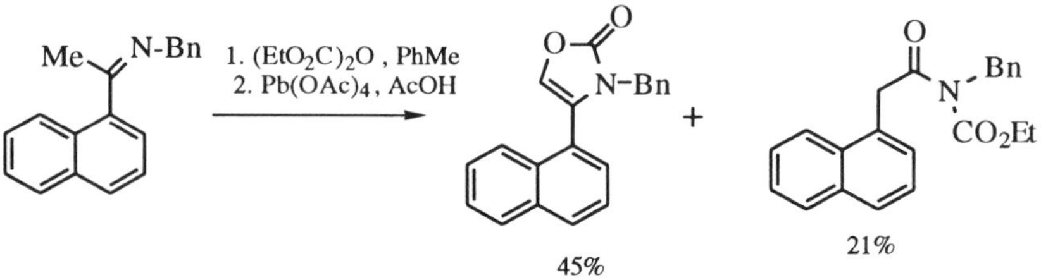
Me
N-Bn
1. (EtO2C)2O , PhMe
2. Pb(OAc)4 , AcOH
O
O
N-Bn
+
O
N
Bn
CO2Et
45%
21%

Lenz, G.R.; Costanza, C.; Lessor, R.A.; Ezell, E.F. *J. Org. Chem.*, ***1990***, *55*, 1753.

Me N Me $ClO_4^{\ominus}$ hv → N Me Me 99%

Cho, I-S.; Tu, C-L.; Mariano, P.S. *J. Am. Chem. Soc.*, ***1990***, *112*, 3594.

=N-OH Me — montmorillonite, KSF; PhMe, reflux, 13 h → O N- H Me 59%

Meshram, H.M. *Synth. Commun.*, ***1990***, *20*, 3253.

=N-OMs — 1.$LiAlH_4$, ether, 0°C; 2. Ac_2O, Py, RT → NHAc 76%

Kočovský, P. *SynLett*, ***1990***, 677.

N OTMS — $SbCl_5$, $AgSbF_6$, MeCN; reflux, 3 h → N H O 77%

Mukaiyama, T.; Harada, T. *Chem. Lett.*, ***1991***, 1653.

=N OAc — piperidine, THF, 25°C; 30 min → N O Me 89%

Fernández, S.; Menéndez, E.; Gotor, V. *Synthesis*, ***1991***, 713.

Ph N Ph — 1. MeC≡N-O; 2. *t*-BuOK, THF; 3. H_2O → O Ph NHPh 51%

Aitken, R.A.; Raut, S.V. *SynLett*, ***1991***, 189.

HO NH_2 — $RuH_2(PPh_3)_4$, DME, 3 h; benzalacetone, 140°C → N H O 59%

Naota, T.; Murahashi, S. *SynLett*, ***1991***, 693.

1. MeOC(O)Cl, NEt_3, MeOH
2. TMSOTf

N-CO_2Me Ph 82%

modified Pictet-Spengler

Cheung, G.K.; Earle, M.J.; Fairhurst, R.A.; Heaney, H.; Shuhaibar, K.F.; Eyley, S.C.; Ince, F. *SynLett*, ***1991***, 721.

MeO OMe Me N Ph Me — Me OH, 110°C → Me O N Ph Me 82%

Bauermeister, S.; Gows, I.D.; Strauss, H.F.; Venter, E.M.M. *J. Chem. Soc., Perkin Trans. I*, ***1991***, 561.

Ph N *t*-Bu — $KMnO_4$, aq. MeCN → Ph N H *t*-Bu O 73%

Larsen, J.; Jørgensen, K.A.; Christensen, D. *J. Chem. Soc., Perkin Trans. I*, ***1991***, 1187.

NH_2 — ethyl butyrate, 40°C; lipase SP 382, 1 h → NHAc quant.

hindered amines gives lower yields

Djeghaba, Z.; Deleuze, H.; De Jeso, B.; Messadi, D.; Maillard, B. *Tetrahedron Lett.*, ***1991***, *32*, 761.

N OH — Et_3SiH, RT; $ClCO_2Me$; overnight → N-CO_2Me OH 74%

Wu, P-L.; Sun, C-J. *Tetrahedron Lett.*, ***1991***, *32*, 4137.

N OMe — $PhSO_2Cl$, Bu_4NI; CH_2Cl_2, 25°C → N O SO_2Ph 76%

Sheu, J.; Smith, M.B.; Oeschger, T.R.; Satchell, J. *Org. Prep. Proceed. Int.*, ***1992***, *24*, 147.

2 *n*-$BuNH_2$ — Ph_3SbO, P_4S_{10}, CO_2; 80°C → BuHN C(O) NHBu 88%

Nomura, R.; Hasegawa, Y.; Ishimoto, M.; Toyosaki, T.; Matsuda, H. *J. Org. Chem.*, ***1992***, *57*, 7339.

$Co_2(CO)_8$, CO (54 atm)
PhH , 72 h

56%

Wang, M.D.; Alper, H. *J. Am. Chem. Soc.*, ***1992***, *114*, 7018.

TEA , $[EtO_2CCH_2Cl/Zn]$

59%

Mulengi, J.K.; Fatmi, N. *Bull. Soc. Chim. Belg.*, ***1992***, *101*, 257.

3 eq. Me_3SiCl , 6 eq. DMF
3 eq. imidazole , RT

84%

Berry, M.B.; Blagg, J.; Craig, D.; Willis, M.C. *SynLett*, ***1992***, 659.

SECTION 83: AMIDES FROM ESTERS

MeHNCHO , THF
NaOMe , MeOH
100°C , 15 min

90%

Jagdmann, G.E. Jr.; Munson, H.R. Jr.; Gero, T.W. *Synth. Commun.*, ***1990***, *20*, 1203.

1. $TiCl_4$, NEt_3 , -78°C
2. 0°C , Ph–CH=N·Bn

(72 : 28) 99%

Cinquini, M.; Cozzi, F.; Cozzi, P.G.; Consolandi, E. *Tetrahedron*, ***1991***, *47*, 8767.

$(TMS)_2N$-SnNHBn
hexane , RT , 12 h

Wang, W-B.; Roskamp, E.J. *J. Org. Chem.*, ***1992***, *57*, 6101.

Me(NH$_2$)CH–CH$_2$CH$_2$CO$_2$iPr → porcine pancreatic lipase, *t*-amyl alcohol, pH 7.8, 6 d → 5-methyl-2-pyrrolidinone (Me, N–H, =O) 40%

Gutman, A.L.; Meyer, E.; Yue, X.; Abell, C. *Tetrahedron Lett.*, ***1992***, *33*, 3943.

2-NH$_2$-C$_6$H$_4$-C(O)OMe → 1. 2 eq. Me$_2$NLi; 2. H$_2$O → 2-NH$_2$-C$_6$H$_4$-C(O)NMe$_2$ 88%

Rivière-Baudet, M.; Morère, A.; Dias, M. *Tetrahedron Lett.*, ***1992***, *33*, 6453.

cyclic carbonate (O, O, =O) with vinyl group → Ph-N=C=O, THF, 8 h; Pd(PPh$_3$)$_4$, RT → cyclic carbamate (O, N–H, =O) with vinyl group 70%

Tamaru, Y.; Bando, T.; Kawamura, Y.; Okamura, K.; Yoshida, Z.; Shiro, M. *J. Chem. Soc., Chem. Commun.*, ***1992***, 1498.

SECTION 84: AMIDES FROM ETHERS, EPOXIDES AND THIOETHERS

NO ADDITIONAL EXAMPLES

SECTION 85: AMIDES FROM HALIDES AND SULFONATES

camphor N-SO$_2$Me imine → 1. 2 BuLi, -78°C; 2. BnBr → Ph–CH$_2$–CH(CH$_3$)–SO$_2$NH$_2$ 72% (13% ee)

Davis, F.A.; Zhou, P.; Lal, G.S. *Tetrahedron Lett.*, ***1990***, *31*, 1653.

CH$_2$=CHCH$_2$–C(I)(CO$_2$Me)(MeO$_2$C) → 1. (Bu$_3$)$_2$Sn$_2$, hν, MeO–C$_6$H$_4$–S(O$_2$)–NH–CH$_2$CH=CH$_2$; 2. NEt$_3$, reflux → bicyclic (MeO$_2$C)$_2$ N–S(O$_2$)–C$_6$H$_4$–OMe 59%

Flynn, D.L.; Zabrowski, D.L. *J. Org. Chem.*, ***1990***, *55*, 3673.

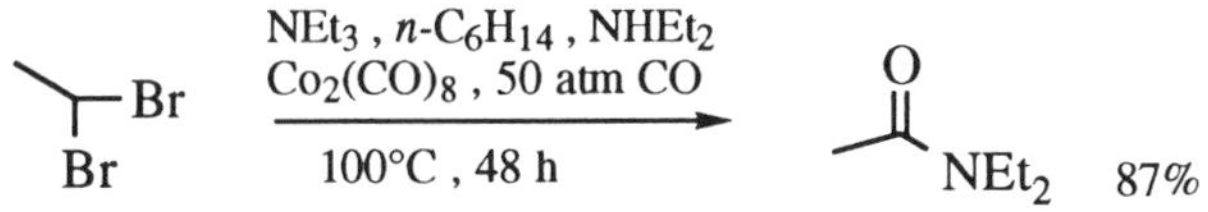

NEt_3, n-C_6H_{14}, $NHEt_2$
$Co_2(CO)_8$, 50 atm CO
100°C, 48 h
Br Br → NEt$_2$ 87%

Miyashita, A.; Kawashima, T.; Kaji, S.; Nomura, K.; Nohira, H. *Tetrahedron Lett.*, ***1991***, *32*, 781.

SECTION 86: AMIDES FROM HYDRIDES

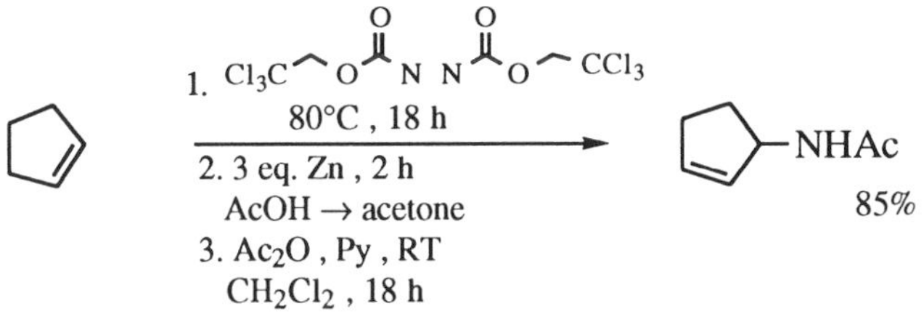

1. Cl_3C–O–C(O)–N=N–C(O)–O–CCl_3, 80°C, 18 h
2. 3 eq. Zn, 2 h, AcOH → acetone
3. Ac_2O, Py, RT, CH_2Cl_2, 18 h

NHAc 85%

Leblanc, Y.; Zamboni, R.; Bernstein, M.A. *J. Org. Chem.*, ***1991***, *56*, 1871.

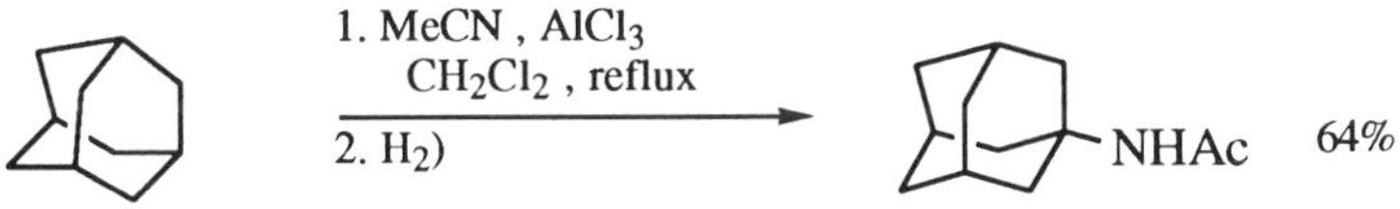

1. MeCN, $AlCl_3$, CH_2Cl_2, reflux
2. H_2)

NHAc 64%

with addition of CO, get a low yield of the aldehyde

Olah, G.A.; Wang, Q. *Synthesis*, ***1992***, 1090.

SECTION 87: AMIDES FROM KETONES

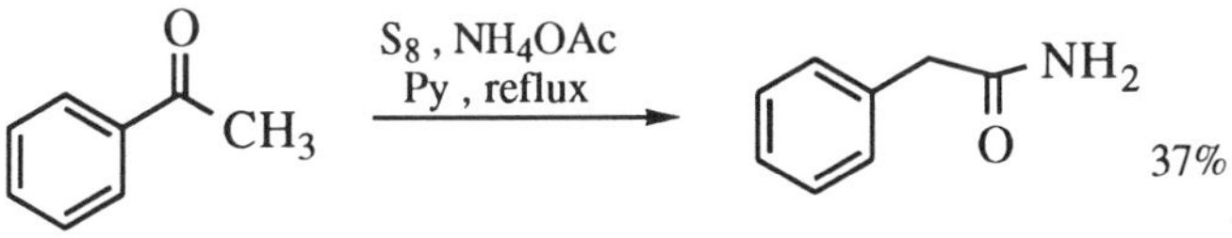

S_8, NH_4OAc
Py, reflux

CH$_3$ → NH$_2$ 37%

You, Q-D.; Zhou, H-Y.; Wang, Q-Z.; Lei, X-H. *Org. Prep. Proceed. Int.*, ***1991***, *23*, 435.

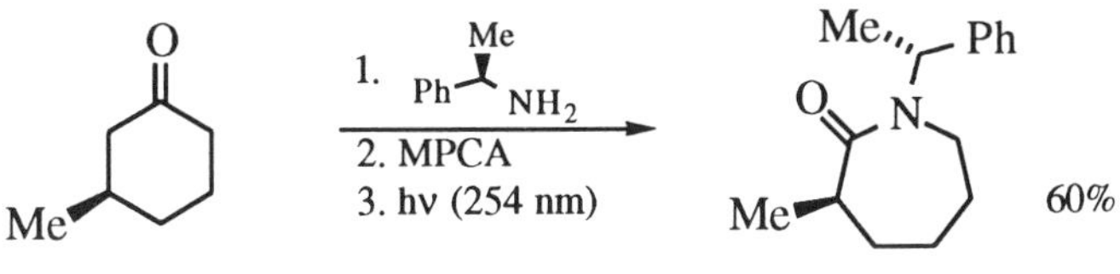

1. Me, Ph, NH$_2$
2. MPCA
3. hν (254 nm)

Me → Me, Ph, N, O, Me 60%

MPCA = monoperoxycamphoric acid

Aubé, J.; Hammond, M.; Gherardini, E.; Takusagawa, F. *J. Org. Chem.*, ***1991***, *56*, 499.

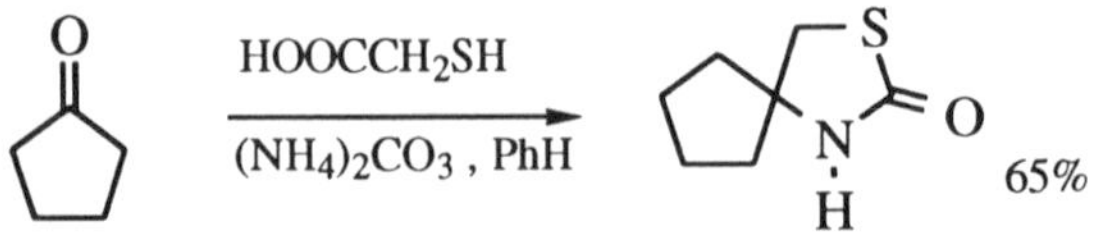

$HOOCCH_2SH$, $(NH_4)_2CO_3$, PhH — 65%

El-Zohry, M.F. *Org. Prep. Proceed. Int.*, ***1992***, *24*, 81.

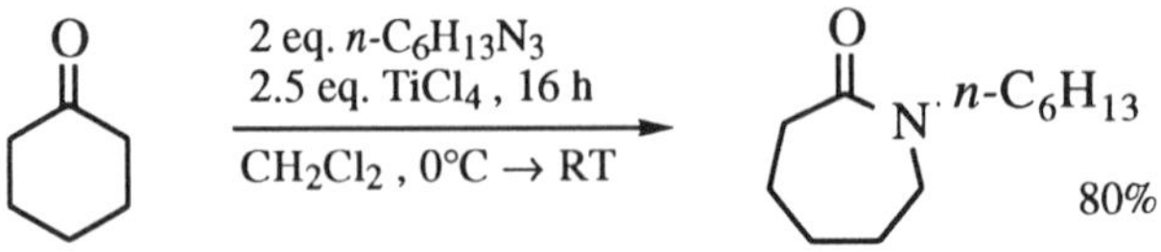

2 eq. n-$C_6H_{13}N_3$, 2.5 eq. $TiCl_4$, 16 h, CH_2Cl_2, 0°C → RT — n-C_6H_{13} — 80%

Aubé, J.; Milligan, G.L.; Mossman, C.J. *J. Org. Chem.*, ***1992***, *51*, 1635.
Aubé, J.; Milligan, G.L. *J. Am. Chem. Soc.*, ***1991***, *113*, 8965.

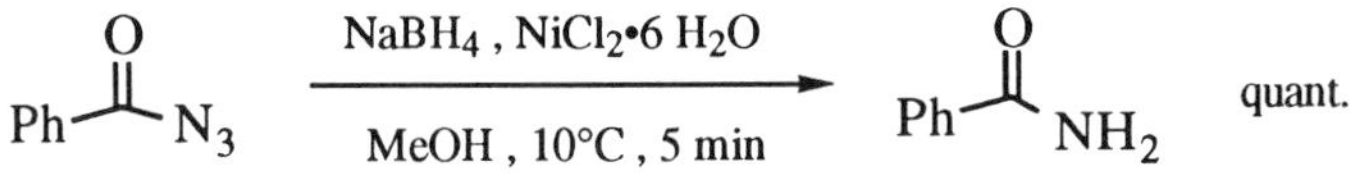

Ph–CO–N_3 → $NaBH_4$, $NiCl_2$•6 H_2O, MeOH, 10°C, 5 min → Ph–CO–NH_2 quant.

Rao, H.S.P.; Reddy, K.S.; Turnbull, K.; Borchers, V. *Synth. Commun.*, ***1992***, *22*, 1339.

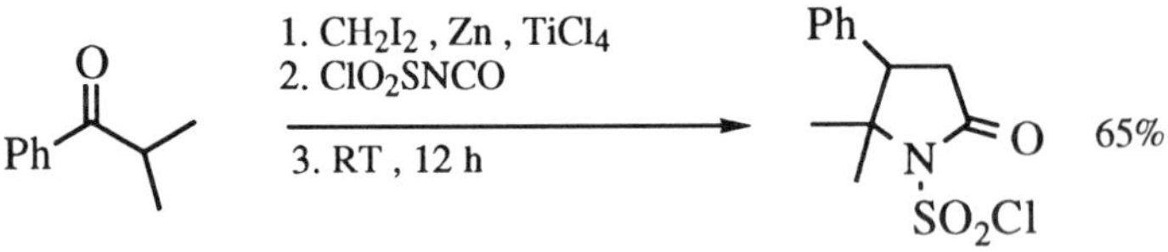

1. CH_2I_2, Zn, $TiCl_4$
2. ClO_2SNCO
3. RT, 12 h

Ph, SO_2Cl — 65%

Black, T.H.; Olson, J.T.; Abt, D.C. *Synth. Commun.*, ***1992***, *22*, 2729.

REVIEWS:

"Carbon-Carbon Bond Cleavage by the Haller-Bauer and Related Reactions"
Gilday, J.P.; Paquette, L.A. *Org. Prep. Proceed. Int.*, ***1990***, *22*, 167.

SECTION 88: AMIDES FROM NITRILES

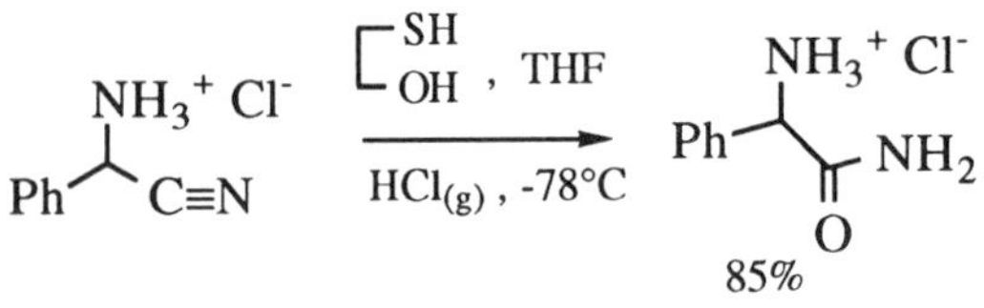

$NH_3^+ Cl^-$, Ph–CH–C≡N → HS–CH$_2$CH$_2$–OH, THF, $HCl_{(g)}$, -78°C → $NH_3^+ Cl^-$, Ph–CH–CO–NH_2 85%

new Pinner type reaction

Lee, Y.B.; Goo, Y.M.; Lee, Y.Y.; Lee, J.K. *Tetrahedron Lett.*, ***1990***, *31*, 1169.

Ph(CH₂)₇C≡N → 1. $MeAl(Cl)NH_2$, 80°C, PhMe; 2. H_2O → Ph(CH₂)₇C(=NH)NH₂ 93%

Garigipati, R.S. *Tetrahedron Lett.*, ***1990***, *31*, 1969.

PhCN → 1. $(Me_3Si)_2Se$, 2.2 BF_3•OEt_2, CH_2Cl_2, 60°C, 8 h; 2. aq. $NaHCO_3$ → PhC(=Se)NH₂ 66%

Shimada, K.; Hikage, S.; Takeishi, Y.; Takikawa, Y. *Chem. Lett.*, ***1990***, 1403.

Br–C₆H₄–C≡N → $NaBO_3$•4 H_2O, H_2O, dioxane, 89°C, 65h → Br–C₆H₄–C(=O)NH₂ 65% (39% "purified")

Reed, K.L.; Gupton, J.T.; Solarz, T.L. *Synth. Commun.*, ***1990***, *20*, 563.

Ph-C≡N → Na_2CO_3•3 H_2O_2, acetone, H_2O, 50°C → PhC(=O)NH₂ 94%

Kabalka, G.W.; Deshpande, S.M.; Wadgaonkar, P.P.; Chatla, N. *Synth. Commun.*, ***1990***, *20*, 1445.

PhCH₂CN → 1. P_4S_{10}, Na_2S, THF, 20°C; 2. H_2O → PhCH₂C(=S)NH₂ 73%

Brillon, D. *Synth. Commun.*, ***1992***, *22*, 1397.

Ph-C≡N → $(Me_3Si)_2S$, NaOMe, DMEU, 30°C → PhC(=S)NH₂ 80%

DMEU = 1,3-dimethyl-2-imidazolidinone

Lin, P-Y.; Ku, W-S.; Shiao, M-J. *Synthesis*, ***1992***, 1219.

SECTION 89: AMIDES FROM ALKENES

Me₂C=CHCH₂SiMe₃ → 1. O=C=N-SO₂Cl, CCl_4, 0°C → 25°C; 2. aq. Na_2So_3 → 3,3-dimethyl-4-(trimethylsilylmethyl)azetidin-2-one (N-H) 60%

Colvin, E.W.; Monteith, M. *J. Chem. Soc., Chem. Commun.*, ***1990***, 1230.

$ClSO_2NCO$, CH_2Cl_2 ; Na_2SO_3 , $NaHCO_3$; -20°C , 4 h — 51%

Nativi, C.; Perrotta, E.; Ricci, A.; Taddei, M. *Tetrahedron Lett.*, ***1991***, *32*, 2265.

1. Cl_2NCO_2t-Bu , PhMe, 50°C , 6 h; 2. Zn , NH_4OAc , 0°C, dioxane , 14 h — NHBoc, 60%

Orlek, B.S.; Stemp, G. *Tetrahedron Lett.*, ***1991***, *32*, 4045.

Ph_3P=CHPh — 1. Se_8; 2. morpholine (O N·H) — 56%

Okuma, K.; Komiya, Y.; Ohta, H. *Bull. Chem. Soc. Jpn.*, ***1991***, *64*, 2402.

SnCl$_4$, -78°C — 58% (26:1 dr , S)

Whitesell, J.K.; Yaser, H.K. *J. Am. Chem. Soc.*, ***1991***, *113*, 3526.

1. NHBoc, $(Bu_3Sn)_2$, hν; 2. NEt_3 — PhO_2S, MeO_2C, N-Boc, 50%

Flynn, D.L.; Zabrowski, D.L.; Nosal, R. *Tetrahedron Lett.*, ***1992***, *33*, 7281.

SECTION 90: AMIDES FROM MISCELLANEOUS COMPOUNDS

CO_2Me, SCN — Bu_3SnH , PhMe, AIBN , 75°C, 40 min — S, CO_2Me, H-N, 94%

Bachi, M.D.; Denenmark, D. *J. Org. Chem.*, ***1990***, *55*, 3442.

$PdCl_2$-$H_5PV_2Mo_{12}O_{40}$, 3 eq. CO, MeOH, DME: $PhNO_2$ → $PhNHCO_2Me$ 96%

Izumi, Y.; Satoh, Y.; Urabe, K. *Chem. Lett.*, ***1990***, 795.

3-Cl-C$_6$H$_4$-NCO, $AlCl_3$, CH_2Cl_2, 20°C: Me-C$_6$H$_4$-$SnBu_3$ → Me-C$_6$H$_4$-C(O)-NH-C$_6$H$_4$-Cl 88%

Kobs, U.; Neumann, W.P. *Chem. Ber.*, ***1990***, *123*, 2191.

BuLi, CH_2Cl_2: t-BuO_2C-N=N-C(O)-N(piperidine) → t-BuO_2C-N(Bu)-NH-C(O)-N(piperidine) + t-BuO_2C-NH-N(Bu)-C(O)-N(piperidine) (74 : 26) 88%

Yamamoto, Y.; Yumoto, M.; Yamada, J. *Tetrahedron Lett.*, ***1991***, *32*, 3079.

1. 3-lithiofuran, THF, -40°C; 2. HCl, MeOH; 3. TMSNCO: nitrone (tetrahydropyranyl, Me) → furyl-CH(Me)-N(OH)-C(O)-NH_2 25%

Basha, A.; Ratajczyk, J.D.; Brooks, D.W. *Tetrahedron Lett.*, ***1991***, *32*, 3783.

BnOH, HCl/CH_2Cl_2, RT, 5 h: Et-CH_2-N=C=O → Et-CH_2-NHCbz 92%

Benalil, A.; Roby, P.; Carboni, B.; Vaultier, M. *Synthesis*, ***1991***, 787.

1. $PhSnMe_3$, 20°C; 2. H_2O: $ClSO_2$-N=C=O → $PhSO_2NH_2$ 82%

Arnswald, M.; Neumann, W.P. *Chem. Ber.*, ***1991***, *124*, 1997.

Zeolite (H-Y), PhMe, reflux, 24 h: $(MeS)_2C$=N-cyclohexyl → MeS-C(O)-NH-cyclohexyl 70%

Reddy, T.I.R.; Bhawal, B.M.; Rajappa, S. *Tetrahedron Lett.*, ***1992***, *33*, 2857.

BuHN, H, Ph --- Rh(Cl)CO → 1. $P(OMe)_3$, HCl CH_2CL_2 , -78°C 2. $P(OMe)_3$, MeOH -78°C – RT → O, N- Bu, Ph 90%

Krafft, M.E.; Yu, X.Y.; Milczanowski, S.E.; Donnelly, K.D. *J. Am. Chem. Soc.*, ***1992***, *114*, 9215.

O, EtO–P, Ph, EtO → 1. NaH , THF 2. NH , $\frac{1}{4}$ S_8 reflux , 4 h → S, Ph, N 46%

Okuma, K.; Ikari, K.; Ohta, H. *Chem. Lett.*, ***1992***, 131.

REVIEWS:

"Organometallic Carboxamidation"

Screttas, C.G.; Steele, B.R. *Org. Prep. Proceed. Int.*, ***1990***, *22*, 271.

SECTION 90A: PROTECTION OF AMIDES

N, H, O → n-C_3H_7CHO , p-TsOH , PhMe 70% → N, O

← 1. Et_3O^+ BF_4^- 2. H_2O 3. pH 8 67%

Smith, M.B.; Wang, C-J.; Keusenkothen, P.F.; Dembofsky, B.T.; Fay, J.G.; Zezza, C.A.; Kwon, T.W.; Sheu, J-L.; Son, Y.C.; Menezes, R.F. *Chem. Lett.*, ***1992***, 247.

CHAPTER 7
PREPARATION OF AMINES

SECTION 91: AMINES FROM ALKYNES

$n\text{-}C_8H_{17}\text{-}C\equiv C$–$CH_2CH_2$–$NH_2$ → 5% $PdCl_2(MeCN)_2$, $MeCN\text{-}H_2O$, 5 h, reflux → (N-ring)–$n\text{-}C_8H_{17}$ 67%

Fukuda, Y.; Matsubara, S.; Utimoto, K. *J. Org. Chem.*, ***1991***, *56*, 5812.

$n\text{-}C_6H_{13}\text{-}C\equiv C$–$(CH_2)_4$–$NH_2$ → cat. $NaAuCl_4$, MeCN, reflux , 1 h → $n\text{-}C_6H_{13}$–(N-ring) 80%

Fukuda, Y.; Utimoto, K. *Synthesis*, ***1991***, 975.

SECTION 92: AMINES FROM ACID DERIVATIVES

PhCOOH → 1. $SOCl_2$, reflux; 2. HOSA , PhMe , reflux; 3. H^+ → $PhNH_2$ 67%

HOSA = hydroxylamine O-sulfonic acid

Wallace, R.G.; Barker, J.M.; Wood, M.L. *Synthesis*, ***1990***, 1143.

Me, N (quinoline) → (cyclohexyl)–CO_2H, lepidine , hν, $Ph(O_2CCF_3)$, 2.5 h, 0°C → RT → Me, N (2-cyclohexylquinoline) 52%

Togo, H.; Aoki, M.; Yokoyama, M. *Tetrahedron Lett.*, ***1991***, *32*, 6559.

SECTION 93: AMINES FROM ALCOHOLS AND THIOLS

Ph⌄OH —(MeNHTf, THF / DEAD, PPh_3)→ Ph⌄N(Tf)Me 70%

Edwards, M.L.; Stemerick, D.M.; McCarthy, J.R. *Tetrahedron Lett.*, ***1990***, *31*, 3417.

Me–CH(Ph)–OH —(ZnN_6•2 Py, PPh_3, RT; $iPrO_2CN{=}NCO_2iPr$ / PhMe)→ Me–CH(Ph)–N_3 81%

Viaud, M.C.; Rollin, P. *Synthesis*, ***1990***, 130.

O_2N–C6H4–OH —(1. Tf_2O, CH_2Cl_2, NEt_3, -15°C; 2. piperidine, MeCN / reflux, 12 h)→ O_2N–C6H4–N(piperidine) quant.

some amines required 10 kbar of pressure

Kotsuki, H.; Kibayashi, S.; Suenaga, H.; Nishizawa, H. *Synthesis*, ***1990***, 1145, 1147.

Ph–CH(OH)–CH(Me)–NMe_2 —(1. MsCl, NEt_3, THF; 2. c-$C_6H_{11}NH_2$, NEt_3 / PhH, reflux)→ Ph–CH(NH-cyclohexyl)–CH(Me)–NMe_2 50%

Dieter, R.K.; Deo, N.; Lagu, B.; Dieter, J.W. *J. Org. Chem.*, ***1992***, *57*, 1663.

SECTION 94: AMINES FROM ALDEHYDES

$PhNH_2$ —(OHC⌄⌄⌄CHO; $HFe(CO)_4^-$, 1M KOH/EtOH, CO, RT, 24h)→ ⌄⌄⌄⌄NHPh 52%

Shim, S.C.; Kwon, Y.G.; Doh, C.H.; Kim, H.S.; Kim, T.J. *Tetrahedron Lett.*, ***1990***, *31*, 105.

$(CH_3)_2CH$–CHO —(1. TsN=S=O, CH_2Cl_2; 2. MeMgBr)→ $(CH_3)_2CH$–CH(NHTs)–CH_3 73%

Sisko, J.; Weinreb, S.M. *J. Org. Chem.*, ***1990***, *55*, 393.

2 benzotriazole + $CHOCHO$ (glutaraldehyde) → 1. $BnNH_2$, 20°C, H_2O; 2. $NaBH_4$ → N-Bn piperidine 91%

Katritzky, A.R.; Fan, W-Q. *J. Org. Chem.*, ***1990***, *55*, 3205.
Katritzky, A.R.; Fan, W-Q.; Fu, C. *J. Org. Chem.*, ***1990***, *55*, 3209.

EtCHO → 1. TsNSO, $SnCl_4$, PhH, 0°C → RT; 2. allyl-$SiMe_3$ → Et–CH(NHTs)–allyl 95%

Ralbovsky, J.L.; Kinsella, M.A.; Sisko, J.; Weinreb, S.M. *Synth. Commun.*, ***1990***, *20*, 573.

Me, morpholino, CHO, Me substrate → 1. $PhNH_2$, PhH, reflux; 2. $NaBH_3CN$, MeOH; THF, 5N HCl → Me, morpholino, NHPh, Me product 65%

Barluenga, J.; Aznar, F.; Fraiz, S. Jr.; Pinto, A.C. *Tetrahedron Lett.*, ***1991***, *32*, 3205.

t-Bu–CHO → Ts–N=Te=N–Ts; PhMe, toluene, 1 h → t-Bu–CH=N–Ts 90%

Trost, B.M.; Marrs, C. *J. Org. Chem.*, ***1991***, *56*, 6468.

OHC–$(CH_2)_3$–CHO → $PhNH_2$, $NaBH_4$, MeOH; 3M H_2SO_4, THF → piperidine N–Ph 90%

Verardo, G.; Giumanini, A.G.; Favret, G.; Strazzolini, P. *Synthesis*, ***1991***, 447.

CH_3CH_2CHO → 1. *t*-$BuNH_2$; 2. LDA, THF; 3. $Cl(CH_2)_3Br$, THF, 0°C; 4. iPrONa, iPrOH → N-*t*-Bu, Me tetrahydropyridine 44%

Stevens, C.; De Kimpe, N. *SynLett*, ***1991***, 351.

Related Methods: Section 102 (Amines from Ketones)

SECTION 95: AMINES FROM ALKYLS, METHYLENES AND ARYLS

NO ADDITIONAL EXAMPLES

SECTION 96: AMINES FROM AMIDES

1. Me_3SiCl , NEt_3 PhMe , 40°C , 4 h
2. BuLi

Bu
70%

Hua, D.H.; Miao, S.W.; Bharathi, S.N.; Katsuhira, T.; Bravo, A.A. *J. Org. Chem.*, ***1990***, *55*, 3682.

O
Et_2N Me
$NaBH_4$, 25°C , 50 h
Br Se Br Br Br
Et_2N Me 68%

Akabori, S.; Takanohashi, Y. *Chem. Lett.*, ***1990***, 251.

Se
Ph N
1. 2.2 eq. PhLi , THF , 0°C
2. H_2O , 0°C
Ph N Ph 94% + PhSeSePh 73%

Sekiguchi, M.; Ogawa, A.; Fujiwara, S.; Ryu, I.; Kambe, N.; Sonoda, N. *Chem. Lett.*, ***1990***, 2053.

O
Me_2N Me
$NaBH_4$, THF , 65°C , 4 h
$(BrCH_2CH_2)_2SeBr_2$
Me_2N Me 82%

Akabori, S.; Takanohashi, Y. *J. Chem. Soc., Perkin Trans. I*, ***1991***, 479.

CF_3
NHAc
1. PCl_5
2. MeLi
3. O N·Li
O N N Me 57%

Strekowski, L.; Petterson, S.E.; Janda, L.; Wydra, R.L.; Harden, D.B.; Lipowska, M.; Cegla, M.T. *J. Org. Chem.*, ***1992***, *57*, 196.

O Me Cl N Bu

NaOEt , DMF
EtOH , reflux

Me CO_2Et N Bu

47%

Bartholomew, D.; Stocks, M.J. *Tetrahedron Lett.*, ***1991***, *32*, 4795.

Ph N Ac Me

$NaBH_4$, I_2 , 70°C , 3 h

Ph N Et Me

74%

Prasad, A.S.B.; Kanth, J.V.B.; Periasamy, M. *Tetrahedron*, ***1992***, *48*, 4623.

N O Ts

1. Dibal , CH_2Cl_2 , -78°C
2. PPTS
3. $TiCl_4$, Me_3Si

N Ts

87%

Åhman, J.; Somfai, P. *Tetrahedron*, ***1992***, *48*, 9537.

S SMe MeO N Bn

$Bu_4N^+ H_2F_3^-$, NBS
$(HF)_9$ / Py , CH_2Cl_2
0°C → RT , 1 h

MeO N CF_3 Bn

97%

Kuroboshi, M.; Hiyama, T. *Tetrahedron Lett.*, ***1992***, *33*, 4177.

Ph N OH O O Ph + HO O Ph O O O

AIBN , xylene
reflux , 2 h

N H Ph

67%

Hirao, K.; Mohri, K.; Yonemitsu, O.; Tabata, M.; Sohma, J. *Tetrahedron Lett.*, ***1992***, *33*, 1459.

O O-S=O N H

1. PhLi , THF
-78°C → RT
2. HCl/H_2O , EtOH

H H N • HCl Ph

62%

Cooper, G.F.; McCarthy, K.E.; Martin, M.G. *Tetrahedron Lett.*, ***1992***, *33*, 5895.

Related Methods: Section 105A (Protection of Amines)

SECTION 97: AMINES FROM AMINES

Ph-N(Me)(Et) → MeOH, 0.25 M NaOH, DCN*, 12 h, hν (>280 nm) → Ph-N(H)(Et) 80%

Pandey, G.; Sudha Rani, K.; Bhalerao, U.T. *Tetrahedron Lett.*, ***1990***, *31*, 1199.

Bu_3SnH, AIBN 60%

Murphy, J.A.; Sherburn, M.S. *Tetrahedron Lett.*, ***1990***, *31*, 1625, 3495.
Murphy, J.A.; Sherburn, M.S.; Dickinson, J.M.; Goodman, C. *J. Chem. Soc., Chem. Commun.*, ***1990***, 1065.
Murphy, J.A.; Sherburn, M.S. *Tetrahedron*, ***1991***, *47*, 4077.

Bu_3SnH, AIBN, PhH, reflux 66%

Rosa, A.M.; Prabhakar, S.; Lobo, A.M. *Tetrahedron Lett.*, ***1990***, *31*, 1881.

Bu_3SnH, AIBN, PhMe, reflux 51%

Takano, S.; Suzuki, M.; Kijima, A.; Ogasawara, K. *Tetrahedron Lett.*, ***1990***, *31*, 2315.

1. 4 BnMgCl, ether, RT, 12h
2. KOH, aq. MeOH
76%

Gottlieb, L.; Meyers, A.I. *Tetrahedron Lett.*, ***1990***, *31*, 4723.

hν , MeCN

Ph N H → Ph N + Ph CHO

65% 15%

Lewis, F.D.; Reddy, G.D. *Tetrahedron Lett.*, ***1990***, *31*, 5293.

MeO NH_2 hν (>280 nm) DCN , MeCN → MeO N H 80%

DCN = 1,4-dicyanonaphthalene

Pandey, G.; Sridhar, M.; Bhalerao, U.T. *Tetrahedron Lett.*, ***1990***, *31*, 5373.

Bn Me_2N O MeO

Ph N OMe → MeLi , PhMe , -78°C 20 min → Ph H N Me OMe 94%

(70% ee, R)

Tomioka, K.; Inoue, I.; Shindo, M.; Koga, K. *Tetrahedron Lett.*, ***1990***, *31*, 6681.

H_2N NH_2 O_2 , H_2O , pH 7 DAO , 30°C → N 78%

DAO = pea seedling diamine oxidase

Cragg, J.E.; Herbert, R.B.; Kgaphola, M.M. *Tetrahedron Lett.*, ***1990***, *31*, 6907.

O_2N N Me Me O $NaHCO_3$, EtOH heat , 3 h → O_2N N Me 77%

Fadda, A.A. *Ind. J. Chem.*, ***1990***, *29B*, 1017.

$RuCl_2(PPh_3)_3$

PhMe , reflux

quant.

Tsuji, Y.; Kotachi, S.; Huh, K-T.; Watanabe, Y. *J. Org. Chem.*, ***1990***, *55*, 580.

1. $Ti(OiPr)_4$, PhCHO , RT

2. $NaBH_3CN$, H_2O , 20 h

70%

Mattson, R.J.; Pham, K.M.; Leuck, D.J.; Cowen, K.A. *J. Org. Chem.*, ***1990***, *55*, 2552.

1. iPrMgBr

2. iPrMgBr

87%

Katritzky, A.R.; Rachwal, S.; Wu, J. *Can. J. Chem.*, ***1990***, *68*, 456.

Katritzky, A.R.; Latif, M.; Urogdi, L. *J. Chem. Soc., Perkin Trans. I*, ***1990***, 667.

Katritzky, A.R.; Noble, G.; Pilarski, B.; Harris, P. *Chem. Ber.*, ***1990***, *123*, 1443.

$NaBH_4$ - $ZrCl_4$, THF , RT

95%

(64% ee , S)

Itsuno, S.; Sakurai, Y.; Shimizu, K.; Ito, K. *J. Chem. Soc., Perkin Trans. I*, ***1990***, 1859.

1. , 80°C, 2 h

2. NaOMe

3. heat , 2 h

35%

Walizei, G.H.; Breitmaier, E. *Liebigs Ann. Chem.*, ***1990***, 605.

1. BuLi

2. *t*-BuLi

3. BuLi , TMEDA , 20°C

4. $(EtO)_2CO$, -50°C → 20°C

5. H_2O

85%

Barluenga, J.; Foubelo, F.; González, R.; Fañanás, F.J.; Yus, M. *J. Chem. Soc., Chem. Commun.*, ***1990***, 1521.

Bn(N=C(Me)Ph) → [$Li[Ir(dppe)_4]$, H_2 (25 bar); THF, CH_2Cl_2, 30°C, 1 h; stainless steel autoclave] → BnHN–CH(Me)Ph quant.

Chan, Y.N.C.; Meyer, D.; Osborn, J.A. *J. Chem. Soc., Chem. Commun.*, ***1990***, 869.

Ph(C=N–OH)Et → [2% $Ru_3(CO)_{12}$, PhH, 4 h; 100°C, CO (20 Kg/cm[2])] → Ph(C=NH)Et quant.

Akazome, M.; Tsuji, Y.; Watanabe, Y. *Chem. Lett.*, ***1990***, 635.

PhCH=N–Ph → [1. $Zn(BH_4)_2$, ether, 0°C → RT; 2. 2N NaOH] → $PhCH_2NHPh$ 98%

Kotsuki, H.; Yoshimura, N.; Kadota, I.; Ushio, Y.; Ochi, M. *Synthesis*, ***1990***, 401.

Me_3SiCH=CHCH$_2$–B(pinacolato) → [1. *n*-C_3H_7-CH=NMe, THF, 65°C, 72 h; 2. 30% H_2O_2, 1M NaOH; 3. 2M HCl ; 2M NaOH] → Me_3SiCH(CH=CH$_2$)CH(*n*-C_3H_7)NHMe 53%

Guyot, B.; Pornet, J.; Miginiac, L. *Synth. Commun.*, ***1990***, *20*, 2409.

$PhCH_2NH_2$ → [$HPhPO_2H$, 120°C, DMSO] → $PhCH_2NMe_2$ 73%

Chaucin, R. *Synth. Commun.*, ***1991***, *21*, 1425.

1,2,3,4-tetrahydroquinoline → [1. BuLi; 2. $Cp_2ZrMeCl$; 3. 60°C, THF, 12 h; 4. cyclohexene; 5. MeOH] → 2-cyclohexyl-1,2,3,4-tetrahydroquinoline 75%

Coles, N.; Whitby, R.J.; Blagg, J. *SynLett*, ***1990***, 271.

indole (N–H) → [1. BuLi, THF; 2. BuOMs] → N-butylindole 92%

28% with BuBr

Epling, G.A.; Kumar, A. *SynLett*, ***1991***, 347.

$$PhCH{=}NOMe \xrightarrow[-78°C]{BF_3{\cdot}OEt_2\ ,\ PhLi\ ,\ PhMe} PhCH_2NHOMe \quad 59\%$$

Uno, H.; Terakawa, T.; Suzuki, H. *SynLett*, ***1991***, 559.

Clostridia thermoaceticum , 0.033 M; 0.1 M phosphate buffer (pH 8.5); 0.3 M sodium formate , 5 h; 0.001 methyl viologen , 40°C

t-Bu, NHOH → t-Bu, NH_2 80%

Braun, H.; Schmidtchen, F.P.; Schneider, A.; Simon, H. *Tetrahedron*, ***1991***, *47*, 3329.

1. 200°C , neat , 1 h; 2. TMSCl , MeOH; 0°C , 1 h

Ph, N, $SiMe_3$, $SiMe_3$ → N, Ph, $SiMe_3$ 65%

Palomo, C.; Aizpurua, J.M.; García, J.M.; Legido, M. *J. Chem. Soc., Chem. Commun.*, ***1991***, 524.

MeO–C6H4–$Pb(OAc)_3$; $Cu(OAc)_2$, CH_2Cl_2 , 2 h

NH_2 → H, N, OMe 95%

Barton, D.H.R.; Donnelly, D.M.X.; Finet, J-P.; Guiry, P.J. *J. Chem. Soc., Perkin Trans. I*, ***1991***, 2095.

1. $HSiCl_3$, $BF_3{\cdot}OEt_2$ PhH; 2. KOH , aq. EtOH

=N–Ph → NHPh 81%

Okamoto, H.; Kato, S. *Bull. Chem. Soc. Jpn.*, ***1991***, *64*, 3466.

1. KOH , DMSO; 2. (2-bromobenzyl bromide) Br, Br; 3. $Pd(PPh_3)_4$, KOAc DMA , 130°C , 10 h

N–H → N 77%

Kozikowski, A.P.; Ma, D. *Tetrahedron Lett.*, ***1991***, *32*, 3317.

H_3BO_3 , 180°C , 1 h

CO_2Et

95%

Delbecq, P.; Bacos, D.; Celerier, J.P.; Lhommet, G. *Can. J. Chem.*, ***1991***, *69*, 1201.

HO

NHPh

DEAD , PPh_3 , 3 h
CH_2Cl_2

Me

N

Ph

80%

Bernotas, R.C.; Cube, R.V. *Tetrahedron Lett.*, ***1991***, *32*, 161.

Ph−C≡C

Me

NH_2

H_2/CO(400 psi)
PPh_3 , $[Rh(OAc)_2]_2$
70°C , 20 h

Ph

Me

N
H

78%

Campi, E.M.; Jackson, W.R.; Nilsson, Y. *Tetrahedron Lett.*, ***1991***, *32*, 1093.

1. 3 eq. BuLi•2 *t*-BuOK
2. MeBr

Me

86%

Naruse, Y.; Ito, Y.; Inagaki, S. *J. Org. Chem.*, ***1991***, *56*, 2256.

Me_3Si

Ph

1. PhMe , reflux , 5 h
2. H_2O

Ph

Ph

53%

Ohno, M.; Miyata, H.; Komatsu, M.; Ohshiro, Y. *Tetrahedron Lett.*, ***1991***, *32*, 5093.

N-OH

Ph Ph

TsCN , NEt_3 , CCl_4
0°C

N-Ts

Ph Ph

69%

Boger, D.L.; Corbett, W.L. *J. Org. Chem.*, ***1992***, *57*, 4777.

N-OH

Ph

hv , MeOH , 2.5 h
1% H_2SO_4

N

73%

Olsen, R.J. *Tetrahedron Lett.*, ***1991***, *32*, 5235.

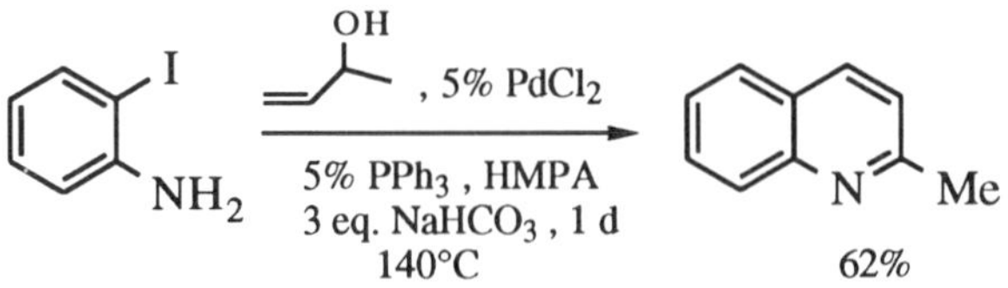

OH, 5% $PdCl_2$; 5% PPh_3, HMPA; 3 eq. $NaHCO_3$, 1 d; 140°C; I; NH_2; N; Me; 62%

Larock, R.C.; Kuo, M-Y. *Tetrahedron Lett.*, ***1991***, *32*, 569.

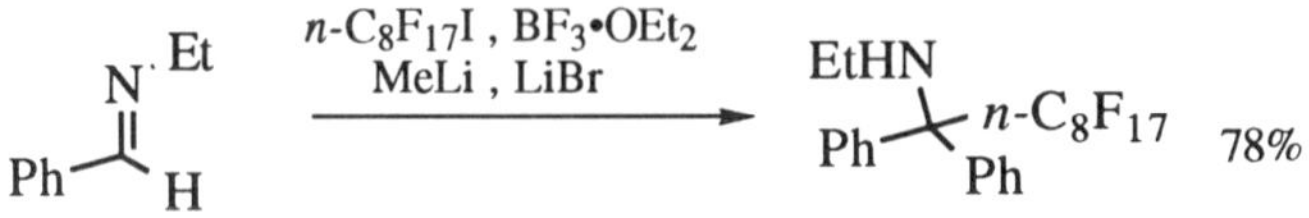

n-$C_8F_{17}I$, BF_3•OEt_2; MeLi, LiBr; N; Et; Ph; H; EtHN; Ph; n-C_8F_{17}; Ph; 78%

Uno, H.; Okada, S.; Ono, T.; Shiraishi, Y.; Suzuki, H. *J. Org. Chem.*, ***1992***, *57*, 1504.

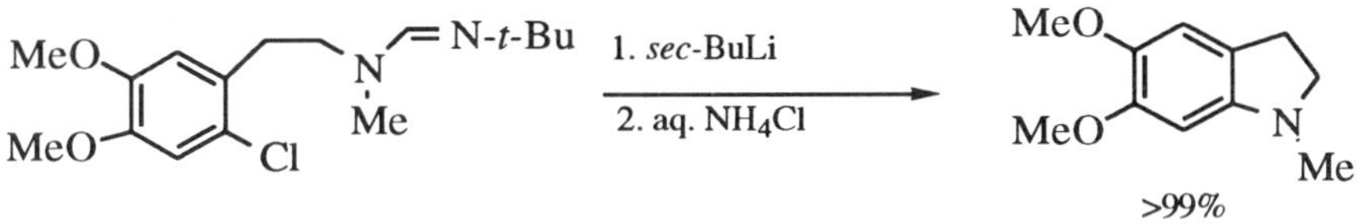

MeO; MeO; N-*t*-Bu; N; Me; Cl; 1. *sec*-BuLi; 2. aq. NH_4Cl; MeO; MeO; N; Me; >99%

Sielecki, T.M.; Meyers, A.I. *J. Org. Chem.*, ***1992***, *57*, 3673.

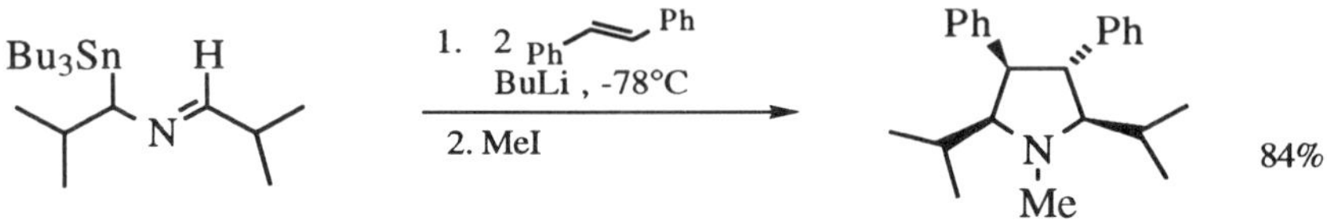

Bu_3Sn; H; N; 1. 2 Ph Ph; BuLi, -78°C; 2. MeI; Ph; Ph; N; Me; 84%

Pearson, W.H.; Postich, M.J. *J. Org. Chem.*, ***1992***, *57*, 6354.

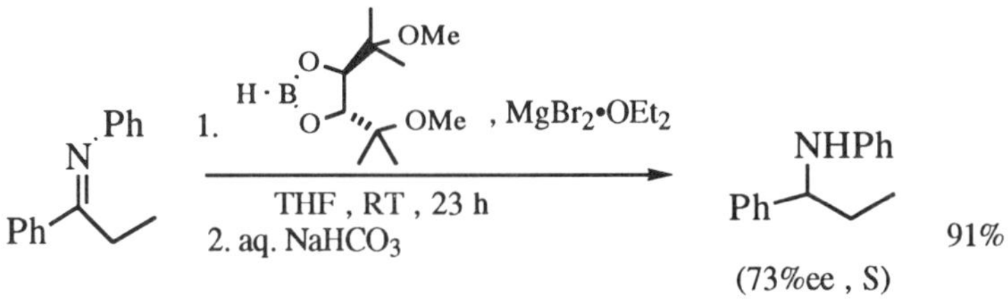

N; Ph; Ph; 1. H·B, O, O, OMe, OMe, $MgBr_2$•OEt_2; THF, RT, 23 h; 2. aq. $NaHCO_3$; NHPh; Ph; 91%; (73%ee, S)

Kawate, T.; Nakagawa, M.; Kakikawa, T.; Hino, T. *Tetrahedron Asymmetry*, ***1992***, *3*, 227.

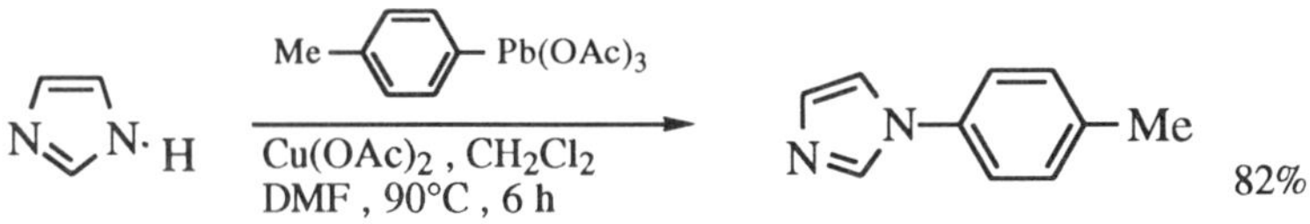

N; N·H; Me; $Pb(OAc)_3$; $Cu(OAc)_2$, CH_2Cl_2; DMF, 90°C, 6 h; N; N; Me; 82%

López-Alvarado, P.; Avendaño, C.; Menéndez, J.C. *Tetrahedron Lett.*, ***1992***, *33*, 659.

Me, Ph, N, Me, OH → CS_2, MeCN → Me, Ph, N, Me, H 87%

Schwartz, M.A.; Gu, J.; Hu, X. *Tetrahedron Lett.*, ***1992***, *33*, 1687.

Me, NHOH, MeO, OMe → 1. acetone 2. EtMgBr 3. CS_2 → Me, Me, Et, N, Me, H, MeO, OMe 84%

Schwartz, M.A.; Hu, X. *Tetrahedron Lett.*, ***1992***, *33*, 1689.

Ph, NHMe, Ph → 0.1 BuLi, THF → Ph, N, Me, Ph 99%

Fujita, H.; Tokuda, M.; Nitta, M.; Siginome, H. *Tetrahedron Lett.*, ***1992***, *33*, 6359.

$SiMe_3$, N → hν, iPrOH, [1]DCN → H, Me, N + H, Me, N

(97 : 3) 90%

Pandey, G.; Reddy, G.D. *Tetrahedron Lett.*, ***1992***, *33*, 6533.

Ph-NMe$_2$ → 1. 2 eq. H_2O_2, MeOH, $RuCl_3$•n H_2O; 2. H_3O^+ → Ph-NHMe 67%

Murahashi, S-I.; Naota, T.; Miyaguchi, N.; Nakato, T. *Tetrahedron Lett.*, ***1992***, *33*, 6991.

OH, Ph, NH → OCO_2Me, $RuH_2(PPh_3)_4$, PhH, 100°C → Ph, N, H 86%

Izumi, T.; Yokota, T. *J. Het. Chem.*, ***1992***, *29*, 1085.

Ph, N=CHPh → 0.5% $RuCl_2(PPh_3)_2$, 82°C, 5% K_2CO_3, iPrOH, 18 h → Ph, N, Ph, H 93%

Wang, G-Z.; Bäckvall, J-E. *J. Chem. Soc., Chem. Commun.*, ***1992***, 980.

N.NHBz
Ph

1. 0.2% $Rh(COD)(Et\text{-}DuPHOS)^+ OTf^-$
iPrOH , H_2 (4 atm)
2. SmI_2

NH_2
Ph

quant.

(89% ee , S)

Et-DuPHOS =
Et
P
Et
Et
P
Et

Burk, M.J.; Feaster, J.E. *J. Am. Chem. Soc.*, ***1992***, *114*, 6266.

N Ph

5%
X·Ti·X
, 2 BuLi , $PhSiH_3$
H_2 (2000 psi)

N Ph
H

68%

(58% ee)

X = 1,1'-binaphth-2,2'-diolate

Willoughby, C.A.; Buchwald, S.L. *J. Am. Chem. Soc.*, ***1992***, *114*, 7562.

N
H

2% $NiSO_4$, $K_2S_2O_8$
NaOH , $H_2O\text{-}CH_2Cl_2$
RT , 3.5 h

N

87%

Yamazaki, S. *Chem. Lett.*, ***1992***, 823.

N
H

$CuBr_2$•LiO*t*-Bu , THF
RT , 20 min

N

89%

Yamaguchi, J.; Takeda, T. *Chem. Lett.*, ***1992***, 1933.

⊕ N
H
⊖ O_2CF_3

$Ph_2Se(O_2C\text{-}c\text{-}C_6H_{11})_2$
hν

N

40%

Togo, H.; Miyagawa, N.; Yokoyama, M. *Chem. Lett.*, ***1992***, 1677.

1. EtI
2. KOH , MeOH

$MeCH=N\text{-}NMe_2$ → $EtNMe_2$ 57%

Smith, R.F.; Marcucci, J.L.; Tingue, P.S. *Synth. Commun.*, ***1992***, *22*, 381.

1. t-BuOOH , PhH, $RuCl_2(PPh_3)_3$
2. $TiCl_4$,

NMe_2 → N-Me, 66%

Murahashi, S.; Naota, T.; Nakato, T. *SynLett*, **1992**, 835.

SECTION 98: AMINES FROM ESTERS

Ph⁀OAc → (Me_3SiN_3 , 50°C , 2h; $Pd_2(dba)_3$, 4 PPh_3) → Ph⁀N_3 85%

Safi, M.; Fahrang, R.; Sinou, D. *Tetrahedron Lett.*, **1990**, *31*, 527.

SECTION 99: AMINES FROM ETHERS, EPOXIDES AND THIOETHERS

$PhNH_2$, TiO_2 , 250-300°C
packed bed-flow reactor

N-Ph, 97% conversion, 96% selectivity

Hargis, D.C.; Shubkin, R.L. *Tetrahedron Lett.*, **1990**, *31*, 2991.

$AlCl_3$, AcOH , 20 h
$NH_2OH{\bullet}HCl$, reflux

84%

Jun, J-G.; Shin, H.S. *Tetrahedron Lett.*, **1992**, *33*, 4593.

SECTION 100: AMINES FROM HALIDES AND SULFONATES

OTBS, OMs, CO_2Me → (NaN_3-CuI, HMPA , 4 h) → OTBS, N_3, CO_2Me + OTBS, N_3, CO_2Me

	(85	: 15)	86%
with NaN_3 , DMSO	(0	: 100)	42%

Yamamoto, Y.; Asao, N. *J. Org. Chem.*, **1990**, *55*, 5304.

Cl → (0.1 AgI, $LiN(SiMe_3)_2$; THF, reflux, 6 h) → $N(SiMe_3)_2$ 54%

Murai, T.; Yamamoto, M.; Kato, S. *J. Chem. Soc., Chem. Commun.*, ***1990***, 789.

$NaN(CHO)_2$ → (1. BuOTs, MeCN, 120°C; 2. HCl) → $BuNH_2$•HCl 88%

Yinglin, H.; Hongwen, H. *Synthesis*, ***1990***, 122.

Cl → (NEt_3, 8 kbar, 4 d, 100°C) → NEt_2 51%

Matsumoto, K.; Hashimoto, S.; Otani, S. *J. Chem. Soc., Chem. Commun.*, ***1991***, 306.

Ph–OTs → (Et_2HNHN_3, DMF, 60°C, 30 min) → Ph–N_3

Saito, S.; Yokoyama, H.; Ishikawa, T.; Niwa, N.; Moriwake, T. *Tetrahedron Lett.*, ***1991***, *32*, 663.
Saito, S.; Takahashi, N.; Ishiwawa, T.; Moriwake, T. *Tetrahedron Lett.*, ***1991***, *32*, 667.

Br → (1. Bn_2NH, K_2CO_3, diglyme, 140°C; 2. $HCl_{(g)}$, iPrOH; 3. NH_4HCO_2, 10% Pd/C, MeOH, 65°C) → NH_2 72%

Purchase, C.F. II.; Goel, O.P. *J. Org. Chem.*, ***1991***, *56*, 457.

O_2N–C6H4–Cl → (pyrrolidine, 50°C, 6 kbar, THF) → O_2N–C6H4–N(pyrrolidine) 91%

Ibata, T.; Isogami, Y.; Toyoda, J. *Bull. Chem. Soc. Jpn.*, ***1991***, *64*, 42.

Ph–Br → (1. NaN_3, PhH-DMF, reflux; 2. $P(OEt)_3$; 3. PhCHO; 4. $NaBH_4$, NaOH; 5. aq. HCl) → Ph–NHBn•HCl 77%

Gajda, T.; Koziara, A.; Osowska-Pacewicka, K.; Zawadzki, S.; Zwierzak, A. *Synth. Commun.*, ***1992***, *22*, 1929.

SECTION 101: AMINES FROM HYDRIDES

MeO–C6H4–Me → 1. $PhI(O_2CCF_3)_2$, $(CF_3)_2CHOH$; 2. Me_3SiN_3 → N_3, MeO, Me 45%

Kita, Y.; Tohma, H.; Inagaki, M.; Hatanka, K.; Yakura, T. *Tetrahedron Lett.*, ***1991***, *32*, 4321.

O → 1. $PhNH_2$, Zn, AcOH, 70°C; 2. NH_4OH → NHPh 85%

Mićović, I.V.; Ivanović, M.D.; Piatak, D.M.; Bojić, V.D. *Synthesis*, ***1991***, 1043.

→ $MeNO_2$ CH_3SO_3H, 16 h, 75°C → Ph, OH, N, Me 75% (E+Z) 33% Z

Coustard, J-M.; Jacquesy, J-C.; Violeau, B. *Tetrahedron Lett.*, ***1992***, *33*, 8085.

OSi(iPr)$_3$ → PhIO, $TMSN_3$ → OSi(iPr)$_3$, N_3 83%

Magnus, P.; Lacour, J. *J. Am. Chem. Soc.*, ***1992***, *114*, 767.

SECTION 102: AMINES FROM KETONES

O, Ph, Me → 1. $PhNH_2$, CH_2Cl_2, $TiCl_4$, EtN(iPr)$_2$; 2. $NaCNBH_4$ → Ph, NHPh, Me 94%

Barney, C.L.; Huber, E.W.; McCarthy, J.R. *Tetrahedron Lett.*, ***1990***, *31*, 5547.

O → n-$C_3H_7NH_2$, DCE, AcOH, NaBH(OAc)$_3$, 3 h → NHn-C_3H_7 88%

also works with aldehydes

Abdel-Magid, A.F.; Maryanoff, C.A.; Carson, K.G. *Tetrahedron Lett.*, ***1990***, *31*, 5595.
Abdel-Magid, A.F.; Maryanoff, C.A. *SynLett*, ***1990***, 537.

1. Me(C=CH2)NO2 , LDA , THF , -78°C
2. Ac_2O
3. H_2 Rh/Al_2O_3 , MeOH

Me–CO–n-C_7H_{15} → Me–(pyrrolidine, N–H)–n-C_7H_{15} 53%

H_2 with PtO_2 and treatment of amine with pTsOH leads to a pyrroline

Miyashita, M.; Awen, B.Z.E.; Yoshikoshi, A. *Chem. Lett.*, ***1990***, 239.

1. 33% KOH/EtOH reflux , 2 h
2. AcOH

Ph, Me, O + NH_2, CHO → Ph, N, Me 76%

Akula, M.R.; Wolowyk, M.W.; Knaus, E.E. *Org. Prep. Proceed. Int.*, ***1991***, *23*, 23.

1. OH, Ph, NH_2
2. $LiAlH_4$-$AlCl_3$
3. Pd/C , H_2

Ph, CO_2H, O → Bn, N, H 41%

Meyers, A.I.; Burgess, L.E. *J. Org. Chem.*, ***1991***, *56*, 2294.

H_2 , Pd-C , EtOH

O, O, NO_2 → H, N, Me, Me 75%

Janowitz, A.; Vavrecka, M.; Hesse, M. *Helv. Chim. Acta*, ***1991***, *74*, 1352.

1. N, B, O, Me, Ph, Ph , BH_3 THF
2. PPh_3 , HN_3 , PhMe DEAD , 0°C → RT
3. Pd/C , TFA

O → NH_2 81% (87% ee)

Chen, C-P.; Prasad, K.; Repic, O. *Tetrahedron Lett.*, ***1991***, *32*, 7155.

t-BuNH_2 , $TiCl_4$-C

O → N-t-Bu 99%

Carlson, R.; Larsson, U.; Hansson, L. *Acta Chem. Scand.*, ***1992***, *46*, 1211.

$n\text{-}C_5H_{11}C(O)CCl_3$ → 1. catecholborane (B·H) 2. oxazaborolidine (H, Ph, Ph, O, N, B, Ph), PhMe, -60°C; 3. NaN_3, DME, aq. NaOH; 4. H_3O^+; 5. H_2, Pd-C → $n\text{-}C_5H_{11}CH(NH_2)CO_2H$ 80% (95% ee)

Corey, E.J.; Link, J.O. *J. Am. Chem. Soc.*, ***1992***, *114*, 1906.

Related Methods: Section 94 (Amines from Aldehydes)

SECTION 103: AMINES FROM NITRILES

$t\text{-}BuCN$ → cat. $Co_2(CO)_8$, PPh_3, $HSiMe_3$, CO (atm); PhMe, 100°C, 20 h → $t\text{-}BuCH_2N(SiMe_3)_2$ quant.

Murai, T.; Sakane, T.; Kato, S. *J. Org. Chem.*, ***1990***, *55*, 449.

Ph-CN → 1. iBu_2AlH 2. BuLi, THF; 3. NH_4Cl, NH_4OH → PhCH(Bu)NH_2 98%

Andreoli, P.; Billi, L.; Cainelli, G.; Panunzio, M.; Martelli, G.; Spunta, G. *J. Org. Chem.*, ***1990***, *55*, 4199.

Ph-CN → 5% Rh on Al_2O_3; H_2, AcOH → $PhCH_2NHCH_2Ph$ 85%

Galán, A.; de Mendoza, J.; Prados, P.; Rojo, J.; Echavarren, A.M. *J. Org. Chem.*, ***1991***, *56*, 452.

PhCN → 1. BH_3 2. BuLi, hexane; -80°C, 2 h → PhCH(Bu)NH_2 95%

Itsuno, S.; Hachisuka, C.; Ito, K. *J. Chem. Soc., Perkin Trans. I*, ***1991***, 1767.
Itsuno, S.; Hachisuka, C.; Kitano, K.; Ito, K. *Tetrahedron Lett.*, ***1992***, *33*, 627.

$n\text{-}C_{11}H_{23}CN$ → 3 eq. $MeCeCl_2$, -65°C; 5 h → $Me_2C(NH_2)\text{-}n\text{-}C_{11}H_{23}$ 64%

Ciganek, E. *J. Org. Chem.*, ***1992***, *57*, 4521.

PhCN → 1. $NaBH_{(4-n)}(O_2CEt)_n$; 2. 2 eq. BuLi , 80°C , 1 h → Ph–CH(NH_2)Bu 59%

Itsuno, S.; Hachisuka, C.; Ushijima, Y. *Synth. Commun.*, ***1992***, *22*, 3229.

SECTION 104: AMINES FROM ALKENES

1. BH_3
2. [$BuNH_2$-NaOCl] , 0°C → RT
3. aq. HCl (pH <1)
4. aq. NaOH

NHBu 67%

Kabalka, G.W.; Wang, Z. *Synth. Commun.*, ***1990***, *20*, 231.

Ph

PhI=NTs , MeCN
10% $Cu(acac)_2$

Ph N- Ts 95%

Evans, D.A.; Faul, M.M.; Bilodeau, M.T. *J. Org. Chem.*, ***1991***, *56*, 6744.

N

$MeNH_2$•HOAc
MeOH , reflux
21 h

N NHMe 82%

Cliffe, I.A.; Ifill, A.D.; Mansell, H.L.; Todd, R.S.; White, A.C. *Tetrahedron Lett.*, ***1991***, *32*, 6789.

Ph

Mn (salen)
PhI=NTs

Ph N. Ts 46%

O'Connor, K.J.; Wey, S-J.; Burrows, C.J. *Tetrahedron Lett.*, ***1992***, *33*, 1001.

1. Me_3SiOTf , CH_2Cl_2
-14°C
2. $HN(OMe)_2$
3. 1M NaOH

N- OMe 64%

Vedejs, E.; Sano, H. *Tetrahedron Lett.*, ***1992***, *33*, 3261.

O_2N

Me_3Si N OMe
Bn
cat. TFA , CH_2Cl_2

O_2N N- Bn 71%

Laborde, E. *Tetrahedron Lett.*, ***1992***, *33*, 6607.

1. HSCN
2. (Me, SH, SH-substituted benzene), MeOH, 25°C

NH_2

50%

Cho, C-G.; Posner, G.H. *Tetrahedron Lett.*, ***1992***, *33*, 3599.

REVIEWS:

"An Overview of the Total Synthesis of Pyrrolizidine Alkaloids via [4+1] Azide-Diene Annulation Methodology"
Hudlicky, T.; Seoane, G.; Price, J.D.; Gadamasetti, K.G. *SynLett*, ***1990***, 433.

SECTION 105: AMINES FROM MISCELLANEOUS COMPOUNDS

Ph, N–S, N, S

n-Bu_3SnH

Ph, Me, N

55%

(cis:trans = 1:5)

Boivin, J.; Fouquet, E.; Zard, S.Z. *Tetrahedron Lett.*, ***1990***, *31*, 85.

O, Ph, N, Ph, NO

BF_3•THF
$NaHCO_3$

O, Ph, N, Ph, H

81%

Jeyaraman, R; Ravindran, T. *Tetrahedron Lett.*, ***1990***, *31*, 2787.

N_3

1. DMF, 120°C, 17 h
2. PhH, NH_4Cl, 4 h
sealed tube, 445°C
3. H_2, PtO_2, ether
25°C

N, Me + N, Me

(1 : 3) 28%

Heidt, P.C.; Bergmeier, S.C.; Pearson, W.H. *Tetrahedron Lett.*, ***1990***, *31*, 5441.

N, OH

bakers yeast, pH 5
12 h, 22°C

NH_2

20% (58% ee)

Gibbs, D.E.; Barnes, D. *Tetrahedron Lett.*, ***1990***, *31*, 5555.

1. PhH, reflux, 12 h
2. $NaBH_4$, MeOH
0°C, 1 h

N_3 Cl

96%

Pearson, W.H.; Lin, K-C. *Tetrahedron Lett.*, ***1990***, *31*, 7571.

1. 2 eq. AlI_3, MeCN, 3 h reflux
2. $Na_2S_2O_3$, H_2O

91%

Konwar, D.; Boruah, R.C.; Sandhu, J.S. *Synthesis*, ***1990***, 337.

$Ph\text{-}N_3 \xrightarrow{MeOH,\ reflux} Ph\text{-}NH_2$ 94%

Ⓟ–C_6H_4–$CH_2NMe_3^{\oplus}$ $BH_4^{\ominus}$

Kabalka, G.W.; Wadgaonkar, P.P.; Chatla, N. *Synth. Commun.*, ***1990***, *20*, 293.

1. BCl_2, DCE
2. H_2O
3. KOH

65% (>99% ee)

Brown, H.C.; Salunkhe, A.M.; Singaram, B. *J. Org. Chem.*, ***1991***, *56*, 1170.

$OSiPh_2t$-Bu

PhMe, reflux

$OSiPh_2t$-Bu + $OSiPh_2t$-Bu

(93 : 7) 85%

Ito, M.; Maeda, M.; Kibayashi, C. *Tetrahedron Lett.*, ***1992***, *33*, 3765.

AMINES FROM NITRO COMPOUNDS

$PhNO_2 \xrightarrow[EtOH,\ reflux]{NH_2NH_2,\ 3\ h} PhNH_2$ 85%

Han, B.H.; Jang, D.G. *Tetrahedron Lett.*, ***1990***, *31*, 1181.

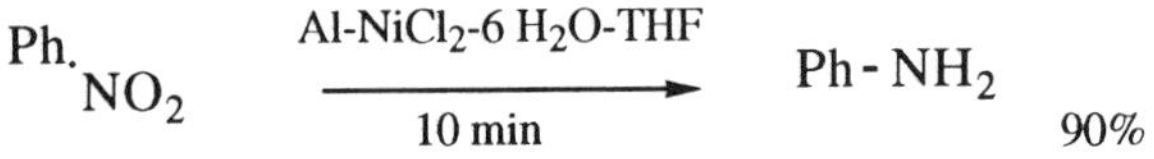

Ph.NO$_2$ $\xrightarrow[\text{10 min}]{\text{Al-NiCl}_2\text{-6 H}_2\text{O-THF}}$ Ph-NH$_2$ 90%

Sarmah, P.; Barua, N.C. *Tetrahedron Lett.*, ***1990***, *31*, 4065.

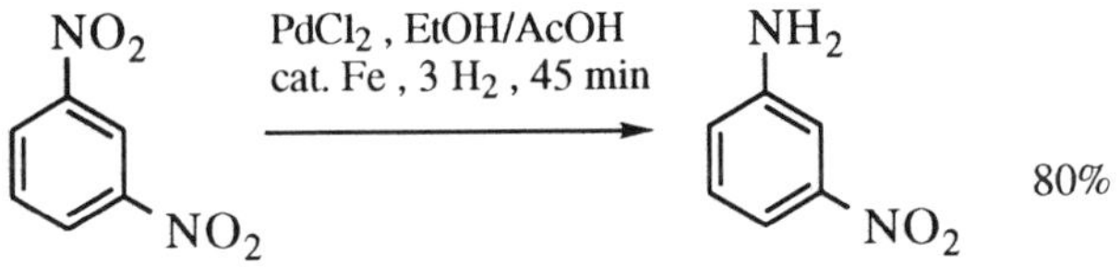

NO$_2$, NO$_2$ $\xrightarrow{\text{PdCl}_2\text{, EtOH/AcOH, cat. Fe, 3 H}_2\text{, 45 min}}$ NH$_2$, NO$_2$ 80%

Theodoridis, G.; Manfredi, M.C.; Krebs, J.D. *Tetrahedron Lett.*, ***1990***, *31*, 6141.

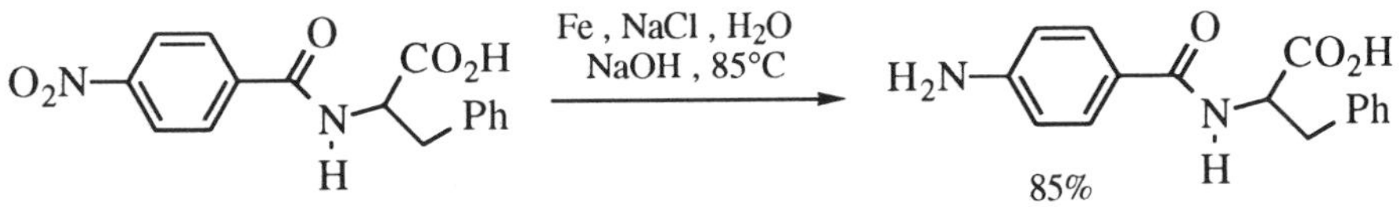

O$_2$N-, O, N-H, CO$_2$H, Ph $\xrightarrow{\text{Fe, NaCl, H}_2\text{O, NaOH, 85°C}}$ H$_2$N-, O, N-H, CO$_2$H, Ph 85%

Reddy, A.P.; Kasi Reddy, C.P.R. *Org. Prep. Proceed. Int.*, ***1990***, *22*, 117.

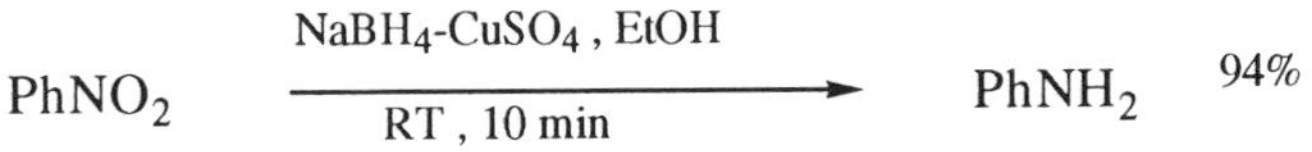

PhNO$_2$ $\xrightarrow[\text{RT, 10 min}]{\text{NaBH}_4\text{-CuSO}_4\text{, EtOH}}$ PhNH$_2$ 94%

Yoo, S.; Lee, S. *SynLett*, ***1990***, 419.

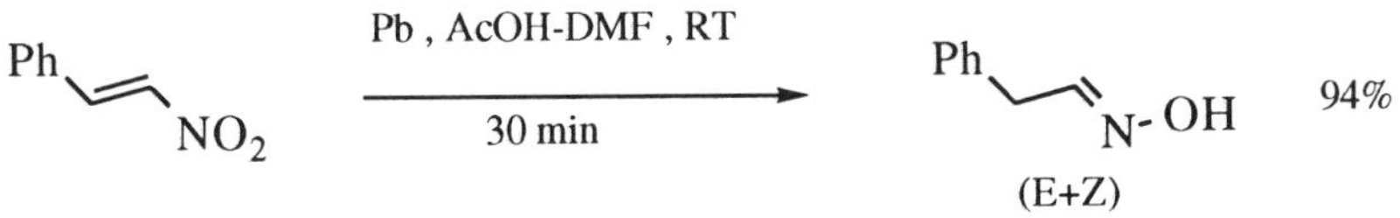

Ph, NO$_2$ $\xrightarrow[\text{30 min}]{\text{Pb, AcOH-DMF, RT}}$ Ph, N-OH (E+Z) 94%

Sera, A.; Yamauchi, H.; Yamada, H.; Itoh, K. *SynLett*, ***1990***, 477.

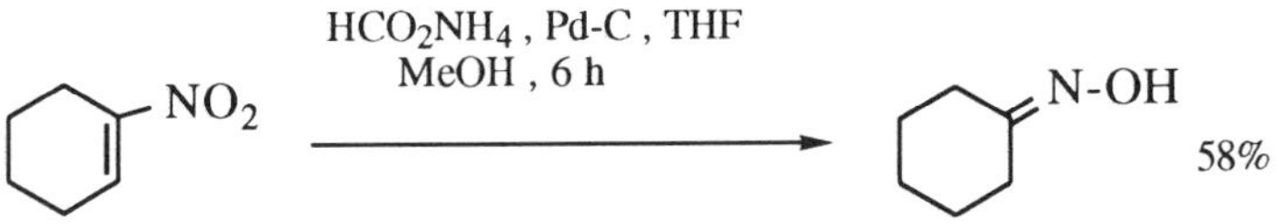

NO$_2$ $\xrightarrow{\text{HCO}_2\text{NH}_4\text{, Pd-C, THF, MeOH, 6 h}}$ N-OH 58%

Kabalka, G.W.; Pace, R.D.; Wadgaonkar, P.P. *Synth. Commun.*, ***1990***, *20*, 2453.

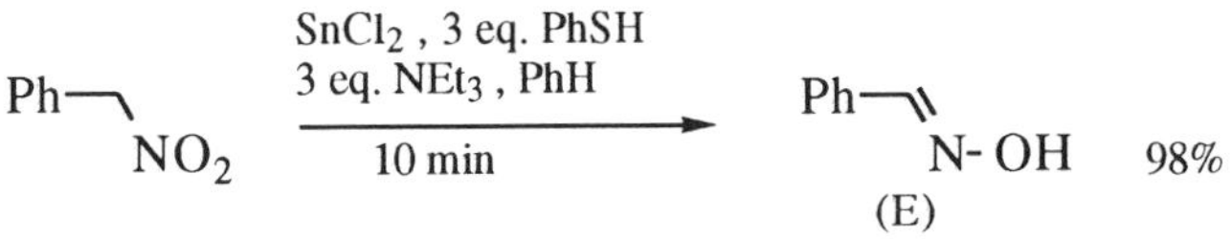

Ph, NO$_2$ $\xrightarrow[\text{10 min}]{\text{SnCl}_2\text{, 3 eq. PhSH, 3 eq. NEt}_3\text{, PhH}}$ Ph, N-OH (E) 98%

Bartra, M.; Romea, P.; Urpí, F.; Vilarrasa, J. *Tetrahedron*, ***1990***, *46*, 587.

$C_6H_5NO_2$ → 1. NH_2NH_2, 160°C, 2 h, acetone, sealed ampule; 2. MeOH → $C_6H_5NH_2$ 95%

Abdel-Baky, S.; Zhuang, M.; Giese, R.W. *Synth. Commun.*, ***1991***, *21*, 161.

1-Nitrocyclohexene → e^-, Hg electrode/C cathode, H_2SO_4, iPrOH, H_2O → cyclohexanone N-OH 87%

Wessling, M.; Schäfer, H.J. *Chem. Ber.*, ***1991***, *124*, 2303.

MeO-C_6H_4-$(CH_2)_4$-NO_2 → 1. 4 eq. SmI_2, THF, THF/MeOH, RT, 3 min; 2. 10% $Na_2S_2O_3$ → MeO-C_6H_4-$(CH_2)_4$-NHOH 72%

Kende, A.S.; Mendoza, J.S. *Tetrahedron Lett.*, ***1991***, *32*, 1699.

1-Nitrocyclohexene → 1. $Zn(BH_4)_2$, DME, 50°C, 1 h; 2. H_2O → cyclohexanone N-OH 90%

Ranu, B.C.; Chakraborty, R. *Tetrahedron Lett.*, ***1991***, *32*, 3579.

O_2N-C_6H_4-C≡N → $Ru_3(CO)_{12}$-$NH(iPr)_2$, CO, H_2O, 150°C, 2 h, 20 atm → H_2N-C_6H_4-C≡N quant.

Nomura, K. *Chem. Lett.*, ***1991***, 1679.

Ph-NO_2 → MoO_3, Na_2SeO_3, $NaBH_4$, H_2O → Ph-NH_2 96%

Yanada, K.; Yanada, R.; Meguir, H. *Tetrahedron Lett.*, ***1992***, *33*, 1463.

2-Nitrostilbene (Ph, NO_2) → 5% $PdCl_2(PPh_3)_2$, $SnCl_2$, 2 eq. CO, 100°C → 2-phenylindole (N-H, Ph) 75%

Akazome, M.; Kondo, T.; Watanabe, Y. *Chem. Lett.*, ***1992***, 769.

$PhNO_2$ → Fe, NH_4Cl, H_2O, reflux, 1.5 h → $PhNH_2$ 82%

Ramadas, K.; Srinivasan, N. *Synth. Commun.*, ***1992***, *22*, 3189.

REVIEWS:

"Electrophilic Aminations with Oxaziridines"
Andreae, S.; Schmitz, E. *Synthesis*, ***1991***, 327.

SECTION 105A: PROTECTION OF AMINES

BnHN~~OBn → H_2N~~OBn (20% $Pd(OH)_2$-C , EtOH; Parr shaker , H_2 (55 psi); 19 h)

Bernotas, R.C.; Cube, R.V. *Synth. Commun.*, ***1990***, *20*, 1209.

Cyclohexyl-NH_2 → Cyclohexyl-N(H)C(O)CF_3 93% (ether , 20°C , 30 min; 2-pyridyl trifluoroacetate)

Keumi, T.; Shimada, M.; Morita, T.; Kitajima, H. *Bull. Chem. Soc. Jpn.*, ***1990***, *63*, 2252.

NH_2•HCl, Me, CO_2Et → Bn, Bn, N, CO_2Et, Me 87% (H-N(Bn)C(O)N(Bn)-H; aq. HCHO; N-Me morpholine; dioxane, toluene)

← aq. HCl , MeOH

Knapp, S.; Hale, J.J.; Bastos, M.; Gibson, F.S. *Tetrahedron Lett.*, ***1990***, *31*, 2109.

Me Me Si-H, Si H, Me Me → Me Me Si, N-Ph, Si, Me Me 72% ($RhCl(PPh_3)_3$, 14 h; $PhNH_2$, 80°C)

← TBAF or TFAA

also with aliphatic amines

Bonar-Law, R.P.; Davis, A.P.; Dorgan, B.J. *Tetrahedron Lett.*, ***1990***, *31*, 6721, 6725.

N-Cbz → N-H 92% (BF_3•OEt_2 , EtSH; ether , 8 h)

Subhas Bose, D.; Thurston, D.E. *Tetrahedron Lett.*, ***1990***, *31*, 6903.

1. HCHO
2. cyclopentadiene , NMM

1. cat. $CuSO_4$, aq. EtOH , 70°C
2. Na_2So_3 , aq. NH_4Cl

NH_2

84%

Grieco, P.A.; Clark, J.D. *J. Org. Chem.*, ***1990***, *55*, 2271.

Ph NH_2 CO_2H → NEt_3 → Ph NHCbz CO_2H 88%

BnO S S Me N N O

also for protection of poly-ols

Allainmat, M.; L'Haridon, P.; Toupet, L.; Plusquellec, D. *Synthesis*, ***1990***, 27.

H N OH → $PhCHN_2$, $SnCl_2$•2 H_2O CH_2Cl_2 , 5.5 d , RT → Bn N OH + H N OBn

40% 8%

Liotta, L.J.; Ganem, B. *Isr. J. Chem.*, ***1991***, *31*, 215.

Ph NH_2•HCl → Boc_2O , $NaHCO_3$ EtOH , RT , 3 h)))))))) → Ph N-Boc H 99%

Einhorn, J.; Einhorn, C.; Luche, J-L. *SynLett*, ***1991***, 37.

BnO CO_2Me NHCbz → , 10% Pd-C Boc_2O , EtOH , RT → BnO CO_2Me NHBoc 92%

Bajwa, J.S. *Tetrahedron Lett.*, ***1992***, *33*, 2955.

TsO-N-Boc H → 1. BuLi 2. BBu_3 , -78°C , 1 h → Bu N Boc H 81%

Genêt, J-P.; Hajicek, J.; Bischoff, L.; Greck, C. *Tetrahedron Lett.*, ***1992***, *33*, 2677.

Ph
H_2N CO_2H

N_3 N_3

2 eq. $NaHCO_3$, 18 h
aq. dioxane

O Ph
O N CO_2H
H
O
N_3
85%

Loubinoux, B.; Gerardin, P. *Tetrahedron Lett.*, ***1991***, *32*, 351.

O
BnHN NHBn, aq. HCHO

n-C_7H_{15} NH_2

aq. HCl

O
n-C_7H_{15} N N·Bn
N
Bn
91%

Knapp, S.; Hale, J.J.; Bastos, M.; Molina, A.; Chen, K.Y. *J. Org. Chem.*, ***1992***, *57*, 6239.

N OBn
O

AcCl, NaI

N. Ac
84%

Ihara, M.; Hirabayashi, A.; Taniguchi, N.; Fukumoto, K. *Heterocycles*, ***1992***, *33*, 851.

H-N NH_2

1. PhCHO
2. Boc_2O
3. 1N $KHSO_4$, NaOH

Boc-N NH_2
92%

Prugh, J.D.; Birchenough, L.A.; Egbertson, M.S. *Synth. Commun.*, ***1992***, *22*, 2357.

Also see Section 82 (Amides from Amines)

CHAPTER 8

PREPARATION OF ESTERS

SECTION 106: ESTERS FROM ALKYNES

$$\text{(CH}_2)_4\text{-linked C}\equiv\text{C-Et / C}\equiv\text{C-SiMe}_3 \xrightarrow[\text{Ni(COD)}_2\text{ , CO}_2\text{, 100°C}]{2\ \text{2-pyridyl-CH}_2\text{CH}_2\text{PBu}_2} \text{tetrahydroisochromenone (Et, SiMe}_3) \quad 74\%$$

Tsuda, T.; Morikawa, S.; Hasegawa, N.; Saegusa, T. *J. Org. Chem.*, ***1990***, *55*, 2978.

SECTION 107: ESTERS FROM ACID DERIVATIVES

The following types of reactions are found in this section:

1. Esters from the reaction of alcohols with carboxylic acids, acid halides and anhydrides.
2. Lactones from hydroxy acids
3. Esters from carboxylic acids and halides, sulfoxides and miscellaneous compounds

$$n\text{-C}_{15}\text{H}_{30}\text{COOH} \xrightarrow[\text{xylene , 9 h}]{\text{Me}_3\text{N}\bullet\text{BH}_3\text{ , reflux}} n\text{-C}_{15}\text{H}_{30}\text{CO}_2\text{CH}_2n\text{-C}_{15}\text{H}_{30} \quad 61\%$$

Trapani, G.; Reho, A.; Latrofa, A.; Liso, G. *Synthesis*, ***1990***, 853.

$$\text{PhCH}_2\text{COCl} \xrightarrow[\text{15 h}]{t\text{-BuOH , activated Al}_2\text{O}_3} \text{PhCH}_2\text{CO}_2t\text{-Bu} \quad 85\%$$

Nagasawa, K.; Yoshitake, S.; Amiya, T.; Ito, K. *Synth. Commun.*, ***1990***, *20*, 2033.

$$n\text{-C}_8\text{H}_{17}\text{Br} \xrightarrow[\text{microwaves}]{\text{dry AcOK , Al}_2\text{O}_3} n\text{-C}_8\text{H}_{17}\text{OAc} \quad 92\%$$

Bram, G.; Loupy, A.; Majdoub, M.; Gutierrez, E.; Ruiz-Hitzky, E. *Tetrahedron*, ***1990***, *46*, 5167.

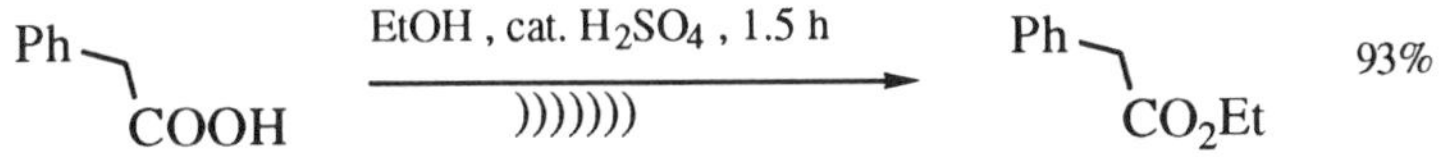

Ph–CH_2–COOH —(EtOH , cat. H_2SO_4 , 1.5 h;))))))))→ Ph–CH_2–CO_2Et 93%

Khurana, J.M.; Sahoo, P.K.; Maikap, G.C. *Synth. Commun.*, ***1990***, *20*, 2267.

$EtCO_2Na$ —(BuBr , Aliquat 336 , reflux)→ $EtCO_2Bu$ 95%

Vinczer, P.; Novak, L.; Szantay, C. *Synth. Commun.*, ***1991***, *21*, 1545.

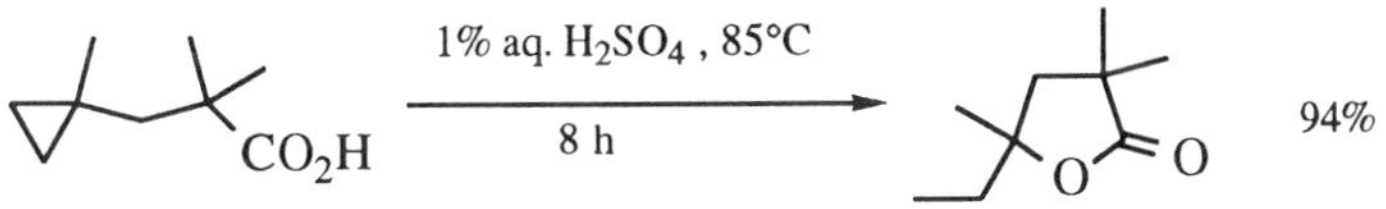

CO_2H —(1% aq. H_2SO_4 , 85°C; 8 h)→ O O 94%

Sugahara, K.; Fujita, T.; Watanabe, S.; Sakamoto, M.; Sugimoto, K. *Synthesis*, ***1990***, 783.

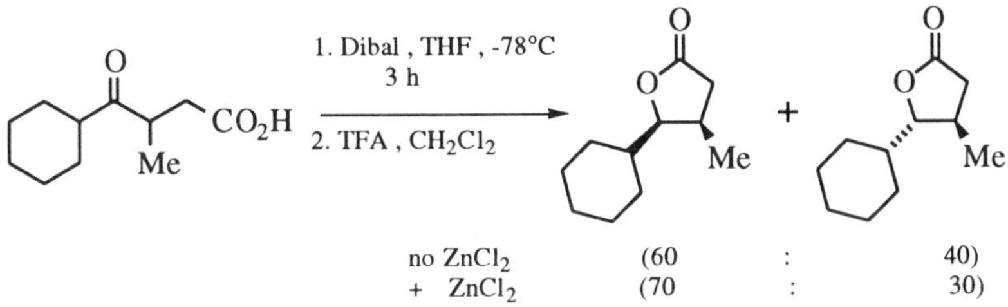

O Me CO_2H —(1. Dibal , THF , -78°C 3 h; 2. TFA , CH_2Cl_2)→ O O Me + O O Me

no $ZnCl_2$ (60 : 40)
\+ $ZnCl_2$ (70 : 30)

Frenette, R.; Monette, M.; Bernstein, M.A.; Young, R.N.; Verhoeven, T.R. *J. Org. Chem.*, ***1991***, *56*, 3083.

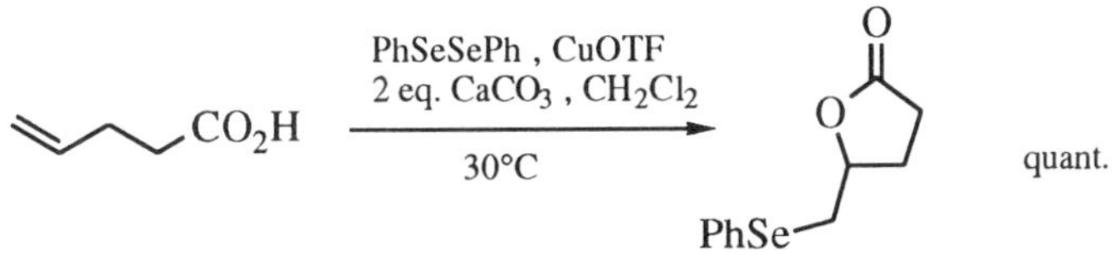

CO_2H —(PhSeSePh , CuOTF; 2 eq. $CaCO_3$, CH_2Cl_2; 30°C)→ O O PhSe quant.

Miyachi, N.; Satoh, H.; Shibasaki, M. *J. Chem. Soc., Perkin Trans. I*, ***1991***, 2049.

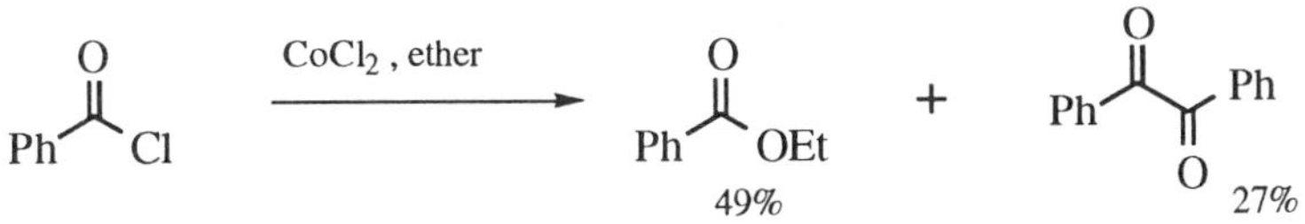

PhC(O)Cl —($CoCl_2$, ether)→ PhC(O)OEt 49% + PhC(O)C(O)Ph 27%

Iqbal, J.; Srivastava, R.R. *Tetrahedron*, ***1991***, *47*, 3155.

PhCOOH —(EtI , CsF , DMF)→ $PhCO_2Et$ 94%

Sato, T.; Otera, J.; Nozaki, H. *J. Org. Chem.*, ***1992***, *57*, 2166.

1.2 eq. PIFA , 3 d

71%

PIFA = phenyliodonium bis-trifluoroacetate

Casey, M.; Manage, A.C.; Murphy, P.J. *Tetrahedron Lett.*, ***1992***, *33*, 965.

$PhCH_2CO_2H \xrightarrow{Fe(ClO_4)_3 \cdot 9\ H_2O\ ,\ EtOH} PhCH_2CO_2Et$ 93%

Kumar, B.; Kumar, H.; Parmar, A. *Synth. Commun.*, ***1992***, *22*, 1087.

BnCl , $NaHCO_3$, TBAB
NaI , MeCN , 80°C , 8 h

98%

Hu, Y.; Pa, W.; Cui, W.; Wang, J. *Synth. Commun.*, ***1992***, *22*, 2763.

$n\text{-}C_5H_{11}COOH \xrightarrow[\text{microwave (560W), 10 min}]{BnBr\ ,\ BnBu_3NCl\ ,\ neat} n\text{-}C_5H_{11}CO_2Bn$ 72%

Yuncheng, Y.; Yulin, J.; Dabin, G. *Synth. Commun.*, ***1992***, *22*, 3109.

Further examples of the reaction RCO_2H + R'OH → RCO_2R' are included in Section 108 (Esters from Alcohols and Phenols) and in Section 30A (Protection of Carboxylic Acids).

SECTION 108: ESTERS FROM ALCOHOLS AND THIOLS

PCC , 8 h

53%

Baskaran, S.; Chandrasekaran, S. *Tetrahedron Lett.*, ***1990***, *31*, 2775.

1. CO , S , cat. Se
NEt_3 , THF , 60°C
autoclave , 1 h
2. cat. S , BnBr
250°C

MeOH → MeO(C=O)SBn 69%

Mizuno, T.; Nishiguchi, I.; Hirashima, T.; Ogawa, A.; Kambe, N.; Sonoda, N. *Tetrahedron Lett.*, ***1990***, *31*, 4773.

1. $EtCO_2NHNH_2$
2. $Pb(OAc)_4$

O, OH → O, CO_2Et 87%

Katritzky, A.R.; Kotali, A. *Tetrahedron Lett.*, ***1990***, *31*, 6781.

Me_3Si, C=O, H , $ZnCl_2$

CH_2Cl_2 , RT , 1 h

OH → O, O, $SiMe_3$ 86%

Kita, Y.; Sekihachi, J.; Hayashi, Y.; Da, Y-Z.; Yamamoto, M.; Akai, S. *J. Org. Chem.*, ***1990***, *55*, 1108.

$Ce(SO_4)_2$, SiO_2 , EtOAc

15 min

n-$C_{11}H_{23}$ OH → n-$C_{11}H_{23}$ OAc 97%

Nishiguchi, T.; Taya, J. *J. Chem. Soc., Perkin Trans I*, ***1990***, *172*.

CO , O_2 , $PdCl_2$, THF
$CuCl_2$, HCl , RT , 1 atm

poly-L-leucine

Me OH → Me, O, O 49%

(61% ee , R)

Alper, H.; Hamel, N. *J. Chem. Soc., Chem. Commun.*, ***1990***, 135.

$KMnO_4$ - $CuSO_4$•5 H_2O
CH_2Cl_2 , H_2O

reflux (5 min) - RT (4.5 h)

OH → O, O 78%

Baskaran, S.; Islam, I.; Vankar, P.S.; Chandrasekaran, S. *J. Chem. Soc., Chem. Commun.*, ***1990***, 1670.

lipase PS-30 , DME , RT

OAc

Ph, O, AcO, OH ± → Ph, O, AcO, OH 47% (>97% ee) + Ph, O, AcO, OH 47% (?97% ee)

Berkowitz, D.B.; Danishefsky, S.J. *Tetrahedron Lett.*, ***1991***, *32*, 5497.

1. PhI, $Pd(OAc)_2$, KOAc, 10 h, $BnEt_3NBr$, iBuCN, 90°C
2. PCC

66%

Mandai, T.; Hasegawa, S.; Fujimoto, T.; Kawada, M.; Nokami, J.; Tsuji, J. *SynLett*, ***1990***, 85.

TPAP - NMO, MS
CH_2Cl_2

76%

TPAP = tetra-*n*-propylammonium perruthenate

Bloch, R.; Brillet, C. *SynLett*, ***1991***, 829.

$SiMe_3$ OH *Pseudomonas AK* 25% conversion OAc

$SiMe_3$ OH 63% (<5% ee) + $SiMe_3$ OAc 23% (80% ee)

Sparks, M.A.; Panek, J.S. *Tetrahedron Lett.*, ***1991***, *32*, 4085.

porcine pancretic lipase, ether, $BuCO_2CH_2CF_3$, 25°C, 48 h

38% (98% ee) + 41% (97% ee)

Chong, J.M.; Mar, E.K. *Tetrahedron Lett.*, ***1991***, *32*, 5683.

PhMe, NEt_3, 80°C, 36 h

general preparation of angelate esters

Hartmann, B.; Kanazawa, A.M.; Deprés, J-P.; Greene, A.E. *Tetrahedron Lett.*, ***1991***, *32*, 5077.

$Pd(dba)_2$•dppb, DME, CO (40 atm), 190°C, 48 h

78%

El Ali, B.; Alper, H. *J. Org. Chem.*, ***1991***, *56*, 5357.

$O_2N-C_6H_4-CO_2H$

DEAD , PPh_3 , PhH

RT , 1 h

84%

modified Mitsunobu - for sterically hindered 2° alcohols

Martin, S.F.; Dodge, J.A. *Tetrahedron Lett.*, ***1991***, *32*, 3017.

NaH , THF

(83 : 1) 96%

Yamada, S. *J. Org. Chem.*, ***1992***, *57*, 1591.
Yamada, S. *Tetrahedron Lett.*, ***1992***, *33*, 2171.

porcine pancreatic lipase

$BuCO_2CH_2CF_3$

ether , 16 h

53% ee 95% ee

36% conversion

Morgan, B.; Oehlschlager, A.C.; Stokes, T.M. *J. Org. Chem.*, ***1992***, *57*, 3231.

1. BuCHO , CH_2Cl_2 , MS
2. $BrZnCH_2CO_2Et$, ether
3. satd. NH_4Cl
4. H_2 , 10% Pd-C , EtOH 9 bar

50%

Andrés, C.; González, A.; Pedrosa, R.; Pérez-Encabo, A. *Tetrahedron Lett.*, ***1992***, *33*, 2895.

3M $LiClO_4$, ether

15 min

86%

Grieco, P.A.; Collins, J.L.; Henry, K.J. Jr. *Tetrahedron Lett.*, ***1992***, *33*, 4735.

$MeCO_3H$, NaBr

40°C , 2 h

92%

Morimoto, T.; Hirano, M.; Hamaguchi, T.; Shimoyama, M.; Zhuang, X. *Bull. Chem. Soc. Jpn.*, ***1992***, *65*, 703.

OH (phenol) → (CO_2Et acrylate), 20 eq. LiH / 180°C, 40 h → dihydrocoumarin 66%

Pickett, J.E.; Van Dort, P.C. *Tetrahedron Lett.*, ***1992***, *33*, 1161.

1. Cr reagent, CH_2Cl_2, 0°C; 2. mCPBA; 3. $NaHCO_3$ — *t*-Bu 98%

Morin-Fox, M.L.; Lipton, M.A. *Tetrahedron Lett.*, ***1992***, *33*, 5699.

Further examples of the reaction ROH → RCO_2R' are included in Section 107 (Esters from Acid Derivatives) and in Section 45A (Protection of Alcohols and Phenols).

SECTION 109: ESTERS FROM ALDEHYDES

Me$(CH_2)_8$OH → *n*-C_3H_7CHO, 1.5 Br_2, H_2O/CH_2Cl_2 / HMPT - $NaHCO_3$ → Me$(CH_2)_8$O(C=O)*n*-C_4H_9 94%

Al Neirabeyeh, M.; Pujol, M.D. *Tetrahedron Lett.*, ***1990***, *31*, 2273.

CHO → KF, 18-crown-6, 1 d / MeCN, 70°C → 85%

Olofson, R.A.; Dang, V.A.; Morrison, D.S.; DeCusati, D.F. *J. Org. Chem.*, ***1990***, *55*, 1.

CHO → *n*-$C_{12}H_{25}$OTMS, NBS / AIBN, 95°C, 15 min → O*n*-$C_{12}H_{25}$ 72%

Markó, I.E.; Mekhalfia, A.; Ollis, W.D. *SynLett*, ***1990***, 347.

2.6 eq. KOH
1.3 eq. I_2 , MeOH , 1 h

PhCHO → $PhCO_2Me$ 98%

Yamada, S.; Morizono, D.; Yamamoto, K. *Tetrahedron Lett.*, ***1992***, *33*, 4329.

1. Na_2CO_2 , MeOH
2. TBPSCl , imidazole
DMF , 40°C , 24 h
3. mCPBA , BF_3•OEt_2
CH_2Cl_2 , -10°C

OAc, H, O, OEt, MeO → t-$BiPh_2SiO$, H, O, O, MeO 68%

Gómez, A.M.; Valverde, S.; López, J.C. *Tetrahedron Lett.*, ***1992***, *33*, 5105.

1. NH_2OH•HCl , NaOAc
MeOH , reflux , 4 h
2. H_2O_2 , 2-NBSeA , MeOH
reflux, 1 h

PhCHO → $PhCO_2Me$ 72%

2-NBSeA = 2-nitrobenzene seleninic acid

Said, S.B.; Skarzewski, J.; Młochowski, J. *Synth. Commun.*, ***1992***, *22*, 1851.

Related Methods: Section 117 (Esters from Ketones)

SECTION 110: ESTERS FROM ALKYLS, METHYLENES AND ARYLS

No examples of the reaction R-R → RCO_2R' or $R'CO_2R$ (R,R' = alkyl, aryl, etc.) occur in the literature. For the reaction R-H → RCO_2R' or $R'CO_2R$, see Section 116 (Esters from Hydrides).

NO ADDITIONAL EXAMPLES

SECTION 111: ESTERS FROM AMIDES

1. BuLi , THF , -78°C
2. (acryloyl N-Me, N-NMe_2 hydrazide: Me, N, NMe_2, O)
3. 1.5 eq. conc. $HClO_4$
CH_2Cl_2 , RT, overnight

OMOM, $SnBu_3$ → (lactone: O, O) 43% (92%ee)

Chong, J.M.; Mar, E.K. *Tetrahedron Lett.*, ***1990***, *31*, 1981.

SECTION 112: ESTERS FROM AMINES

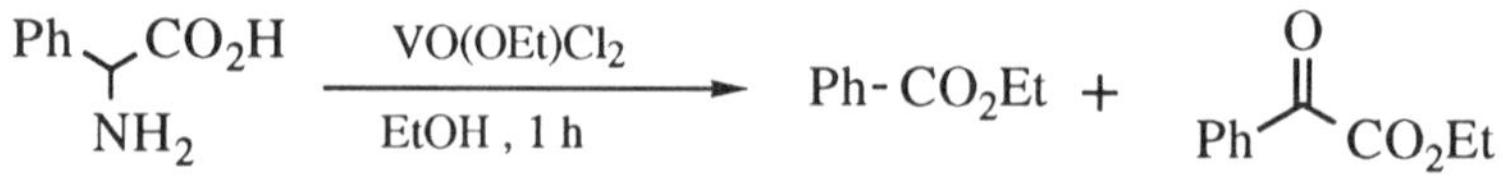

Ph(NH$_2$)CH–CO$_2$H —[VO(OEt)Cl$_2$, EtOH, 1 h]→ Ph-CO$_2$Et + PhCOCO$_2$Et

(89 : 11) 74%

Hirao, T.; Ohshiro, Y. *Tetrahedron Lett.*, ***1990***, *31*, 3917.

SECTION 113: ESTERS FROM ESTERS

Conjugate reductions and conjugate alkylations of unsaturated esters are found in Section 74 (Alkyls from Alkenes).

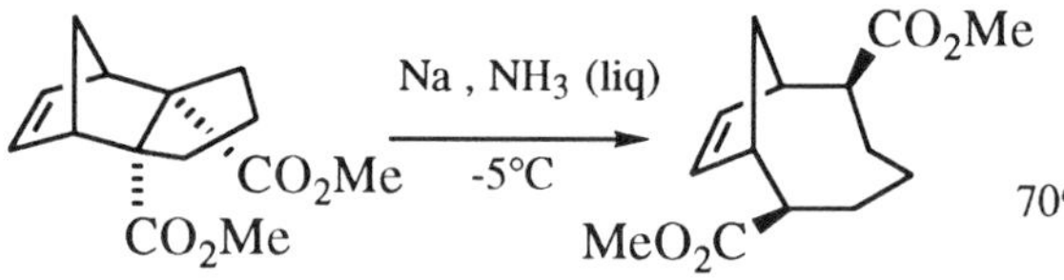

Na, NH$_3$ (liq), -5°C

CO$_2$Me, CO$_2$Me → CO$_2$Me, MeO$_2$C 70%

Saha, G.; Bhattacharya, A.; Roy, S.S.; Ghosh, S. *Tetrahedron Lett.*, ***1990***, *31*, 1483.

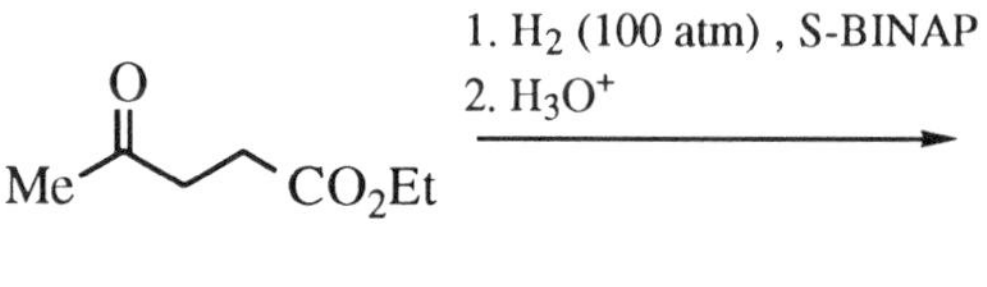

Me–CO–CH$_2$CH$_2$–CO$_2$Et —[1. H$_2$ (100 atm), S-BINAP; 2. H$_3$O$^+$]→ lactone (Me) 96% (99.5% ee)

Ohkuma, T.; Kitamura, M.; Noyori, R. *Tetrahedron Lett.*, ***1990***, *31*, 5509.

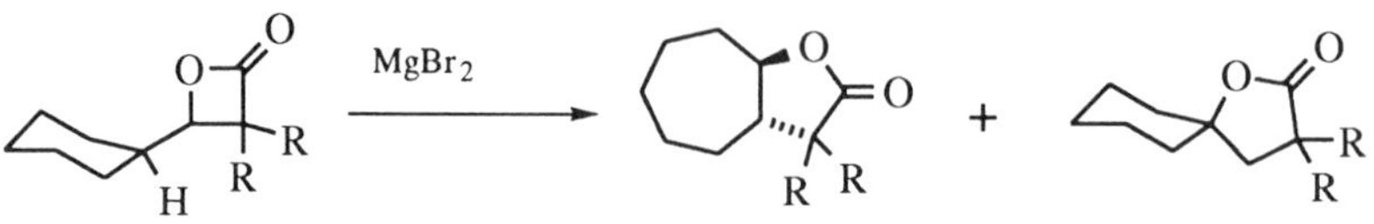

MgBr$_2$

	fused		spirocycle	yield
R = Me	(69	:	31)	72%
R = Et	(75	:	25)	31%
R = Pr	(>1	:	99)	64%

using TiCl$_4$ gave fused > spirocycle in all cases

Black, T.H.; McDermott, T.S.; Eisenbeis, S.A. *Tetrahedron Lett.*, ***1990***, *31*, 6617.

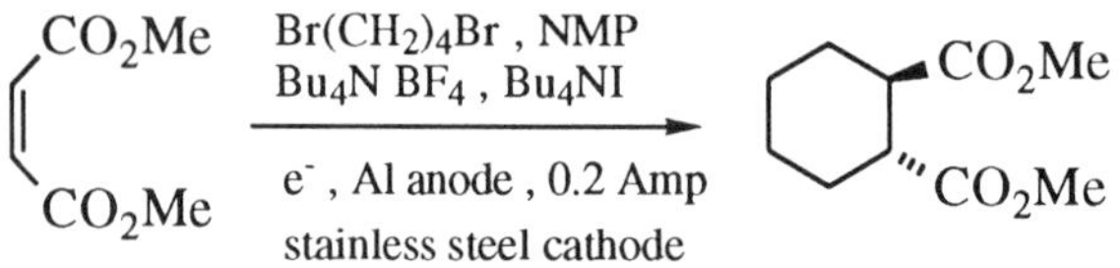

CO$_2$Me / CO$_2$Me —[Br(CH$_2$)$_4$Br, NMP; Bu$_4$N BF$_4$, Bu$_4$NI; e$^-$, Al anode, 0.2 Amp; stainless steel cathode]→ CO$_2$Me / CO$_2$Me 40%

Lu, Y-W.; Nédeléc, J-Y.; Folest, J-C.; Périchon, J. *J. Org. Chem.*, ***1990***, *55*, 2503.

O, O — 1. LDA, Me_3SiCl 2. $MeCHCl_2$, BuLi 3. xylene, reflux, 10 h 4. H_2, Pd-C, 4 atm. — Me, O, O 77%

Fouque, E.; Rousseau, G.; Seyden-Penne, J. *J. Org. Chem.*, ***1990***, *55*, 4807.

O, O — 1. Br, CO_2Et, Zn, 10% $TiCl_4$, CH_2Cl_2, 40°C, 10 min 2. 10% aq. HCl — CO_2Et (1:1 erythro:threo) 97%

Hayashi, M.; Sugiyama, M.; Toba, T.; Oguni, N. *J. Chem. Soc., Chem. Commun.*, ***1990***, 767.

Et, CO_2Me — 1. EtMgBr 2. EtCHO — Et, CO_2Me, Et, Et, O, Et, O + Et, CO_2Me, Et, Et, O, Et, O (2.3 : 1) 77%

Yamaguchi, K.; Kurasawa, Y.; Yokota, K. *Chem. Lett.*, ***1990***, 719.

n-C_5H_{11} SEt (O) — e^-, Pt electrodes, MeCN, Ph OH, Bu_4NI, RT — n-C_5H_{11} O Ph (O) 83%

Yamaguchi, M.; Tsukamoto, Y.; Minami, T. *Chem. Lett.*, ***1990***, 1223.

CO_2Et, CO_2Et — Ph, $Fe(ClO_4)_3 \cdot 9\ H_2O$ — Me, EtO_2C, O, O, Ph 90%

Citterio, A.; Sebastiano, R.; Nicolini, M.; Santi, R. *SynLett*, ***1990***, 42.

On-C_6H_{13}, $OSiMe_3$ — Me_3Al, CH_2Cl_2, 3 h, -78°C — Me_3Si On-C_6H_{13} (O) 88%

other R_3Al and R_2AlCl reagents used

Maruoka, K.; Banno, H.; Yamamoto, H. *SynLett*, ***1991***, 253.

OMe, OSiMe$_3$, 1 h, 25°C

(η^3-$C_4H_7PdOAc)_2$, dppf

Me, OAc → Me, CO_2Me 85%

Carfagna, C.; Marian, L.; Musco, A.; Sallese, G.; Santi, R. *J. Org. Chem.*, ***1991***, *56*, 3924.

Bu_3SnH, AIBN, PhH (0.02 M), reflux

O, O, SePh → Me, O, O 92%

Bachi, M.D.; Bosch, E. *J. Org. Chem.*, ***1992***, *57*, 4696.

1. 2.2 eq. CH_2Br_2 2.2 eq. LiTMP

2. 2 eq. LiHMDS

3. 4 eq. *sec*-BuLi

4. 2 eq. *n*-BuLi

5. EtOH, HCl

CO_2Et → CO_2Et 79%

Kowalski, C.J.; Reddy, R.E. *J. Org. Chem.*, ***1992***, *57*, 7194.

t-BuOK, THF, air 24°C

Ph, CO_2Me, OMe → $PhCO_2Me$ 73%

Heckmann, B.; Alayrac, C.; Mioskowski, C.; Chandrasekhar, S.; Falck, J.R. *Tetrahedron Lett.*, ***1992***, *33*, 5205.

Ph, $OSiMe_3$, $Sn(OTf)_2$, RT

$(F_3C-C_6H_4-CO)_2O$, CH_2Cl_2

RT, 2 h

O, $OSiMe_3$ → O, O, Ph 95%

Mukaiyama, T.; Shiina, I.; Miyashita, M. *Chem. Lett.*, ***1992***, 625.

SECTION 114: ESTERS FROM ETHERS, EPOXIDES AND THIOETHERS

$OSiMe_3$, OMe, $TiCl_4$, -78°C

CH_2Cl_2

O, O, Me, OEt → O, O, Me, CO_2Me 65%

Collins, D.J.; Choo, G.L.P.; Obrist, H. *Aust. J. Chem.*, ***1990***, *43*, 617.

CO_2-O_2 (65 psi), 70°C
Rh(NBD)(*p*-$Me_2Ph)_3$ BF_4
7 d
150 turnovers

Fazlur-Rahman, A.K.; Tsai, J-C.; Nicholas, K.M. *J. Chem. Soc., Chem. Commun.*, ***1992***, 1334.

SECTION 115: ESTERS FROM HALIDES AND SULFONATES

$CHBr_2$, $CHBr_2$, O_2N —H_2SO_4→ O_2N + O_2N

(2.6 : 1) 61%

Guo, W.; Duann, Y.F. *Org. Prep. Proceed. Int.*, ***1990***, *22*, 85.

$ClCH_2CO_2Me$, $NiBr_2$
e^- (Al anode, C cathode)
DMF, Bu_4NBr, 2,2-bipy
CO_2Me 70%

Conan, A.; Sibille, S.; d'Incan, E.' Périchon, J. *J. Chem. Soc., Chem. Commun.*, ***1990***, 48.

$PhCH_2Br$ —$Co(CO)_4$, Bu_4NI, CO, NaOEt, PhH→ $PhCH_2CO_2Et$ 95%

Kantam, M.L.; Reddy, N.P.; Choudary, B.M. *Synth. Commun.*, ***1990***, *20*, 2631.

MeO(C=O)OMe —PhMgBr, THF, 10°C, 2 h→ Ph(C=O)OMe 87%

Satyanarayama, G.; Sivaram, S. *Synth. Commun.*, ***1990***, *20*, 3273.

Ph OMs —10% $Co_2(CO)_8$, NaI, CO (50 atm), TMU, EtOH, 100°C, 24 h→ Ph CO_2Et 61%

TMU = tetramethyl urea

Urata, H.; Goto, D.; Fuchikami, T. *Tetrahedron Lett.*, ***1991***, *32*, 3091.

I —AcOH, NaOMe, 40°C, autoclave, N_2 (15 bar), $PdCl_2(PPh_3)_2$, 4 h→ CO_2Me 98%

Carpentier, J-F.; Castanet, Y.; Brocard, J.; Mortreux, A.; Petit, F. *Tetrahedron Lett.*, ***1991***, *32*, 4705.

PhOTf → (Me, OMe, OSiMe$_3$ ketene silyl acetal), THF; [η^3-C_4H_7•$Pd(OAc)_2$, LiOAc, dppf, reflux, 6 h → Me–CH(Ph)–CO_2Me 73%

dppf = 1,1'=bis(diphenylphosphino)ferrocene

Carfagna, C.; Musco, A.; Sallese, G.; Santi, R.; Fiorani, T. *J. Org. Chem.*, ***1991***, *56*, 261.

C_6H_5–OSO_2F → ($Pd(OAc)_2$, dppp, NEt_3, DMSO, MeOH, CO; 60°C, 2 h) → C_6H_5–CO_2Me 88%

Roth, G.P.; Thomas, J.A. *Tetrahedron Lett.*, ***1992***, *33*, 1959.

$I(CH_2)_4I$ → (TBAF•3 H_2O, $Mo(CO)_6$, THF; reflux, 20 h) → δ-valerolactone 69%

Imbeaux, M.; Mestdagh, H.; Moughamir, K.; Rolando, C. *J. Chem. Soc., Chem. Commun.*, ***1992***, 1678.

PhI → (1. $Bu_3SnC{\equiv}COEt$, $PdCl_2(PPh_3)_2$, Et_4NCl, DMF; 2. H_2SO_4, aq. acetone, reflux) → $PhCH_2CO_2Et$ 60%

Sakamoto, T.;Yasuhara, A.; Kondo, Y.; Yamanaka, H. *SynLett*, ***1992***, 502.

Related Methods: Section 25 (Acid Derivatives from Halides).

SECTION 116: ESTERS FROM HYDRIDES

This section contains examples of the reaction R-H → RCO_2R' or $R'CO_2R$ (R = alkyl, aryl, etc.).

cyclohexene → (5% $Pd(OAc)_2$, AcOH, 20% benzoquinone; 2 eq. MnO_2, 60°C, 50°C) → 3-acetoxycyclohexene (OAc) 77%

Hansson, S.; Heumann, A.; Rein, T.; Åkermark, B. *J. Org. Chem.*, ***1990***, *55*, 975.

5% $Pd(OAc)_2$, AcOH
5% $Cu(OAc)_2$, 50°C
10% hydroquinone
1 atm. O_2, 22 h

OAc

>85%

Byström, S.E.; Larsson, E.M.; Åkermark, B. *J. Org. Chem.*, ***1990***, *55*, 5674.

O
$Mn(OAc)_3$, PhCOOH
PhH, reflux, 46 h
O
OBz
72%
MeO
MeO

Demir, A.S.; Sayrac, T.; Watt, D.S. *Synthesis*, ***1990***, 1119.

Also via: Section 26 (Acid Derivatives) and Section 41 (Alcohols).

SECTION 117: ESTERS FROM KETONES

$Pb(OAc)_4$, BF_3•OEt_2
MeOH, PhH, 12 h

O
Ph
Ph
CO_2Me
68%

Mathew, F.; Myrbok, B. *Tetrahedron Lett.*, ***1990***, *31*, 3757.

1. Ce, THF, RT
2. acetophenone

Br
CO_2Et
Ph
Me
O
O
70%
+
OH Ph
Ph
Me
Me OH
20%

Fukuzawa, S.; Sumimoto, N.; Fujinami, T.; Sakai, S. *J. Org. Chem.*, ***1990***, *55*, 1628.

Ph
C=O
Me
1. THF, O^- Me
2. H_2O
Me
Ph
O
O
50%

Gong, L.; Leung-Toung, R.; Tidwell, T.T. *J. Org. Chem.*, ***1990***, *55*, 3634.

O
Me
PhI(OH)OTs, MeOH
36 h
CO_2Me
57%

Prakash, O.; Goyal, S.; Moriatry, R.M.; Khosrowshah. J.S. *Ind. J. Chem.*, ***1990***, *29B*, 304.

Acinetobacter calcoaceticus NC1B 9871, tetraethyl pyrophosphate

n-$C_{11}H_{23}$ → n-$C_{11}H_{23}$

25% (74% ee)

+ 65% recovered ketone (36% ee)

Alphand, V.; Archelas, A.; Furstoss, R. *J. Org. Chem.*, ***1990***, *55*, 347.

mCPBA, CH_2Cl_2, 60°C

n-C_6H_{13} SnBu$_3$ → n-$C_6H_{13}CO_2Bn$

t-Bu, *t*-Bu, HO, OH, S, Me, Me

Bakale, R.P.; Scialdone, M.A.; Johnson, C.R. *J. Am. Chem. Soc.*, ***1990***, *112*, 6729.

Me, MeO, MeO, CO_2Et

1. O; $TiCl_4$, MeCN, -45°C
2. H_2O
3. cat. *p*-TsOH, PhH reflux

O, Me, n-C_3H_7, CO_2Et, n-C_3H_7 + O, Me, n-C_3H_7, CO_2Et, n-C_3H_7

(94 : 6) 68%

Saigo, K.; Shimada, S.; Hashimoto, Y.; Nagashima, T.; Hasegawa, M. *Chem. Lett.*, ***1990***, 1101.

1. [O SPh, LDA] THF, -78°C → 0°C
2. aaq. NH_4Cl

92%

Danheiser, R.L.; Nowick, J.S. *J. Org. Chem.*, ***1991***, *56*, 1176.

BnO, O

Acinetobacter NC1B 9871

BnO, O, O + BnO, OH

26% (85% ee)

35% endo (95% ee)
8% exo (89% ee)

Königsberger, K.; Alphand, V.; Furstoss, R.; Griengl, H. *Tetrahedron Lett.*, ***1991***, *32*, 499.

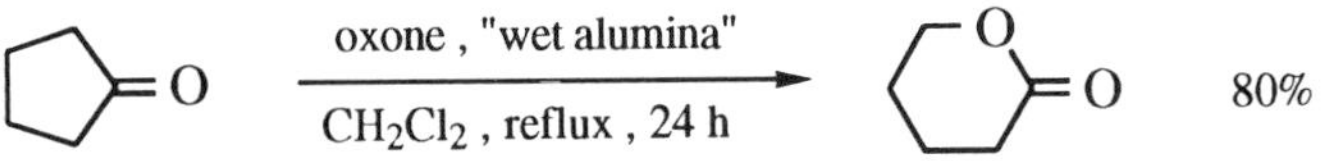

oxone , "wet alumina"; CH_2Cl_2 , reflux , 24 h — 80%

Hirano, M.; Oose, M.; Morimoto, T. *Chem. Lett.*, ***1991***, 331.

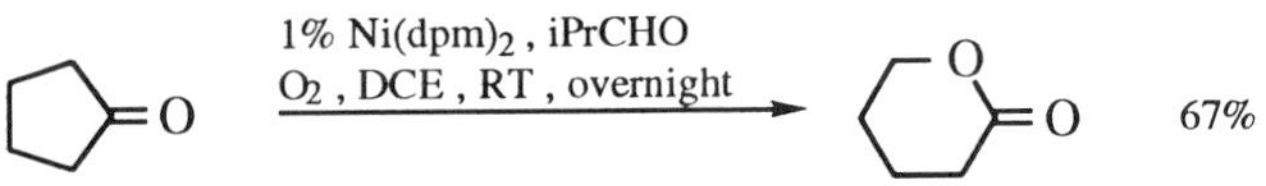

1% $Ni(dpm)_2$, iPrCHO; O_2 , DCE , RT , overnight — 67%

$Ni(dpm)_2$ = bis-(dipivaloylmethanato) nickel (II)

Yamada, T.; Takahashi, K.; Kato, K.; Takai, T.; Inoki, S.; Mukaiyama, T. *Chem. Lett.*, ***1991***, 641.

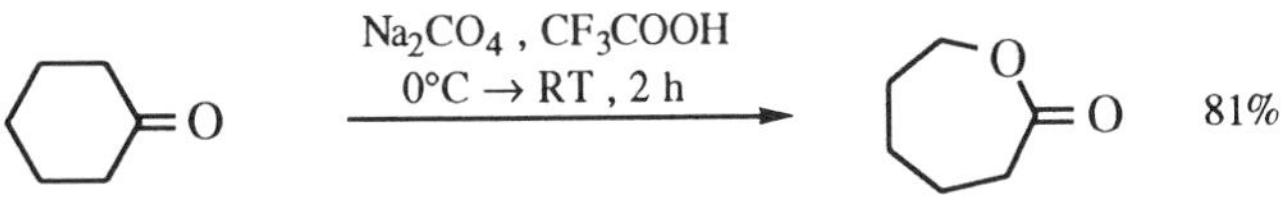

Na_2CO_4 , CF_3COOH; 0°C → RT , 2 h — 81%

Olah, G.A.; Wang, Q.; Trivedi, N.J.; Surya Prakash, G.K. *Synthesis,* ***1991***, 739.

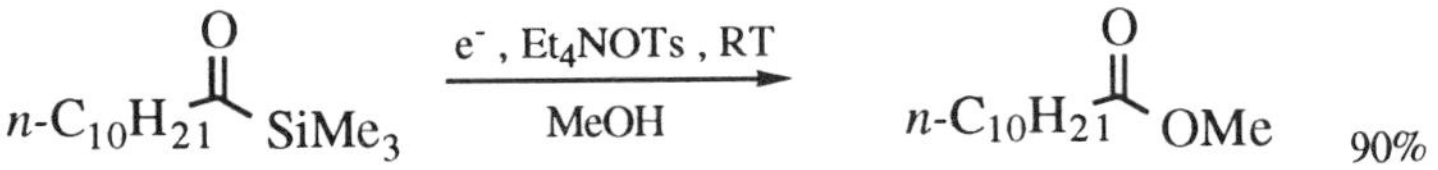

n-$C_{10}H_{21}C(O)SiMe_3$ → n-$C_{10}H_{21}C(O)OMe$; e^- , Et_4NOTs , RT; MeOH — 90%

Yoshida, J.; Itoh, M.; Matsunaga, S.; Isoe, S. *J. Org. Chem.*, ***1992***, *57*, 4877.

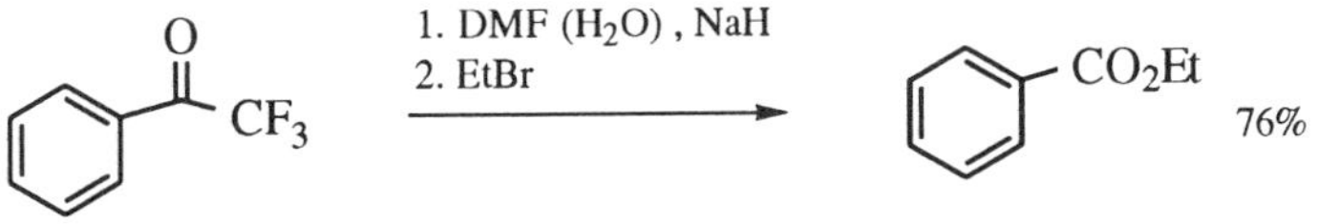

$PhC(O)CF_3$ → $PhCO_2Et$; 1. DMF (H_2O) , NaH; 2. EtBr — 76%

Delgado, A.; Clardy, J. *Tetrahedron Lett.*, ***1992***, *33*, 2789.

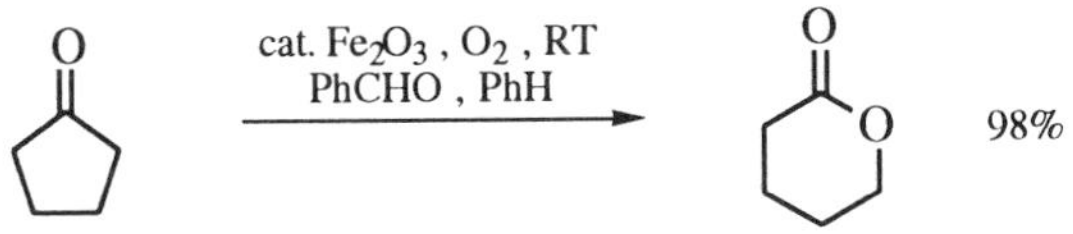

cat. Fe_2O_3 , O_2 , RT; PhCHO , PhH — 98%

Murahashi, S-I.; Oda, Y.; Naota, T. *Tetrahedron Lett.*, ***1992***, *33*, 7557.

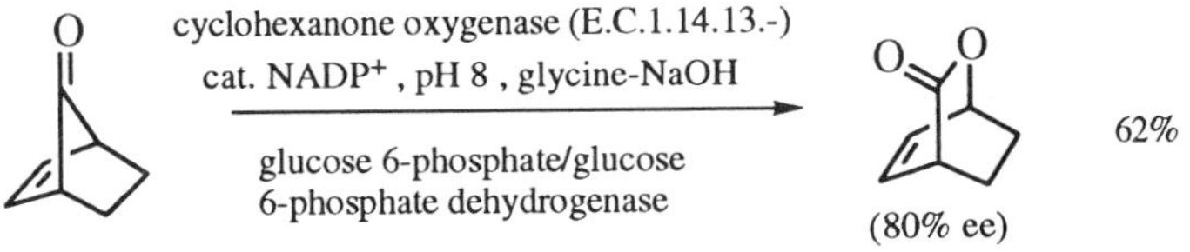

cyclohexanone oxygenase (E.C.1.14.13.-); cat. $NADP^+$, pH 8 , glycine-NaOH; glucose 6-phosphate/glucose 6-phosphate dehydrogenase — (80% ee) 62%

Taschner, M.J.; Peddada, L. *J. Chem. Soc., Chem. Commun.*, ***1992***, 1384.

Also via: Section 27 (Acid Derivatives).

SECTION 118: ESTERS FROM NITRILES

PPE-Me , MeOH , 150°C , 24 h

Ph–CN → Ph–CO_2Me 97%

PPE-Me = polyphosphate methyl ester - see *J. Org. Chem.*, ***1969***, *34*, 2665

Mills, F.D.; Brown, R.T. *Synth. Commun.*, ***1990***, *20*, 3131.

SECTION 119: ESTERS FROM ALKENES

PCC , CH_2Cl_2

28°C

90%

Baskaran, S.; Islam, I.; Chandrasekaran, S. *J. Org. Chem.*, ***1990***, *55*, 891.

Me BzO → BzO Me CO_2Me

1. B_2H_6; 2 eq. $PhCHCO_2^{\ominus\ominus}$

18 h , 0°C → 66°C

2. NaOH , H_2O_2 , 0°C

75%

Hara, S.; Kishimura, K.; Suzuki, A.; Dhillon, R.S. *J. Org. Chem.*, ***1990***, *55*, 6356.

OH SiMe$_3$

1. 2 eq. PhI(OTs)OH CH_2Cl_2 , 0°C

2. 2.2 eq. imidazole

H O H O + H O H OTs

(3 : 1) 87%

Schaumann, E.; Kirshning, A. *J. Chem. Soc., Perkin Trans. I*, ***1990***, 1481.

n-C_9H_{19}

Pd-C , 3 $CuCl_2$

CO (1 atm) , MeOH

25°C , 8 d

Me n-C_9H_{19} CO_2Me + MeO_2C n-C_9H_{19}

(81 : 19) quant.

Inomata, K.; Toda, S.; Kinoshita, H. *Chem. Lett.*, ***1990***, 1567.

OH

1. $Me_2S{\bullet}BH_3$, ether -78°C → 25°C

2. $Na_2Cr_2O_7{\bullet}2\ H_2O$

H_2SO_4 , H_2O , 10°C → reflux

O O

40%

Mandal, A.K.; Mahajan, S.W. *Synthesis*, ***1991***, 311.

Me Me OBn

1. Thex-BH_2 , THF NaCN, TFAA NaOOH
2. mCPBA , CH_2Cl_2

O O OBn H Me H Me 57%

Yokoyama, Y.; Kawashima, H.; Kohno, M.; Ogawa, Y.; Uchida, S. *Tetrahedron Lett.*, ***1991***, *32*, 1479

Bu $SiMe_3$

2% $PdCl_2(PPh_3)_2$, THF
90°C , CO (60 kg/cm^2)
EtOH

CO_2Et Bu $SiMe_3$ + Bu CO_2Et $SiMe_3$

(96 : 4) 70%

with $Co_2(CO)_8$ (0 : 100) 18%

Takeuchi, R.; Ishii, N.; Sato, N. *J. Chem. Soc., Chem. Commun.*, ***1991***, 1247.

Also via: Section 44 (Alcohols).

SECTION 120: ESTERS FROM MISCELLANEOUS COMPOUNDS

SO_2Ph

$NaCH(CO_2Me)_2$
$Mo(CO)_6$, PhMe

CO_2Me CO_2Me + CO_2Me CO_2Me

(90 : 10) 61%

Trost, B.M.; Merlic, C.A. *J. Org. Chem.*, ***1990***, *55*, 1127.

$(CO)_5Cr$ OEt Ph

, acetone , 20°C
3 h

Ph- CO_2Et 97%

Lluch, A-M.; Jordes, L.; Sánchez-Baeza, F.; Ricart, S.; Camps, F.; Messeguer, A.; Moretó, J.M. *Tetrahedron Lett.*, ***1992***, *33*, 3021.

$(CO)_5Cr$ OMe Ph

1. OMe OMe , THF , CO
2. aq, HCl , THF

Ph O O 68%

Wang, S.L.B.; Su, J.; Wulff, W.D. *J. Am. Chem. Soc.*, ***1992***, *114*, 10665.

Me_3Si =C=O H

EtCHO , MABR
-78°C , 2 h

Me_3Si O O Et 82%

MABR = methyl aluminum bis(4-bromo-2,6-di*t*-butylphenoxide)

Maruoka, K.; Concepcion, A.B.; Yamamoto, H. *SynLett*, ***1992***, 31.

CHAPTER 9

PREPARATION OF ETHERS, EPOXIDES AND THIOETHERS

SECTION 121: ETHERS, EPOXIDES AND THIOETHERS FROM ALKYNES

$n\text{-}C_5H_{11}C\equiv C\text{-}n\text{-}C_5H_{11}$ → (1. $TaCl_5$, Zn, DMF/PhH; 2. $n\text{-}C_8H_{17}CHO$, 25°C; 3. 2,6-dimethylphenyl-N≡C, 25°C) → 3,4-di-$n\text{-}C_5H_{11}$-2-$n\text{-}C_8H_{17}$-furan

Takai, K.; Tezuka, M.; Kataoka, Y.; Utimoto, K. *J. Org. Chem.*, ***1990***, *55*, 5310.

SECTION 122: ETHERS, EPOXIDES AND THIOETHERS FROM ACID DERIVATIVES

NO ADDITIONAL EXAMPLES

SECTION 123: ETHERS, EPOXIDES AND THIOETHERS FROM ALCOHOLS AND THIOLS

n-alkyl-OH → (Me_2SiCHN_2, 42% aq. HBF_4; CH_2Cl_2, 0°C) → n-alkyl-OMe 92%

Aoyama, T.; Shiori, T. *Tetrahedron Lett.*, ***1990***, *31*, 5507.

cyclohexane-1,2-diol (OH, OH) → ($CDCl_3$, RT, Tf_2O; $Ph_3P{=}O$, 20 min) → cyclohexene oxide (O) quant.

Hendrickson, J.B.; Hussoin, Md.S. *SynLett*, ***1990***, 423.

n-$C_{13}H_{27}$ (Et)(OH) CH_2SePh —[mCPBA, MeOH / 25°C]→ n-$C_{13}H_{27}$(Et) epoxide 97%

Uemura, S.; Ohe, K.; Sugita, N. *J. Chem. Soc., Perkin Trans. I*, ***1990***, 1697.

H-C≡C-n-C_4H_9 —[HS⌒SH, BPr_3, MeOH / O_2, PhH]→ n-C_4H_9-1,4-dithiane 49%

Demchuk, D.V.; Lutsenko, A.I.; Troyanskii, E.I.; Nikishin, G.I. *Izv. Akad. Nauk. SSSR*, ***1990***, *39*, 2801 (Engl., p. 2542).

2-[Ph(SiPr)CH]phenol (OH) —[=/Ph, 0.15 BF_3•OEt_2 / PhH, RT]→ 2,4-diphenylchroman 64% (90:10 Z:E)

Inoue, T.; Inoue, S.; Sato, K. *Chem. Lett.*, ***1990***, 55.

4-bromo-2-butanol (OH, Br) —[Ph_4SbOMe, DCE, 60°C; sealed ampule, 24 h]→ 2-methyloxetane 74%

Fujiwara, M.; Hitomi, K.; Baba, A. *Synthesis*, ***1990***, 106.

BuSH —[1. e^-, Pt electrodes, MeCN; Et_4BNr / 2.BuBr]→ Bu-S-Bu 86%

Petrosyan, V.A.; Niyazymbetov, M.E.; Konyushkin, L.D.; Litvinov, V.P. *Synthesis*, ***1990***, 841.

HO-C_6H_4-OH —[EtI, 9% TBAB, KOH / neat, 60°C, 24 h]→ EtO-C_6H_4-OEt 90%

Loupy, A.; Sansoulet, J.; Díez-Barra, E.; Carrillo, J.R. *Synth. Commun.*, ***1991***, *21*, 1465.

$(CH_3)_3CCH_2$OH —[Bu_3P-PhSSPh]→ $(CH_3)_3CCH_2$SPh

1 atm, 16 h 90%

10 kbar, 62°C, 3 h quant.

Kotsuki, H.; Matsumoto, K.; Nishizawa, H. *Tetrahedron Lett.*, ***1991***, *32*, 4155.

HO, OH + OH — TFA, RT, 30 min → HO, O 63%

Ismail, F.M.D.; Hilton, M.J.; Štefinović, M. *Tetrahedron Lett.*, **1992**, *33*, 3795.

OH — 1. O·Et, RT, $Hg(OAc)_2$; 2. OH, 2.5% TsOH, CH_2Cl_2, RT → O O 57%

Guy, R.K.; DiPietro, A. *Synth. Commun.*, **1992**, *22*, 687.

2 eq. OH — OMe Br, 140°C; 2 eq. K_2CO_3, 2 h; 5% CuI,)))))) → O OMe (71% ee, R) 63%

Smith, K.; Jones, D. *J. Chem. Soc., Perkin Trans. I*, **1992**, 407.

PhS Ph Me OH — 1.2 eq. $(CF_3CO_2)_2Hg$; CH_2Cl_2, 20°C, 18 h → Ph Me O 80%

Krief, A.; Bousbaa, J.; Hobe, M. *SynLett*, **1992**, 320.

SECTION 124: ETHERS, EPOXIDES AND THIOETHERS FROM ALDEHYDES

PhCHO — BnO, OBn, ⊕S Bn; 50% NaOH, cat. $BnEt_3NHCl$; CH_2Cl_2, 0°C → O Ph Ph 53% (60% ee)

Breau, L.; Ogilvie, W.W.; Durst, T. *Tetrahedron Lett.*, **1990**, *31*, 35.

⊕ Ph_3As OH X ⊖ — PhCHO, MeCN; $KF{\bullet}Al_2O_3$ → Ph O OH 78%

Wang, W-B.; Shi, L-L.; Li, Z-Q.; Huang, Y-Z. *Tetrahedron Lett.*, **1991**, *32*, 3999.

PhCHO → (allyl bromide, iBu_2Te; Cs_2CO_3, THF/ether/H_2O, 50°C, 19 h) → vinyl epoxide (Ph) 75% (71:29 cis:trans)

Zhou, Z-L.; Shi, L-L.; Huang, Y-Z. *Tetrahedron Lett.*, ***1990***, *31*, 7657.

$Ph_2Te^{\oplus}$-CH_3 $BPh_4^{\ominus}$ → (1. LiTMP, THF, -78°C; 2. PhCHO, -78°C → RT) → Ph epoxide 67%

also reacts with ketones

Shi, L-L.; Zhou, Z-L.; Huang, Y-Z. *Tetrahedron Lett.*, ***1990***, *31*, 4173.

crotonaldehyde (CHO) → (1. Ph_3As, THF, -78°C, iPr_3SiOTf; 2. $KNTMS_2$, -78°C; 3. PhCHO, -78°C; 4. TBAF, -78°C → RT) → 3-Me-2-Ph-furan

Kim, S.; Kim, Y.G. *Tetrahedron Lett.*, ***1991***, *32*, 2913.

PHCHO + Ph(C=CH₂)OSiMe₃ → (hν) → oxetane Ph, Ph, $OSiMe_3$ + oxetane $OSiMe_3$, Ph, Ph (>95 : 5) 39%

Bach, T. *Tetrahedron Lett.*, ***1991***, *32*, 7037.

benzotriazole (N, N, N-H) → (1. *n*-C_3H_7CHO, PhSH, H^+; 2. BuLi; 3. MeI; 4. $PhCH_2MgX$) → Me, SPh, Ph 46%

Katritzky, A.R.; Perumal, S.; Kuzmierkiewicz, W.; Lue, P.; Greenhill, J.V. *Helv. Chim. Acta*, ***1991***, *74*, 1924.

Me, Me, CO_2Me, $SiMe_2Ph$ → (BnO⌄CHO, -78°C; CF_3•OEt_2, CH_2Cl_2, 15 h) → tetrahydrofuran Me, $SiMe_2Ph$, CO_2Me, H, O, H, Me 85% (30:1 4,5-syn:anti ; 96% dr)

Panek, J.S.; Yang, M. *J. Am. Chem. Soc.*, ***1991***, *113*, 9868.

PhCHO $\xrightarrow[\text{TMSI, pentane, -78°C} \rightarrow \text{RT}]{t\text{-BuOSiMe}_3\text{, Me}_3\text{SiH}}$ $PhCH_2Ot$-Bu 71%

Hartz, N.; Surya Prakash, G.K.; Olah, G.A. *SynLett*, ***1992***, 569.

PhCHO + SiMe3 / OSiMe3 $\xrightarrow[\text{CCl}_4\text{, 20°C}]{\text{20\% Me}_3\text{SiOTf}}$ Ph, O 85%

Markó, I.E.; Mekhalfia, A. *Tetrahedron Lett.*, ***1992***, *33*, 1799.

SECTION 125: ETHERS, EPOXIDES AND THIOETHERS FROM ALKYLS, METHYLENES AND ARYLS

NO ADDITIONAL EXAMPLES

SECTION 126: ETHERS, EPOXIDES AND THIOETHERS FROM AMIDES

NO ADDITIONAL EXAMPLES

SECTION 127: ETHERS, EPOXIDES AND THIOETHERS FROM AMINES

Cl, NH_2 $\xrightarrow[\text{2. HO}_2\text{CH}_2\text{SH}]{\text{1. NaNO}_2\text{, HCl}}$ Cl, S–CO_2H 35%

Yildirir, Y.; Okay, G. *Org. Prep. Proceed. Int.*, ***1991***, *23*, 198.

SECTION 128: ETHERS, EPOXIDES AND THIOETHERS FROM ESTERS

S, O $\xrightarrow[\text{3. Ph}_3\text{SnH}]{\text{1. MeLi; 2. MeI}}$ Me, O 85%

Nicolaou, K.C.; McGarry, D.G.; Somers, P.K.; Kim, B.H.; Ogilvie, W.W.; Yiannikouros, G.; Prasad, C.V.V.; Veale, C.A.; Hark, R.R. *J. Am. Chem. Soc.*, ***1990***, *112*, 6263.

PhSeSePh , THF , 15 h
2.5 eq. SmI_2 , 5% $PdCl_2$
20% PPh_3 , 25°C

OAc → SePh 84%

Fukuzawa, S.; Fujinami, T.; Sakai, S. *Chem. Lett.*, ***1990***, 927.

5% $Pd(OAc)_2$, THF
20% PPh_3 , 25°C , 1 h

OPh 87%

Larock, R.C.; Lee, N.H. *Tetrahedron Lett.*, ***1991***, *32*, 6315.

2 eq. $NiCl_2$•6 H_2O
2 eq. Zn* , 8 eq. PPh_3
DMF , 23°C , 20 h

SEt → SEt 73%

Wenkert, E.; Chianelli, D. *J. Chem. Soc., Chem. Commun.*, ***1991***, 627.

3 eq. DBH , CH_2Cl_2
$(HF)_9$ / Py

n-$C_{10}H_{21}O$ SMe → n-$C_{10}H_{21}$-O-CF_3 80%

DBH = 1,3-dibromo-5,5-demethyl hydantoin

Juroboshi, M.; Suzuki, K.; Hiyama, T. *Tetrahedron Lett.*, ***1992***, *33*, 4173.

$Pd_2(dba)_3$, dppb , PhOH
THF , 60°C , 12 h

OMe → OPh 75%

Goux, C.; Lhoste, P.; Sinou, D. *SynLett*, ***1992***, 725.
Goux, C.; Lhoste, P.; Sinou, D. *Tetrahedron Lett.*, ***1992***, *33*, 8099 [with thiols].

SECTION 129: ETHERS, EPOXIDES AND THIOETHERS FROM ETHERS, EPOXIDES AND THIOETHERS

1. 5% $Pd(OAc)_2$
3 KOAc , DMF
Bu_4NCl , 80°C
MeO, MeO, I
2. 4% $Pd(OAc)_2$
9% PPh_3 , MeCN
2 Ag_2CO_3 , 80°C
I
3. H_2 , PtO_2 , EtOAc , 100min

OMe, OMe

68% overall

Larock, R.C.; Gong, W.H. *J. Org. Chem.*, ***1990***, *55*, 407.

PhOSiMe$_3$, 1 h
2% CsF , 130°C
PhO OPh
OSiMe$_3$ quant.

Nambu, Y.; Endo, T. *Tetrahedron Lett.*, ***1990***, *31*, 1723.

t-BuLi , TMEDA
Me$_3$SiCl , -90°C
Ph
SiMe$_3$ 96%

Eisch, J.J.; Galle, J.E. *J. Org. Chem.*, ***1990***, *55*, 4835.

OMe
e$^-$, S$_x$, Pt electrodes
OMe
S
OMe
37%

Le Guillanton, G.; Do, Q.T.; Simonet, J. *J. Chem. Soc., Chem. Commun.*, ***1990***, 393.

Br
t-BuOK , 18-crown-6
PhH , reflux , 4 h
O
84%

Dulcère, J-P.; Rodriguez, J. *Synth. Commun.*, ***1990***, *20*, 1893.

iBu$_2$AlSePh , PhMe
EtO OEt
50°C , 2 h
EtO SePh 87%

Nishiyama, Y.; Nakata, S.; Hamanaka, S. *Chem. Lett.*, ***1991***, 1775.

SmI$_2$-MeCN , MeCN
t-BuOH , HMPA
Bn
O Ph
Ph
O
89%

with amines as well as ethers

Inanaga, J.; Ujikawa, O.; Yamaguchi, M. *Tetrahedron Lett.*, ***1991***, *32*, 1737.

O
Zn(BH$_4$)$_2$ on SiO$_2$
THF , -10°C , 8h
Me
OH
Me
80%

Rawal, V.H.; Singh, S.P.; Dufour, C.; Michoud, C. *J. Org. Chem.*, ***1991***, *56*, 5245.

$$\text{Ph-CH(OSiMe}_3\text{)(SEt)} \xrightarrow[\text{InCl}_3\text{ , CH}_2\text{Cl}_2\text{ , RT}]{\text{Et}_3\text{SiH , TMSCl}} \text{PhCH}_2\text{SEt} \quad 83\%$$

Mukaiyama, T.; Ohno, T.; Nishimura, T.; Han, J.S.; Kobayashi, S. *Bull. Chem. Soc. Jpn.*, ***1991***, *64*, 2524.

BnOTMS , Et_3SiH , 0°C
5% $SbCl_5$-$AgSbF_6$, 2 h
83%

Harada, T.; Mukaiyama, T. *Chem. Lett.*, ***1992***, 1901.

SECTION 130: ETHERS, EPOXIDES AND THIOETHERS FROM HALIDES AND SULFONATES

Me_2GaCl , MeCN
90 h
(20 : 80) quant.

Kobayashi, S.; Koide, K.; Ohno, M. *Tetrahedron Lett.*, ***1990***, *31*, 2435.

$$\text{Ph-I} \xrightarrow[\text{250°C , 8 h}]{\text{PhSSPh , PhOPh}} \text{Ph-S-Ph} \quad \text{quant.}$$

Wang, Z.Y.; Hay, A.S. *Tetrahedron Lett.*, ***1990***, *31*, 5685.

$$\text{PhCH}_2\text{Br} \xrightarrow{\text{Ph}_3\text{Sn-Te-SnPh}_3\text{ , 45°C; 4 CsF , CD}_3\text{CN , 24 h}} \text{PhCH}_2\text{TeCH}_2\text{Ph} \quad \text{quant.}$$

Li, C.J.; Harpp, D.N. *Tetrahedron Lett.*, ***1990***, *31*, 6291.

$$\text{PhF} \xrightarrow[\text{140°C , 2 h}]{\text{BnOH , }t\text{-BuOK , TDA-1}} \text{PhOBn} \quad 98\%$$

TDA-1 = see *J. Org. Chem.*, ***1985***, *50*, 3717

Loupy, A.; Philippon, N.; Pigeon, P.; Sansoulet, J.; Galons, H. *Synth. Commun.*, ***1990***, *20*, 2855.

BnS–C(=N-H)–NH_2•HCl —(1. NaOEt; 2. MeI)→ Bn-S-Me 91%

Hengchang, C.; Zhenzhong, L.; Wei, L. *Synth. Commun.*, ***1990***, *20*, 3313.

AgOAc, MeOH; RT, 5 h — N-iPr imine (OMe) → N-iPr imine (OMe) 94% + ketone (OMe) 3%

De Kimpe, N.; Stevens, C. *Tetrahedron*, ***1990***, *46*, 6753.

Cl–C6H4–OTf; $Pd(OAc)_2$•(R)-BINAP; iPr_2NEt, PhH, 22 h — Cl (83 : 17) 86%

Ozawa, F.; Kubo, A.; Hayashi, T. *J. Am. Chem. Soc.*, ***1991***, *113*, 1417.

SMe, Ph, Ph, Br — hv, PhSMe; MeCN → SMe, Ph, Ph, SPh 83%

Kitamura, T.; Kabashima, T.; Taniguchi, H. *J. Org. Chem.*, ***1991***, *56*, 3739.

Se —(1. CO (5 atm), DBU, H_2O, MeCN, 80°C, 1 h; 2. BuCl, 80°C, 1 h)→ Bu-Se-Bu 75%

Nishiyama, Y.; Katsuura, A.; Negoro, A.; Hamanaka, S.; Miyoshi, N.; Yamana, Y.; Ogawa, A.; Sonoda, N. *J. Org. Chem.*, ***1991***, *56*, 3776.

PhBr —(NaOMe, DMF, reflux; 10% CuBr, 30 min)→ PhOMe 95%

Keegstra, M.A.; Peters, T.H.A.; Brandsma, L. *Tetrahedron*, ***1992***, *48*, 3633.

allyl Br —($S(MgBr)_2$, ether)→ diallyl S 86%

Nedugov, A.N.; Pavlova, N.N. *Zhur. Org. Khim.*, ***1992***, *28*, 1401 (Engl., p. 1103).

OBn OBn / OMs — $NaCN$, $EtOH$, H_2O, reflux → BnO (tetrahydrofuran) 78%

Dehmlow, H.; Muler, J.; Seilz, C.; Strecker, A.R.; Kohlmann, A. *Tetrahedron Lett.*, ***1992**, 33,* 3607.

$Br(CH_2)_6Br$ — Al_2O_3, Na_2S, 1.69×10^{-3} M → (thiepane) 79%

Tan, L.C.; Pagni, R.M.; Kabalka, G.W.; Hillmyer, M.; Woosley, J. *Tetrahedron Lett.*, ***1992**, 33*, 7709.

Related Methods: Section 123 (Ethers from Alcohols).

SECTION 131: ETHERS, EPOXIDES AND THIOETHERS FROM HYDRIDES

NO ADDITIONAL EXAMPLES

SECTION 132: ETHERS, EPOXIDES AND THIOETHERS FROM KETONES

O / Me / CO_2Et — 1. I_2, MeOH; 2. PhH, $NaHCO_3$, $Na_2S_2O_3$, aq.NaOH → OMe / Me / CO_2Et 90%

see Tamara Y.; Yoshimoto, Y. *Chem. Ind.*, ***1980***, 888.

Kotnis, A.S. *Tetrahedron Lett.*, ***1990**, 31,* 481.

O — $Me_3S{=}O^+ I^-$, *t*-BuOK, DMSO, RT, 16 h → O 86%

Ng, J.S. *Synth. Commun.*, ***1990**, 20,* 1193.

Ph / Br — $SnBu_3$, 80°C, 2h; *p*-nititrobenzene, Bu_2SnBr_2 - HMPT → Ph / O quant.

Yano, K.; Hatta, Y.; Baba, A.; Matsuda, H. *SynLett*, ***1991***, 555.

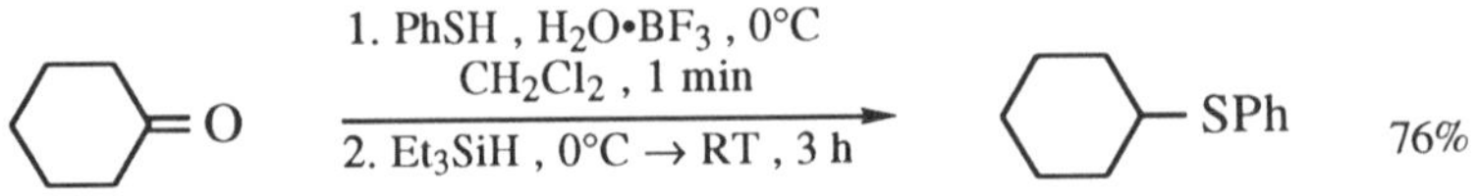

1. PhSH , $H_2O \cdot BF_3$, 0°C
CH_2Cl_2 , 1 min
2. Et_3SiH , 0°C → RT , 3 h

SPh 76%

Olah, G.A.; Wang, Q.; Trivedi, N.J.; Surya Prakash, G.K. *Synthesis*, ***1992***, 465.

Related Methods: Section 124 (Epoxides from Aldehydes).

SECTION 133: ETHERS, EPOXIDES AND THIOETHERS FROM NITRILES

NO ADDITIONAL EXAMPLES

SECTION 134: ETHERS, EPOXIDES AND THIOETHERS FROM ALKENES

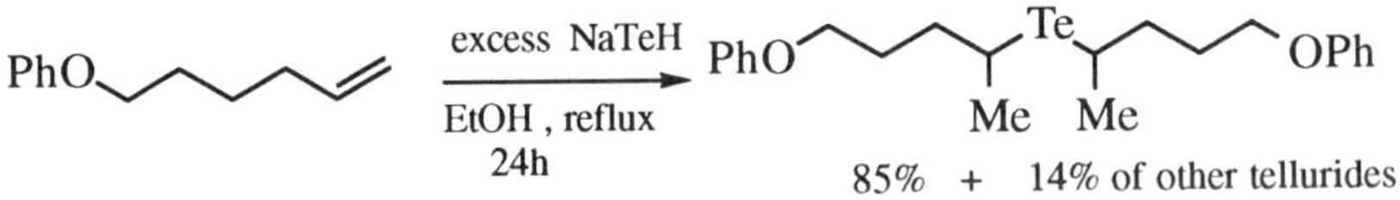

PhO

excess NaTeH
EtOH , reflux
24h

PhO Te OPh
Me Me

85% + 14% of other tellurides

Barton, D.H.R.; Bohé, L.; Lusinchi, X. *Tetrahedron Lett.*, ***1990***, *31*, 93.

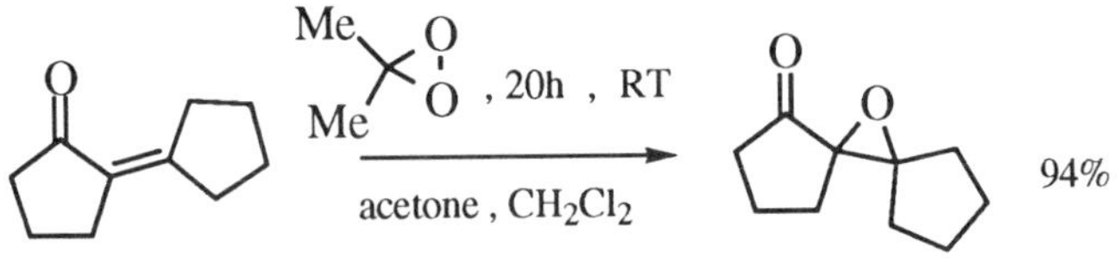

O

Me Me O O , 20h , RT
acetone , CH_2Cl_2

O O

94%

Adam, W.; Hadjiarapoglou, L.; Nestler, B. *Tetrahedron Lett.*, ***1990***, *31*, 331.

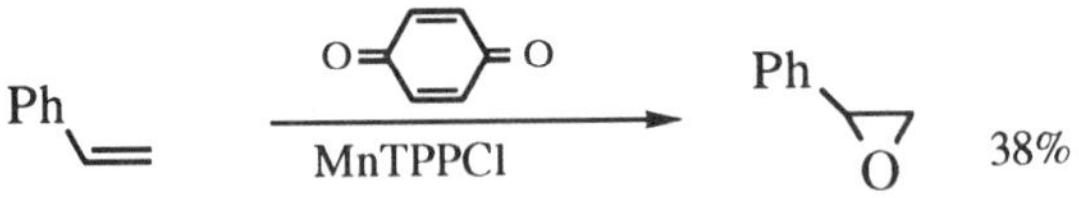

Ph

O O
MnTPPCl

Ph O 38%

TPPCl was undefined; other quinones led to higher yields

Hirao, T.; Ohno, M.; Ohshiro, Y. *Tetrahedron Lett.*, ***1990***, *31*, 6039.

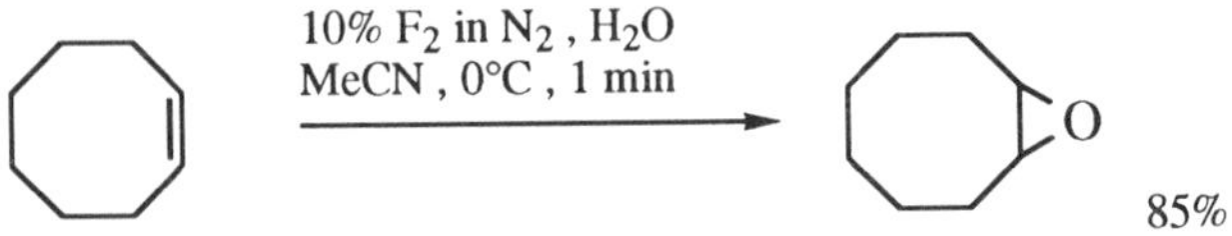

10% F_2 in N_2 , H_2O
MeCN , 0°C , 1 min

O

85%

Rozen, S.; Kol, M. *J. Org. Chem.*, ***1990***, *55*, 5155.

PhSeSePh , hemin-NaBH$_4$, PhH - EtOH , 24 h

Ph-CH=CH$_2$ → PhCH(Me)SPh 47%

Kano, K.; Takeuchi, M.; Hashimoto, S.; Yoshida, Z. *Chem. Lett.*, ***1990***, 1381.

3% VO(acp)$_2$, DCE; 3 atm O_2 , MS 4Å , 75°C → 78%

acp = 2-acetylcyclopentanone

Takai, T.; Yamada, T.; Mukaiyama, T. *Chem. Lett.*, ***1990***, 1657.

NaOCl , 0.05 M , RT; 2% Ni(bpy)$_2$Cl$_2$ → 74%

Yamazaki, S.; Yamazaki, Y. *Bull. Chem. Soc. Jpn.*, ***1991***, *64*, 3185.

aq. NOCl , CH_2Cl_2; 5% Mn (salen) derivative → 88% (84% ee , 1R2S)

Jacobsen, E.N.; Zhang, W.; Muci, A.R.; Ecker, J.R.; Deng, L. *J. Am. Chem. Soc.*, ***1991***, *113*, 7063.

Zhang, W.; Jacobsen, E.N. *J. Org. Chem.*, ***1991***, *56.*, 2296.

Lee, N.H.; Muci, A.R.; Jacobsen, E.N. *Tetrahedron Lett.*, ***1991***, *32*, 5055.

1. $MeOCH_2OSO_2OMe$, -20°C; 2. MeOH → Ph-CH(OMe)-CH$_2$-CH$_2$-OMe 56%

Lebedev, M.Yu.; Balenkova, E.S. *Zhur. Org. Khim.*, ***1991***, *27*, 1388 (Engl., p. 1214).

4.5 eq. $ZnBr_2$, CH_2Br_2; e^- (Zn anode) → 33%

Durandetti, S.; Sibille, S.; Périchon, J. *J. Org. Chem.*, ***1991***, *56*, 3255.

V_2O_5 , 5 *t*-BuOOH; 45°C , 3 h → 45%

Laszlo, P.; Levart, M.; Singh, G.P. *Tetrahedron Lett.*, ***1991***, *32*, 3167.

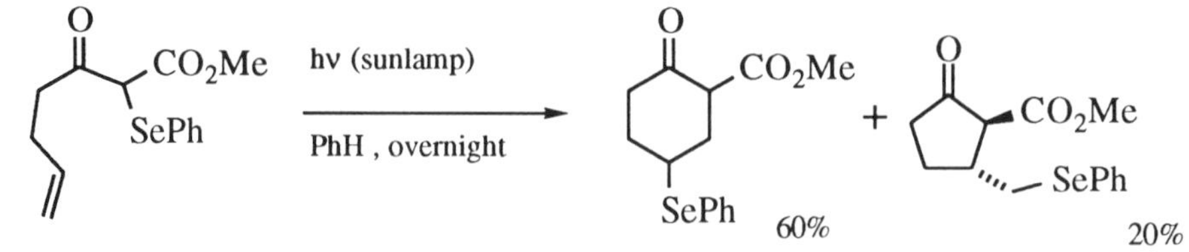
CO2Me
SePh
hν (sunlamp)
PhH , overnight
CO2Me
SePh
60%
+
CO2Me
SePh
20%

Byers, J.H.; Gleason, T.G.; Knight, K.S. *J. Chem. Soc., Chem. Commun.*, ***1991***, 354.

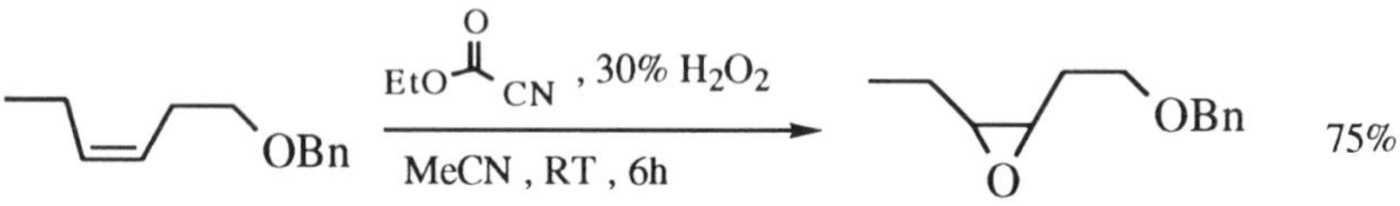
OBn
EtO CN , 30% H2O2
MeCN , RT , 6h
OBn
75%

yields are poor (16%) with 1-decene as a substrate

Masaki, Y.; Miura, T.; Mukai, I.; Itoh, A.; Oda, H. *Chem. Lett.*, ***1991***, 1937.

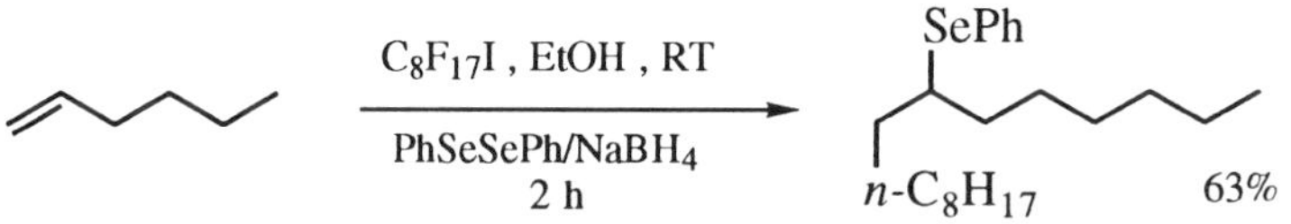
C8F17I , EtOH , RT
PhSeSePh/NaBH4
2 h
SePh
n-C8H17
63%

Uneyama, K.; Kitagawa, K. *Tetrahedron Lett.*, ***1991***, *32*, 375.

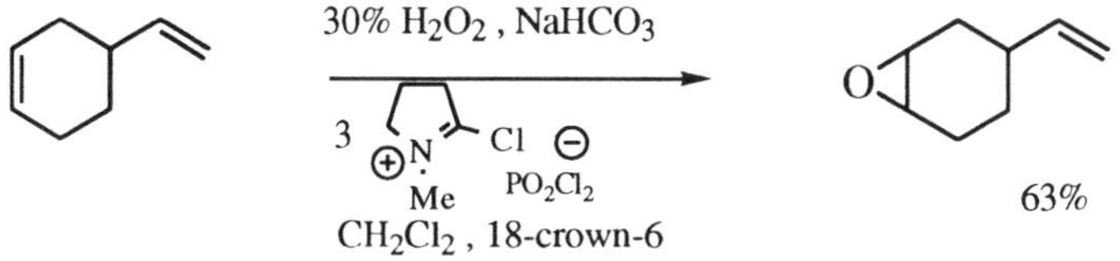
30% H2O2 , NaHCO3
3
Cl
Me
PO2Cl2
CH2Cl2 , 18-crown-6
63%

Rodriguez, T.; Dulcère, J-P. *J. Org. Chem.*, ***1991***, *56*, 469.

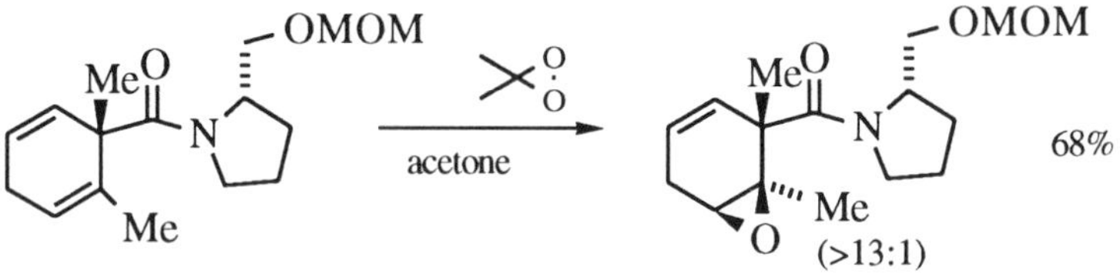
OMOM
Me
Me
acetone
OMOM
Me
Me
(>13:1)
68%

Schultz, A.G.; Harrington, R.E.; Tham, F.S. *Tetrahedron Lett.*, ***1992***, *33*, 6097.

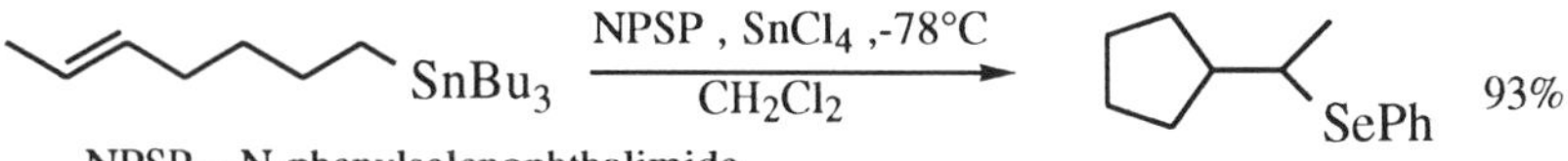
SnBu3
NPSP , SnCl4 ,-78°C
CH2Cl2
SePh
93%

NPSP = N-phenylselenophthalimide

Herndon, J.W.; Harp, J.J. *Tetrahedron Lett.*, ***1992***, *33*, 6243.

Bu$_3$SnCl , NaCNBH$_3$
AIBN , t-BuOH
reflux , 8 h

OSiMe$_2$CH$_2$Br

(1 : 1) 86%

Koreeda, M.; Visger, D.C. *Tetrahedron Lett.*, ***1992***, *33*, 6603.

25 eq. MeCHO , O$_2$
40°C , 6 h

80%

Kaneda, K.; Haruna, S.; Imanaka, T.; Hamamoto, M.; Nishiyama, Y.; Ishii, Y. *Tetrahedron Lett.*, ***1992***, *33*, 6827.

2.5% Ni(dmp)$_2$, O$_2$ (1 atm)
2 eq. iPrCHO , RT

quant

Ni(dmp)$_2$ = [1,3-bis(p-methoxyphenyl)-1,3-propanedionato] nickel (II)

Yamada, T.; Takai, T.; Rhode, O.; Mukaiyama, T. *Bull. Chem. Soc. Jpn.*, ***1991***, *64*, 2109.
Takai, T.; Hata, E.; Yamada, T.; Mukaiyama, T. *Bull. Chem. Soc. Jpn.*, ***1991***, *64*, 2513 [with Fe(dmp)$_3$].
Mukaiyama, T.; Takai, T.; Yamada, T.; Rhode, O. *Chem. Lett.*, ***1990***, 1661.
Yamada, T.; Takai, T.; Rhode, O.; Mukaiyama, T. *Chem. Lett.*, ***1991***, 1.

N-methylimidazole , t-BuCHO
Mn (salen) derivative
PhF, RT , O$_2$

78%

(63% ee , 1S,2R)

Yamada, T.; Imagawa, K.; Nagata, T.; Mukaiyama, T. *Chem. Lett.*, ***1992***, 2231.
Takai, T.; Hata, E.; Yorozu, K.; Mukaiyama, T. *Chem. Lett.*, ***1992***, 2077.

Mn (salen) derivative
PhIO , MeCN , RT

83%

(34% ee , RR)

Hatayama, A.; Hosoya, N.; Irie, R.; Ito, Y.; Katsuki, T. *SynLett*, ***1992***, 407.
Irie, R.; Noda, K.; Ito, Y.; Matsumoto, N.; Katsuki, T. *Tetrahedron Lett.*, ***1990***, *31*, 7345.
Irie, R.; Noda, K.; Ito, Y.; Katsuki, T. *Tetrahedron Lett.*, ***1991***, *32*, 1055.
Irie, R.; Noda, K.; Ito, Y.; Matsumoto, N.; Katsuki, T. *Tetrahedron Asymmetry*, ***1991***, *2*, 481.

PhIO , Mn porphyrin / imidazole

45% (48% ee)

Konishi, K.; Oda, K.; Nishida, K.; Aida, T.; Inoue, S. *J. Am. Chem. Soc.*, ***1992***, *114*, 1313.

2 eq. iPrCHO , O_2 , DCE / 2% Ni (salen) derivative , 10 h

88%

Oda, T.; Irie, R.; Katsuki, T.; Okawa, H. *SynLett*, ***1992***, 641.
Irie, R.; Ito, Y.; Katsuki, T. *Tetrahedron Lett.*, ***1991***, *32*, 6891.

SECTION 135: ETHERS, EPOXIDES AND THIOETHERS FROM MISCELLANEOUS COMPOUNDS

n-$C_{12}H_{25}$-S(O)-n-$C_{12}H_{25}$ → (7:2:1) NaH:t-AmONa:Ni(OAc)$_2$, THF , 16.5 h → n-$C_{12}H_{25}$-S-n-$C_{12}H_{25}$ 67%

Becker, S.; Fort, Y.; Caubère, P. *J. Org. Chem.*, ***1990***, *55*, 6194.

PhCH$_2$S(O)Me → BnBr , 3 eq. DMF , 155°C → PhCH$_2$SMe 70%

Bernard, A.M.; Caredda, M.C.; Piras, P.P.; Serra, E. *Synthesis*, ***1990***, 329.

Bu-S(O)-Bu → $TiCl_4$, NaI , MeOH , RT, 5 min → Bu-S-Bu 92%

Balicki, R. *Synthesis*, ***1991***, 155.

$PhN_2^+ \ BF_4^-$ → 4 eq. Me_3SiOMe , 60°C, Freon 113 , 17 h,)))))))) → PhOMe 51%

Olah, G.A.; Wu, A. *Synthesis*, ***1991***, 204.

Ph-S(O)-Ph → Lawesson's reagent, THF , RT → Ph-S-Ph

Bartsch, H.; Erker, R. *Tetrahedron Lett.*, ***1992***, *33*, 199.

$$\text{Bu-S(=O)-Bu} \xrightarrow[\text{30 min}]{\text{2 eq. TsOH , 2 eq. NaI}} \text{Bu-S-Bu} \quad \text{quant.}$$

Drabowicz, J.; Dudziński, B.; Mokołajczk, M. *SynLett*, ***1992***, 252.

REVIEWS:

"Recent Developments in the Area of Annulated Furans"
Padwa, A.; Murphree, S.S. *Org. Prep. Proceed. Int.*, ***1991***, *23*, 547.

CHAPTER 10

PREPARATION OF HALIDES AND SULFONATES

SECTION 136: HALIDES AND SULFONATES FROM ALKYNES

NO ADDITIONAL EXAMPLES

SECTION 137: HALIDES AND SULFONATES FROM ACID DERIVATIVES

NO ADDITIONAL EXAMPLES

SECTION 138: HALIDES AND SULFONATES FROM ALCOHOLS AND THIOLS

OH → PPh_3 , imidazole , I_2 / CH_2Cl_2 , RT → I 60-98%

Lange, G.L.; Gottardo, C. *Synth. Commun.*, ***1990***, *20*, 1473.

OH → $CFCl_3$, -78°C → 20°C ; m ; $\oplus$N H $F(HF)_{n-1}^{\ominus}$ → F 67%

Olah, G.A.; Li, X-Y. *SynLett*, ***1990***, 267.

HO–(C6H4)–OMe (m-) $\xrightarrow[H_2O,\ 160°C]{[PhPCl_2/Cl_2,\ 80°C]}$ Cl–(C6H4)–OMe (m-) 80%

Bay, E.; Bak, D.A.; Timony, P.E.; Leone-Bay, A. *J. Org. Chem.*, ***1990***, *55*, 3415.

allyl–OH $\xrightarrow{P_2I_4,\ 10^{-2}\ torr,\ 85°C}$ allyl–I 86%

Lasne, M-C.; Cairon, P.; Villemin, D. *Synth. Commun.*, ***1990***, *20*, 41.

Ph(Me)2C–OTHP $\xrightarrow{LiBr•BF_3,\ ether}$ Ph(Me)2C–Br 53%

also with NaI•BF_3 to give iodides
and with 1° and 2° OTHP derivatives

Vankar, Y.D.; Shah, K. *Tetrahedron Lett.*, ***1991***, *32*, 1081.

$n\text{-}C_8H_{17}OH \xrightarrow[7\ h]{(BrCH_2CH_2)_2SeBr_2,\ 80°C} n\text{-}C_8H_{17}Br$ 98%

Akabori, S.; Takanohashi, Y. *Bull. Chem. Soc. Jpn.*, ***1991***, *64*, 3482.

HOCH2–CH(OH)–CH(OH)–CH(OH)–CH2OH $\xrightarrow{3\ eq.\ Me_2N{=}CCl_2^{\oplus}\ Cl^{\ominus},\ dioxane,\ 40°C,\ 90\ min}$ ClCH2–CH(OH)–CH(OH)–CH(OH)–CH2Cl 73%

Benazza, M.; Uzan, R.; Beaupère, D.; Demailly, G. *Tetrahedron Lett.*, ***1992***, *33*, 3129, 4901.

$n\text{-}C_5H_{11}$–CH=CH–CH2CH2–OH $\xrightarrow{Ph_3P•Br_2,\ MeCN,\ Py,\ 0°C - RT}$ $n\text{-}C_5H_{11}$–CH=CH–CH2CH2–Br 96%

Sandri, J.; Viala, J. *Synth. Commun.*, ***1992***, *22*, 2945.

SECTION 139: HALIDES AND SULFONATES FROM ALDEHYDES

$n\text{-}C_6H_{13}CHO \xrightarrow[-15°C \rightarrow 0°C]{"(PhO)_3PBr_2",\ CH_2Cl_2} n\text{-}C_6H_{13}CHBr_2$ 70%

Hoffmann, R.W.; Bovicelli, P. *Synthesis*, ***1990***, 657.

CHO → 1. Tf_2O 2. $Bu_4^+ Ph_3SnF_2^-$ → F, F

68%

Martínez, A.G.; Barcina, J.O.; Rys, A.Z.; Subramanian, L.R. *Tetrahedron Lett.*, ***1992***, *33*, 7787.

SECTION 140: HALIDES AND SULFONATES FROM ALKYLS, METHYLENES AND ARYLS

For the conversion R-H → R-Halogen, see Section 146 (Halides from Hydrides).

NO ADDITIONAL EXAMPLES

SECTION 141: HALIDES AND SULFONATES FROM AMIDES

NO ADDITIONAL EXAMPLES

SECTION 142: HALIDES AND SULFONATES FROM AMINES

O_2N, N, S, NH_2 → *t*-BuS-N=O , MeCN 25°C , 5 h → O_2N, N, S, F

quant.

Kim, Y.H.; Lee, C.H.; Chang, K.Y. *Tetrahedron Lett.*, ***1990***, *31*, 3019.

NH_2 → 1. ONO , CH_2I_2 RT → reflux 2. MeCN , N-H → I

70%

Smith, W.B.; Ho, O-C. *J. Org. Chem.*, ***1990***, *55*, 2543.

NH_2, Bu → 1. HCl , $NaNO_2$ 2. PhSH 3. 6 eq. $AgNO_3$, PhMe HF-Py , 90°C → F, Bu + Bu

39% 54%

Haroutounian, S.A.; Di Zio, J.P.; Katzenellenbogen, J.A. *J. Org. Chem.*, ***1991***, *56*, 4993.

N-NH$_2$ / Ph–C–Ph → (NBS, Py/(HF)$_n$) → Ph_2CF_2 96%

Surya Prakash, G.K.; Reddy, V.P.; Li, X-Y.; Olah, G.A. *SynLett*, ***1990***, 594.

PhNH$_2$ → (Me_3SiCl, $NaNO_2$, $BnEt_3NCl$, CCl_4, 10 h) → PhCl 86%

Lee, J.G.; Cha, H.T. *Tetrahedron Lett.*, ***1992***, *33*, 3167.

SECTION 143: HALIDES AND SULFONATES FROM ESTERS

1-naphthyl–C(=S)SMe → (4 eq. DBH, $Bu_4N^+ H_2F_3^-$) → 1-naphthyl–CF_3 63%

DBH = 1,3-dibromo-5,5-dimethyl hydantoin

Kuroboshi, M.; Hiyama, T. *Chem. Lett.*, ***1992***, 827.

SECTION 144: HALIDES AND SULFONATES FROM ETHERS, EPOXIDES AND THIOETHERS

NO ADDITIONAL EXAMPLES

SECTION 145: HALIDES AND SULFONATES FROM HALIDES AND SULFONATES

n-octyl–Br → (HI, 130°C, 5 h) → n-octyl–I 90%

Namavari, M.; Satyamurthy, N.; Phelps, M.E.; Barrio, J.R. *Tetrahedron Lett.*, ***1990***, *31*, 4973.

BrMg–C$_6$H$_4$–Me + N-[^{18}F]fluoro-2-pyrrolidinone → (freon, 0°C) → ^{18}F–C$_6$H$_4$–Me 30%

Satyamurthy, N.; Bida, G.T.; Phelps, M.E.; Barrio, J.R. *J. Org. Chem.*, ***1990***, *55*, 3373.

Br
Me

1. Mg, ether
2. CS_2
3. XeF_2, CH_2Cl_2
0°C → RT (72 h)

CF_3
Me
73%

+

O F
Me
21%

Zupan, M.; Bregar, Z. *Tetrahedron Lett.*, ***1990***, *31*, 3357.

Me
Br Br
Me

1. Mg
2. Cp_2ZrCl_2
3. Br_2

Me
Br Br
Me

+

Me
Br Br
Me

(99 : 1)

Takahashi, T.; Fujimori, T.; Seki, T.; Saburi, M.; Uchida, Y.; Rousset, C.J.; Negishi, E. *J. Chem. Soc., Chem. Commun.*, ***1990***, 182.

TfO

MgI_2, CS_2, 48°C, 5 h
))))))))

I

75%

Martínez, A.G.; Vilar, E.T.; López, J.C.; Alonso, J.M.; Hanack, M.; Subramanian, L.R. *Synthesis*, ***1991***, 353.

MeO–C_6H_4–I

CF_3SiEt_3, KF
CuI, DMF/NMP
80°C, 24 h

MeO–C_6H_4–CF_3

82%

Urata, H.; Fuchikami, T. *Tetrahedron Lett.*, ***1991***, *32*, 91.

Br

NaOCl (pH 9), $CHCl_3$
6% Ni TPP, $BnBu_3NBr$

Cl

quant.

TPP = 5,10,15,20-tetraphenyl porphyrin

O'Connor, K.J.; Burrows, C.J. *J. Org. Chem.*, ***1991***, *56*, 1344.

Ph Br

CCF_2CO_2Me, KF
CuI, 120°C, 8 h

Ph CF_3

84%

Su, D-B.; Duan, J-X.; Chen, Q-Y. *Tetrahedron Lett.*, ***1991***, *32*, 7689.

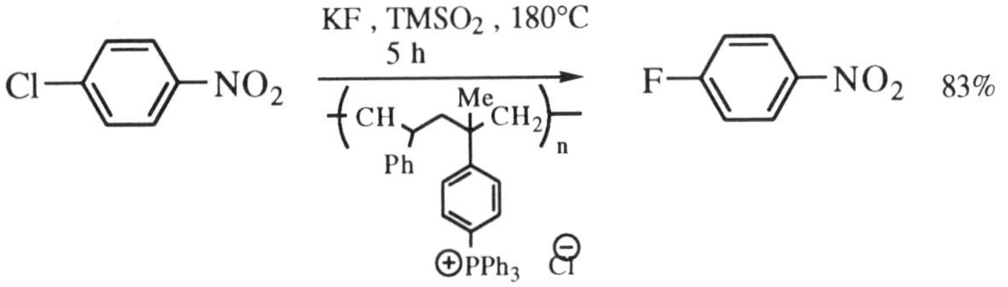

KF , $TMSO_2$, 180°C, 5 h

Yoshida, Y.; Kimura, Y.; Tomoi, M. *Chem. Lett.*, ***1990***, 769.

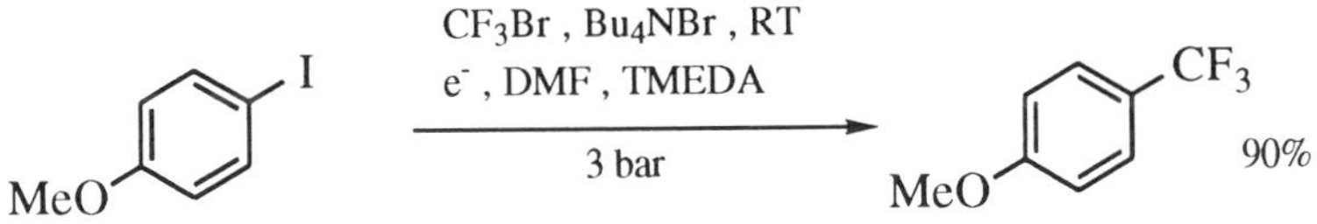

CF_3Br , Bu_4NBr , RT, e^- , DMF , TMEDA, 3 bar

90%

Paratian, J.M.; Sibille, S.; Périchon, J. *J. Chem. Soc., Chem. Commun.*, ***1992***, 53.

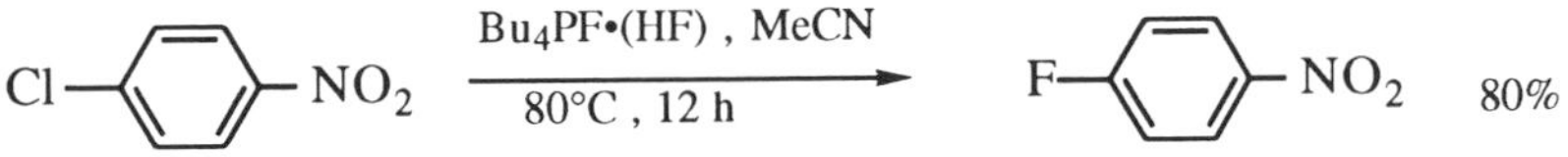

$Bu_4PF{\bullet}(HF)$, MeCN, 80°C , 12 h

80%

Uchibori, Y.; Umeno, M.; Seto, H.; Qian, Z.; Yoskioka, H. *SynLett*, ***1992***, 345.

SECTION 146: HALIDES AND SULFONATES FROM HYDRIDES

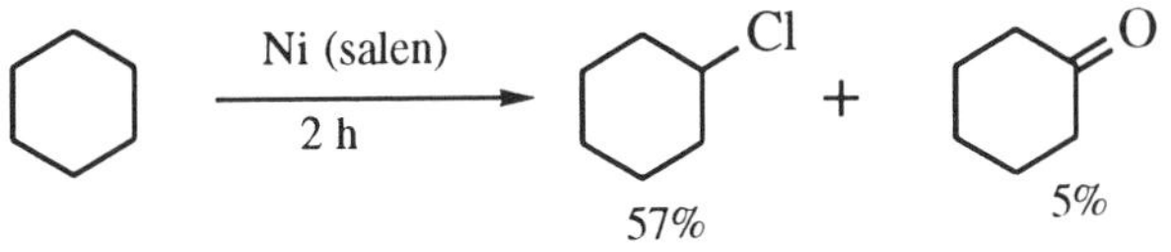

Ni (salen), 2 h

57% + 5%

Querci, C.; Strologo, S.; Ricci, M. *Tetrahedron Lett.*, ***1990***, *31*, 6577.

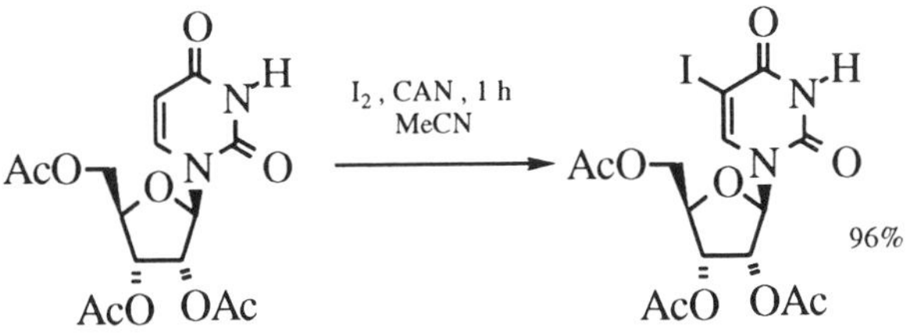

I_2 , CAN , 1 h, MeCN

96%

Asakura, J.; Robins, M.J. *J. Org. Chem.*, ***1990***, *55*, 4928.

$BnMe_3NBr$, RT, CH_2Cl_2 - MeOH, 2 h

98%

Kajigaeshi, S.; Moriwaki, M.; Tanaka, T.; Fujisaki, S.; Kakinami, T.; Okamoto, T. *J. Chem. Soc., Perkin Trans. I*, ***1990***, 897.

HCl-mCPBA , DMF
RT , 20 min

OH Cl Cl Cl

30% 68% 2%

Chung, K.H.; Kim, H.J.; Kim, H.R.; Ryu, E.K. *Synth. Commun.*, ***1990***, *20*, 2991.

I_2, Ag_2SO_4 , EtOH
RT , 15 min

NMe_2 I NMe_2 84%

Sy, W-W. *Synth. Commun.*, ***1992***, *22*, 3215.
Sy, W-W.; Lodge, B.A.; By, A.W. *Synth. Commun.*, ***1990***, 877.

CuBr , TFA , MeCN
Et_4NOTs , e^-
C electrodes , 20 F/mol

O Br 80%

Mitani, M.; Kobayashi, T.; Koyama, K. *J. Chem. Soc., Chem. Commun.*, ***1991***, 1418.

I_2 , $CuCl_2$, Al_2O_3
120°C , 14 h

Me Me I + Me I

75%

Kodomari, M.; Amanokura, N.; Takeuchi, K.; Yoshitomi, S. *Bull. Chem. Soc. Jpn.*, ***1992***, *65*, 306.

Amberlyst-A26•Br_3 , 24 h
CH_2Cl_2 (dark)

OH Br OH + OH Br

86% 7%

Smith, K.; James, D.M.; Matthews, I.; Bye, M.R. *J. Chem. Soc., Perkin Trans. I*, ***1992***, 1877.

IPy_2BF_4 , 2 HBF_4
CH_2Cl_2 , RT

I 48%

Barluenga, J.; González, J.M.; García-Martín, M.A.; Campos, P.J.; Asensio, G. *J. Chem. Soc., Chem. Commun.*, ***1992***, 1016.

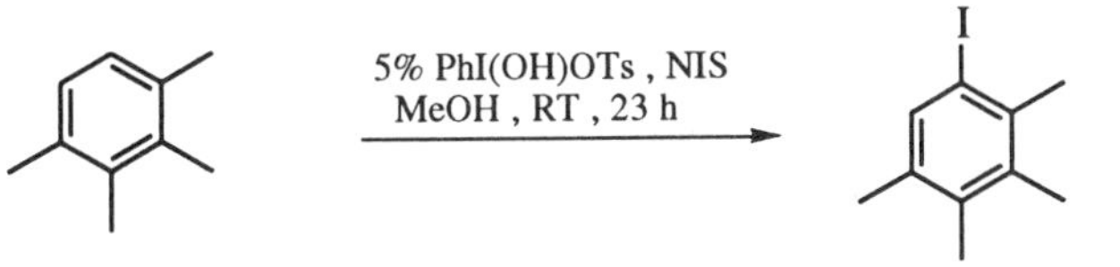
5% PhI(OH)OTs , NIS
MeOH , RT , 23 h
I

84% conversion

Bovonsombat, P.; Angara, G.J.; McNelis, E. *SynLett*, ***1992***, 131.

α-Halogenations of aldehydes, ketones and acids are found in Sections 338 (Halide-Aldehyde), 369 (Halide-Ketone), 359 (Halide-Esters) and 319 (Halide-Acids).

SECTION 147: HALIDES AND SULFONATES FROM KETONES

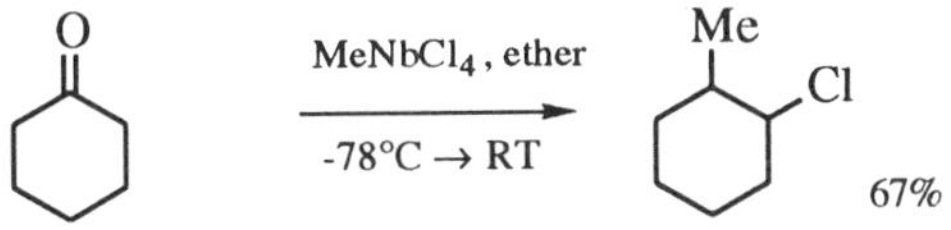
O
$MeNbCl_4$, ether
-78°C → RT
Me
Cl
67%

also with aldehydes

also with Me_2NbCl_3

Kauffmann, T.; Abel, T.; Neiteler, G.; Schreer, M. *Tetrahedron Lett.*, ***1990***, *31*, 493.

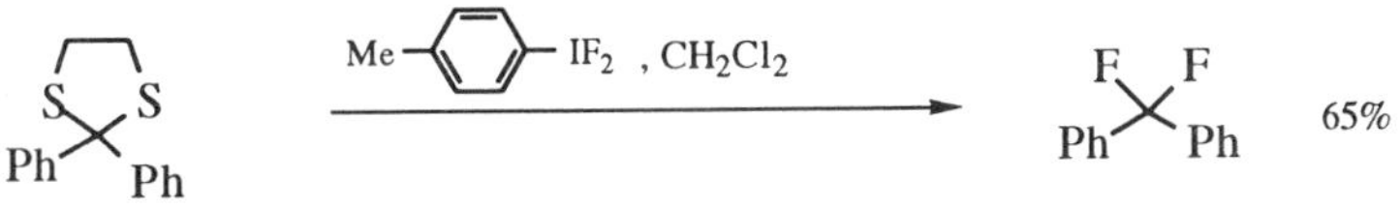
S
S
Ph
Ph
Me
IF_2 , CH_2Cl_2
F
F
Ph
Ph
65%

Motherwell, W.B.; Wilkinson, J.A. *SynLett*, ***1991***, 191.

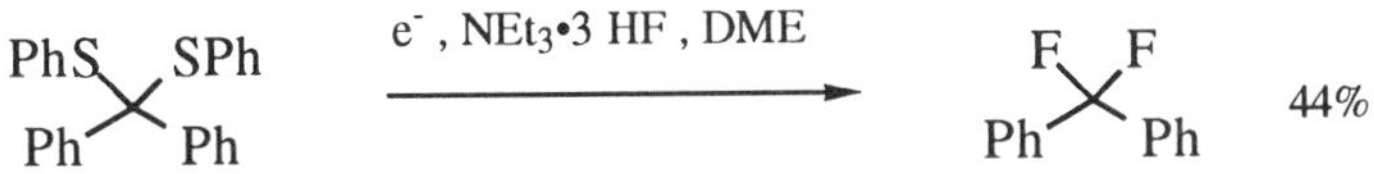
PhS
SPh
Ph
Ph
e^- , NEt_3•3 HF , DME
F
F
Ph
Ph
44%

Yoshiyama, T.; Fuchigami, T. *Chem. Lett.*, ***1992***, 1995.

SECTION 148: HALIDES AND SULFONATES FROM NITRILES

NO ADDITIONAL EXAMPLES

SECTION 149: HALIDES AND SULFONATES FROM ALKENES

For halocyclopropanations, see Section 74E (Alkyls from Alkenes).

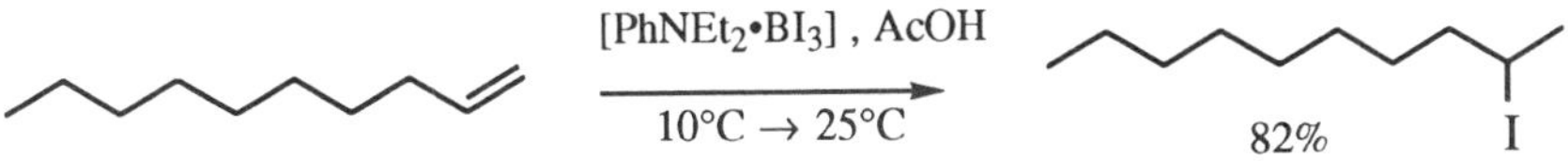
[$PhNEt_2$•BI_3] , AcOH
10°C → 25°C
82%
I

a new HI surrogate

Reddy, Ch.K.; Periasamy, M. *Tetrahedron Lett.*, ***1990***, *31*, 1919.

$PhMe_3NBr \cdot Br_2$, trace Py

$\Delta^{14,14}$ 70%
$\Delta^{3,4}$ 20%
$\Delta^{4,5}$ 5%

Collado, I.G.; Madero, J.G.; Massanet, G.M.; Luis, F.R. *Tetrahedron Lett.*, ***1990***, *31*, 563.

n-C_6H_{13} → $BnEt_3N\ ICl_2$, RT, 2 h, CH_2Cl_2 → 82%

Kajigaeshi, S.; Moriwaki, M.; Fujisaki, S.; Kakinami, T.; Okamoto, T. *Bull. Chem. Soc. Jpn.*, ***1990***, *63*, 3033.

n-C_5H_{11} → CH_2Cl_2 → n-C_5H_{11} (Br) + n-C_5H_{11} Br

HBr, 20 min	11%	83%
HBr, SiO_2, 0.7 h	96%	0%
$(COBr)_2$, Al_2O_3, 20 min	99%	0%

Kropp, P.J.; Daus, K.A.; Crawford, S.D.; Tubergen, M.W.; Kepler, K.D.; Craig, S.L.; Wilson, V.P. *J. Am. Chem. Soc.*, ***1990***, *112*, 7433.

MnO_2, Me_3SiCl, THF, 30°C, 6h → 84%

Bellesia, F.; Ghelfi, F.; Pagnoni, U.M.; Pinetti, A. *Synth. Commun.*, ***1991***, *21*, 489.

$Bu_4NBr \cdot Br_2$, $CHCl_3$

700 min	78%
))))))), 120 min	98%

Berthelot, J.; Benammar, Y.; Lange, C. *Tetrahedron Lett.*, ***1991***, *32*, 4135.

CCl_4, DCE, $Cu(iPrNH_2)_2Cl$, 72°C, microwaves (600 W) → Ph–CHCl–CH_2CCl_3 90% conversion

Adámek, F.; Hájek, M. *Tetrahedron Lett.*, ***1992***, *33*, 2039.

SECTION 150: HALIDES AND SULFONATES FROM MISCELLANEOUS COMPOUNDS

NO ADDITIONAL EXAMPLES

CHAPTER 11

PREPARATION OF HYDRIDES

This chapter lists hydrogenolysis and related reactions by which functional groups are replaced by hydrogen: e.g. $RCH_2X \rightarrow RCH_2$-H or R-H.

SECTION 151: HYDRIDES FROM ALKYNES

NO ADDITIONAL EXAMPLES

SECTION 152: HYDRIDES FROM ACID DERIVATIVES

This section lists examples of decarboxylations ($RCO_2H \rightarrow$ R-H) and related reactions.

CO_2Me 1. 2 H_3BO_3 , 220°C, 1 h; 2. MeOH , HCl ; HCl

N H O — N H O 50%

Delbecq, P.; Celerier, J-P.; Lhommet, G. *Tetrahedron Lett.*, ***1990***, *31*, 4873.

O Cl — $(Me_3Si)_3SiH$, AIBN, PhMe , 80°C , 10 h — 92%

Ballestri, M.; Chatgilialoglu, C.; Cardi, N.; Sommazzi, A. *Tetrahedron Lett.*, ***1992***, *33*, 1787.

SECTION 153: HYDRIDES FROM ALCOHOLS AND THIOLS

This section lists examples of the hydrogenolysis of alcohols and phenols (ROH $\rightarrow$ R-H).

OH — 5.5% β-cyclodextrin, 0.4 N KOH , KCN/KCl, $CoCl_2$•6 H_2O — 91%

Lee, J-T.; Alper, H. *Tetrahedron Lett.*, ***1990***, *31*, 4101.

6 Li°, 10% DBB, 48 h,))))))))) — 72%

DBB = 4,4'-di-*t*-butylbiphenyl

Karaman, R.; Kohlman, D.T.; Fry, J.L. *Tetrahedron Lett.*, ***1990***, *31*, 6155.

$WCl_2(PMePh_2)_4$, <1 min — 93% + 7%

Jang, S.; Atagi, L.M.; Mayer, J.M. *J. Am. Chem. Soc.*, ***1990***, *112*, 6413.

Et_3SiH, CH_2Cl_2, $TMSCl{\bullet}InCl_3$, RT — 83%

Mukaiyama, T.; Ohno, T.; Nishimura, T.; Han, J.S.; Kobayashi, S. *Chem. Lett.*, ***1990***, 2239.

$B(iPr)_3$, CF_3SO_3H, ClF_2CCCl_2F, -30°C → RT — 80%

Olah, G.A.; Wu, A. *Synthesis*, ***1991***, 407.

$NaBH_4$, ether, TFA, 25°C, 1 h

Nutaitis, C.F.; Patragnoni, R.; Goodkin,G.; Neighbour, B.; Obaza-Nutaitis, J. *Org. Prep. Proceed. Int.*, ***1991***, *23*, 403.
Nutaitis, C.F.; Bernardo, J.E. *Synth. Commun.*, ***1990***, *20*, 487.

5% SmI_2, HMPA, ethylene glycol, RT, 3 h — 90%

Hanessian, S.; Girard, C.; Chiara, J.L. *Tetrahedron Lett.*, ***1992***, *33*, 573.

$CH_3(CH_2)_{16}CH_2OH \xrightarrow[\text{[PhCO]}_2\text{O , reflux}]{\text{MeO}\cdot\text{P(O)H / MeO , dioxane , 3 h}} CH_3(CH_2)_{16}CH_3$ 93%

Barton, D.H.R.; Jang, D.O.; Jaszberenyi, J.Cs. *Tetrahedron Lett.*, ***1992***, *33*, 2311.

PhH , TsOH
hν (λ = 254 nm)
75%

Piva, O. *Tetrahedron Lett.*, ***1992***, *33*, 2459.

1. TsCl , NaH , THF
2. $NaBH_4$, $NiCl_2$•6 H_2O
MeOH
92%

Wang, F.; Chiba, K.; Tada, M. *J. Chem. Soc., Perkin Trans. I*, ***1992***, 1897.

3 eq. $[VCl_3(THF)_6]_2$ $[Zn_2Cl_6]$
CH_2Cl_2
59%

using acetate as starting material led to 91% yield of ketone

Inokuchi, T.; Kawafuchi, H.; Torii, S. *Chem. Lett.*, ***1992***, 1895.

Also via: Section 160 (Halides and Sulfonates).

SECTION 154: HYDRIDES FROM ALDEHYDES

For the conversion RCHO → R-Me, etc., see Section 64 (Alkyls from Aldehydes).

0.05 eq. Rh(S-BINAP) ClO_4
CH_2Cl_2 , RT
87%
(>99% ee , R)

Wu, X-M.; Funakoshi, K.; Sakai, K. *Tetrahedron Lett.*, ***1992***, *33*, 6331.

SECTION 155: HYDRIDES FROM ALKYLS, METHYLENES AND ARYLS

NO ADDITIONAL EXAMPLES

SECTION 156: HYDRIDES FROM AMIDES

NO ADDITIONAL EXAMPLES

SECTION 157: HYDRIDES FROM AMINES

This section lists examples of the conversion RNH_2 (or R_2NH) $\rightarrow$ R-H.

1. $LiAlH_4$, RT
2. H_3O^+

59%

Leone, C.L.; Chamberlin, A.R. *Tetrahedron Lett.*, ***1991***, *32*, 1691.

1. KOH
2. 2 eq. NH_2Cl

MeO NHTs → MeO 56%

Guziec, F.S. Jr.; Wei, D. *J. Org. Chem.*, ***1992***, *57*, 3772.

HCO_2NH_4, Pd-C, MeOH
RT, 2 h

98%

Balicki, R. *Gazz. Chim. Ital.*, ***1992***, *122*, 133.

SECTION 158: HYDRIDES FROM ESTERS

This section lists examples of the reactions $RCO_2R' \rightarrow$ R-H and $RO_2CR' \rightarrow$ RH.

BEt_3, $PhSiH_2$
PhH, hexane
air, 80°C, 10 min

80%

Barton, D.H.R.; Jang, D.O.; Jaszberenyi, J.C. *Tetrahedron Lett.*, ***1990***, *31*, 4681.

2% t-$C_{12}H_{25}SH$
2% dicumyl peroxide
Et_3SiH
SMe
O
S
92%

Kirwan, J.N.; Roberts, B.P.; Willis, C.R. *Tetrahedron Lett.*, ***1990***, *31*, 5093.

1. 1M NaOH , EtOH RT
2. DMSO:H_2O reflux , 30 min
NO_2 Ph CO_2Et → NO_2 Ph H 79% + H Ph NO_2 11%

O'Bannon, P.E.; Dailey, W.P. *J. Org. Chem.*, ***1990***, *55*, 353.

$RhCl(PPh_3)_3$, Et_3SiH
PhH , reflux
Ph O O → Ph Me 90%

Liu, H-J.; Zhu, B-Y. *Synth. Commun.*, ***1990***, *20*, 557.

$PhSiH_3$, AIBN , 1 h
PhMe , reflux
O SMe S → quant.

Barton, D.H.R.; Jang, D.O.; Jaszberenyi, J.Cs. *SynLett*, ***1991***, 435.

1. MeLi , ether 0°C → RT
2. OH CO_2me -100°C → RT
O O Ph → O Ph 82% (82% ee , R)

Matsumoto, K.; Ohta, H. *Tetrahedron Lett.*, ***1991***, *32*, 4729.

SECTION 159: HYDRIDES FROM ETHERS, EPOXIDES AND THIOETHERS

This section lists examples of the reaction R-O-R' → R-H.

1. Et_3SiH , $SnBr_2$-AcBr CH_2Cl_2 , 0°C
2. Bu_3SnH , AIBN PhH , reflux
MeO OMe → Me 85%

Oriyama, T.; Ichimura, Y.; Koga, G. *Bull. Chem. Soc. Jpn.*, ***1991***, *64*, 2581.

REVIEWS:

"Transition Metal Mediated C-S Bond Cleavage Reactions"
Luh, T-Y.; Ni, Z-J. *Synthesis*, ***1990***, 89.

SECTION 160: HYDRIDES FROM HALIDES AND SULFONATES

This section lists the reductions of halides and sulfonates, R-X → R-H.

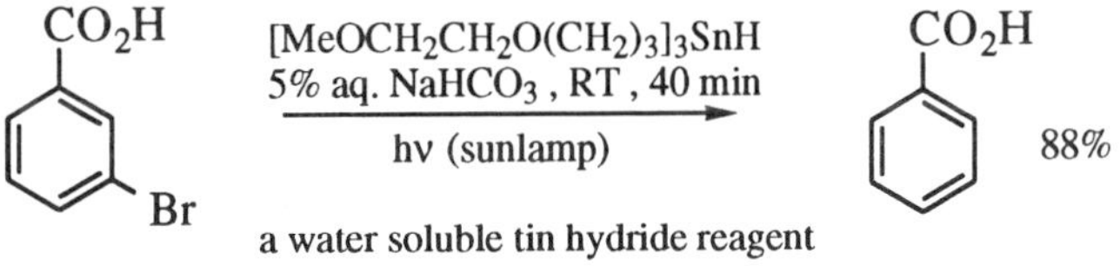

$[MeOCH_2CH_2O(CH_2)_3]_3SnH$, 5% aq. $NaHCO_3$, RT, 40 min; hν (sunlamp) — 88%

a water soluble tin hydride reagent

Light, J.; Breslow, R. *Tetrahedron Lett.*, ***1990***, *31*, 2957.

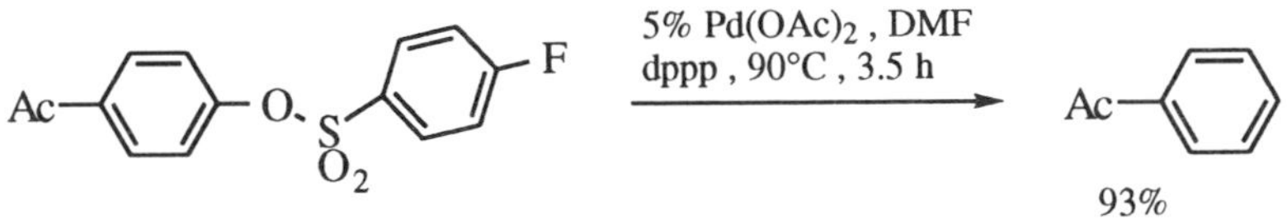

5% $Pd(OAc)_2$, DMF, dppp, 90°C, 3.5 h — 93%

Cabri, W.; De Bernardinis, S.; Francalanci, F.; Panco, S.; Santi, R. *J. Org. Chem.*, ***1990***, *55*, 350.

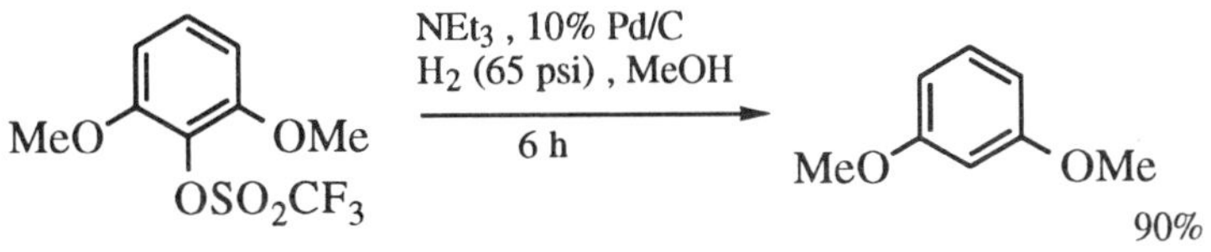

NEt_3, 10% Pd/C, H_2 (65 psi), MeOH, 6 h — 90%

Saá, J.M.; Dopico, M.; Martorell, G.; García-Raso, A. *J. Org. Chem.*, ***1990***, *55*, 991.

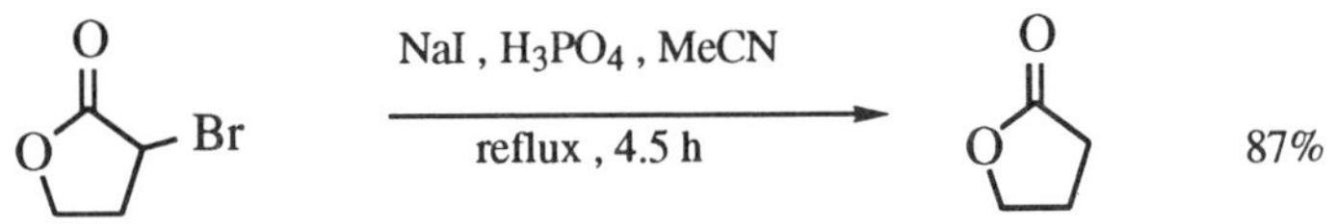

NaI, H_3PO_4, MeCN, reflux, 4.5 h — 87%

Mandal, A.K.; Nijasure, A.M. *SynLett*, ***1990***, 554.

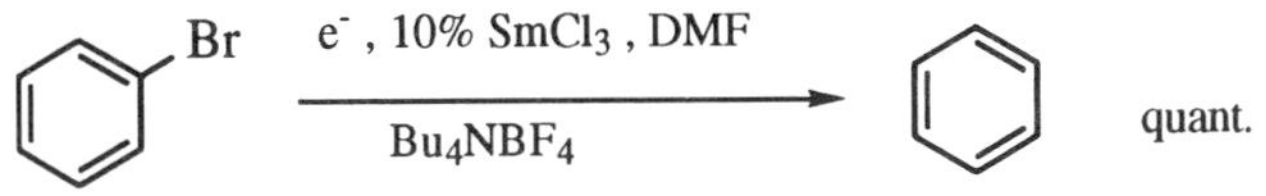

e^-, 10% $SmCl_3$, DMF, Bu_4NBF_4 — quant.

Hebri, H.; Duñach, E.; Périchon, J. *Synth. Commun.*, ***1991***, *21*, 2377.

Cl (n-octyl chloride) → (Et_3Si, hexane, reflux; 5% TBHN; 2% *t*-dodecanethiol) → octane 96%

TBHN = di-*t*-butyl hyponitrite

Cole, S.J.; Kirwan, J.N.; Roberts, B.P.; Willis, C.R. *J. Chem. Soc., Perkin Trans. I*, ***1991***, 103.

Ph–CH(OMe)–C(Me)(I)–CO_2Me → (Bu_3SnH, $MgBr_2$•OEt_2) → Ph–CH(OMe)–CH(Me)–CO_2Me + Ph–CH(OMe)–CH(Me)–CO_2Me

(>25 : 1) 84%

no Lewis acid (1 : >25) 90%

Guindon, Y.; Lavallée, J-F.; Llinas-Brunet, M.; Horner, G.; Rancourt, J. *J. Am. Chem. Soc.*, ***1991***, *113*, 9701.

PhCOCH$_2$Br → (1. AlI_3, MeCN, 80°C, 1 h; 2. H_2O) → PhCOCH$_3$ 92%

Borah, H.N.; Boruah, R.C.; Sandhu, J.S. *J. Chem. Soc., Chem. Commun.*, ***1991***, 154.

2-bromononane → ($(Me_3Si)_2SiMeSH$, 80°C; *t*-BuONNO*t*-Bu, 1 h) → nonane 92%

Daroszewski, J.; Lusztyk, J.; Degueil, M.; Navarro, C.; Maillard, B. *J. Chem. Soc., Chem. Commun.*, ***1991***, 586.

Me–C$_6$H$_4$–OTf → ($NiBr_2$(dppp), PPh_3, Zn; DMF, MeOH, KI; 20°C, 16 h) → Me–C$_6$H$_5$ 97%

Sasaki, K.; Sakai, M.; Sakaibara, Y.; Takagi, K. *Chem. Lett.*, ***1991***, 2017.

Ph–CH(OMe)–C(Me)(Br)–CO_2Et → (Bu_3SnH, AIBN; PhMe, -78°C; hν (sunlamp); 30 min) → Ph–CH(OMe)–CH(Me)–CO_2Et + Ph–CH(OMe)–CH(Me)–CO_2Et

(32 : 1) 90%

Guindon, Y.; Lavallée, J-F.; Boisvert, L.; Chabot, C.; Delorme, D.; Yoakim, C.; Hall, D.; Lemieux, R.; Simoneau, B. *Tetrahedron Lett.*, ***1991***, *32*, 27.

Guindon, Y.; Yoakim, C.; Lemieux, R.; Boisvert, L.; Delorme, D.; Lavallée, J-F. *Tetrahedron Lett.*, ***1990***, *31*, 2845.

$Br-CH_2-n-C_{11}H_{23}$ —(*1*, *2*, PhMe, 24 h; 100°C, 10 eq. $NaBH_4$)→ $H_3C-n-C_{11}H_{23}$ 64%

1 =

2 =

Ph OH O O O O O Ph $SnCl_2Ph$ n n

Blanton, J.R.; Salley, J.M. *J. Org. Chem.*, ***1991***, *56*, 490.

$(Me_3Si)_3SiH$, AIBN, PhMe, 90°C

Cl → 82%

Ballestri, M.; Chatgilialoglu, C.; Clark, K.B.; Griller, D.; Giese, B.; Kopping, B. *J. Org. Chem.*, ***1991***, *56*, 678.

n-$C_{10}H_{21}$ I —(Bu_3SnH, $Pd(PPh_3)_4$; 25°C, 3 h)→ n-$C_{10}H_{21}$ 92%

Taniguchi, M.; Takeyama, Y.; Fugami, K.; Oshima, K.; Utimoto, K. *Bull. Chem. Soc. Jpn.*, ***1991***, *64*, 2593.

I —($NiBr_2$, Zn, EtOH; 60°C, 20 h)→ 90%

Sakai, M.; Lee, M-S.; Yamaguchi, K.; Kawai, Y.; Sasaki, K.; Sakakibara, Y. *Bull. Chem. Soc. Jpn.*, ***1992***, *65*, 1739.

OTs Bu —(BuZnCl, PPh_3, $Pd(dba)_2$)→ Bu 85%

Ollivier, J.; Piras, P.P.; Stolle, A.; Aufranc, P.; de Meijere, A.; Salaün, J. *Tetrahedron Lett.*, ***1992***, *33*, 3307.

O Ph Br —(Ph_3PHI, MeCN, RT; 3 h)→ O Ph 85%

Kamiya, N.; Tanmatu, H.; Ishii, Y. *Chem. Lett.*, ***1992***, 293.

REVIEWS:

"New Mechanistic Insight into Reductions of Halides and Radicals with Samarium (II) Iodide"
Curran, D.P.; Fevig, T.L.; Jasperse, C.P.; Totleben, M.J. *SynLett*, ***1992***, 943.

SECTION 161: HYDRIDES FROM HYDRIDES

NO ADDITIONAL EXAMPLES

SECTION 162: HYDRIDES FROM KETONES

This section lists examples of the reaction $R_2C\text{-}(C{=}O)R \rightarrow R_2C\text{-}H$.

1. $TsNHNH_2$, MeOH
2. $LiAlH_4$, THF, 60°C

Ph (ketone, NO_2) → Ph 66%

Ballini, R.; Petrini, M.; Rosini, G. *J. Org. Chem.*, ***1990***, *55*, 5159.

$P(OiPr)_3$, reflux

Ph → Ph 51%

Olah, G.A.; Wu, A. *SynLett*, ***1990***, 54.

1. Dibal, heptane 90°C, 6 h
2. 2 eq. $AlBr_3$, 90°C 20 h

Ph → Ph 65%

Eisch, J.J; Liu, Z-R.; Boleslawski, M.P. *J. Org. Chem.*, ***1992***, *57*, 2143.

Ph SEt Ph SEt — P_2I_4, CH_2Cl_2, 25°C, 2 h → Ph Ph + Ph Ph =O

(85 : 15) 85%

Saimoto, H.; Kanzaki, A.; Miyazaki, K.; Sashiwa, H.; Shigemasa, Y. *Bull. Chem. Soc. Jpn.*, ***1992***, *65*, 2842.

Et_3SiH, PPHF, 0°C → RT

Ph Ph (O) → Ph Ph 96%

PPHF = pyridinium poly(hydrogen fluoride)

Olah, G.A.; Wang, Q.; Surya Prakash, G.K. *SynLett*, ***1992***, 647.

SECTION 163: HYDRIDES FROM NITRILES

This section lists examples of the reaction, R-C≡N → R-H (includes reactions of isonitriles (R-N≡C).

3 eq. BH_3 , THF , 3 h

71%

Ogura, K.; Shimamura, Y.; Fujita, M. *J. Org. Chem.*, ***1991***, *56*, 2920.

1. Bu_3SnH , AIBN PhH , reflux
2. DBU

81%

Curran, D.P.; Seong, C.M. *SynLett*, ***1991***, 107.

1. MeLi
2. H_2O

79%

"normal" nitrile addition with RMgX

Kulp, S.S.; Romanelli, A.*Org. Prep. Proceed. Int.*, ***1992***, *24*, 7.

SECTION 164: HYDRIDES FROM ALKENES

NO ADDITIONAL EXAMPLES

SECTION 165: HYDRIDES FROM MISCELLANEOUS COMPOUNDS

NaHTe , DMF , RT

64%

Huang, X.; Pi, J-H. *Synth. Commun.*, ***1990***, *20*, 2297.

SmI_2 , THF , HMPA
-20°C , 30 min

74%

Künzer, H.; Stahnke, M.; Sauer, G.; Wiechert, R. *Tetrahedron Lett.*, ***1991***, *32*, 1949.

CHAPTER 12

PREPARATION OF KETONES

SECTION 166: KETONES FROM ALKYNES

1. NBS, $AgNO_3$, acetone
2. $Hg(O_2CCF_3)_2$, CH_2Cl_2, RT
3. alkaline $NaBH_4$

Ph O C≡C-H → Ph O 81%

Subramanian, R.S.; Balasubramanian, K.K. *J. Chem. Soc., Chem. Commun.*, ***1990***, 1469.

HCO_2H, 100°C, 6 h

92% O

Menashe, N.; Reshef, D.; Shvo, Y. *J. Org. Chem.*, ***1991***, *56*, 2912.

$n\text{-}C_6H_{13}\text{-}C\equiv C\text{-}Me$ → ($NaAuClO_4$, aq. MeOH, 5 h) O, $n\text{-}C_6H_{13}$, Me + $n\text{-}C_6H_{13}$, Me, O

(40 : 60) 94%

Fukuda, Y.; Utimoto, K. *J. Org. Chem.*, ***1991***, *56*, 3729.

$Ph\text{-}C\equiv C\text{-}H$ → (PhSePh, MeOH, 1 h; $(NH_4)_2S_2O_8$, reflux) O, Ph, OMe, OMe 75%

Tiecco, M.; Testaferri, L.; Tingoli, M.; Chianelli, D.; Bartoli, D. *J. Org. Chem.*, ***1991***, *56*, 4529.

SECTION 167: KETONES FROM ACID DERIVATIVES

O, $n\text{-}C_9H_{19}$, Cl → (Me_3Tl) O, $n\text{-}C_9H_{19}$, Me 85%

Markó, I.E.; Southern, J.M. *J. Org. Chem.*, ***1990***, *55*, 3368.

Ph₂CHCO₂H → [2 $Cu(NO_3)_2/H_2O_2/2\ H_2O$], MeCN → $Ph_2C{=}O$ 55%

Capdevielle, P.; Maumy, M. *Tetrahedron Lett.*, ***1990***, *31*, 3891.

Ethylbenzene → 1. Me_3SiCl, Li/THF; 2. air; 3. BzCl, $AlCl_3$; 4. KF/DMF → 2-ethylbenzophenone 41%

Bennetau, B.; Krempp, M.; Dunogues, J.; Ratton, S. *Tetrahedron Lett.*, ***1990***, *31*, 6179.

2-(2-Phenylethyl)benzoyl chloride → Nafion-H, PhH, reflux, 30 min → dibenzosuberone 90%

Yamato, T.; Hideshima, C.; Surya Prakash, G.K.; Olah, G.A. *J. Org. Chem.*, ***1991***, *56*, 3955.

$PhCH_2COCl$ → allyl-$SnBu_3$, Et_4NCl, Bu_2SnCl_2, 30 min → $PhCH_2COCH_2CH{=}CH_2$ 82%

Yano, K.; Baba, A.; Matsuda, H. *Chem. Lett.*, ***1991***, 1181.

PhCOCl → Bu_4SbCH_2Ph, THF, -78°C → RT, 1 h → $PhCOCH_2Ph$ 87%

Zhang, L-J.; Huang, Y-Z.; Jiang, H-X.; Duan-Mu, J.; Liao, Y. *J. Org. Chem.*, ***1992***, *57*, 774.

n-$C_7H_{15}COCl$ → 3% $MnCl_4Li_2$, THF, BuMgCl, 100°C → n-$C_7H_{15}COBu$ 87%

Cahiez, G.; Laboue, B. *Tetrahedron Lett.*, ***1992***, *33*, 4439.
Cahiez, G.; Chavant, P-Y.; Metais, E. *Tetrahedron Lett.*, ***1992***, *33*, 5245.

$CH_2{=}C(SnMe_3)CH_2SiMe_3$ → $CH_3CH_2CH_2COCl$, $AlCl_3$, CH_2Cl_2, -78°C, 1 h → $CH_2{=}C(SiMe_3)CH_2COCH_2CH_2CH_3$ 85%

Kang, K-T.; Lee, J.C.; SunU, J. *Tetrahedron Lett.*, ***1992***, *33*, 4953.

$n\text{-}C_{15}H_{31}COCl$ —(TFAA, Py, ether; 20°C, 0.8 h)→ $n\text{-}C_{15}H_{31}COCF_3$ 81%

Boivin, J.; El Kaim, L.; Zard, S.Z. *Tetrahedron Lett.*, ***1992***, *33*, 1285.

PhCOOH —(1. $Ph_3P{=}CHCO_2t\text{-}Bu$, DMAP, $Me_2N(CH_2)_3N{=}C{=}N\text{-}Et$, CH_2Cl_2; 2. oxone)→ $PhCOCOCO_2t\text{-}Bu$ 79%

Wasserman, H.H.; Ennis, D.S.; Blum, C.A.; Rotello, V.M. *Tetrahedron Lett.*, ***1992***, *33*, 6003.

Allyl chloride —(1. Cu*; 2. BzCl)→ allyl phenyl ketone 65%

Cu* = activated zerovalent copper

Stack, D.E.; Dawson, B.T.; Rieke, R.D. *J. Am. Chem. Soc.*, ***1992***, *114*, 5110.

SECTION 168: KETONES FROM ALCOHOLS AND THIOLS

cis-Cyclohexane-1,2-diol —(1. MsCl, CH_2Cl_2, NEt_3, 0°C, 3h; 2. 2 eq. NaOEt, EtOH, 12h, 25°C)→ cyclohexanone 67%

a Grob-like reaction

Pestchanker, M.J.; Giordano, O.S. *Tetrahedron Lett.*, ***1990***, *31*, 463.

Octane-1,7-diol —(Cr-PILC/TBHP, CH_2Cl_2, RT, 24 h)→ 8-hydroxy-2-octanone 94%

Cr-PILC = chromia-pillared montmorillonite

Choudary, B.M.; Durgaprasad, A.; Valli, V.L.K. *Tetrahedron Lett.*, ***1990***, *31*, 5785.

Cyclohexanol —($Ba(MnO_4)_2$)→ cyclohexanone 55%

Firouzabadi, H.; Seddighi, M.; Mottaghinejad, E.; Bolourchian, M. *Tetrahedron*, ***1990***, *46*, 6869.

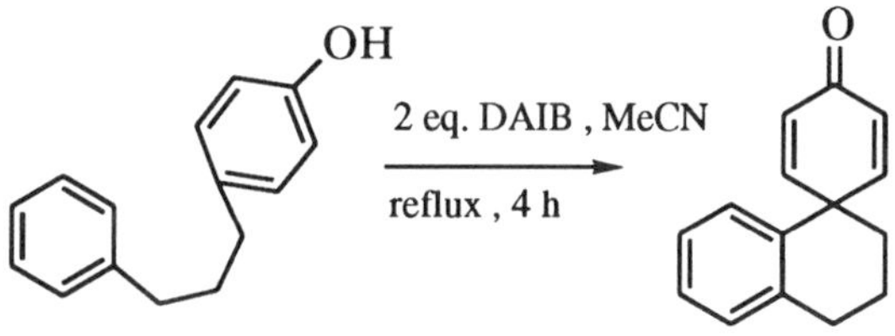

OH

2 eq. DAIB , MeCN

reflux , 4 h

O

DAIB = (diacetoxyiodo)benzene 30%

Rama Krishna, K.V.; Sujatha, K.; Kapil, R.S. *Tetrahedron Lett.*, ***1990***, *31*, 1351.

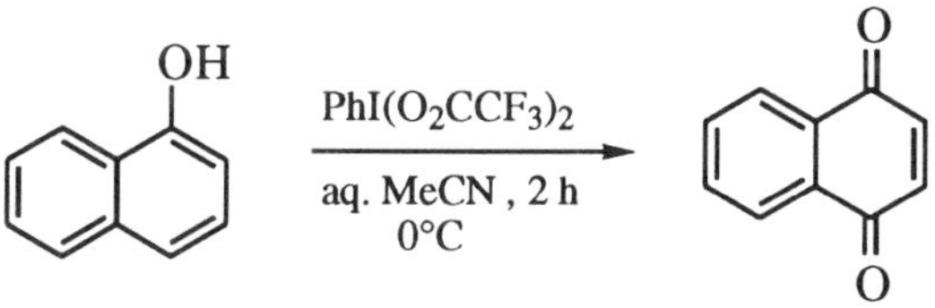

OH

$PhI(O_2CCF_3)_2$

aq. MeCN , 2 h
0°C

O

O 73%

Barret, R.; Daudon, M. *Tetrahedron Lett.*, ***1990***, *31*, 4871.

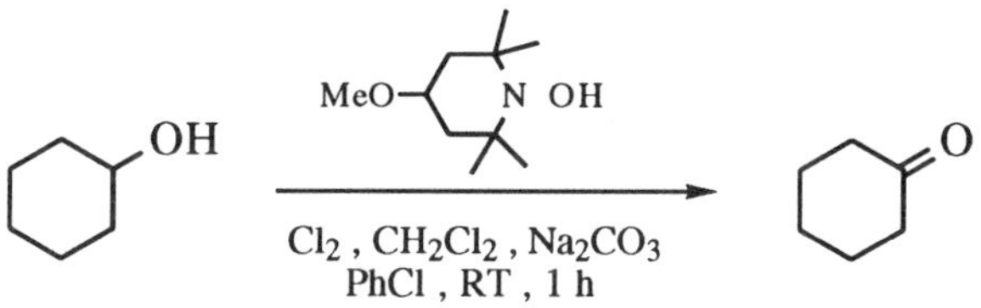

MeO N OH

OH

Cl_2 , CH_2Cl_2 , Na_2CO_3
PhCl , RT , 1 h

O 63%

Yamaguchi, M.; Takata, T.; Endo, T. *Bull. Chem. Soc. Jpn.*, ***1990***, *63*, 947.

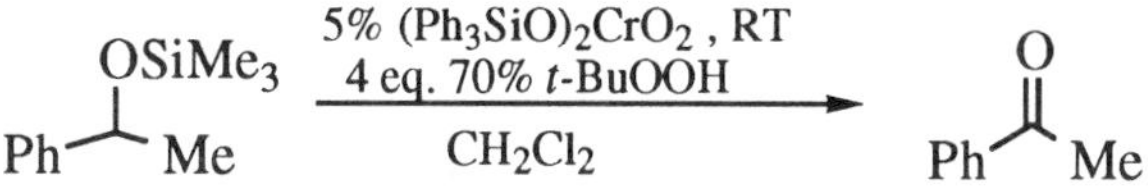

$OSiMe_3$

Ph Me

5% $(Ph_3SiO)_2CrO_2$, RT
4 eq. 70% *t*-BuOOH

CH_2Cl_2

O

Ph Me 97%

Muzart, J.; Ajjou, A.N. *SynLett*, ***1991***, 497.

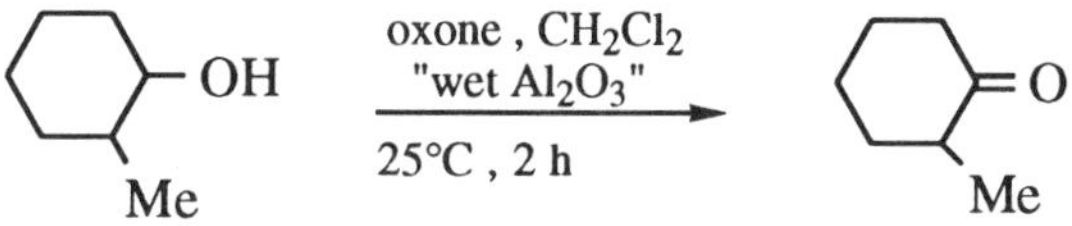

OH

Me

oxone , CH_2Cl_2
"wet Al_2O_3"

25°C , 2 h

O

Me 80%

Hirano, M.; Oose, M.; Morimoto, T. *Bull. Chem. Soc. Jpn.*, ***1991***, *64*, 1046.

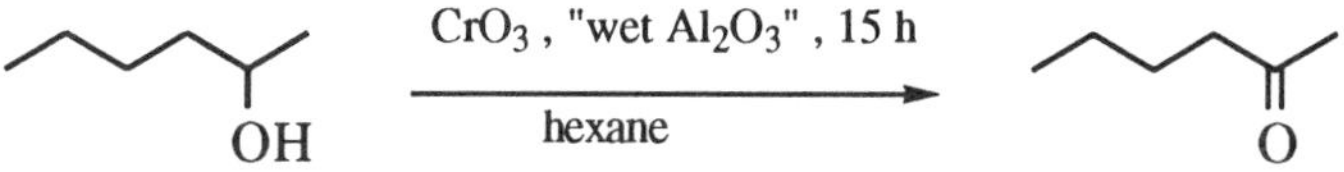

OH

CrO_3 , "wet Al_2O_3" , 15 h

hexane

O 95%

Hirano, M.; Nagasawa, S.; Morimoto, T. *Bull. Chem. Soc. Jpn.*, ***1991***, *64*, 2857.

OH → (dimethyldioxirane, TFP; CH_2Cl_2, 20 min, -20°C) → O 99%

Mello, R.; Cassidei, L.; Fiorentino, M.; Fusco, C.; Hümmer, W.; Jäger, V.; Curci, R. *J. Am. Chem. Soc.*, ***1991***, *113*, 2205.

NMO, CH_2Cl_2
MS 4Å, TPAP
))))))))

HO, O, O, O 75-80%

TPAP = tetra-*n*-butylammonium perruthenate

Miranda Moreno, M.J.S.; Sáe Melo, M.L.; Campos Neves, A.S. *Tetrahedron Lett.*, ***1991***, *32*, 3201.

$OSiMe_3$ → (hv (254 nm), DDQ; CH_2Cl_2, MeCN) → O 85%

also - aldehydes from $R\text{-}CH_2OTMS$

Piva, O.; Amougay, A.; Pete, J-P. *Tetrahedron Lett.*, ***1991***, *32*, 3993.

OH → (oxaziridine: O, N, $n\text{-}C_3F_7$, $n\text{-}C_4F_9$, F; $CFCl_3$, 0°C → RT, 9 h) → O 98%

DesMarteau, D.D.; Petrov, V.A.; Montanari, V.; Pregnolato, M.; Resnati, G. *Tetrahedron Lett.*, ***1992***, *33*, 7245.

Ph, OH → ($PdCl_2(PPh_3)_2$, DME; $Ti(OiPr)_4$, $Zn(acac)_2$; 120°C, 18 h) → Ph, O, O 40% + Ph, O 51%

Itoh, K.; Hamaguchi, N.; Miura, M.; Nomura, M. *J. Chem. Soc., Perkin Trans. I*, ***1992***, 2833.

$n\text{-}C_6H_{13}$(OH)Me → ($Zn(ClCrO_3)_2 \cdot 9\ H_2O$; CH_2Cl_2, RT) → $n\text{-}C_6H_{13}$(O)Me 90%

Firouzabadi, H.; Sharifi, A. *Synthesis*, ***1992***, 999.

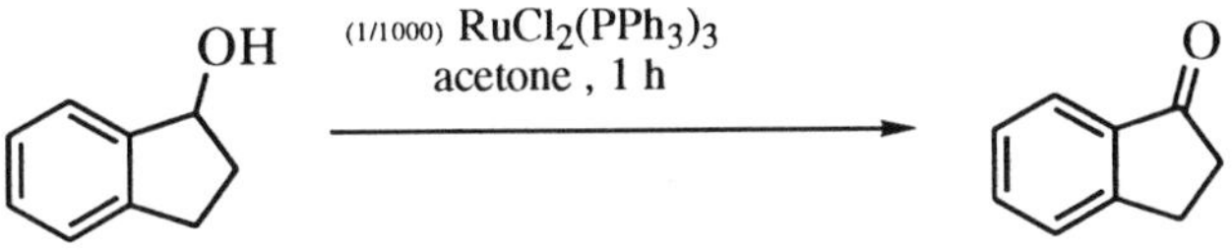

OH (1/1000) $RuCl_2(PPh_3)_3$ acetone , 1 h O 91%

Wang, G-Z.; Bäckvall, J-E. *J. Chem. Soc., Chem. Commun.*, ***1992***, 337.

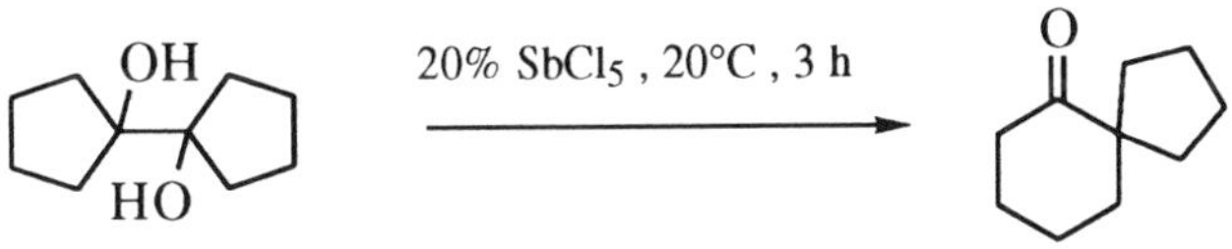

OH HO 20% $SbCl_5$, 20°C , 3 h O 83%

Harada, T.; Mukaiyama, T. *Chem. Lett.*, ***1992***, 81.

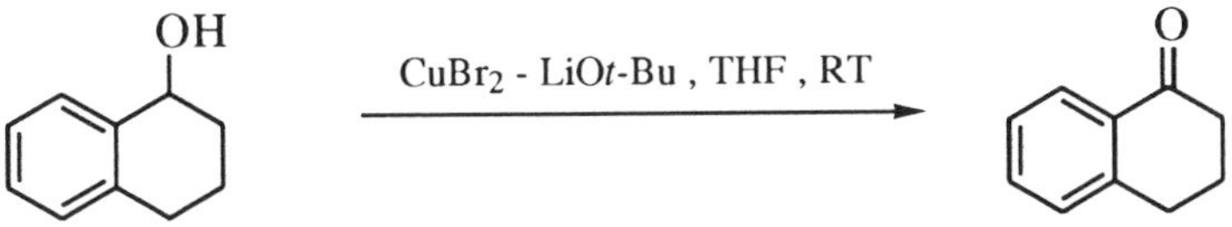

OH $CuBr_2$ - LiO*t*-Bu , THF , RT O 94%

also for preparation of aldehydes from 1° alcohols

Yamaguchi, J.; Yamamoto, S.; Takeda, T. *Chem. Lett.*, ***1992***, 1185.

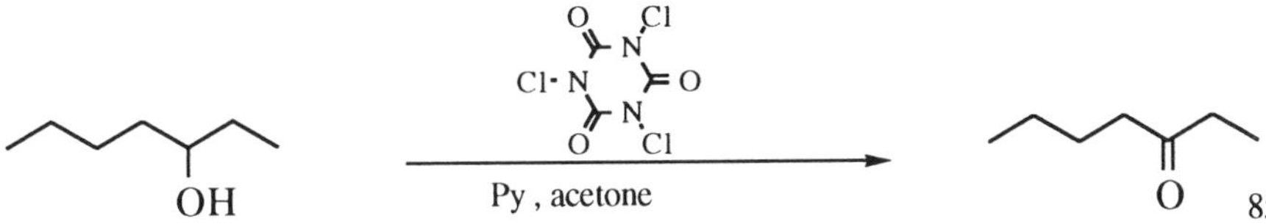

O Cl N Cl- N O N O Cl

OH Py , acetone O 85%

Hiegel, G.A.; Nalbandy, M. *Synth. Commun.*, ***1992***, *22*, 1589.

REVIEWS:

"Synthetic Oxidations with Hypochlorites"
Skarzewski, J.; Siedlecka, R. *Org. Prep. Proceed. Int.*, ***1992***, *24*, 625.

Related Methods: Section 48 (Aldehydes from Alcohols and Phenols).

SECTION 169: KETONES FROM ALDEHYDES

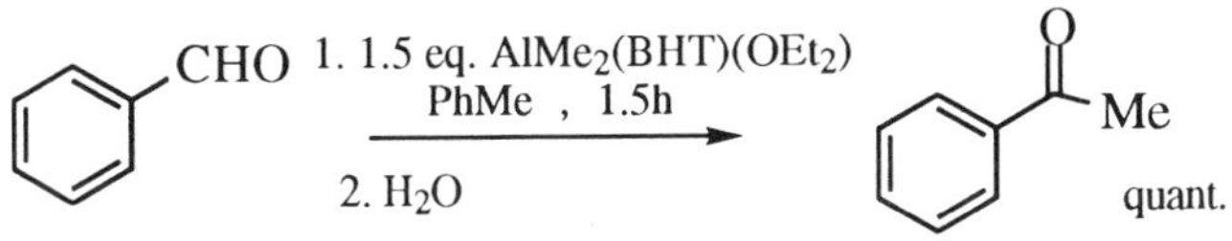

CHO 1. 1.5 eq. $AlMe_2(BHT)(OEt_2)$ PhMe , 1.5h 2. H_2O O Me quant.

BHT = 2,6-di-*t*-butyl-4-methylphenol

Power, M.B.; Barron, A.R. *Tetrahedron Lett.*, ***1990***, *31*, 323.

$Co_2(CO)_8$, Py , 12 h
stainless steel autoclave
(CO:H_2 - 10 MPa)

Ph CHO → Ph, O, Ph

60%

Fontaine, M.; Noels, A.F.; Demonceau, A.; Hubert, A.J. *Tetrahedron Lett.*, ***1990***, *31*, 3117.

1. BuLi , *t*-BuONa
THF , -78°C
2. MeI , -78°C

MEMO, S, S, OBn → MEMO, S, S, Me, OBn

also reacts with aldehydes

97%

Lipshutz, B.H.; Garcia, E. *Tetrahedron Lett.*, ***1990***, *31*, 7261.

PhCHO → , CO , 200°C; $Ru_3(CO)_{12}$, 48 h → O

50%

Kondo, T.; Akazome, M.; Tsuji, Y.; Watanabe, Y. *J. Org. Chem.*, ***1990***, *55*, 1286.

n-C_9H_{19}CHO

1. $(Me_3Si)_2C(Li)OMe$
THF
2. HCl , aq. THF

→ O, n-C_9H_{19}, $SiMe_3$

75%

Yoshida, J.; Matsunaga, S.; Ishichi, Y.; Maekawa, T.; Isoe, S. *J. Org. Chem.*, ***1991***, *56*, 1307.

Me_3SiO, Ph, Me

PhCHO

20% iPrO, O, CO_2H, O, CO_2H•BH_3-THF, OH, OiPr

→ O, $OSiMe_3$, Ph, Ph, Ph

86%

(95:5 erythro:threo)
(85% ee , R)

Furuta, K.; Maruyama, T.; Yamamoto, H. *J. Am. Chem. Soc.*, ***1991***, *113*, 1041.

SECTION 170: KETONES FROM ALKYLS, METHYLENES AND ARYLS

This section lists examples of the reaction, R-CH_2-R' → R(C=O)-R'.

O

CO_2Et

1. $(O_2N-C_6H_4-SO_2-O)_2$

CH_2Cl_2, 0°C

2. TEA, PhH, 25°C

O

CO_2Et

O good yield

Hoffman, R.V.; Kim, H-O.; Wilson, A.L. *J. Org. Chem.*, ***1990***, *55*, 2820.

SECTION 171: KETONES FROM AMIDES

O

n-C_6H_{13} N OMe

Me

1. Li

ether, RT

2. H_2O

3. wet silica

O

n-C_6H_{13}

86% O

Bergman, R.; Nilsson, B.; Wickberg, B. *Tetrahedron Lett.*, ***1990***, *31*, 2783.

Me

O N

Me

1. TFAA, DCE

2,6-di-*t*-butyl-4-methyl pyridine

2. 83°C

3. H_2O/CCl_4, reflux

H O

H 88%

(98% ee)

Chen, L.; Ghosez, L. *Tetrahedron Lett.*, ***1990***, *31*, 4467.

O

I N OMe

Me

t-BuLi, THF

-78°C → -20°C

O

>90%

Souchet, M.; Clark, R.D. *SynLett*, ***1990***, 151.

O

Ph NEt_2

1. *sec*-BuLi, TMEDA

2. Br CHO

3. *t*-BuLi ; -78°C → RT

4. air

O

O 60%

Wang, X.; Snieckus, V. *SynLett*, ***1990***, 313.

MeO N Me O O N OMe Me
1.5 eq. PhMgBr , THF
0°C , 1 h
Ph O O N OMe Me 95%

4 eq. PhMgBr , 65°C , 7 h
21% + Ph O O Ph 62%

Sibi, M.P.; Sharma, R.; Paulson, K.L. *Tetrahedron Lett.*, ***1992***, *33*, 1941.

SECTION 172: KETONES FROM AMINES

Me
1. H_2 , TFA , PtO_2
2. MeI
3. Na , EtOH
4. H_2O
O
85%

Stevens, R.V.; Canary, J.W. *J. Org. Chem.*, ***1990***, *55*, 2237.

N-NMe$_2$
MMPP•6 H_2O , MeOH
pH 7 buffer , 0°C , 5 min
O
76%

MMPP = magnesium monoperoxyphthalate

Enders, D.; Plant, A. *SynLett*, ***1990***, 725.

N Ph Ph
1. PhCH=NPh , DMF
t-BuOK , 75°C
2. aq. H^+
N-H Ph + Ph Ph O
94%

Paventi, M.; Hay, A.S. *Tetrahedron Lett.*, ***1991***, *32*, 1617.

Ph Ph N OH
$Fe(ClO_4)_3$, MeCN
Ph Ph O
79%

Kumar, H.; Kaur, B.; Kumar, B. *Ind. J. Chem.*, ***1991***, *30B*, 869.

Ph N-OH
$PhICl_2$, Py , $CHCl_3$
10°C , 3 h
Ph O
75%

Radhakrishna, A.S.; Augustine, B.; Sivaprakash, K.; Singh, B.B. *Synth. Commun.*, ***1991***, *21*, 1473.

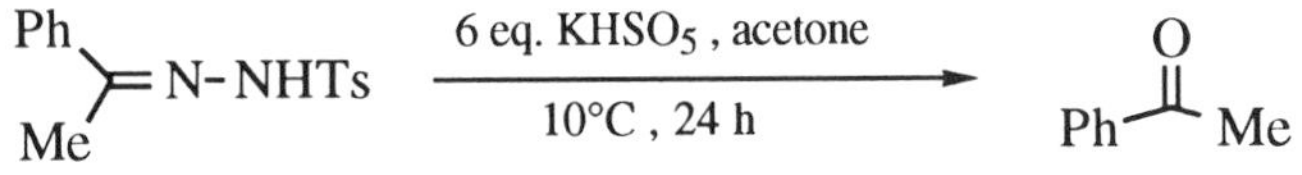

Ph(Me)C=N-NHTs → 6 eq. $KHSO_5$, acetone; 10°C, 24 h → PhC(O)Me 99%

Jung, J.C.; Kim, K.S.; Kim, Y.H. *Synth. Commun.*, ***1992***, *22*, 1583.

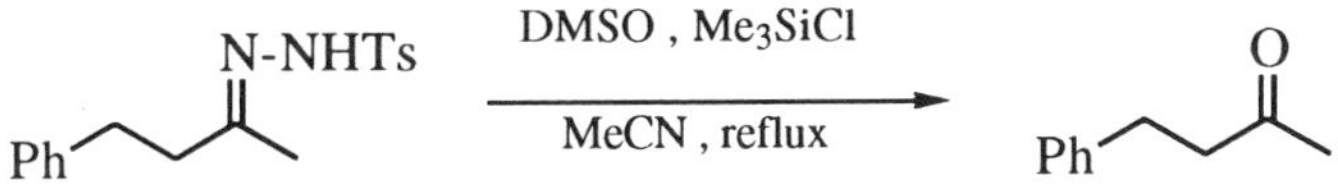

DMSO, Me_3SiCl; MeCN, reflux → 98%

Ghelfi, F.; Grandi, R.; Pagnoni, U.M. *Synth. Commun.*, ***1992***, *22*, 1845.

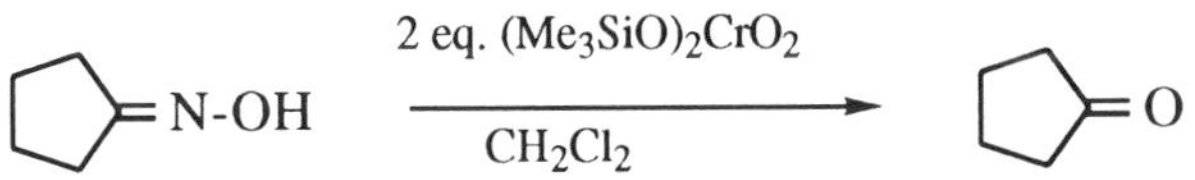

2 eq. $(Me_3SiO)_2CrO_2$; CH_2Cl_2 → quant.

Lee, J.G.; Kwak, K.H.; Hwang, J.P. *Synth. Commun.*, ***1992***, *22*, 2425.

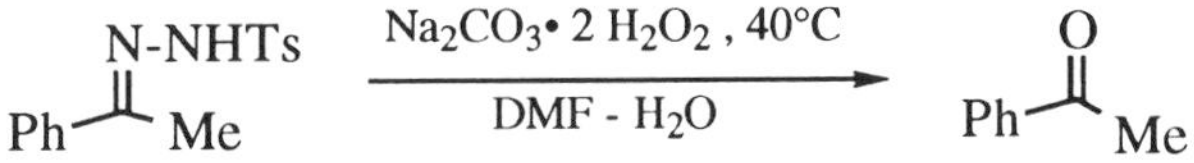

Na_2CO_3• 2 H_2O_2, 40°C; DMF - H_2O → 82%

Narayana, C.; Reddy, N.K.; Kabalka, G.W. *Synth. Commun.*, ***1992***, *22*, 2587.

SECTION 173: KETONES FROM ESTERS

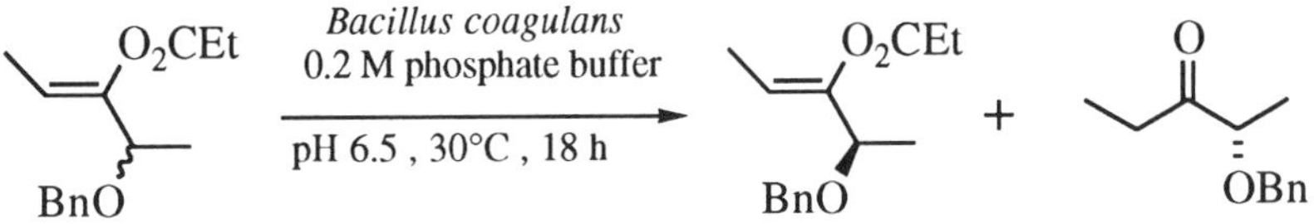

Bacillus coagulans, 0.2 M phosphate buffer; pH 6.5, 30°C, 18 h → 26% (95% ee) + 57% (42% ee)

Matsumoto, K.; Suzuki, N.; Ohta, H. *Tetrahedron Lett.*, ***1990***, *31*, 7159.

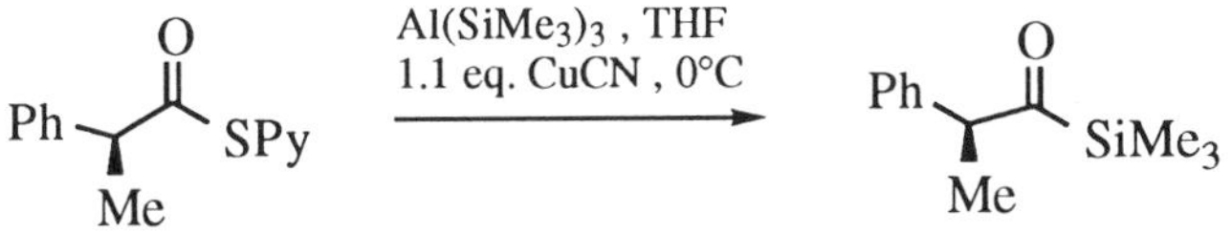

$Al(SiMe_3)_3$, THF; 1.1 eq. CuCN, 0°C → 82%

Nakada, M.; Nakamura, S.; Kobayashi, S.; Ohno, M. *Tetrahedron Lett.*, ***1991***, *32*, 4929.

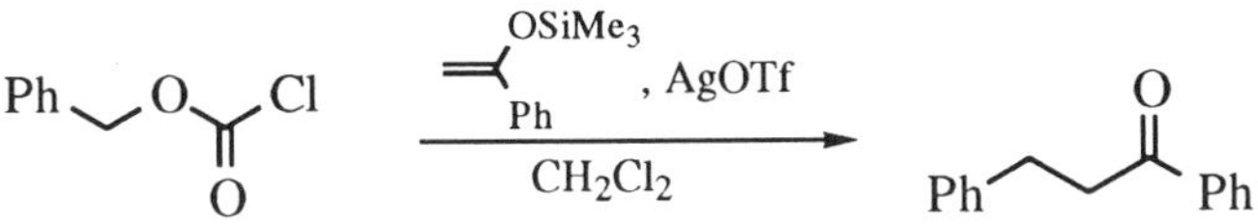

$CH_2=C(OSiMe_3)Ph$, AgOTf; CH_2Cl_2 → 87%

Takeda, K.; Ayabe, A.; Kawashima, H.; Harigaya, Y. *Tetrahedron Lett.*, ***1992***, *33*, 951.

OAc

CO_2Et

1. 1.4 $Fe_2(CO)_9$, ether
2. BF_3•OEt_2 , CH_2Cl_2
OSiMe$_3$
3. Me_3N-O

O

CO_2Et

71%

(1:1 diastereomer ratio)

Green, J.R.; Carroll, M.K. *Tetrahedron Lett.*, ***1991***, *32*, 1141.

Me_3SiO

Ph

DCA (λ = 450 nm) MeCN
[photoinduced electronic transfer]

O

Ph

38%

DCA = 9,10-dicyanoanthracene

Heidbreder, A.; Mattay, J. *Tetrahedron Lett.*, ***1992***, *33*, 1973.

O

O

S

Pichia farinosa
0.2 phosphate buffer
pH 6.5 , 30 min

O

S

74% (55% ee)

+

OH

S

23%

Kume, Y.; Ohta, H. *Tetrahedron Lett.*, ***1992***, *33*, 6367.

SECTION 174: KETONES FROM ETHERS, EPOXIDES AND THIOETHERS

MeO

PhS SPh
SPh

AgOTf
CH_2Cl_2

MeO

O

95%

Gramage, S.A.; Smith, R.A.J. *Aust. J. Chem.*, ***1990***, *43*, 815.

OSiMe$_3$

Ph_2I^+ F^- , THF
-40°C → RT

O

Ph

88%

Chen, K.; Koser, G.F. *J. Org. Chem.*, ***1991***, *56*, 5764.

t-$BuMe_2SiO$ Ph $\xrightarrow{Cu(OTf)_2, Cu_2O, MeCN, 0°C}$ H O H Et + H O H Et

(>20 : 1) 87%

Snider, B.B.; Kwon, T. *J. Org. Chem.*, ***1990***, *55*, 4786.

$OSiMe_3$ $\xrightarrow{BnBr, THF, 20°C; Bu_4N^+(Ph_3SnF_2)^-; 6 h}$ O Bn 99%

Gingras, M. *Tetrahedron Lett.*, ***1991***, *32*, 7381.

$OSiMe_3$ $\xrightarrow{AgO_2CCF_3, BuI; CH_2Cl_2; -78°C \rightarrow 25°C}$ O Bu 56%

Jefford, C.W.; Sledeski, A.W.; Lelandais, P.; Boukouvalas, J. *Tetrahedron Lett.*, ***1992***, *33*, 1855.

SECTION 175: KETONES FROM HALIDES AND SULFONATES

F_3C Cl NO_2 $\xrightarrow{1.\ EtNO_2, DBU, EtOAc;\ 2.\ H_2O_2, K_2CO_3}$ F_3C O Me NO_2 91%

Reid, J.G.; Reny Runge, J.M. *Tetrahedron Lett.*, ***1990***, *31*, 1093.

Br $\xrightarrow{=\!\!/OBu,\ NEt_3;\ TlOAc, DMF;\ 100°C, dppe}$ O Me 91%

Cabri, W.; Candiani, I.; Bedeschi, A.; Santi, R. *Tetrahedron Lett.*, ***1991***, *32*, 1753.

Br $\xrightarrow{1.\ Zn^*; CuCN•2 LiBr;\ 2.\ BzCl}$ Ph O 92%

$Zn^* = ZnCl_2$/Li naphth.

also for conjugate addition reactions and reaction with ArX

Zhu, L.; Wehmeyer R.M.; Rieke, R.D. *J. Org. Chem.*, ***1991***, *56*, 1445.

OMs

Me–C6H4–OMe

$TiCl_4$, -78°C

O, OMe, Me

82%

Kraus, G.A.; Hansen, J. *SynLett*, ***1990***, 483.

Ph-OTf

1. =/OBu , $Pd(OAc)_2$

dppp , NEt_3 , DMF

80°C , 3 h

2. H^+

Ph, Me, O

90%

Cabri, W.; Candiani, I.; Bedeschi, A.; Pence, S.; Santi, R. *J. Org. Chem.*, ***1992***, *57*, 1481.

O, SO_2Tol

t-BuOK , THF , -78°C

Cl, O, N, H, S, O_2, Cl

O, O

81%

Williams, D.R.; Robinson, L.A.; Amato, G.S.; Osterhout, M.H. *J. Org. Chem.*, ***1992***, *57*, 3740.

Br

1. Li , $ZnCl_2$, 0°C

)))))))

2. Cl_3CCOCl

0°C → 23°C

CCl_3, O

82%

Corey, E.J.; Link, J.O. *Tetrahedron Lett.*, ***1992***, *33*, 3435.

I

1. *n*-C_8H_{17}, B , *t*-Bu-N≡C , K_3PO_4

$Pd(PPh_3)_4$

2. H_3O^+

O, Ph, *n*-C_8H_{17}

97%

Ishiyama, T.; Oh-e, T.; Miyaura, N.; Suzuki, A. *Tetrahedron Lett.*, ***1992***, *33*, 4465.

Related Methods: Section 177 (Ketones from Ketones).
Section 55 (Aldehydes from Halides).

SECTION 176: KETONES FROM HYDRIDES

This section lists examples of the replacement of hydrogen by ketonic groups, R-H → R(C=O)-R'. For the oxidation of methylenes, R_2CH_2 → $R_2C=O$, see section 170 (Ketones from Alkyls).

Me; 3-pyridyl trifluoroacetate ($OC(O)CF_3$), $AlCl_3$ → $F_3C(C=O)$-aryl-Me 61% + 2-pyridone (N-H, O)

Keumi, T.; Shimada, M.; Takahashi, M.; Kitajima, H. *Chem. Lett.*, ***1990***, 783.

Me-Ph; $PhCCl_3$, $AlCl_3$, 25°C, DCE → Me-C_6H_4-C(O)Ph 95%

Ayyangar, N.R.; Lahoti, R.J.; Srinivasan, K.V.; Daniel, T. *Synthesis*, ***1991***, 322.

Ethylbenzene; $Ag(Py)_4S_2O_8$, MeCN, reflux → acetophenone 98%

ArMe gives ArCHO

Firouzabadi, H.; Salehi, P.; Sardarian, A.R.; Seddighi, M. *Synth. Commun.*, ***1991***, *21*, 1121.

MeO-Ph; n-C_5H_{11}C(O)OC(O)n-C_5H_{11}, 10% $GaCl_3$, 20% $AgClO_4$, CH_2Cl_2 , RT , 20 h → MeO-C_6H_4-C(O)n-C_5H_{11} 91%

Mukaiyama, T.; Ohno, T.; Nishimura, T.; Suda, S.; Kobayashi, S. *Chem. Lett.*, ***1991***, 1059.

Ph-CH_2-Ph; t-BuOOH, Cr-pillared montmorillonite → Ph-C(O)-Ph 90%

Choudary, B.M.; Prasad, A.D.; Bhuma, V.; Swapna, V. *J. Org. Chem.*, ***1992***, *57*, 5841.

Ph-OMe; Ac_2O , CH_2Cl_2, 4% $SbCl_5$, 30 min, 1 eq. $LiClO_4$ → Me-C(O)-C_6H_4-OMe 89%

Mukaiyama, T.; Suzuki, K.; Han, J.S.; Kobayashi, S. *Chem. Lett.*, ***1992***, 435.

SECTION 177: KETONES FROM KETONES

This section contains alkylations of ketones and protected ketones, ketone transpositions and annulations, ring expansions and ring openings and dimerizations. Conjugate reductions and Michael alkylations of enone are listed in Section 74 (Alkyls from Alkenes).

For the preparation of enamines or imines from ketones, see Section 356 (Amine-Alkene).

excess MeI, DMSO; KOH (solid); 60°C — Me, Me, O, Me, Me 90%

Langhals, E.; Langhals, H. *Tetrahedron Lett.*, ***1990***, *31*, 859.

O, $SiMe_3$, Bu, SPh — PhMe, 4 h; $Mo(CO)_6$; also with $Cr(CO)_6$ — O, Bu 74%

Kang, J.; Choi, Y.R.; Kim, B.J.; Jeong, J.U.; Lee, S.; Lee, J.H.; Pyun, C. *Tetrahedron Lett.*, ***1990***, *31*, 2713.

$PhB(OH)_2$ — 1. $Pb(OAc)_4$, $Hg(OAc)_2$, $CHCl_3$; 2. Py, O, CO_2Et — O, Ph, CO_2Et 77%

Morgan, J.; Pinhey, J.T. *J. Chem. Soc., Perkin Trans. I*, ***1990***, 715.

O, O, Me — 1. hv, hexane; 2. 150°C — O 45% + O

Yoshioka, M.; Arai, M.; Nishizawa, K.; Hasegawa, T. *J. Chem. Soc., Chem. Commun.*, ***1990***, 374.

O, $SnBu_3$ — 1. LDA, THF, -78°C; 2. BuLi, -78°C → 0°C; 3. n-$C_5H_{11}Br$, -78°c → -20°C; 4. H_2O — O 71%

Nakahira, H.; Ryu, I.; Ikebe, M.; Kanbe, N.; Sonoda, N. *Angew. Chem. Int. Ed. Engl.*, ***1991***, *30*, 177.

$OSiMe_3$ 1. MeLi-LiBr , ether 2. PhMeO N H 3. BnBr , PhMe — O Bn 68% (92% ee , R)

Murakata, M.; Nakajima, M.; Koga, K. *J. Chem. Soc., Chem. Commun.*, ***1990***, 1657.

O Me OR 1. 2.5 LDA , THF 10 HMPA , 0°C 2. MeI , -78°C — O Me OR Me + O Me OR Me

R =				
R = Li	>95	:	5	(34%)
= Bn	80	:	20	(90%)
= Bz	>95		5	(79%)

McGarvey, G.J.; Andersen, M.W. *Tetrahedron Lett.*, ***1990***, *31*, 4569.

O Cl CD_3CN , $KF \cdot 2\ H_2O$ $(Ph_3Sn)_2Te$, 5 h — O 63%

Li, C-J.; Harpp, D.N. *Tetrahedron Lett.*, ***1991***, *32*, 1545.

O anhydrous $FeCl_3$ on SiO_2 , CH_2Cl_2 ethanedithiol RT , <1 min — S 99%

Patney, H.K. *Tetrahedron Lett.*, ***1991***, *32*, 2259.

Me Me H O CHN_2 $Rh_2(OAc)_4$, RT CH_2Cl_2 , 12 h — H Me O 85%

Sonawane, H.R.; Bellur, N.S.; Ahuja, J.R.; Kulkarni, D.G. *J. Org. Chem.*, ***1991***, *56*, 1434.

N 1. Cl NO_2 , DMF NEt_3 , 3 h , hν (sunlamp) 2. H_3O^+ — O NO_2 90%

Russell, G.A.; Wang, K. *J. Org. Chem.*, ***1991***, *56*, 3475.

SmI_2 , THF , DMPA

39%

DMPU = 1,3-dimethyl-3,4,5,6-tetrahydro-2(1H)-pyrimidinone

Batey, R.A.; Motherwell, W.B. *Tetrahedron Lett.*, ***1991***, *32*, 6211, 6649.

Br

Cl

O

reflux

slow addition (7-10 h) of 3.3 eq. Bu_3SnH with AIBN to PhH

O 74% + 18% O

Dowd, P.; Zhang, W. *J. Am. Chem. Soc.*, ***1991***, *113*, 9875.
Zhang, W.; Dowd, P. *Tetrahedron Lett.*, ***1992***, *33*, 3285.

Ph CF_3 O

1. $POCl_3$, DMF
2. Bu_3SnH , $Pd(PPh_3)_4$

O Ph CF_3

82%

Laurent, A.J.; Lesniak, S. *Tetrahedron Lett.*, ***1992***, *33*, 3311.

O Me H

1. $EtAlCl_2$, CH_2Cl_2 RT , 16 h
2. H_2O

Me O H

(93:7 cis:trans)

77%

Fujiwara, T.; Tomaru, J.; Suda, A.; Takeda, T. *Tetrahedron Lett.*, ***1992***, *33*, 2583.

O Ph O

Se , H_2O , DBU
CO (30 atm) , 120°C
24 h

O Ph 70% + Ph 3%

Nishiyama, Y.; Inoue, J.; Teranishi, K.; Moriwaki, M.; Hamanaka, S. *Tetrahedron Lett.*, ***1992***, *33*, 6347.

O

1. Tol-S(O) n-C_5H_{11} Cl
2. 5 eq. t-BuLi , -60°C

O n-C_5H_{11} 66%

Satoh, T.; Itoh, N.; Gengyo, K.; Yamakawa, K. *Tetrahedron Lett.*, ***1992***, *33*, 7543.

Related Methods: Section 49 (Aldehydes from Aldehydes).

SECTION 178: KETONES FROM NITRILES

CN; N—Ts (allyl)

1. Cp_2Zr-propylene/"Cp_2Zr"
2. H_3O^+

O; Me; N; Ts

54%

Mori, M.; Uesaka, N.; Shibasaki, M. *J. Chem. Soc., Chem. Commun.*, ***1990***, 1222.

SECTION 179: KETONES FROM ALKENES

2% $Pd(OAc)_2$, aq. MeCN
90% benzoquinone
0.24M $HClO_4$

82% O

Miller, D.G.; Wayner, D.D.M. *J. Org. Chem.*, ***1990***, *55*, 2924.

CO_2Et

$[RhCl(COD)]_2$, aq. MeCN
CO (40 bar) , 165°C

CO_2Et; O

81%d

Eilbracht, P.; Hüttmann, G-E. *Chem. Ber.*, ***1990***, *123*, 1053.
Eilbracht, P.; Hüttmann, G-E.; Deußen, R. *Chem. Ber.*, ***1990***, *123*, 1063.

cat. RuO_2 , MeCHO , 3 h
acetone , O_2 , 40°C
pyromellitic acid

O

77%

pyromellitic acid = 1,2,4,5-benzenetetracarboxylic acid

Kaneda, K.; Haruna, S.; Imanaka, T.; Kawamoto, K. *J. Chem. Soc., Chem. Commun.*, ***1990***, 1467.

Br

CO (75 atm) , 80°C , 3 h
Bu_3SnH , AIBN , PhH
0.05 M

O

77%

(32:68 Z:E)

Ryu, I.; Kusano, K.; Hasegawa, M.; Kambe, N.; Sonoda, N. *J. Chem. Soc., Chem. Commun.*, ***1991***, 1018.

1. $BH_3 \cdot THF$, THF , 2°C
2. PCC , CH_2Cl_2 , reflux

77%

Parish, E.J.; Parish, S.; Honda, H. *Synth. Commun.*, ***1990***, *20*, 3265.

1. 9-BBN
2. CO , n-$C_6H_{13}I$
$Pd(PPh_3)_4$, hν
K_3PO_4 , PhH , RT
24 h

n-C_6H_{13} n-C_8H_{17}

67%

Ishiyama, T.; Miyama, N.; Suzuki, A. *Tetrahedron Lett.*, ***1991***, *32*, 6923.

$[PdCl(NO_2)(MeCN)_2]$
$CuCl_2$, O_2 , DPA
CH_2Cl_2

DPA = N,N-diethyl pivaloyl chloride

15% CHO + 48%

Kiers, N.H.; Feringa, B.L.; van Leeuwen, P.W.N.M. *Tetrahedron Lett.*, ***1992***, *33*, 2403.

See also: Section 134 (Ethers from Alkenes). Section 174 (Ketones from Ethers).

SECTION 180: KETONES FROM MISCELLANEOUS COMPOUNDS

Conjugate reductions and reductive alkylations of enones are listed in Section 74 (Alkyls from Alkenes). Ketones from oximes are listed here.

S N-N N N Ph + $OSiMe_3$ Ph

2 eq. AgOTf
$ClCH_2CH_2Cl$
reflux , 6h

O Ph

62%

Takeda, K.; Torii, K.; Ogura, H. *Tetrahedron Lett.*, ***1990***, *31*, 265.

N .OH Me

TMSCl , $NaNO_2$, CCl_4
5% Aliquat 336 , RT , 5 h

O Me

95%

Lee, J.G.; Kwak, K.H.; Hwang, J.P. *Tetrahedron Lett.*, ***1990***, *31*, 6677.

Me_3Si H C=O

1. $PhCeCl_2$, THF , -78°C
2. aq. NH_4Cl , -78°C → RT

Me_3Si O Ph

79%

Kita, Y.; Matsuda, S.; Kitagaki, S.; Tsuzuki, Y.; Akai, S. *SynLett*, ***1991***, 401.

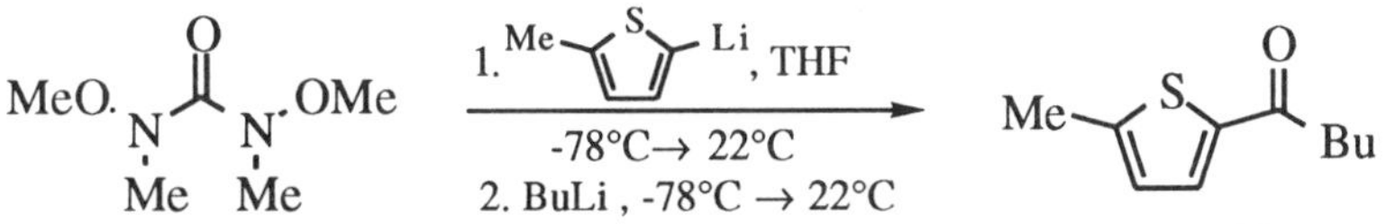

1. 5-methyl-2-thienyllithium, THF, -78°C→ 22°C
2. BuLi, -78°C → 22°C

99%

Whipple, W.L.; Reich, H.J. *J. Org. Chem.*, ***1991***, *56*, 2911.

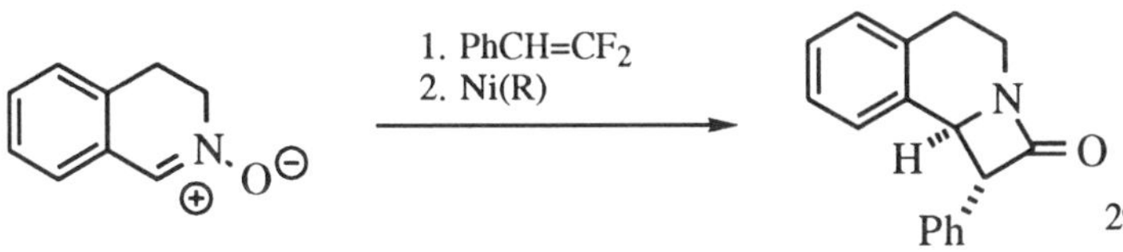

1. $PhCH=CF_2$
2. Ni(R)

29%

Purrington, S.T.; Sheu, K-W. *Tetrahedron Lett.*, ***1992***, *33*, 3289.

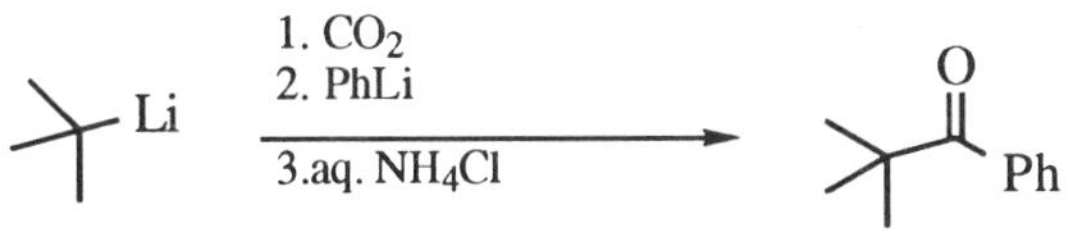

1. CO_2
2. PhLi
3. aq. NH_4Cl

92%

Zadel, G.; Breitmaier, E. *Angew. Chem. Int. Ed. Engl.*, ***1992***, *31*, 1035.

SECTION 180A: PROTECTION OF KETONES

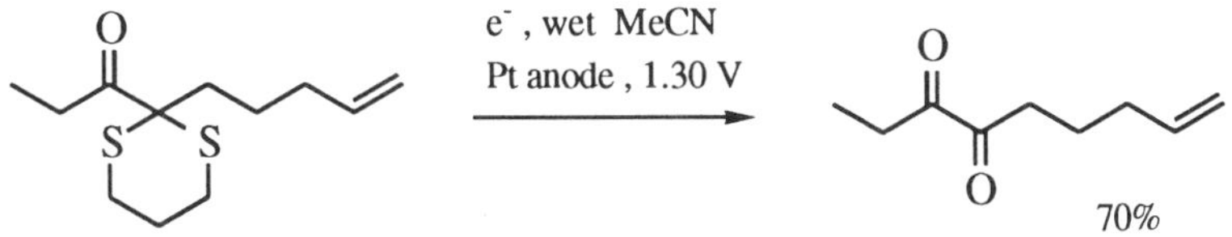

e^-, wet MeCN
Pt anode, 1.30 V

70%

Martre, A-M.; Mousset, G.; Bel Rhlid, R.; Veschambre, H. *Tetrahedron Lett.*, ***1990***, *31*, 2599.

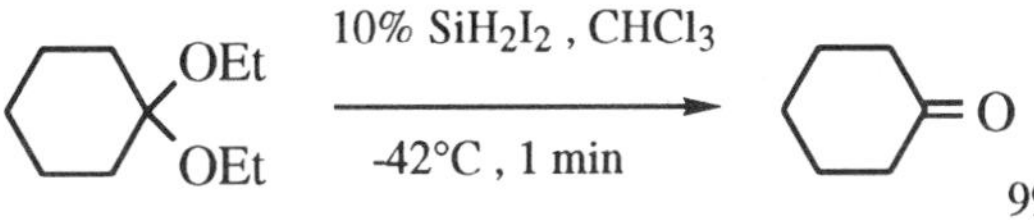

10% SiH_2I_2, $CHCl_3$
-42°C, 1 min

99%

Keinan, E.; Perez, D.; Sahai, M.; Shvily, R. *J. Org. Chem.*, ***1990***, *55*, 2927.

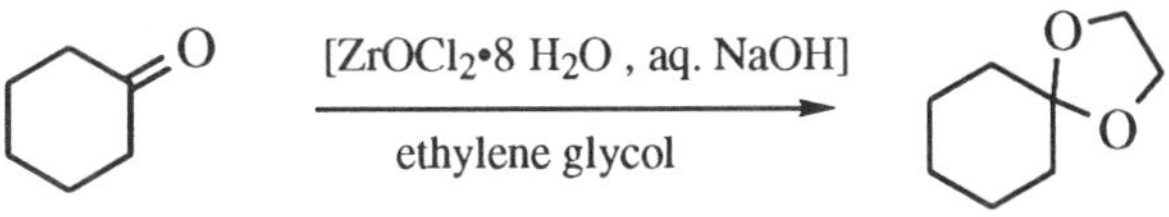

[$ZrOCl_2 \cdot 8\ H_2O$, aq. NaOH]
ethylene glycol

74%

Shibagaki, M.; Takahashi, K.; Kuno, H.; Matsushita, H. *Bull. Chem. Soc. Jpn.*, ***1990***, *63*, 1258.

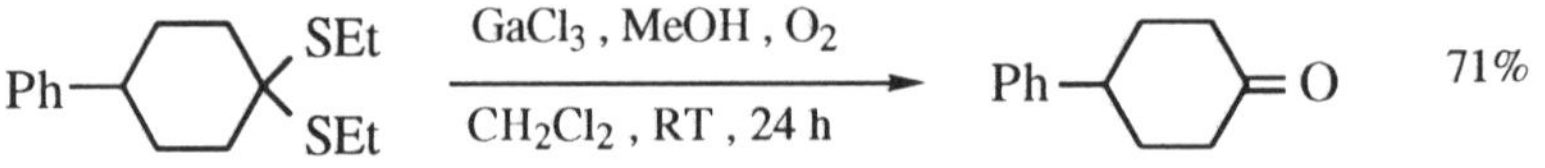
Ph
SEt
SEt
GaCl3 , MeOH , O2
CH2Cl2 , RT , 24 h
Ph
O
71%

Saigo, K.; Hashimoto, Y.; Kihara, N.; Umehara, H.; Hasegawa, M. *Chem. Lett.*, ***1990***, 831.

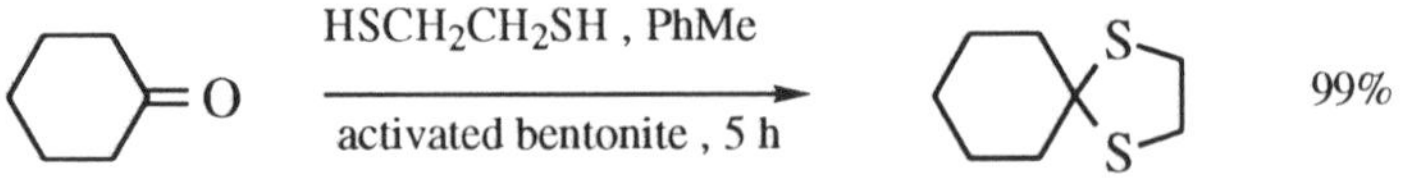
O
HSCH2CH2SH , PhMe
activated bentonite , 5 h
S
S
99%

Miranda, R.; Cervantes, H.; Joseph-Nathan, P. *Synth. Commun.*, ***1990***, *20*, 153.

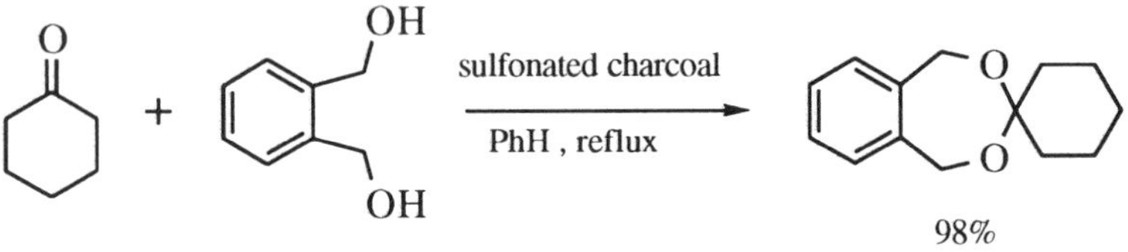
O
+
OH
OH
sulfonated charcoal
PhH , reflux
O
O
98%

Patney, H.K. *Tetrahedron Lett.*, ***1991***, *32*, 413.

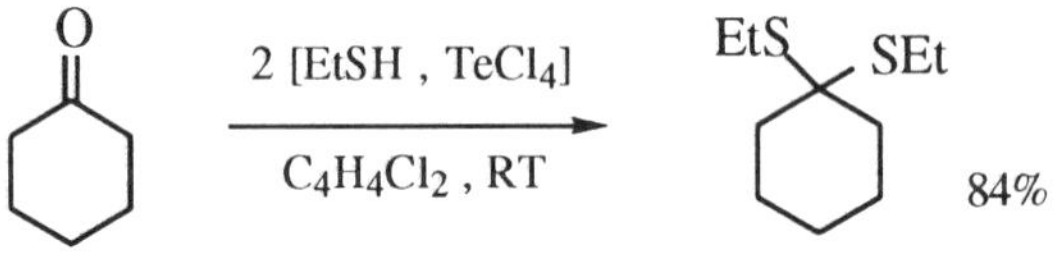
O
2 [EtSH , TeCl4]
C4H4Cl2 , RT
EtS
SEt
84%

Tani, H.; Masumoto, K.; Inamasu, T.; Suzuki, H. *Tetrahedron Lett.*, ***1991***, *32*, 2039.

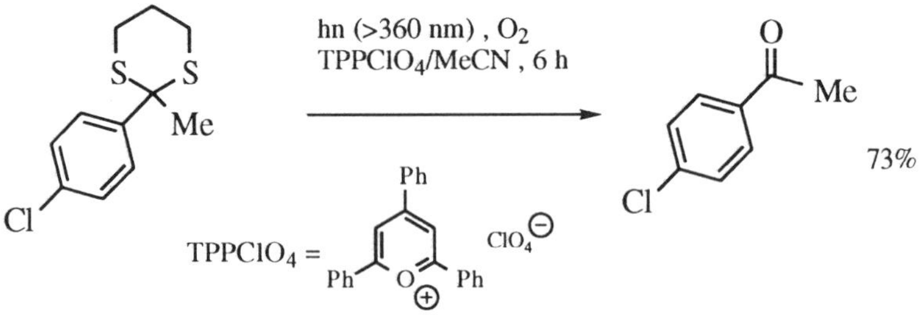
S
S
Me
Cl
hn (>360 nm) , O2
TPPClO4/MeCN , 6 h
TPPClO4 =
Ph
Ph
O
Ph
ClO4
O
Me
Cl
73%

Kamata, M.; Kato, Y.; Hasegawa, E. *Tetrahedron Lett.*, ***1991***, *32*, 4349.

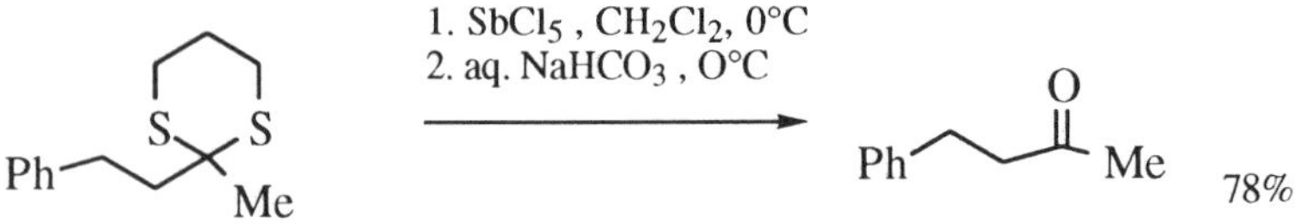
S
S
Ph
Me
1. SbCl5 , CH2Cl2, 0°C
2. aq. NaHCO3 , 0°C
O
Ph
Me
78%

Kamata, M.; Otogawa, H.; Hasegawa, E. *Tetrahedron Lett.*, ***1991***, *32*, 7421.

Cyclohexanone → hydroxy apatite, 1 h; $(EtO)_3SiH$, heptane → cyclohexyl–$OSi(OEt)_3$ 99%

hydroxy apatite = $[Ca_{10}(PO_4)_6(OH)_2]$

Izumi, Y.; Nanami, H.; Higuchi, K.; Onanaka, M. *Tetrahedron Lett.*, ***1991***, *32*, 4741.

1. TMSOTf, Me_2S, CH_2Cl_2, -78°C
2. TMSOTf, $TMSOCH_2CH_2OTMS$
3. aq. K_2CO_3

(CHO → CHO) 80%

Kim, S.; Kim, Y.G.; Kim, D. *Tetrahedron Lett.*, ***1992***, *33*, 2565.

acetone-CH_2Cl_2, 0°C, 24 h; 98% conversion; O >95%

Curci, R.; D'Accolti, L.; Fiorentino, M.; Fusco, C.; Adam, W.; González-Nuñez, M.E.; Mello, R. *Tetrahedron Lett.*, ***1992***, *33*, 4225.

Ph, $SiMe_3$, S S → e^- (Pt electrodes); $NaClO_4$, aq. MeCN → Ph, $SiMe_3$, O 95%

Suda, K.; Watanabe, J.; Takanami, T. *Tetrahedron Lett.*, ***1992***, *33*, 1355.

S S, Ph Me → hν (visible), $MeCN/H_2O$; methylene green → O, Ph Me 91%

Epling, G.A.; Wang, Q. *Tetrahedron Lett.*, ***1992***, *33*, 5909.
Epling, G.A.; Wang, Q. *SynLett*, ***1992***, 335.

Ph SEt, Me SEt → DMSO, 140°C, 4 h → Ph, O, Me 94%

Rao, Ch.S.; Chandrasekharam, M.; Ila, H.; Junjappa, H. *Tetrahedron Lett.*, ***1992***, *33*, 8163.

See Section 362 (Ester-Alkene) for the formation of enol esters and Section 367 (Ether-Alkenes) for the formation of enol ethers. Many of the methods in Section 60A (Protection of Aldehydes) are also applicable to ketones.

CHAPTER 13
PREPARATION OF NITRILES

SECTION 181: NITRILES FROM ALKYNES

NO ADDITIONAL EXAMPLES

SECTION 182: NITRILES FROM ACID DERIVATIVES

NO ADDITIONAL EXAMPLES

SECTION 183: NITRILES FROM ALCOHOLS AND THIOLS

$$PhCH_2OH \xrightarrow[\;]{(NiSO_4/K_2S_2O_8\text{-}NaOH)\; NH_3,\ H_2O} PhCN \quad 82\%$$

Yamazaki, S.; Yamazaki, Y. *Chem. Lett.*, ***1990***, 571.

SECTION 184: NITRILES FROM ALDEHYDES

$$PhCH_2C(CH_3)_2CH{=}N\text{-}NMe_2 \xrightarrow{1.\ MeI,\ 150\ min;\ 2.\ DBU,\ 25°C} PhCH_2C(CH_3)_2CN \quad 83\%$$

Moore, J.S.; Stupp, S.I. *J. Org. Chem.*, ***1990***, *55*, 3374.

$$PhCHO \xrightarrow[2.5\ K_2S_2O_8]{2\ eq.\ Cu(NO_3)_2} \underset{82\%}{PhCN} + \underset{11\%}{PhCOOH}$$

Arora, P.K.; Sayre, L.M. *Tetrahedron Lett.*, ***1991***, *32*, 1007.

$$PhCHO \xrightarrow{(NiSO_4/K_2S_2O_8\text{-}NaOH)\; NH_3,\ H_2O} PhCN \quad 76\%$$

Yamazaki, S.; Yamazaki, Y. *Chem. Lett.*, ***1990***, 571.

$$PhCHO \xrightarrow[\text{microwaves (2450 MHz)}]{NH_2OH \cdot HCl,\ \text{Mexican Bentonite}} PhCN \quad 50\%$$

Delgado, F.; Cano, A.C.; García, O.; Alvarado, J.; Velasco, L.; Alvarez, C.; Rudler, H. *Synth. Commun.*, ***1992***, *22*, 2125.

SECTION 185: NITRILES FROM ALKYLS, METHYLENES AND ARYLS

NO ADDITIONAL EXAMPLES

SECTION 186: NITRILES FROM AMIDES

Ph_3PBr_2, NEt_3, RT; CH_2Cl_2, 10 h — 89%

Walters, M.A.; McDonough, C.S.; Brown, P.S. Jr.; Hoem, A.B. *Tetrahedron Lett.*, ***1991***, *32*, 179.

1. CBr_4, PPh_3, MeCN, RT → reflux; 2. Zn, DMF, AcOH, reflux — 69%

Sakamoto, T.; Mori, H.; Takizawa, M.; Kikygawa, Y. *Synthesis*, ***1991***, 750.

SECTION 187: NITRILES FROM AMINES

1. EtO(C=O)CN, MeCN, cat. NEt_3, 25°C; 2. heat → Et-CN 95%

Thomas, H.G.; Greyn, H.D. *Synthesis*, ***1990***, 129.

4 eq. PPh_3, 2 eq. CCl_4, MeCN, RT — 97%

Kim, J.N.; Chung, K.H.; Ryu, E.K. *Synth. Commun.*, ***1990***, *20*, 2785.

Ph–CH=N·OH —[KSF, montmorillonite, PhMe; reflux, 15 h]→ Ph-C≡N 70%

Meshram, H.M. *Synthesis*, ***1992***, 943.

SECTION 188: NITRILES FROM ESTERS

NO ADDITIONAL EXAMPLES

SECTION 189: NITRILES FROM ETHERS, EPOXIDES AND THIOETHERS

MeO OMe / Ph–C–CH=CH–Ph —[TMS-CM, RT, 5 h; $[Rh(COD)Cl]_2$]→ MeO CN / Ph–C–CH=CH–Ph 86%

Soga, T.; Takenoshita, H.; Yamada, M.; Han, J.S.; Mukaiyama, T. *Bull. Chem. Soc. Jpn.*, ***1991***, *64*, 1108.
Mukaiyama, T.; Takenoshita, H.; Yamada, M.; Soga, T. *Chem. Lett.*, ***1990***, 229 (Z product].

SECTION 190: NITRILES FROM HALIDES AND SULFONATES

PhOTf —[2 eq. KCN, dppf, NMP; 2% $Pd_2(dba)_3$•$CHCl_3$]→ PhCN 96%

Takagi, K.; Sasaki, K.; Sakakibara, Y. *Bull. Chem. Soc. Jpn.*, ***1991***, *64*, 1118.

SECTION 191: NITRILES FROM HYDRIDES

EtO–C_6H_5 —[$AlCl_3$, CS_2; -70°C → 0°C; N≡CO-anthraquinone-OC≡N]→ —[$HCl_{(g)}$]→ EtO–C_6H_4–C≡N 51%

Buttke, K.; Reiher, T.; Niclas, H-J. *Synth. Commun.*, ***1992***, *22*, 2237.

SECTION 192: NITRILES FROM KETONES

Preparations of nitriles from oximes and hydrazones, which are ketone derivatives, are found in Section 195.

NO ADDITIONAL EXAMPLES

SECTION 193: NITRILES FROM NITRILES

Conjugate reductions and Michael alkylations of alkene nitriles are found in Section 74D (Alkyls from Alkenes).

HCHO , H_2 (CO) , DMF
cat. $Ru_3(CO)_{12}$, 230°C
16 h
86%

Abe, F.; Hayashi, T.; Tanaka, M. *Chem. Lett.*, ***1990***, 765.

Rhodococcus butanica
phosphate buffer
30°C , 3 h
0.1% substrate
56% (27% ee) 30% (95% ee)

Kakeya, H.; Sakai, N.; Sugai, T.; Ohta, H. *Tetrahedron Lett.*, ***1991***, *32*, 1343.

K_2CO_2 , EtBr
58% 16%

Xu, C.Z.; Mao, X.J.; Shen, H.B.; Chen, W.Q. *Org. Prep. Proceed. Int.*, ***1991***, *23*, 153.

SECTION 194: NITRILES FROM ALKENES

NO ADDITIONAL EXAMPLES

SECTION 195: NITRILES FROM MISCELLANEOUS COMPOUNDS

The preparation of nitriles from oximes and other compounds are presented here.

1. AlI_3 , MeCN
82°C , 2 h
2. H_2O
Ph-C≡N 92%

Konwar, D.; Boruah, R.C.; Sandhu, J.S. *Tetrahedron Lett.*, ***1990***, *31*, 1063.

$n\text{-}C_{11}H_{23}CH_2NO_2$ $\xrightarrow[\text{DMAP, CH}_2\text{Cl}_2\text{, 0°C, 10 min}]{\text{Sn(SPh)}_4\text{, PBu}_3\text{, DEAD}}$ $n\text{-}C_{11}H_{23}-CN$ 98%

Urpí, F.; Vilarrasa, J. *Tetrahedron Lett.*, ***1990***, *31*, 7499.

Cyclohexylidene=N-NHBz $\xrightarrow[\text{Pt electrodes}]{\text{e}^- \text{(0.7 V), 2.2 F/mol; NaCN, MeOH}}$ Cyclohexyl-CN 70%

Okimoto, M.; Chiba, T. *J. Org. Chem.*, ***1990***, *55*, 1070.

$Ph\text{-}N{=}CCl_2$ $\xrightarrow[\text{15°C}]{\text{e}^-\text{, Hg cathode, Pt anode; DMF-LiClO}_4\text{, Na}_2\text{CO}_3}$ $Ph-N{\equiv}C$ 91%

Guirado, A.; Zapata, A.; Fenor, M. *Tetrahedron Lett.*, ***1992***, *33*, 4779.

REVIEWS:

"The Isocyanide-Cyanide Rearrangement; Mechanism and Preparative Applications" Rüchardt, C.; Meier, M.; Haaf, K.; Pakusch, J.; Wolber, E.K.A.; Müller, B. *Angew. Chem. Int. Ed. Engl.*, ***1991***, *30*, 893.

CHAPTER 14

PREPARATION OF ALKENES

SECTION 196: ALKENES FROM ALKYNES

$n\text{-}C_5H_{11}-C\equiv C-n\text{-}C_5H_{11}$ → 1. $NbCl_5$, HMPA/THF or $TaCl_5$, Zn, DME/PhH; 2. aq. NaOH → $n\text{-}C_5H_{11}$CH=CH$n\text{-}C_5H_{11}$

Nb 74% (>99:1 Z:E)
Ta 85% (>99:1 Z:E)

Kataoka, Y.; Takai, K.; Oshima, K.; Utimoto, K. *Tetrahedron Lett.*, ***1990***, *31*, 365.

1. *t*-BuLi, hexane/ether, -78°C → 25°C; 2. H_2O

SiMe$_3$ — SiMe$_3$ 95%

Wu, G.; Cederbaum, F.E.; Negishi, E. *Tetrahedron Lett.*, ***1990***, *31*, 493.

1. *t*-BuLi, -78°C pentane/ether; 2. -78°C, 1h

Ph, I — Ph 88% + Ph 6% + Ph 6%

Bailey, W.F.; Ovaska, T.V. *Tetrahedron Lett.*, ***1990***, *31*, 627.

OH → $PhMe_2SiH$, THF, 2 h, cat. $Cy_3P\text{-}Pt[\text{-}CH_2{=}CH_2]_2$ → $PhMe_2Si$CH=CHCH$_2$OH + OH, SiMe$_2$Ph (5.2 : 1)

82%

Murphy, P.J.; Spencer, J.L.; Procter, G. *Tetrahedron Lett.*, ***1990***, *31*, 1051.

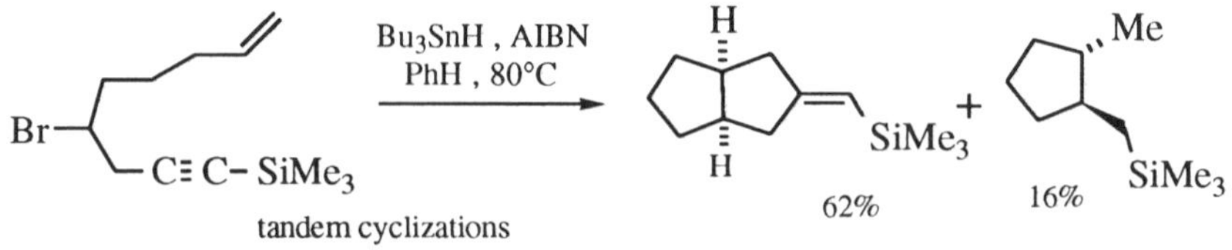
Br
C≡C–SiMe3
Bu3SnH , AIBN
PhH , 80°C
tandem cyclizations
H
H
SiMe3
62%
+
Me
SiMe3
16%

<u>Kilburn, J.D.</u> *Tetrahedron Lett.*, ***1990***, *31*, 2193.

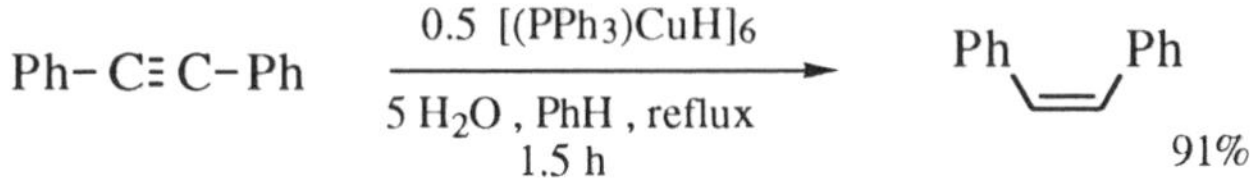
Ph–C≡C–Ph
0.5 [(PPh3)CuH]6
5 H2O , PhH , reflux
1.5 h
Ph
Ph
91%

Daeuble, J.F.; McGettigan, C.; <u>Stryker, J.M.</u> *Tetrahedron Lett.*, ***1990***, *31*, 2397.

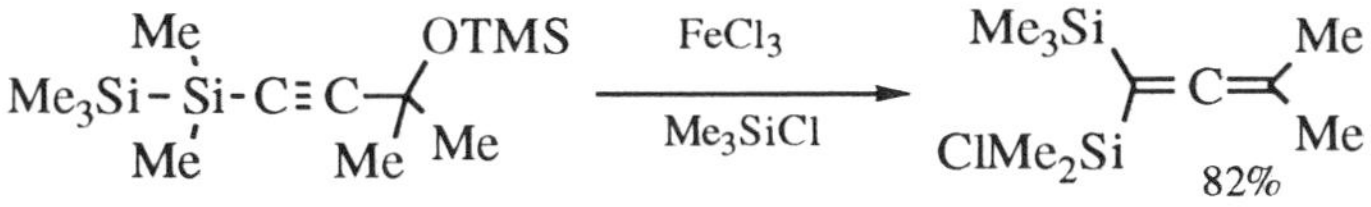
Me
Me3Si–Si–C≡C
Me
OTMS
Me
Me
FeCl3
Me3SiCl
Me3Si
ClMe2Si
C
Me
Me
82%

<u>Cunico, R.F.</u> *Tetrahedron Lett.*, ***1990***, *31*, 5607.

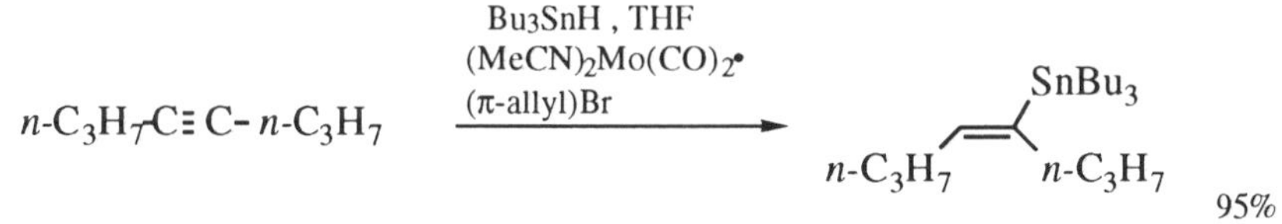
n-C3H7-C≡C-n-C3H7
Bu3SnH , THF
(MeCN)2Mo(CO)2•
(π-allyl)Br
SnBu3
n-C3H7
n-C3H7
95%

Zhang, H.X.; <u>Guibé, F.</u>; Balavoine, G. *J. Org. Chem.*, ***1990***, *55*, 1857.

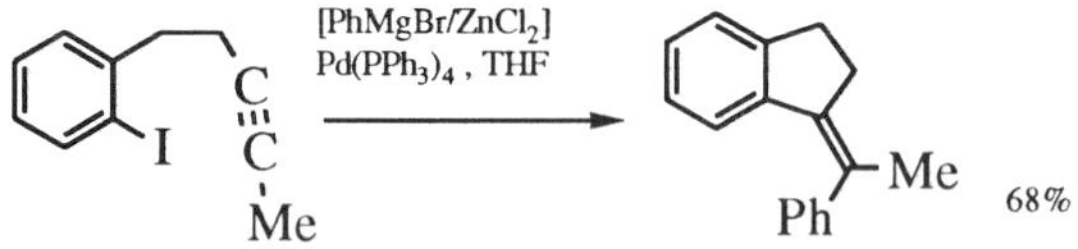
I
C
C
Me
[PhMgBr/ZnCl2]
Pd(PPh3)4 , THF
Me
Ph
68%

Wang, R-T.; Chou, F-L.; <u>Luo, F-T.</u> *J. Org. Chem.*, ***1990***, *55*, 4846.

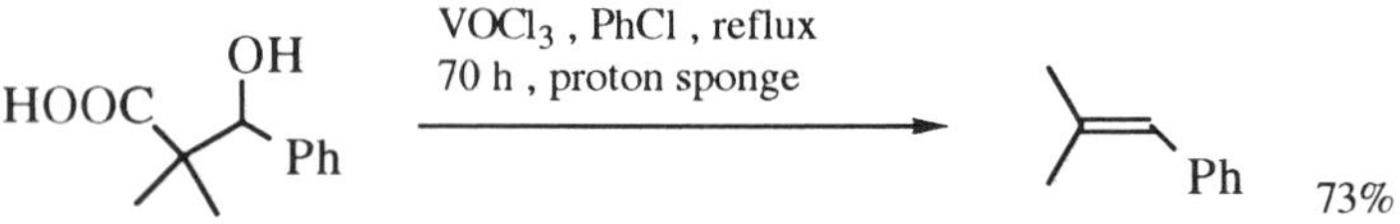
OH
HOOC
Ph
VOCl3 , PhCl , reflux
70 h , proton sponge
Ph
73%

Meier, I.K.; <u>Schwartz, J.</u> *J. Org. Chem.*, ***1990***, *55*, 5619.

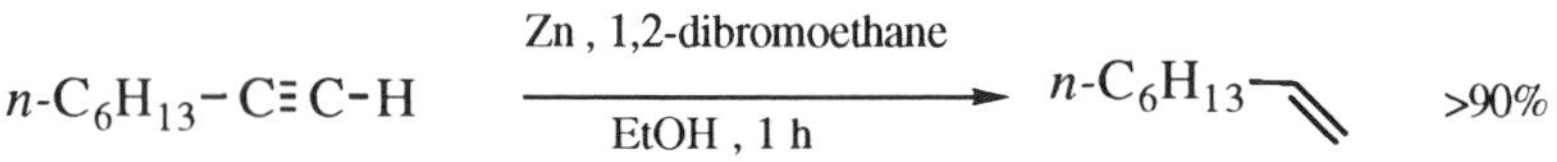
n-C6H13–C≡C–H
Zn , 1,2-dibromoethane
EtOH , 1 h
n-C6H13
>90%

Aerrsens, M.H.P.J.; van der Heiden, R.; Heus, M.; <u>Brandsma, L.</u> *Synth. Commun.*, ***1990***, *20*, 3421.

n-$C_7H_{15}C{\equiv}C$–CH(X)Me ⟶ (n-C_7H_{15})(R_3Sn)C=C=CH(Me) + n-$C_7H_{15}C{\equiv}C$–CH(SnR_3)Me

Bu_3SnLi , THF , 78°C	X=Br, R=Bu	(50	:	50)	67%
$Me_3SnCuBrLi$, THF	X = OTs, R=Me	(100	:	0)	75%

Marshall, J.A.; Wang, X *J. Org. Chem.*, ***1990***, *55*, 6246.

H- C≡C- n-C_5H_{11} —[1. 2 eq. Dibal; 2. BuLi , THF; 3. n-$C_5H_{11}CHO$]→ n-C_5H_{11}CH=CHn-C_5H_{11} 45%

(80:20 Z:E)

Kuchin, A.V.; Markova, S.A.; Gorobets, E.V.; Tolstikov, G.A. *Izv. Akad. Nauk. SSSR*, ***1990***, *39*, 1151 (Engl., p. 1034).

Ph(CH₂)₄C≡C-H —[1. $Bu_3SnSiMe_3$, THF, cat. $Pd(PPh_3)_4$, 60°C; 2. HI , Bu_4NI , PhMe, 0°C]→ Ph(CH₂)₄CH=CH-$SiMe_3$ 90%

Mori, M.; Watanabe, N.; Kaneta, N.; Shibasaki, M. *Chem. Lett.*, ***1991***, 1615.

Ph- C≡C- Ph —[$CoCl_2$•4 PPh_3 , 2 h, iPrOH]→ (Z)-PhCH=CHPh + (E)-PhCH=CHPh

(>99 : 1) 59%

Inanaga, J.; Yokoyama, Y.; Baba, Y.; Yamaguchi, M. *Tetrahedron Lett.*, ***1991***, *32*, 5559.

2-(OTf)C₆H₄CH₂CH₂C≡CH —[Bu_3SnH , LiCl , DMF, 5% $Pd(PPh_3)_4$, 60°C, 8 h]→ 1-(pentylidene)indane (=CH-Bu) 57%

Luo, F-T.; Wang, R-T. *Tetrahedron Lett.*, ***1991***, *32*, 7703.

Ph- C≡C- $SiMe_3$ —[1. HB(c-C6H11)$_2$, THF, 0°C → 25°C; 2. PhBr , THF , NaOH, $Pd(PPh_3)_4$, reflux; 3. NaOH , H_2O_2]→ PhCH=C(Ph)$SiMe_3$

40%

Soderquist, J.A.; León-Colón, G. *Tetrahedron Lett.*, ***1991***, *32*, 43.

Ph·C≡C–(CH2)3–Br → (1.5 eq. SmI_2, THF, 24 h, reflux) → benzylidenecyclopentane (Ph) 60%

Bennett, S.M.; Larouche, D. *SynLett*, ***1991***, 805.

7% $Co(acac)_3$, dppe; 6 eq. Et_2AlCl — C≡C-H — 78%

Lautens, M.; Tam, W.; Edwards, L.G. *J. Org. Chem.*, ***1992***, *57*, 8.

n-C_3H_7-C≡C-H → 1. cat. Zr porphyrin, Et_3Al, 55 h; 2. H^+ → n-C_3H_7, Et 26%

Shibata, K.; Aida, T.; Inoue, S. *Tetrahedron Lett.*, ***1992***, *33*, 1077.

Me–C≡C–$SnBu_3$ → $Cp_2Zr(H)Cl$, THF → Me, $SnBu_3$ 92%

Lipshutz, B.H.; Keil, R.; Barton, J.C. *Tetrahedron Lett.*, ***1992***, *33*, 5861.
Lipshutz, B.H.; Keil, R.; Ellsworth, E.L. *Tetrahedron Lett.*, ***1990***, *31*, 7257.

Ph–C≡C, Br → 1. 4 eq. Mg, THF; 2. Cp_2TiCl_2, CH_2Cl_2; 3. 6 eq. $EtAlCl_2$, PhMe, -78°C → Ph + Ph–C≡C (96 : 4) 42%

Harms, A.E.; Stille, J.R. *Tetrahedron Lett.*, ***1992***, *33*, 6565.

H-C≡C-Ph → 1. TMSCl, NaI, H_2O, MeCN; 2. BuZnCl, cat. $Pd(PPh_3)_4$ → Ph, Bu 54%

Luo, F-T.; Fwu, S-L.; Huang, W-S. *Tetrahedron Lett.*, ***1992***, *33*, 6839.

n-C_6H_{13}-C≡C-n-C_6H_{13} → 1% $RhCl_3$•3 H_2O, 15% $Cu(NO_3)_2$, Sol-Gel matrix → n-C_6H_{13} n-C_6H_{13} (8:1 Z:E) 83%

Tour, J.M.; Pendalwar, S.L.; Kafka, C.M.; Cooper, J.P. *J. Org. Chem.*, ***1992***, *57*, 4786.

SECTION 197: ALKENES FROM ACID DERIVATIVES

Ph–C(=O)–Cl → (10 eq. $TiCl_3$, THF; 5 eq. $LiAlH_4$, 11 h) → (Z)-PhCH=CHPh 74% + (E)-PhCH=CHPh 18%

Dang, Y.; Giese, H.J. *Bull. Soc. Chim. Belg.*, ***1991***, *100*, 375.

SECTION 198: ALKENES FROM ALCOHOLS AND THIOLS

1. TMSCl, HMDS
2. Li, NH_3
3. NH_4Cl

69%

Ballester, P.; Capó, M.; Saá, J.M. *Tetrahedron Lett.*, ***1990***, *31*, 1339.

SmI_2, THF

82%

(8:1 E:Z)

Kende, A.S.; Mendoza, J.S. *Tetrahedron Lett.*, ***1990***, *31*, 7105.

Ph–CH₂CH₂–CHO

1. $Me_3GeCH_2CO_2t$-Bu/LDA THF, -78°C
2. H_2O, -78°C
3. BF_3•OEt_2, CH_2Cl_2, 4 h

CO_2t-Bu 80%

(79:21 E:Z)

also with ketones

Inoue, S.; Sato, Y. *J. Org. Chem.*, ***1991***, *56*, 437.

n-$C_{16}H_{33}$ ... OH

BF_3•OEt_2, CH_2Cl_2

25°C, 10 min

n-$C_{16}H_{33}$ + n-$C_{16}H_{33}$

(95 : 5) 90%

Posner, G.H.; Shulman-Roskes, E.M.; Oh, C.H.; Carry, J-C.; Green, J.V.; Clark, A.B.; Dai, H.; Anjeh, T.E.N. *Tetrahedron Lett.*, ***1991***, *32*, 6489.

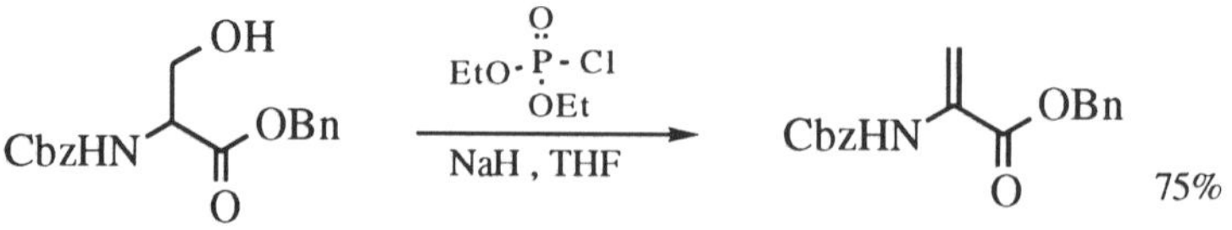
OH
CbzHN
OBn
O
EtO-P-Cl
OEt
NaH, THF
CbzHN
OBn
O
75%

Berti, F.; Ebert, C.; Gardossi, L. *Tetrahedron Lett.*, ***1992***, *33*, 8145.

SECTION 199: ALKENES FROM ALDEHYDES

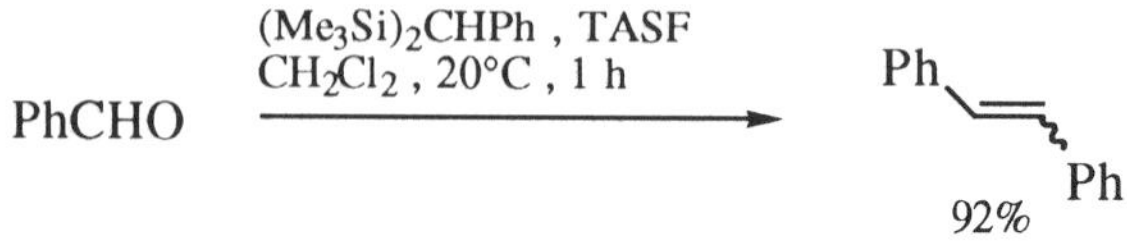
$(Me_3Si)_2CHPh$, TASF
CH_2Cl_2, 20°C, 1 h
PhCHO
Ph
Ph
92%

TASF = tris-(dimethylamino)-sulfonium difluorotrimethyl siliconate

(E:Z = 1:1)

Palomo, C.; Aizpurua, J.M.; García, J.M.; Ganboa, I.; Cossio, F.P.; Lecea, B.; López, C. *J. Org. Chem.*, ***1990***, *55*, 2498.

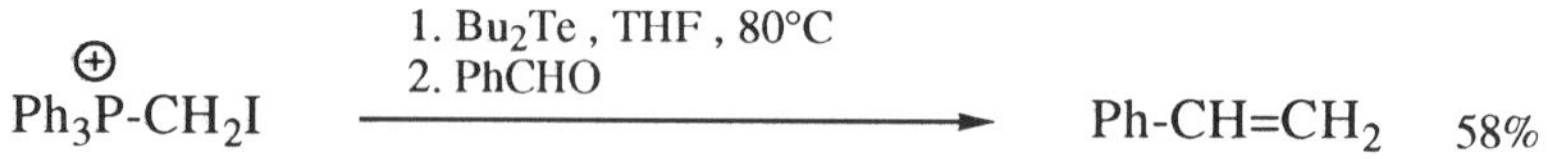
$Ph_3\overset{\oplus}{P}-CH_2I$
1. Bu_2Te, THF, 80°C
2. PhCHO
$Ph-CH=CH_2$
58%

Li, S-W.; Huang, Y-Z.; Shi, L-L. *Chem. Ber.*, ***1990***, *123*, 1441.

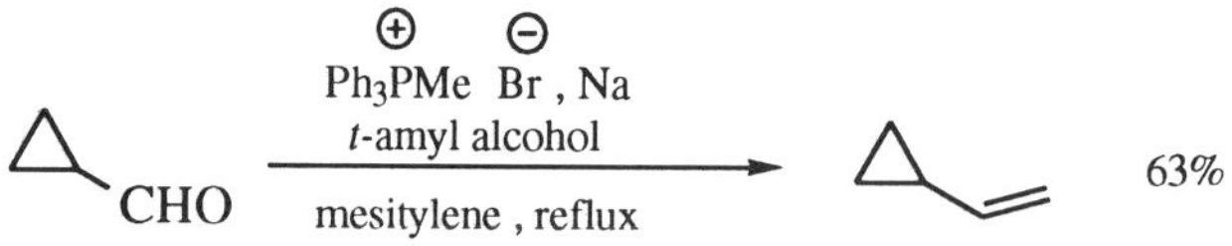
$Ph_3\overset{\oplus}{P}Me$ $\overset{\ominus}{Br}$, Na
t-amyl alcohol
CHO
mesitylene, reflux
63%

Tsunoda, T.; Hudlicky, T. *SynLett*, ***1990***, 322.

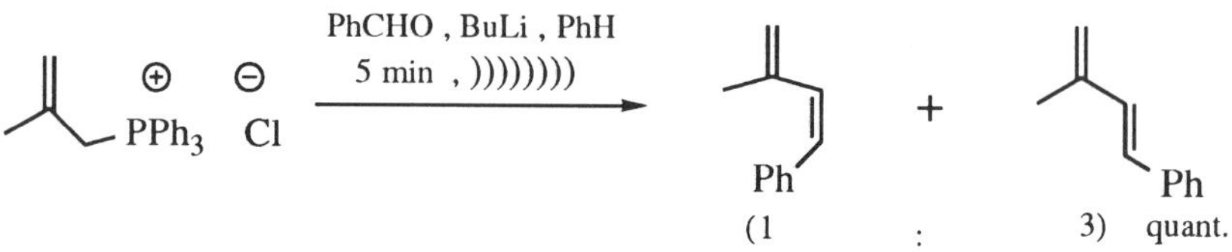
PhCHO, BuLi, PhH
5 min,))))))))
PPh_3
Cl
Ph
+
Ph
(1 : 3) quant.

Low, C.M.R. *SynLett*, ***1991***, 123.

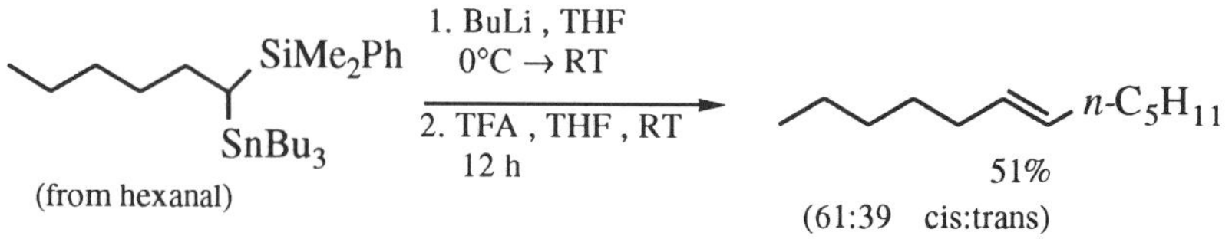
$SiMe_2Ph$
$SnBu_3$
(from hexanal)
1. BuLi, THF
0°C → RT
2. TFA, THF, RT
12 h
n-C_5H_{11}
51%
(61:39 cis:trans)

Barrett, A.G.M.; Hill, J.M. *Tetrahedron Lett.*, ***1991***, *32*, 3285.

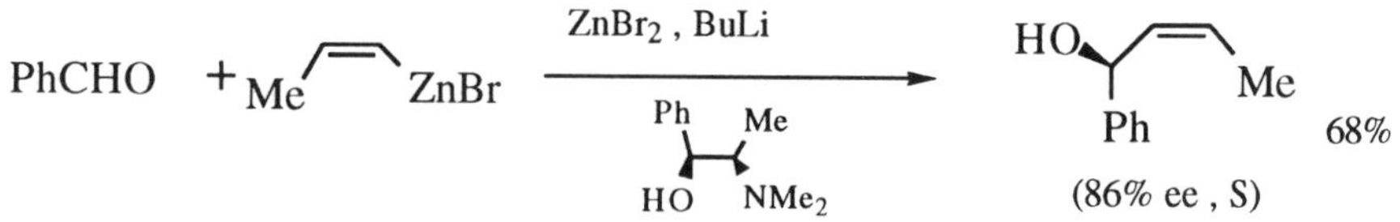

PhCHO + Me–CH=CH–ZnBr —($ZnBr_2$, BuLi; Ph, Me, HO, NMe_2)→ HO–CH(Ph)–CH=CH–Me (86% ee, S) 68%

Oppolzer, W.; Radinov, R.N. *Tetrahedron Lett.*, ***1991***, *32*, 5777.

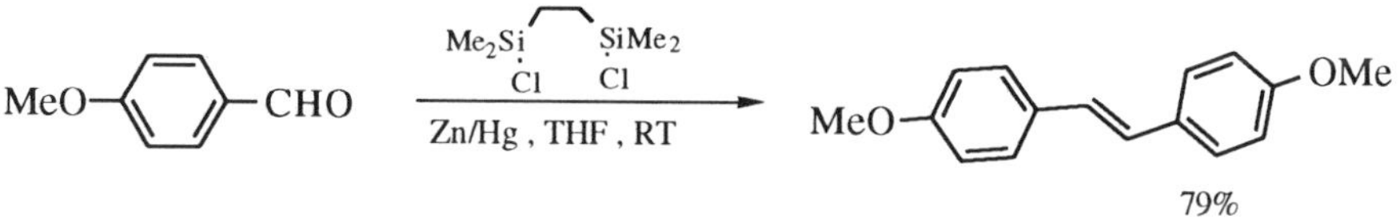

MeO–C_6H_4–CHO —(Me_2Si $SiMe_2$, Cl Cl; Zn/Hg, THF, RT)→ MeO–C_6H_4–CH=CH–C_6H_4–OMe 79%

Afonso, C.A.M.; Motherwell, W.B.; O'Shea, D.M.; Roberts, L.R. *Tetrahedron Lett.*, ***1992***, *33*, 3899.

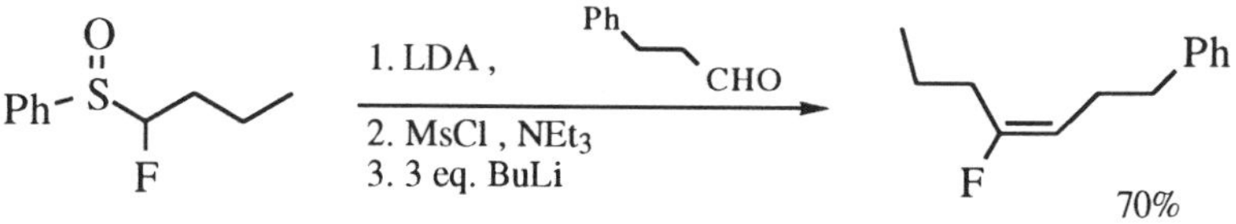

Ph–S(=O)–CH(F)–$CH_2CH_2CH_3$ —(1. LDA, Ph–CH_2CH_2–CHO; 2. MsCl, NEt_3; 3. 3 eq. BuLi)→ F, Ph 70%

Satoh, T.; Itoh, N.; Onda, K.; Kitoh, Y.; Yamakawa, K. *Tetrahedron Lett.*, ***1992***, *33*, 1483.

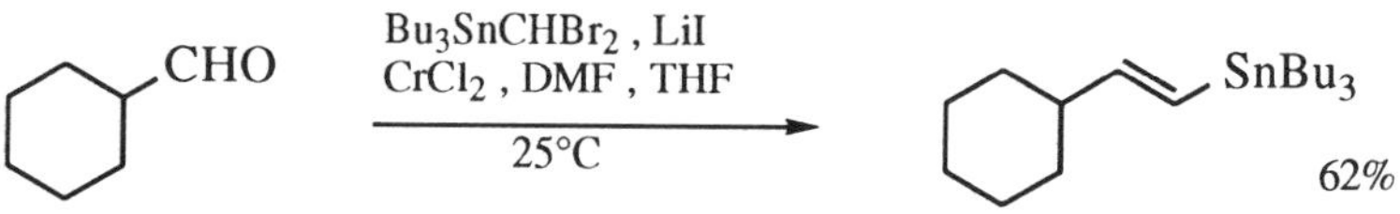

C_6H_{11}–CHO —($Bu_3SnCHBr_2$, LiI; $CrCl_2$, DMF, THF; 25°C)→ C_6H_{11}–CH=CH–$SnBu_3$ 62%

Hodgson, D.M. *Tetrahedron Lett.*, ***1992***, *33*, 5603.

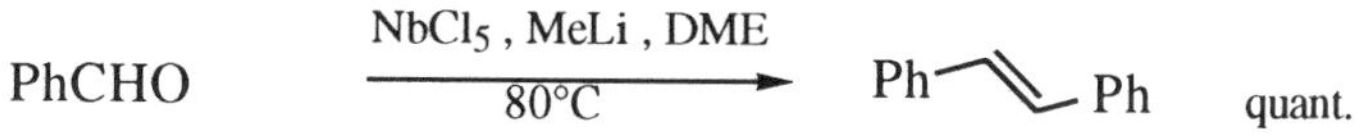

PhCHO —($NbCl_5$, MeLi, DME; 80°C)→ Ph–CH=CH–Ph quant.

Kauffmann, T.; Kallweit, H. *Chem. Ber.*, ***1992***, *125*, 149.

Related Methods: Section 207 (Alkenes from Ketones).

SECTION 200: ALKENES FROM ALKYLS, METHYLENES AND ARYLS

This section contains dehydrogenations to form alkenes and unsaturated ketones, esters and amides. It also includes the conversion of aromatic rings to alkenes. Reduction of aryls to dienes is found in Section 377 (Alkene-Alkene). Hydrogenation of aryls to alkanes and dehydrogenations to form aryls are included in Section 74 (Alkyls from Alkenes).

DCE , $MeCCl_3$, $AlCl_3$

50°C , 30 min

Ph Ph

98%

Sonawane, H.R.; Sudalai, A.; Daniel, T.; Ayyangar, N.R. *SynLett*, ***1991***, 925.

SECTION 201: ALKENES FROM AMIDES

Related Methods: Section 65 (Alkyls from Alkyls).
Section 74 (Alkyls from Alkenes).

NO
N
Ac

2% $Rh_2(OAc)_4$

PhH , reflux

78%

Godfrey, A.G.; Ganem, B. *J. Am. Chem. Soc.*, ***1990***, *112*, 3717.

SECTION 202: ALKENES FROM AMINES

PPh_3

PPh_3

PhMe

N
O
S
O_2

84%

Davis, F.A.; Chen, B. *J. Org. Chem.*, ***1990***, *55*, 360.

n-$C_{10}H_{21}$

Me_3Si NH_2

1. iAmONO , gl. AcOH

2. BF_3•OEt_2

n-$C_{10}H_{21}$

62%

Cunico, R.F. *J. Org. Chem.*, ***1990***, *55*, 4474.

EtO_2C CO_2Et

$=N$-$NHSO_2MES$

Br

Bu_3SnH , AIBN
PhH , reflux

EtO_2C CO_2Et

Me

60%

MES = 2-mesityl

Kim, S.; Cho, J.R *SynLett*, ***1992***, 629.

SECTION 203: ALKENES FROM ESTERS

1. MeLi, ether, 0°C
2. MeCu, 0°C, 23 h
3. 0°C→ 24°C

66% (91% ee)

Denmark, S.E.; Marble, L.K. *J. Org. Chem.*, ***1990***, *55*, 1984.

1. Bu-C≡C-Pb(OAc)$_3$ CH_2Cl_2-THF 23°C, 30 min
2. H_2, Lindlar, PhH 23°C

76% (95:5 Z:E)

Hashimoto, S.; Miyazaki, Y.; Shinoda, T.; Ikegami, S. *J. Chem. Soc., Chem. Commun.*, ***1990***, 1100.

Bu_3SnH, AIBN
PhH, 6 h

63%

Crimmins, M.T.; Dudek, C.M.; Cheung, A.W-H. *Tetrahedron Lett.*, ***1992***, *33*, 181.

$Pd(acac)_2$, PBu_3, PhH
1. [HCOOH/NEt_3/PhH] RT, 30 min
2. H_2O

88%

Mandai, T.; Suzuki, S.; Murakami, T.; Fujita, M.; Kawada, M.; Tsuji, J. *Tetrahedron Lett.*, ***1992***, *33*, 2987.

SECTION 204: ALKENES FROM ETHERS, EPOXIDES AND THIOETHERS

Cp_2Nb, THF, 12 h
-78°C → RT

86% (97:3 Z:E)

Schobert, R.; Höhlein, U. *SynLett*, ***1990***, 465.

1. PhMe, hν, Bu_3Sn—allyl, pyrex, 24 h
2. 10% KF overnight

(60 : 40) 40%

(1.2:1 α:β) (1:2 α:β)

Marco-Contelles, J.L.; Fernández, C.; Gómez, A.; Martín-León, N. *Tetrahedron Lett.*, ***1990***, *31*, 1467.

BuLi, DME, -78°C → RT

3% 85% (98:2 E:Z)

Ukaji, Y.; Yoshida, A.; Fujisawa, T. *Chem. Lett.*, ***1990***, 157.

1. *t*-BuLi
2. *t*-BuLi

93% (16:84 E:Z)

Santiago, B.; Lopez, C.; Soderquist, J.A. *Tetrahedron Lett.*, ***1991***, *32*, 3457.
Soderquist, J.A.; Lopez, C. *Tetrahedron Lett.*, ***1991***, *32*, 6305.

SECTION 205: ALKENES FROM HALIDES AND SULFONATES

2 Et-I

1. *t*-BuLi, ether, -78°C → 25°C
2. Cl_2ZrCp_2
3. 1-vinylcyclohexene
4. 2N HCl

91%

general method - preparation of 1° organolithium reagents

Negishi, E.; Swanson, D.R.; Rousset, C.J. *J. Org. Chem.*, ***1990***, *55*, 5406.

cysteine, Na_2SeO_3, H_2O, 0°C, 5 min

98%

the R,R-diastereomer gave 0% yield of alkene

Yanada, K.; Yanada, R.; Meguri, H. *J. Chem. Soc., Chem. Commun.*, ***1990***, 730.

$PhCHCl_2$ —[iron (II) oxalate , DMF , 1 h; 160°C]→ PhCH=CHPh 92%

Khurana, J.M.; Maikap, G.C.; Mehta, S. *Synthesis*, ***1990***, 731.

—[$=\!\!/^{SnBu_3}$, $Pd_2(dba)_3$; $P(2\text{-furyl})_3$, NMP , RT; 16 h]→ 89%

Farina, V.; Hauck, S.I. *SynLett*, ***1991***, 157.

—[1. 3 eq. $NaNH_2$; 2. NH_3 , ether; < -50°C; 3. aq. NH_4Cl]→ 69%

Isaka, M.; Ando, R.; Morinaka, Y.; Nakamura, E. *Tetrahedron Lett.*, ***1991***, *32*, 1339.

—[1. 2.2 *t*-BuLi , ether , -100°C; 2. $CuBr{\cdot}SMe_2$, AcCl; 3. BuLi; 4. X]→

X = TFA/TBAF 94:6 E:Z) 59%
X = KH , TBAF 5:95 E:Z) 57%

Barrett, A.G.M.; Flygare, J.A. *J. Org. Chem.*, ***1991***, *56*, 638.

—[DMF , 160°C , 210 min]→ 66%

Khurana, J.M.; Maikap, G.C. *J. Org. Chem.*, ***1991***, *56*, 2582.
Khurana, J.M.; Maikap, G.C.; Sahoo, P.K. *Synthesis*, ***1991***, 827.

—[cyclohexylidene=C=CH_2 , 80°C; Na_2CO_3 , 2 d]→ 82%

Larock, R.C.; Berrios-Peña, N.G.; Fried, C.A. *J. Org. Chem.*, ***1991***, *56*, 2615.

n-C_5H_{11}CH(Cl)$SiMe_2Ph$ → 1. Mg* (ether) 2. $CuBr{\cdot}SMe_2$ 3. BuCOCl 4. MeLi 5. X → n-C_5H_{11}CH=C(Me)Bu

Mg* = activatedMag

X = TsOH 60% (5:95 E:Z)
X = KH 61% (95:5 E:Z)

Barrett, A.G.M.; Hill, J.M.; Wallace, E.M. *J. Org. Chem.*, ***1992***, *57*, 386.

PhCH(Br)CH(Br)Ph → Cu (powder), MeOH, $Cu(ClO_4)_2{\cdot}(H_2O)_6$, RT, 4 h → PhCH=CHPh 88%

Vijayashree, N.; Samuelson, A.G *Tetrahedron Lett.*, ***1992***, *33*, 559.

(bis-OMs) → NaI, Zn, HMPA, heat → 82%

Ghosh, S.; Karpa, A.; Saha, G.; Patra, D. *Tetrahedron Lett.*, ***1992***, *33*, 2363.

n-C_6H_{11}-Br → CH_2=CHCH$_2SO_2Ph$, Zn, DMF; 0.05 Co (porphyrin), 20°C → CH_2=CHCH$_2$-n-C_6H_{11} 85%

Giese, B.; Erdmann, P.; Göbel, T.; Springer, R. *Tetrahedron Lett.*, ***1992***, *33*, 545.

SECTION 206: ALKENES FROM HYDRIDES

For conversions of methylenes to alkenes (RCH_2R' → $RR'C=CH_2$), see Section 200 (Alkenes from Alkyls).

cyclooctane → $RhCl(CH_2{=}CH_2)(PMe_3)_2$ 0.7 mmol/dm^3, 70°C; hn (l > 375 nm), $CH_2{=}CH_2$ (1 atm) → cyclooctene

turnover rate = 160

Sakakura, T.; Abe, F.; Tanaka, M. *Chem. Lett.*, ***1991***, 297, 359.

SECTION 207: ALKENES FROM KETONES

$(MeO)_2P(O)CHN_2$, t-BuOK, THF, 5h, -78°C — 68%

Ohira, S.; Ishi, S.; Shinohara, K.; Nozaki, H. *Tetrahedron Lett.*, ***1990***, *31*, 1039.

1. t-BuLi 2. $SOCl_2$ — 76%

Olah, G.A.; Wu, A.; Farooq, O.; Surya Prakash, G.K. *J. Org. Chem.*, ***1990***, *55*, 1792.

$Ph_3P{=}CHCO_2Et$, 70°C, 4 h — EtO_2C — Me — 51% [43% ee, (-)]

enantioselective Wittig-Horner in **solid state**

Toda, F.; Akai, H. *J. Org. Chem.*, ***1990***, *55*, 3446.

Cp_2TiMe_2, PhMe, 65°C — 90%

Petasis, N.A.; Bzowej, E.I. *J. Am. Chem. Soc.*, ***1990***, *112*, 6392.

$TiCl_4$, Hg, Mg, THF, 0°C → reflux, 24h — 87%

Carroll, A.R.; Taylor, W.C. *Aust. J. Chem.*, ***1990***, *43*, 1439.

1. Me_3SiCH_2Li, ether 2. aq. H^+ 3. Nafion-H, CH_2Cl_2, RT, 1 h — 90%

Olah, G.A.; Reddy, V.P.; Surya Prakash, G.K. *Synthesis*, ***1991***, 29.

1. LDA, THF; Tf_2NPh -78°C → 0°C
2. Cl–C_6H_4–$B(OH)_2$, $Pd(PPh_3)_4$, LiCl, Na_2CO_3, DME, H_2O, reflux

O= (piperidine) N–Boc → Cl–C_6H_4–(tetrahydropyridine) N–Boc 83%

Wustrow, D.J.; Wise, L.D. *Synthesis*, ***1991***, 993.

1. LDA, THF, -78°C
2. acetone

SO_2Me 51%

poor yields with saturated aldehydes

Baudin, J.B.; Hareau, G.; Julia, S.A.; Ruel, O. *Tetrahedron Lett.*, ***1991***, *32*, 1175.

Me N

1. 9-BBN
2. NaOH

Me 70%

Singaram, B.; Rangaishenvi, M.V.; Brown, H.C.; Goralski, C.T.; Hasha, D.L *J. Org. Chem.*, ***1991***, *56*, 1543.

O Ph Me

DME, $TiCl_3Li$, 16 h reflux

Ph Me Me Ph

83% E; 9% Z

Nayak, S.K.; Banerji, A. *J. Org. Chem.*, ***1991***, *56*, 1940.

S S Ph Ph

MeMgI, 18 h
6% $NiCl_2(PPh_3)_2$

Ph Ph 70%

Ni, Z-J.; Mei, N-W.; Shi, X.; Tzeng, Y-L.; Wang, M.C.; Luh, T-Y. *J. Org. Chem.*, ***1991***, *56*, 4035.

O

Cp_2TiBn_2, PhMe 55°C

Ph 75% (1:1 E:Z)

Petasis, N.A.; Bzowej, E.I. *J. Org. Chem.*, ***1992***, *57*, 1327.

1. BuLi , [chiral phosphonamide, Me, N, O, P, Ph, N, Me]
2. H_2O
3. AcOH

O, t-Bu → Ph, t-Bu (91% ee ,R)

Hanessian, S.; Beaudoin, S. *Tetrahedron Lett.*, ***1992***, *33*, 7655, 7659.

$n\text{-}C_6H_{13}$, O

$Me_3SiC(N_2)Li$, THF
-78°C → 0°C

$n\text{-}C_6H_{13}$ 49%

Ohira, S.; Okai, K.; Moritani, T. *J. Chem. Soc., Chem. Commun.*, ***1992***, 721.

O, Ph

$(Me_3SiCH_2)_3TiCp$, PhMe
100°C

$SiMe_3$, Ph 97%

Petasis, N.A. Akritopoulou, I. *SynLett*, ***1992***, 665.

Related Methods: Section 199 (Alkenes from Aldehydes).

SECTION 208: ALKENES FROM NITRILES

NO ADDITIONAL EXAMPLES

SECTION 209: ALKENES FROM ALKENES

Bu_3Sn, OEt, OEt

PhBr (sealed tube)
PhH , 80°C , 20 h
2% $Pd(PPh_3)_4$

Ph, OEt, OEt 75% (E/Z 85/15)

Parrain, J-L.; Duchene, A.; Quintard, J-P. *Tetrahedron Lett.*, ***1990***, *31*, 1857.

OTHP, OTHP

1. $n\text{-}C_3H_7MgBr$, ether
$NiCl_2(dppe)$
2. aq. NH_4Cl

OTHP 73%

Malanga, C.; Menicagli, R.; Lardicci, L. *Gazz. Chim. Ital.*, ***1990***, *120*, 217.

1. 15% bpy•Fe(0) PhH, 25°C, 30 min
2. cat. TsOH ethylene glycol

85%

Takacs, B.E.; Takacs, J.M. *Tetrahedron Lett.*, ***1990***, *31*, 2865.

β-cyclodextrin, $CeCl_3$•7 H_2O
$CoCl_2$, KCN, KCS, 24 h
H_2 (1 atm), RT, 3N NaOH

(98 : 2) 95%

Lee, J-T.; Alper, H. *J. Org. Chem.*, ***1990***, *55*, 1854.

$ClSiMe_3$, 10°C 2 min

MeO MeO 84%

Verma, P.; Ray, S. *Ind. J. Chem.*, ***1990***, *29B*, 652.

Me , 25°C, 14 h SiO_2, glass ampule Me 85%

Veselovskii, V.V.; Gybini, A.S.; Lozanova, A.V.; Moiseenkov, A.M.; Smit, V.A. *Izv. Akad. Nauk. SSSR*, ***1990***, *39*, 107 (Engl., p. 94).

CH_2=C=CH_2

1. Bu_6Sn_2/BuLi, CuCN THF, -78°C
2. MeI, -78°C → 0°C

$SnBu_3$ Me 96%

Barbero, A.; Cuadrado, P.; Fleming, I.; González, A.M.; Pulido, F.J. *J. Chem. Soc., Chem. Commun.*, ***1990***, 1030.

n-C_6H_{13} =C=

2 eq. Cl_2AlH
5% $PhB(OH)_2$

n-C_6H_{13} 92%

Nagahara, S.; Maruoka, K.; Doi, Y.; Yamamoto, H. *Chem. Lett.*, ***1990***, 1595.

Me_3Si → (1. BuLi, THF; 2. t-BuOH, -50°C, 24 h; 3. BuI) → Me_3Si ... Bu + Me_3Si ... Bu

(83 : 17) 74%

Degl'Innocenti, A.; Mordini, A.; Pagliai, L.; Ricci, A. *SynLett*, ***1991***, 155.

$Et_4N^+[(CO)_5Cr(\mu\text{-}H)Cr(CO)_5]^-$
H_2 (50 atm), DME, 100°C

90%

Fuchikami, T.; Ubukata, Y.; Tanaka, Y. *Tetrahedron Lett.*, ***1991***, *32*, 1199.

Ph, Me, Me, Si, N, MeO → (1. *sec*-BuLi, ether; 2. MeI) → Ph, Me, Me, Si, Me, N, MeO + Ph, Me, Me, Si, Me, N, MeO

(81 : 15) 92%

(94% ee, S)

Lamothe, S.; Chan, T.H. *Tetrahedron Lett.*, ***1991***, *32*, 1847.

Ph ... Ph → (1. Mg*, THF, RT; 2. -78°C → RT, Br⌒⌒Br) → Ph ... Ph

65%

Mg* = [$MgCl_2$/THF/Li naphth.]

Rieke, R.D.; Xiong, H. *J. Org. Chem.*, ***1991***, *56*, 3109; *J. Org. Chem.*, ***1992***, *57*, 6560. Xiong, H.; Rieke, R.D. *Tetrahedron Lett.*, ***1991***, *32*, 5269.

TsO, Ph, $SnBu_3$ → (BuLi, THF; -20°C → 0°C) → Ph

61%

Barbero, A.; Cuadrado, P.; González, A.M.; Pulido, F.J.; Rubio, R.; Fleming, I. *Tetrahedron Lett.*, ***1992***, *33*, 5841.

n-C_6H_{13} → (1. $Cp_2Zr(H)Cl$, THF; 2. Ph⌒⌒O·P(O)(OEt)(OEt), cat. CuCN) → Ph, n-C_8H_{17} + Ph ... n-C_8H_{17}

(93 : 7) 85%

Venanzi, L.M.; Lehmann, R.; Keil, R.; Lipshutz, B.H. *Tetrahedron Lett.*, ***1992***, *33*, 5857.

1. 9-BBN
2. 3% $Pd(PPh_3)_4$
dioxane , THF
K_3PO_4 , 60°C

Br n-C_8H_{17} → n-C_8H_{17} 80%

Miyaura, N.; Ishikawa, N.; Suzuki, A. *Tetrahedron Lett.*, ***1992***, *33*, 2571.

1. 180°C , 10 h
2. TBAF-DMF

Me Me O Si Ph Me → OH Ph Me 68%

(85:15 cis-4,5:trans-4,5)

Stork, G.; Chan, T.Y.; Breault, G.A. *J. Am. Chem. Soc.*, ***1992***, *114*, 7578.

Bu_3SnH , $Pd(PPh_3)_4$

Me → Me $SnBu_3$ 45%

(36:64 E:Z)

Miyake, H.; Yamamura, K. *Chem. Lett.*, ***1992***, 473.

REVIEWS:

"Ketene Dithioacetals in Organic Synthesis: Recent Developments"
Kolb, M. *Synthesis*, ***1990***, 171.

"Acyclic Stereocontrol via [2,3]-Wittig Sigmatropic Rearrangement"
Mikami, K.; Nakai, T. *Synthesis*, ***1991***, 594.

SECTION 210: ALKENES FROM MISCELLANEOUS COMPOUNDS

FVP (350°C)
10^{-5} mm Hg

N_2 Ph P OMe OMe O → Ph 60%

Tomioka, H.; Watanabe, M.; Kobayashi, N.; Hirai, K. *Tetrahedron Lett.*, ***1990***, *31*, 5061.

THF, 1 mM / reflux

R = Ac (7.3 : 1) 98%
36.4:1 cis anti:cis syn 1.7:1 exo:endo

R = CH_2OH (1 : 1.9) 91%
13.2:1 cis anti:cis syn 1:1.3 exo:endo

Masjededizadeh, M.R.; Dannecker-Doerig, I.; Little, R.D. *J. Org. Chem.*, ***1990***, *55*, 2742.

n-C_8H_{17} — $C_{10}H_8Na$, THF — n-C_8H_{17} 76%

Beels, C.M.D.; Coleman, M.J.; Taylor, R.J.K. *SynLett*, ***1990***, 479.

$(CF_3SO_2)_2O$, CH_2Cl_2 / -78°C → -50°C

(93 : 7) 90%

Creary, X.; Wang, Y-X.; Gill, W. *Tetrahedron Lett.*, ***1991***, *32*, 729.

$Fe(CO)_3$ — 1. 2 eq. Ph_2CHLi, THF HMPA / 2. TFA, -78°C, 1 h / 3. aq. $NaHCO_3$ — 67%

Donaldson, W.A.; Hossain, M.A. *Tetrahedron Lett.*, ***1992***, *33*, 4107.

Ph_3COTf, MeCN / 2,6-lutidine / 60°C — 68% (>99% ee, S)

Denmark, S.E.; Chen, C-T. *J. Am. Chem. Soc.*, ***1992***, *114*, 10674.

$Fe(CO)_4$ — BnCu(CN)ZnI / THF, 0°C — 60%

Yeh, M-C.P.; Tau, S-I. *J. Chem. Soc., Chem. Commun.*, ***1992***, 13.

PhO_2S (Li) n-$C_{11}H_{23}$ —[$ClCH_2MgCl$, THF; -78°C → RT]→ $CH_2{=}CH$-n-$C_{11}H_{23}$ 82%

DeLima, C.; Julia, M.; Verpeaux, J-N. *SynLett, **1992***, 133.

REVIEWS:

"Methods for the Synthesis of Allyl Silanes"
Sarkar, T.K. *Synthesis, **1990***, 969, 1101.

"Vinyl Phosphonates in Organic Synthesis"
Minami, T.; Motoyoshiya, J. *Synthesis, **1992***, 333.

CHAPTER 15

PREPARATION OF OXIDES

This chapter contains reactions which prepare the oxides of nitrogen, sulfur and selenium. Included are *N*-oxides, nitroso and nitro compounds, nitrile oxides, sulfoxides, selenoxides and sulfones. Oximes are considered to be amines and appear in those sections. Preparation of sulfonic acid derivatives are found in Chapter Two and the preparation of sulfonates in Chapter Ten.

SECTION 211: OXIDES FROM ALKYNES

NO ADDITIONAL EXAMPLES

SECTION 212: OXIDES FROM ACID DERIVATIVES

(p-tolyl)SO_2Cl → (p-tolyl)SO_2Me

1. Na_2SO_3, $NaHCO_3$, H_2O, 75°C
2. $ClCH_2COOH$, 19 h, aq. NaOH, reflux

Brown, R.W. *J. Org. Chem.*, ***1991***, *56*, 4974.

SECTION 213: OXIDES FROM ALCOHOLS AND THIOLS

PhSH → $PhSO_2Cl$ 82%

2 eq. SO_2Cl_2, MeCN, 0°C; KNO_3

Park, Y.J.; Shin, H.H.; Kim, Y.H. *Chem. Lett.*, ***1992***, 1483.

SECTION 214: OXIDES FROM ALDEHYDES

CHO → (N-oxide: Me, ⊕, O ⊖, N, Me) good yield

MeNHOH, 25°C

Ciganek, E. *J. Org. Chem.*, ***1990***, *55*, 3007.

Ph CHO → Ph P(=O)(Ph)Ph 92%

1. Ph_2PH, $NbCl_5$, CH_2Cl_2, -78°C
2. BF_3•OEt_2, -78°C → RT

Suzuki, K.; Hashimoto, T.; Maeta, H.; Matsumoto, T. *SynLett*, ***1992***, 125.

SECTION 215: OXIDES FROM ALKYLS, METHYLENES AND ARYLS

NO ADDITIONAL EXAMPLES

SECTION 216: OXIDES FROM AMIDES

NO ADDITIONAL EXAMPLES

SECTION 217: OXIDES FROM AMINES

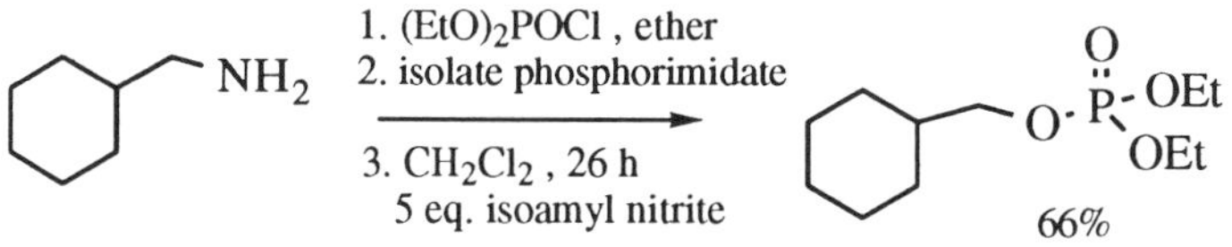

NH_2

1. $(EtO)_2POCl$, ether
2. isolate phosphorimidate
3. CH_2Cl_2, 26 h
5 eq. isoamyl nitrite

O-P(O)(OEt)OEt

66%

Nikolaides, N.; Ganem, B. *Tetrahedron Lett.*, ***1990***, *31*, 1113.
Luo, J.; Ganem, B. *Tetrahedron Lett.*, ***1991***, *32*, 3145 [with $(BnO_2)POH$].

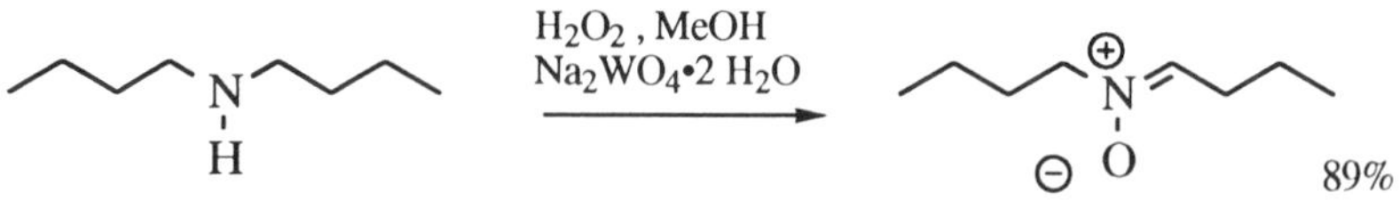

N-H

H_2O_2, MeOH
Na_2WO_4•2 H_2O

$N^{\oplus}$ $O^{\ominus}$

89%

Murahashi, S-I.; Mitsui, H.; Shiota, T.; Tsuda, T.; Watanabe, S. *J. Org. Chem.*, ***1990***, *55*, 1736.

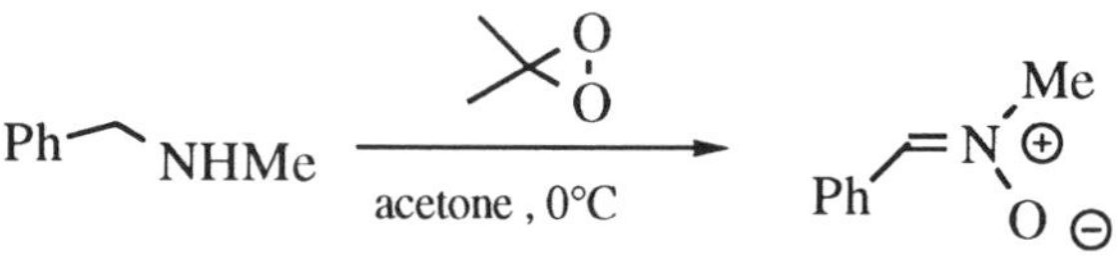

Ph NHMe

acetone, 0°C

Ph N Me O

Murray, R.W.; Singh, M. *J. Org. Chem.*, ***1990***, *55*, 2954.

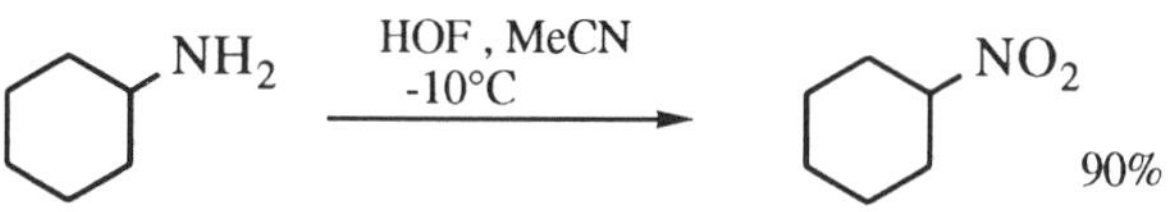

NH_2

HOF, MeCN
-10°C

NO_2

90%

Rozen, S.; Kol, M. *J. Org. Chem.*, ***1992***, *57*, 7342.

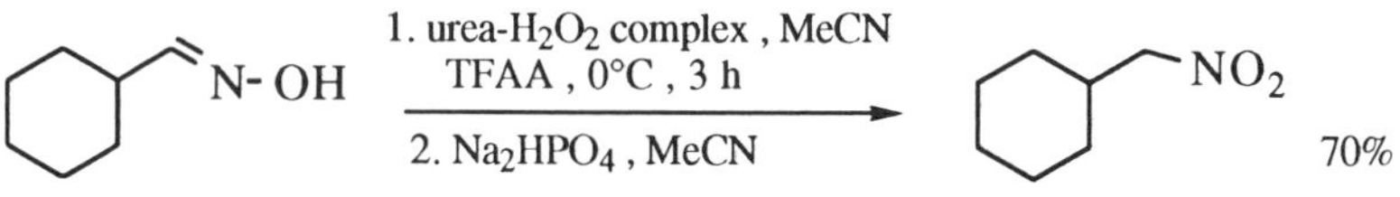

N-OH

1. urea-H_2O_2 complex, MeCN
TFAA, 0°C, 3 h
2. Na_2HPO_4, MeCN

NO_2

70%

Ballini, R.; Marcantoni, E.; Petrini, M. *Tetrahedron Lett.*, ***1992***, *33*, 4835.

$NaBO_3 \cdot 4\ H_2O$
AcOH , 60°C , 2 h

=N-OH → $-NO_2$ 58%

Olah, G.A.; Ramaiah, P.; Lee, G.K.; Surya Prakash, G.K. *SynLett,* ***1992***, 337.

SECTION 218: OXIDES FROM ESTERS

NO ADDITIONAL EXAMPLES

SECTION 219: OXIDES FROM ETHERS, EPOXIDES AND THIOETHERS

$HP(=O)[OH]_2$, Ac_2O
dioxane , reflux
2 h

O → O P=O O H 74%

Klosinski, P. *Tetrahedron Lett.,* ***1990***, *31*, 2025.

0.1 TeO_2 , 0.01 HCl
H_2O_2 , 2 h

Ph S → O Ph S 92%

Kim, K.S.; Hwang, H.J.; Cheong, C.S.; Hahn, C.S. *Tetrahedron Lett.,* ***1990***, *31*, 2893.

O O

CH_2Cl_2

S → S O_2 93%

Miyahara, Y.; Inazu, T. *Tetrahedron Lett.,* ***1990***, *31*, 5955.

Bu_4N^+ $AuCl_4^-$, $MeNO_2$
aq. HNO_3 , 6 h

Ph S → Ph S O 92%

Gasparrini, F.; Giovannoli, M.; Misiti, D.; Natile, G.; Palmieri, G. *J. Org. Chem.,* ***1990***, *55*, 1323.

Me_3PhNBr , Py , H_2O , 3 h
10°C - RT

S → S=O 69%

Rábai, J.; Kapovits, I.; Tanács, B.; Tamás, J. *Synthesis,* ***1990***, 847.

30% H_2O_2 , MeOH , RT
cat. iPrOH/H_2SO_4

85%

Drabowicz, J.; Łyżwa, P.; Popielarczyk, M.; Mikołajczyk, M. *Synthesis*, ***1990***, 937.

$PhSO_2Cl$ → $RuCl_2(PPh_3)_3$, PhH; SPh , 140°C , 24 h → SO_2Ph (71%) + PhS SPh, Cl (54%)

Kamigata, N.; Ishii, K.; Ohtsuka, T.; Matsuyama, H. *Bull. Chem. Soc. Jpn.*, ***1991***, *64*, 3479.

Ph-S-Me → oxone , "wet" kaolin; CH_2Cl_2 , reflux , 4 h → $PhSO_2Me$ 97%

Hirano, M.; Tomaru, J.; Morimoto, T. *Bull. Chem. Soc. Jpn.*, ***1991***, *64*, 3752.
Hirano, M.; Tomaru, J.; Morimoto, T. *Chem. Lett.*, ***1991***, 523 [with montmorillonite].

Mortierella isabellina (ATCC 42613)

48%

(58% ee , R)

Holland, H.L.; Rand, C.G.; Viski, P.; Brown, F.M. *Can. J. Chem.*, ***1991***, *69*, 1989.

$Bu(MnO_4)_2$, MeCN
reflux , 4 h

88%

Firouzabadi, H.; Seddighi, M. *Synth. Commun.*, ***1991***, *21*, 211.

Ph-I=O , 0.1 $PhSeO_2H$
45°C , 1 h

86%

Roh, K.R.; Kim, K.S.; Kim, Y.H. *Tetrahedron Lett.*, ***1991***, *32*, 793.

Ti(OiPr)$_4$, *t*-BuOOH
, H_2O , 45 h

88%

(73% ee , R)

Komatsu, N.; Nishibayashi, Y.; Sugita, T.; Uemura, S. *Tetrahedron Lett.*, ***1992***, *33*, 5391.

Ph-S-Me —(1% OsO_4, NMO)→ Ph-S(O_2)-Me 99%

Kaldor, S.W.; Hammond, M. *Tetrahedron Lett.*, ***1991***, *32*, 5043.

Ph-S-Ph —(H_5IO_6-RuO_4, 3 h; CCl_4/MeCN/H_2O)→ Ph-SO_2-Ph 76%

Rodríguez, C.M.; Ode, J.M.; Palazón, J.M.; Martín, V.S. *Tetrahedron*, ***1992***, *48*, 3571.

Ph-S-Me —(3% Mn (salen) derivative; HOOH, MeOH)→ Ph-S(O)-Me 90%

(47% ee, S)

Palucki, M.; Hanson, P.; Jacobsen, E.N. *Tetrahedron Lett.*, ***1992***, *33*, 7111.

Ph-S-Ph —(NaOBr, hexane/EtOAc; "wet" montmorillonite; 15 min)→ Ph-S(O)-Ph 78%

Hirano, M.; Kudo, H.; Morimoto, T. *Bull. Chem. Soc. Jpn.*, ***1992***, *65*, 1744.

Tol-S-Me —(CCl_4, 4 h; Me, Me, Cl, Cl, Me, O, N, SO_2Ph, Me)→ Tol-S(O)-Me 95%

(>95% ee, S)

Davis, F.A.; Reddy, R.T.; Han, W.; Carroll, P.J. *J. Am. Chem. Soc.*, ***1992***, *114*, 1428.

Ph⁀S⁀Ph —(PhIO, HCl-silica gel; "pulverization")→ Ph⁀SO_2Cl 98%

Sohmiya, H.; Kimura, T.; Fujita, M.; Ando, T. *Chem. Lett.*, ***1992***, 891.

Ph-S-Me —(1 eq. oxone, Al_2O_3; CH_2Cl_2, reflux)→ Ph-S(O)-Me 96%

3 eq. of oxone led to the sulfone

Greenhalgh, R.P. *SynLett*, ***1992***, 235.

REVIEWS:

"Some Routes to Chiral Sulfoxides with Very High Enantiomeric Excesses"
Kagan, H.B.; Reviere, F. *SynLett*, ***1990***, 643.

SECTION 220: OXIDES FROM HALIDES AND SULFONATES

Ph⌒Cl → [1. $P(OEt)_3$, 130°C; 2. Me_3SiCl, KI, MeCN, 60°C; 3. H_2O, RT, 5 min] → $PhCH_2P(O)(OH)_2$ 87%

Katritzky, A.R.; Pilarski, B.; Johnson, J.W. *Org. Prep. Proceed. Int.*, ***1990***, *22*, 209.

$PhCH_2Cl$ → [$TosSO_2Na$, aq. DMF, 3 min,)))))))] → $PhCH_2SO_2Tol$ 85%

Biswas, G.K.; Jash, S.S.; Bhattacharyya, P. *Ind. J. Chem.*, ***1990***, *29B*, 491.

$PhCH_2Cl$ → [$PhSO_2Na$, Al_2O_3,))))))))] → $PhCH_2SO_2Ph$ 97%

Villemin, D.; Ben Alloum, A. *Synth. Commun.*, ***1990***, *20*, 925.

SECTION 221: OXIDES FROM HYDRIDES

Adamantane → [PCl_3, $AlCl_3$, 16 h, CH_2Cl_2, reflux] → 1-adamantyl-$P(O)Cl_2$ 64%

Olah, G.A.; Farooq, O.; Wang, Q.; Wu, A. *J. Org. Chem.*, ***1990***, *55*, 1224.

Ethylbenzene → [NO_2/O_3, CH_2Cl_2, -10°C, 3 h] → nitroethylbenzene (NO_2) 79%

(*o:m:p* 44:3:53)

Suzuki, H.; Murashima, T.; Shimizu, K.; Tsukamoto, K. *Chem. Lett.*, ***1991***, 817.
Suzuki, H.; Murashima, T.; Shimizu, K.; Tsukamoto, K. *J. Chem. Soc., Chem. Commun.*, ***1991***, 1049.

SECTION 222: OXIDES FROM KETONES

Ph–C(=S)–$SiMe_3$ → (1. mCPBA; 2. TBAF , aq. THF, -50°C) → Ph–CH=S→O (90% E) 65%

Barbaro, G.; Battaglia, A.; Giorgianni, P.; Bonini, B.F.; Maccagnani, G.; Zani, P. *J. Org. Chem.*, ***1990***, *55*, 3744.

$CH_3C(=S)$–n-C_5H_{11} → (mCPBA , CH_2Cl_2, 0°C , 5 min) → (Z)-sulfine + (E)-sulfine (100 : 0) quant.

with MeC(=S)SBu - obtain 80:20 mixture

Le Nocher, A.M.; Metzner, P. *Tetrahedron Lett.*, ***1991***, *32*, 747.

SECTION 223: OXIDES FROM NITRILES

NO ADDITIONAL EXAMPLES

SECTION 224: OXIDES FROM ALKENES

NO ADDITIONAL EXAMPLES

SECTION 225: OXIDES FROM MISCELLANEOUS COMPOUNDS

This section includes oxides prepared from other oxides

$CH_2=C(SO_2Ph)CH_2SO_2Ph$ → (1. $Br(CH_2)_3CO_2R$, hν; 2. Bu_3SnH , AIBN) → cyclohexyl–SO_2Ph 82%

R = thiohydroxamic acid

Padwa, A.; Murphree, S.S.; Yeske, P.E. *Tetrahedron Lett.*, ***1990***, *31*, 2983.

$Ph_2C=N_2$ → (H–P(O)(OMe)OMe , PhH; $Cu(acac)_2$) → Ph_2CH–P(O)(OMe)OMe 52%

Polozov, A.M.; Polezhaeva, N.A.; Mustaphin, A.H.; Khotinen, A.V.; Arbuzov, B.A. *Synthesis*, ***1990***, 515.

RMgBr, ether

R = Me (95 : 5) 96%
R = Ph (3 : 97) 79%

Chang, Z-Y.; Coates, R.M. *J. Org. Chem.*, ***1990***, *55*, 3464.

1. BuLi
2. MeI

67%

Chou, T.; Tsai, C-Y.; Huang, L-J. *J. Org. Chem.*, ***1990***, *55*, 5410.

1. 3 eq. UDP, PhMe
2. MeI

92%

UDP = ultrasonically dispersed potassium

Chou, T-s; Hung, S-H.; Peng, M-L.; Lee, S-J. *Tetrahedron Lett.*, ***1991***, *32*, 3551.

1. $AlMe_3$, RT, CH_2Cl_2 30 min
2. PhMgBr, 5 h -70°C → RT

71%

(>99% ee)

Benson, S.C.; Snyder, J.K. *Tetrahedron Lett.*, ***1991***, *32*, 5885.

$AlCl_3$, 90°C, 14 h

46%

Katritzky, A.R.; Wu, J.; Rachwal, S.; Rachwal, B.; Macomber, D.W.; Smith, T.P. *Org. Prep. Proceed. Int.*, ***1992***, *24*, 463.

BuLi, THF
-78°C

92%

Frye, L.L.; Sullivan, E.L.; Cusak, K.P.; Funaro, J.M. *J. Org. Chem.*, ***1992***, *57*, 697.

BuMgBr , THF , -70°C

Br Tol → Tol–S(O)–Bu (99% ee , S) 81%

Cardellicchio, C.; Fiandanese, V.; Naso, F.; Scilimati, A. *Tetrahedron Lett.*, ***1992***, *33*, 5121.

DEAD , PPh_3
PhH , RT

NO_2 Ph OH → O N O Ph 93%

Falck, J.R.; Yu, J. *Tetrahedron Lett.*, ***1992***, *33*, 6723.

1. $NaBH_3CN$, pH 3-4
2. PhMe , reflux

C C N- OH → $O^{\ominus}$ $N^{\oplus}$ 94%

Fox, M.E.; Holmes, A.B.; Forbes, I.T.; Thompson, M. *Tetrahedron Lett.*, ***1992***, *33*, 7421.

REVIEWS:

"Asymmetric Carbon-Carbon Bond Formation Using Sulfoxide-Stabilized Carbanions"
Walker, A.J. *Tetrahedron Asymmetry,* ***1992***, *3*, 961.

"Cyclic Sulfites and Cyclic Sulfates"
Lohray, B.B. *Synthesis,* ***1992***, 1035.

CHAPTER 16

PREPARATION OF DIFUNCTIONAL COMPOUNDS

SECTION 300: ALKYNE - ALKYNE

MeO–CH=CH–C≡C–CH₂–n-C_6H_{13} → (1. 2 LDA; 2. aq. NH_4Cl) → H–C≡C–C≡C–CH₂–n-C_6H_{13} 82%

Stracker, E.C.; Zweifel, G. *Tetrahedron Lett.*, ***1990***, *31*, 6815.

Ph–C≡C–H → (1. $CuCl_2$•2 H_2O; 2. "solid state", RT, 3 h) → Ph–C≡C–C≡C–Ph 60%

A solid-state Glaser reaction

Toda, F.; Tokumaru, Y. *Chem. Lett.*, ***1990***, 987.

t-Bu–O–C≡C–H → (cat. CuI, 2 TMEDA; O_2, acetone, 1 h) → t-Bu–O–C≡C–C≡C–O–t-Bu 77%

Valentí, E.; Pericàs, M.A.; Serratosa, F. *J. Am. Chem. Soc.*, ***1990***, *112*, 7405.

2 eq. Ph–C≡C–Li → (1. CuCN; 2. BuC≡C-I-Ph^+ OTs^-, THF, -70°C → RT) → Ph–C≡C–C≡C–Bu (71%) + Bu–C≡C–C≡C–Bu (8%)

Kitamura, T.; Tanaka, T.; Taniguchi, H.; Stang, P.J. *J. Chem. Soc., Perkin Trans. I*, ***1991***, 2892.

n-C_3H_7–C≡C–H → (1. CuBr, Py, DBU, DMF, O_2, 45°C; 2. aq. KCN) → n-C_3H_7–C≡C–C≡C–n-C_3H_7 85%

Brandsma, L.; Verkruijsse, H.D.; Walda, B. *Synth. Commun.*, ***1991***, *21*, 137.

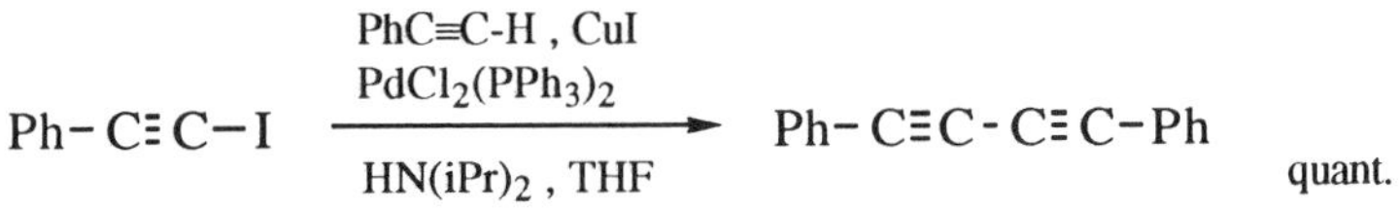

$$Ph-C\equiv C-I \xrightarrow[HN(iPr)_2,\ THF]{PhC\equiv C-H,\ CuI,\ PdCl_2(PPh_3)_2} Ph-C\equiv C-C\equiv C-Ph$$ quant.

Wityak, J.; Chan, J.B. *Synth. Commun.*, ***1991***, *21*, 977.

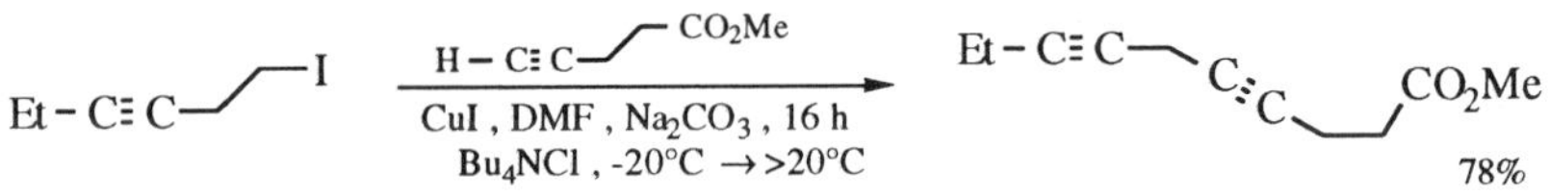

Et-C≡C-CH$_2$CH$_2$-I → (H-C≡C-CH$_2$CH$_2$-CO$_2$Me; CuI, DMF, Na_2CO_3, 16 h, Bu_4NCl, -20°C → >20°C) Et-C≡C-CH$_2$-C≡C-CH$_2$CH$_2$-CO$_2$Me 78%

Jeffery, T.; Gueugnot, S.; Linstrumelle, G. *Tetrahedron Lett.*, ***1992***, *33*, 5757.

SECTION 301: ALKYNE - ACID DERIVATIVES

NO ADDITIONAL EXAMPLES

SECTION 302: ALKYNE - ALCOHOL, THIOL

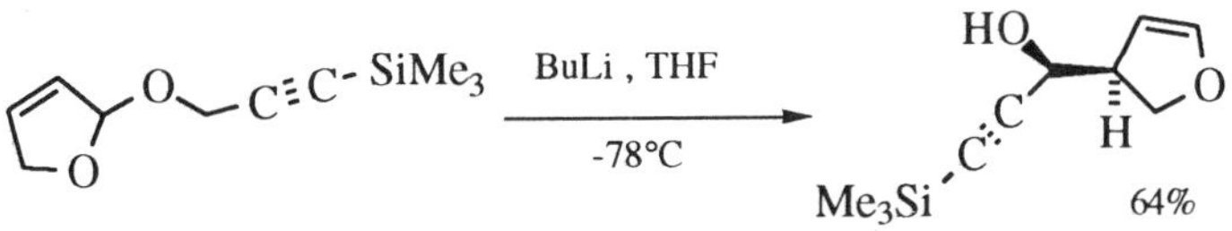

BuLi, THF, -78°C; product Me$_3$Si-C≡C-CH(OH)-(dihydrofuranyl), H 64%

Tomooka, K.; Watanabe, M.; Naka, T. *Tetrahedron Lett.*, ***1990***, *31*, 7353.

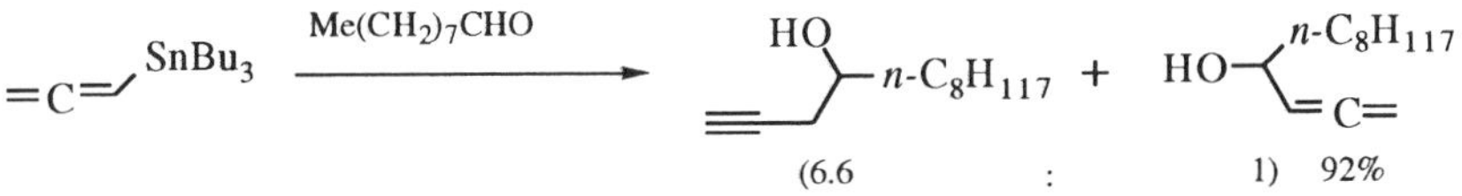

=C=CH-SnBu$_3$ → Me(CH$_2$)$_7$CHO; HO / n-C$_8$H$_{117}$ + HO / n-C$_8$H$_{117}$ =C= (6.6 : 1) 92%

Suzuki, M.; Morita, Y.; Noyori, R. *J. Org. Chem.*, ***1990***, *55*, 441.

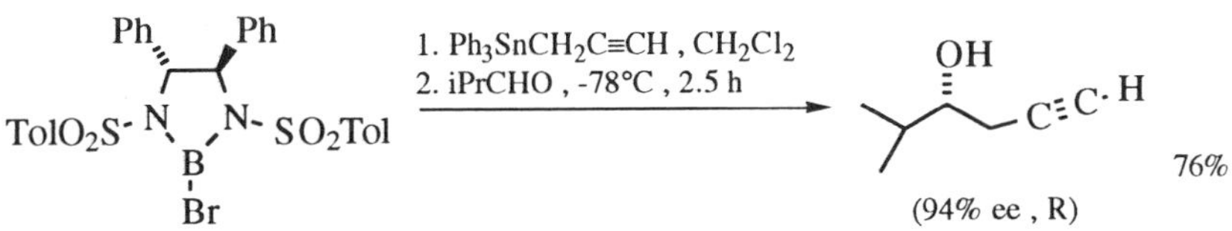

Ph Ph TolO$_2$S-N N-SO$_2$Tol B Br; 1. $Ph_3SnCH_2C\equiv CH$, CH_2Cl_2 2. iPrCHO, -78°C, 2.5 h; OH C≡C-H (94% ee, R) 76%

Corey, E.J.; Yu, C-M.; Lee, D-H. *J. Am. Chem. Soc.*, ***1990***, *112*, 878.

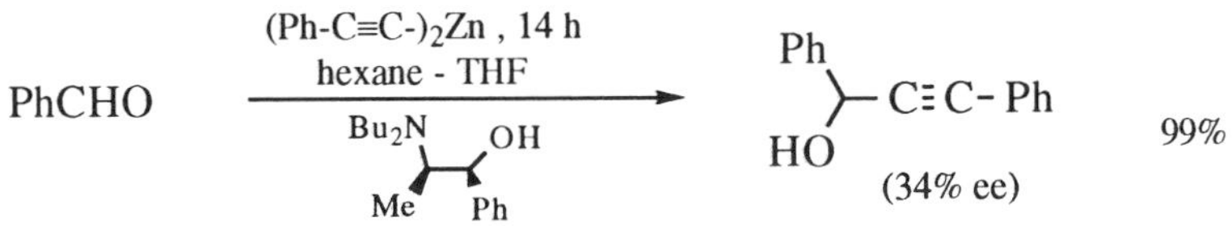

PhCHO → (Ph-C≡C-)$_2$Zn, 14 h, hexane - THF; Bu$_2$N OH Me Ph → Ph(HO)CH-C≡C-Ph (34% ee) 99%

Niwa, S.; Soai, K. *J. Chem. Soc., Perkin Trans. I*, ***1990***, 937.

n-C_6H_{13}-C≡C–Li → (propylene oxide (Me), THF; 8% Me_3Ga, RT, 1 h) → n-C_6H_{13}-C≡C–CH$_2$CH(OH)Me 87%

only 3% yield without Me_3Ga

Fukuda, Y.; Matsubara, S.; Lambert, C.; Shiragami, H.; Nanko, T.; Utimoto, K.; Nozaki, H. *Bull. Chem. Soc. Jpn.*, ***1991***, *64*, 1810.

1-methoxycyclohexyl-C≡C–Cu(CN)Li → (1. PhCHO, THF; 2. ICH_2ZnI) → 1-methoxycyclohexyl-C≡C–CH$_2$CH(OH)Ph 95%

Rozema, M.J.; Knochel, P. *Tetrahedron Lett.*, ***1991***, *32*, 1855.

Ph–C≡C–H → ($Sn(OTf)_2$, CH_2Cl_2; 1,8-bis(dimethylamino)naphthalene (Me_2N, NMe_2), t-BuCHO) → Ph–C≡C–CH(OH)t-Bu 81%

Yamaguchi, M.; Hayashi, A.; Minami, T. *J. Org. Chem.*, ***1991***, *56*, 4091.

H–C≡C–CH$_2$Br → (1. Bu_3Sb, RT, 12 h; 2. BuMgBr, THF, 0°C; 3. PhCHO, 1 h, 0°C → RT) → Ph–CH(OH)–CH$_2$–C≡C–H 91%

Zhang, L.J.; Huang, Y-Z.; Huang, Z-H. *Tetrahedron Lett.*, ***1991***, *32*, 6579.

n-C_6H_{13}-epoxide → (Ph-C≡C-Li, THF; $LiClO_4$, RT, 24 h) → n-C_6H_{13}–CH(OH)–CH$_2$–C≡C–Ph + n-C_6H_{13}–CH(C≡C–Ph)–CH$_2$OH (>99 : <1) 96%

Chini, M.; Crotti, P.; Favero, L.; Macchia, F. *Tetrahedron Lett.*, ***1991***, *32*, 6617.

Ph–C≡C–CH(OBn)Me → (Cp_2ZrCl_2, BuLi, THF, -78°C; BF_3•OEt_2, PhCHO, THF, overnight, 0°C → RT) → Ph–CH(OH)–CH(Me)–C≡C–Ph + Ph–CH(OH)–C(Ph)=C=CHMe (86 : 14) 80%

(3:1 anti:syn)

Ito, H.; Nakamura, T.; Taguchi, T.; Hanzawa, Y. *Tetrahedron Lett.*, ***1992***, *33*, 3769.

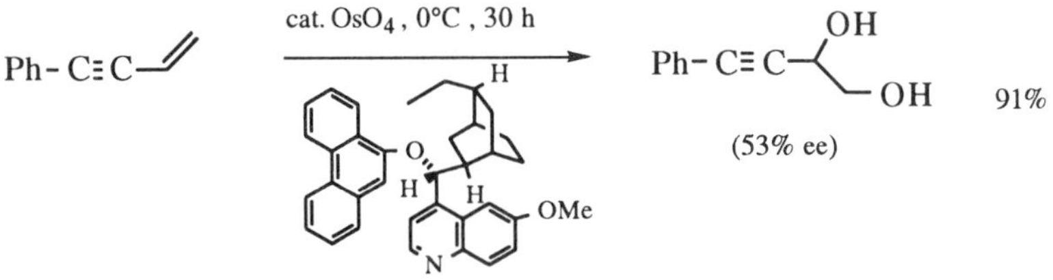

$Ph-C\equiv C-CH=CH_2$ → cat. OsO_4, 0°C, 30 h → $Ph-C\equiv C-CH(OH)CH_2OH$ (53% ee) 91%

Jeong, K-S.; Sjö, P.; Sharpless, K.B. *Tetrahedron Lett.*, ***1992***, *33*, 3833.

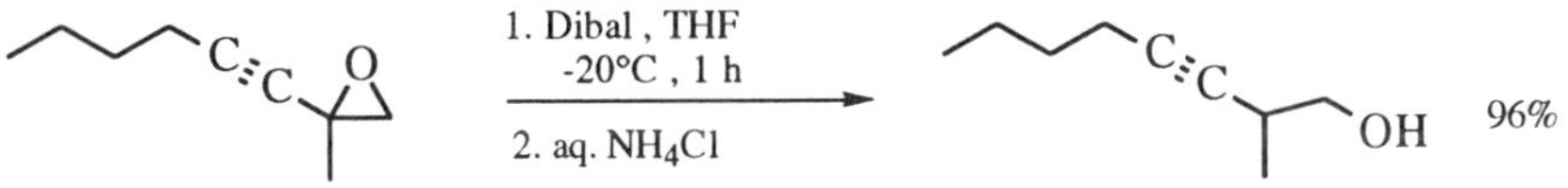

1. Dibal, THF, -20°C, 1 h; 2. aq. NH_4Cl — 96%

Nussbaumer, P.; Stütz, A. *Tetrahedron Lett.*, ***1992***, *33*, 7507.

SECTION 303: ALKYNE - ALDEHYDE

NO ADDITIONAL EXAMPLES

SECTION 304: ALKYNE - AMIDE

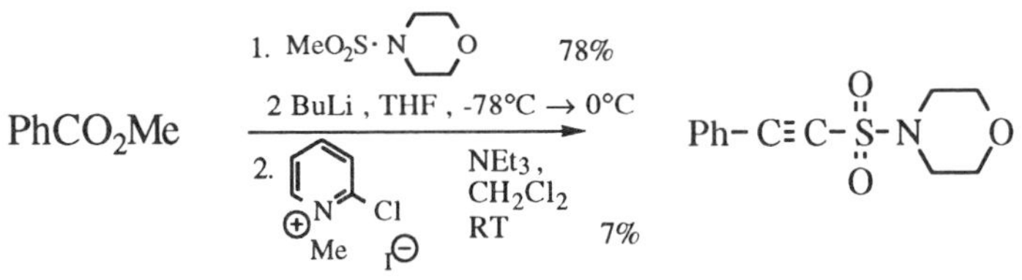

$PhCO_2Me$ → 1. $MeO_2S\cdot N$(morpholine), 2 BuLi, THF, -78°C → 0°C, 78%; 2. 2-chloro-1-methylpyridinium iodide, NEt_3, CH_2Cl_2, RT, 7% → $Ph-C\equiv C-SO_2-N$(morpholine)

LeClercq, M.; Briennc, M.J. *Tetrahedron Lett.*, ***1990***, *31*, 3875.

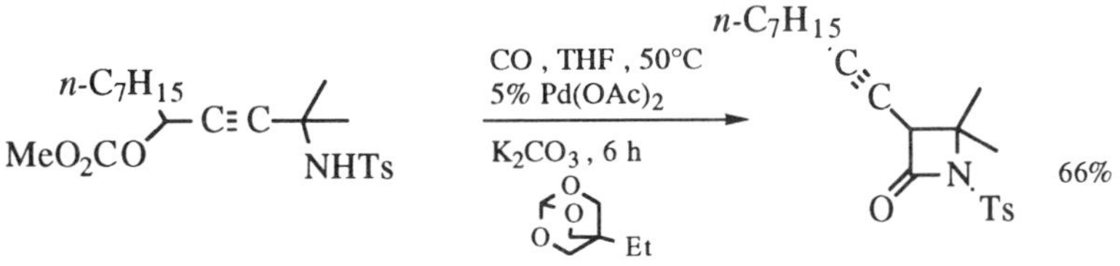

CO, THF, 50°C, 5% $Pd(OAc)_2$, K_2CO_3, 6 h — 66%

Mandai, T.; Ryoden, K.; Kawada, M.; Tsuji, J. *Tetrahedron Lett.*, ***1991***, *32*, 7683.

SECTION 305: ALKYNE - AMINE

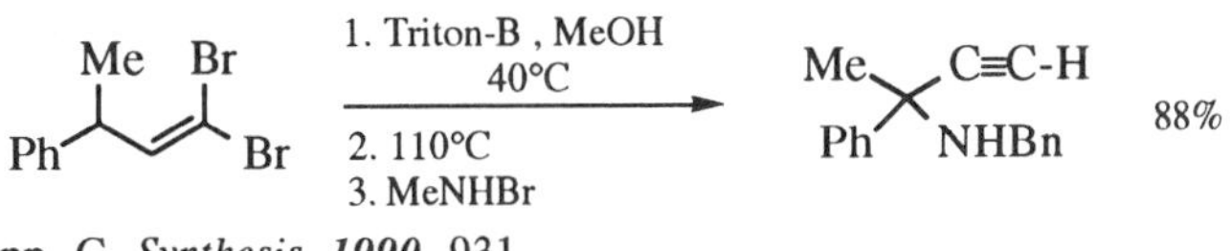

1. Triton-B, MeOH, 40°C; 2. 110°C; 3. MeNHBr — 88%

Frey, H.; Kaupp, G. *Synthesis*, ***1990***, 931.

Ph-C≡C-H → 1. (Br)(Cl)C=C(Cl), $PdCl_2(PPh_3)_2$, CuI, PPh_3, NEt_3, 14 h reflux; 2. Et_2NLi, ether, 14 h, RT → Ph-C≡C-C≡C-NEt_2 15%

Löffler, A.; Himbert, G. *Synthesis*, ***1990***, 125.

Ph(Cl)C=N-Bu → Bu-C≡C-H, $PdCl_2(PPh_3)_2$ / CuI, NEt_3, RT → Ph(C≡C-Bu)C=N-Bu 60%

Lin, S-Y.; Sheng, H-Y.; Huang, Y-Z. *Synthesis*, ***1990***, 235.

C=C=CHBr → $BuNH_2$, 3% CuBr, MeCN, 20°C, 5 h → C≡C-H / NHBu 92%

Caporusso, A.M.; Geri, R.; Polizzi, C.; Lardicci, L. *Tetrahedron Lett.*, ***1991***, *32*, 7471.

SECTION 306: ALKYNE - ESTER

Ph-C≡C-I$^{\oplus}$-Ph $OTs^{\ominus}$ → $Pd(OAc)_2$, NEt_3, EtOH / CO, RT → Ph-C≡C-CO_2Et 80%

Kitamura, T.; Mihara, I.; Taniguchi, H.; Stang, P.J. *J. Chem. Soc., Chem. Commun.*, ***1990***, 614.

Bu-C≡C-CHO → Me-CH=C(OSiMe$_3$)(SEt), $Sn(OTf)_2$ / 20% N-Me pyrrolidine-N-H-naphthyl amine / CH_2Cl_2, -78°C → EtS-C(=O)-CH(Me)-CH(OSiMe$_3$)-C≡C-Bu 67% (93:7 syn:anti; 91% ee)

Mukaiyama, T.; Furuya, M.; Ohsubo, A.; Kobayashi, S. *Chem. Lett.*, ***1991***, 989.

H-C≡C-CH$_2$CH$_2$-CO_2H → 1. *t*-BuOK; 2. Bu-C≡C-Br, THF-MeCN / Pd(P(*o*-tolyl)$_3$, 25°C, 48 h → Bu-C≡C-CH= (lactone)=O 41%

Bouyssi, D.; Gore, J.; Balme, G. *Tetrahedron Lett.*, ***1992***, *33*, 2811.

SECTION 307: ALKYNE - ETHER, EPOXIDE, THIOETHER

Cl−C≡C−Cl → (2 EtSH, 2 KH, THF; 0°C → RT, 20 h) EtS−C≡C−SEt 87%

Riera, A.; Cabré, F.; Moyano, A.; Pericàs, M.A.; Santamaría, J. *Tetrahedron Lett.*, ***1990***, *31*, 2169.

Ph / O−C≡C−I → [$BrCH_2CH_2Br$, Zn; THF, Me_3SiCl; $C_5H_{11}Br$]; CuCN, THF → Ph / O−C≡C−n-C_5H_{11} 56%

Sörensen, H.; Greene, A.E. *Tetrahedron Lett.*, ***1990***, *31*, 7597.

C≡C−$SiMe_3$ / OMe → 1. n-BuLi; 2. $ZnBr_2$; 3. H_2O → Me / C≡C-$SiMe_3$ / OMe 80%

Courtemanche, G.; Normant, J-F. *Tetrahedron Lett.*, ***1991***, *32*, 5317.

n-C_6H_{13} / C / C / Me → 4% Mn (salen) derivative; NaOCl, CH_2Cl_2, RT; pH 11.3, 5 h → n-C_6H_{13} / C / C / O / Me 85%

(2:1 trans:cis)
trans (93% ee, SR)
cis (58% ee)

Lee, N.H.; Jacobsen, E.N. *Tetrahedron Lett.*, ***1991***, *32*, 6533.

AcO / O / AcO / OAc → Me_3Si-C≡C-$SiMe_3$; $SnCl_4$, -20°C → AcO / O / AcO 99%

Tsukiyama, T.; Isobe, M. *Tetrahedron Lett.*, ***1992***, *33*, 7911.

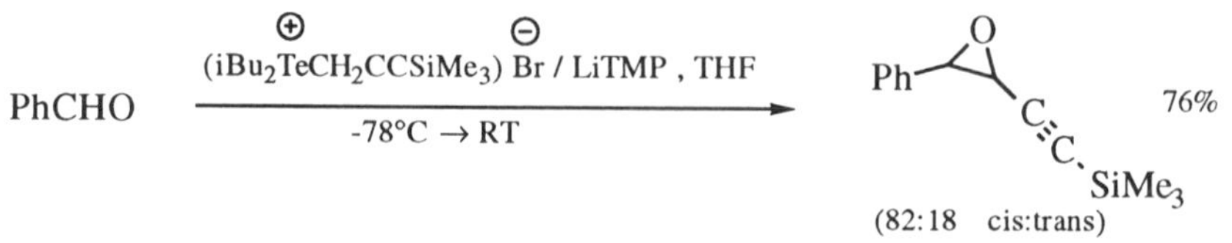

PhCHO → $(iBu_2TeCH_2CCSiMe_3)^{\oplus} Br^{\ominus}$ / LiTMP , THF ; -78°C → RT → Ph-epoxide-C≡C-$SiMe_3$ 76% (82:18 cis:trans)

Mukai, C.; Uchiyama, M.; Hanaoka, M. *J. Chem. Soc., Chem. Commun.*, ***1992***, 1014.

SECTION 308: ALKYNE - HALIDE

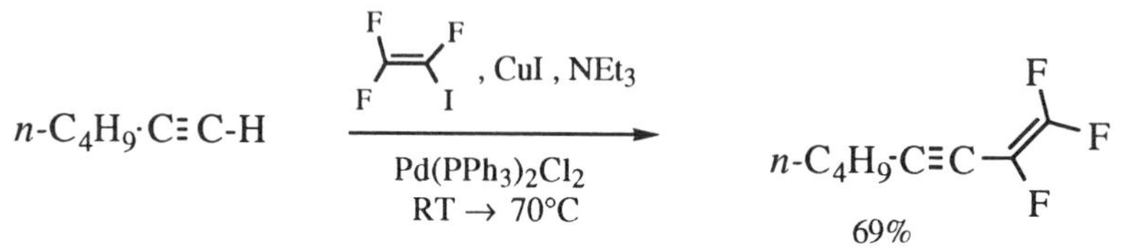

n-C_4H_9·C≡C-H → F,F,F,I-alkene , CuI , NEt_3 ; $Pd(PPh_3)_2Cl_2$; RT → 70°C → n-C_4H_9·C≡C–C(F)=C(F)F 69%

Yang, Z-Y.; Burton, D.J. *Tetrahedron Lett.*, ***1990***, *31*, 1369.

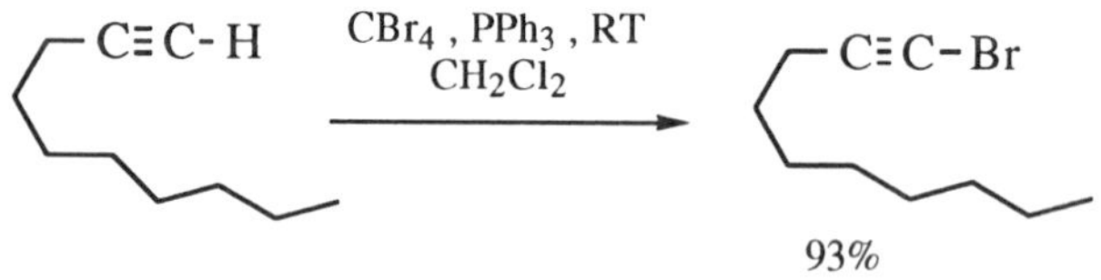

C≡C-H → CBr_4 , PPh_3 , RT ; CH_2Cl_2 → C≡C-Br 93%

Wagner, A.; Heitz, M.P.; Mioskowski, C. *Tetrahedron Lett.*, ***1990***, *31*, 3141.

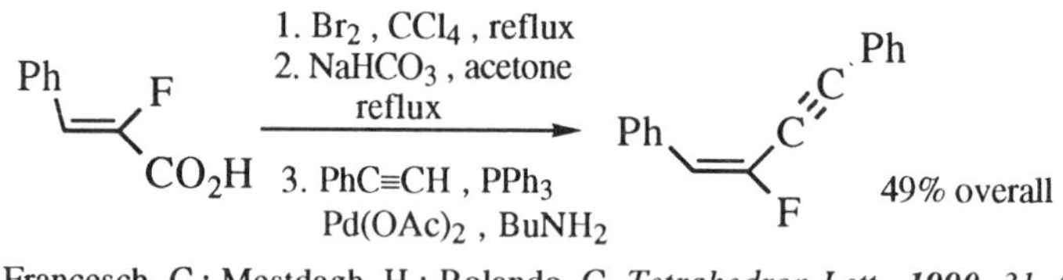

Ph, F, CO_2H → 1. Br_2 , CCl_4 , reflux ; 2. $NaHCO_3$, acetone reflux ; 3. PhC≡CH , PPh_3 ; $Pd(OAc)_2$, $BuNH_2$ → Ph, C≡C-Ph, F 49% overall

Eddarir, S.; Francesch, C.; Mestdagh, H.; Rolando, C. *Tetrahedron Lett.*, ***1990***, *31*, 4449.

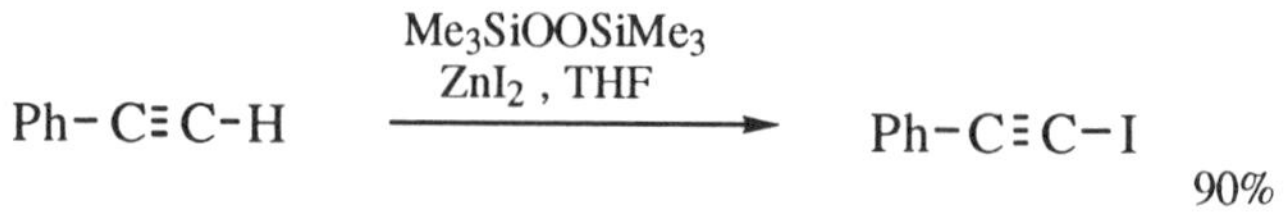

Ph-C≡C-H → $Me_3SiOOSiMe_3$; ZnI_2 , THF → Ph-C≡C-I 90%

Casarini, A.; Dembech, P.; Reginato, G.; Ricci, A.; Seconi, G. *Tetrahedron Lett.*, ***1991***, *32*, 2165.

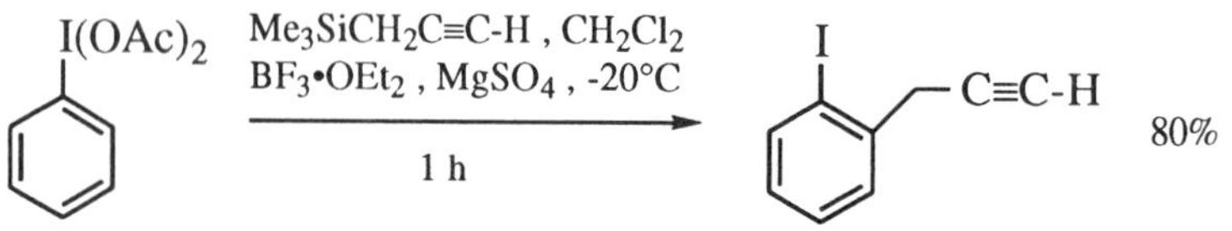

$I(OAc)_2$ → Me_3SiCH_2C≡C-H , CH_2Cl_2 ; BF_3•OEt_2 , $MgSO_4$, -20°C ; 1 h → I, C≡C-H 80%

Ochiai, M.; Ito, T.; Takaoka, Y.; Masaki, Y. *J. Am. Chem. Soc.*, ***1991***, *113*, 1319.

SECTION 309: ALKYNE - KETONE

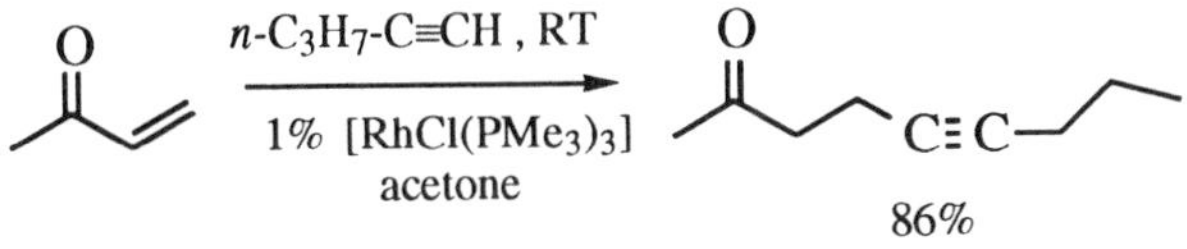

n-C_3H_7-C≡CH , RT
1% [$RhCl(PMe_3)_3$]
acetone
C≡C
86%

Nikishin, G.I.; Kovalev, I. *Tetrahedron Lett.*, ***1990***, *31*, 7063.

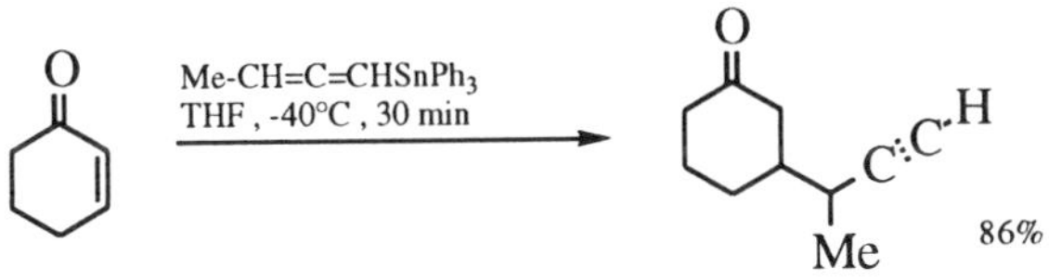

Me-CH=C=CHSnPh$_3$
THF , -40°C , 30 min
H
Me
86%

Haruta, J.; Nishi, K.; Matsuda, S.; Akai, S.; Tamura, Y.; Kita, Y. *J. Org. Chem.*, ***1990***, *55*, 4853.

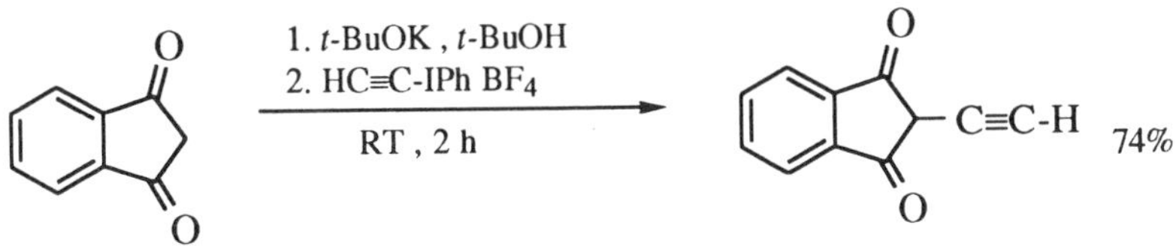

1. t-BuOK , t-BuOH
2. HC≡C-IPh BF_4
RT , 2 h
C≡C-H 74%

Ochiai, M.; Ito, T.; Takaoka, Y.; Masaki, Y.; Kunishima, M.; Tani, S.; Nagao, Y. *J. Chem. Soc., Chem. Commun.*, ***1990***, 118.

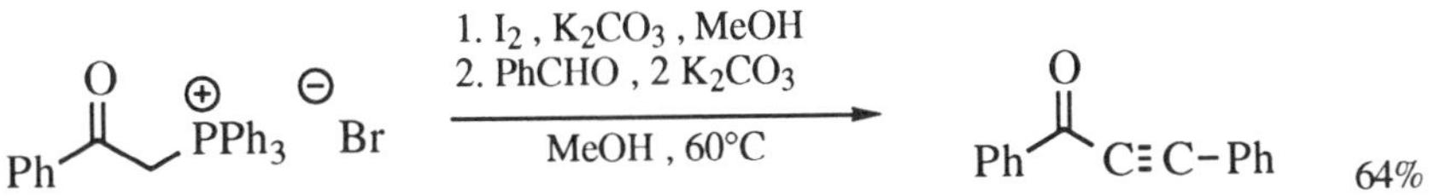

1. I_2 , K_2CO_3 , MeOH
2. PhCHO , 2 K_2CO_3
MeOH , 60°C
Ph PPh$_3$ Br
Ph C≡C-Ph 64%

Iman, M.; Bouyssou, P.; Chenault, J. *Synthesis*, ***1990***, 631.

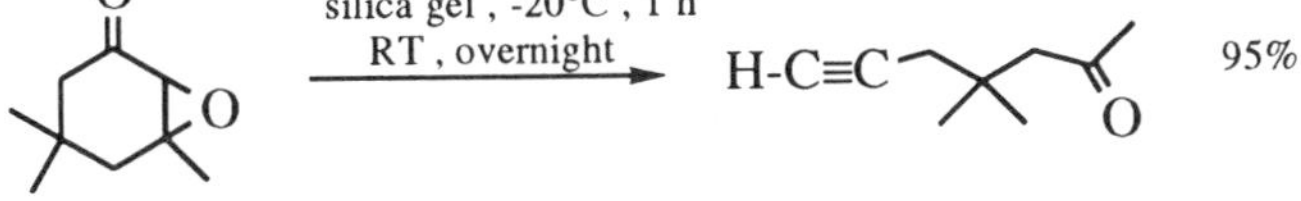

silica gel , -20°C , 1 h
RT , overnight
H-C≡C
95%

silica gel catalyzed Eschenmoser reaction

Abad, A.; Agulló, C.; Arnó, M.; Cuñat, A.C.; Zaragozá, R.J. *SynLett*, ***1991***, 787.

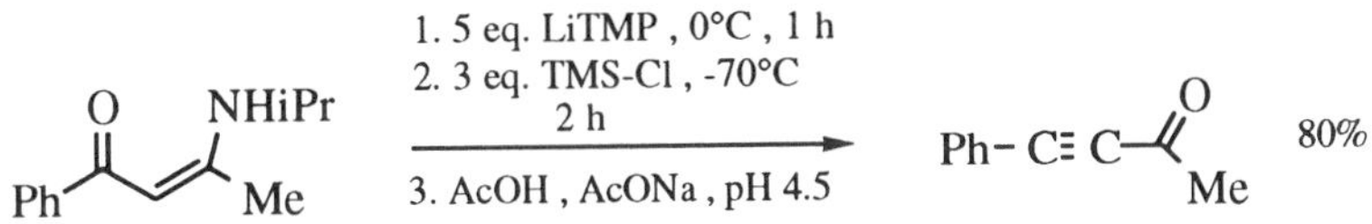

1. 5 eq. LiTMP , 0°C , 1 h
2. 3 eq. TMS-Cl , -70°C
2 h
3. AcOH , AcONa , pH 4.5
Ph NHiPr Me
Ph-C≡C Me
80%

Bartioli, G.; Cimarelli, C.; Palmieri. G. *Tetrahedron Lett.*, ***1991***, *32*, 7091.

t-BuMe$_2$SiO,,, O Et$_2$N; Et$_2$Al· C≡C OSiMe$_2t$-Bu; PhH, RT; t-BuMe$_2$SiO,,, O C≡C t-BuMe$_2$SiO,,, 90% + t-BuMe$_2$SiO,,, O C≡C t-BuMe$_2$SiO,,, 4%

Yoshino, T.; Okamoto, S.; Sato, F. *J. Org. Chem.*, ***1991***, *56*, 3205.

TfO– –Ph; Ph-C≡C-H, CO, DMF; Pd(OAc)$_2$, dppp; NEt$_3$, 60°C, 2.5 h; Ph-C≡C-C(=O)– –Ph; 75%

Ciattini, P.G.; Morera, E.; Ortar, G. *Tetrahedron Lett.*, ***1991***, *32*, 6449.

n-C$_8$H$_{17}$·C≡C-H; 1. HB(OiPr)$_2$; 2. BF$_3$•OEt$_2$, CH$_2$Cl$_2$; Ph, O, Ph, reflux, 8 h; n-C$_8$H$_{17}$-C≡C-CH(Ph)-CH$_2$-C(=O)Ph; 81%

Fujishima, H.; Takada, E.; Hara, S.; Suzuki, A. *Chem. Lett.*, ***1992***, 695.

SECTION 310: ALKYNE - NITRILE

NO ADDITIONAL EXAMPLES

SECTION 311: ALKYNE - ALKENE

Cl, Cl; 1. 6 HC≡CMgCl, THF; 0.065 NaI, 0.15 CuCl; 15 h; 2. 2 BuLi, THF, -78°C; 3. THF, -78°C, 6 h; Cl Cl; 60%; C≡C, C≡C; 5%

Gleiter, R.; Merger, R. *Tetrahedron Lett.*, ***1990***, *31*,1845.

C≡C-H, OH; Cl; CuI, NEt$_3$, 60°C; C≡C, OH; 85%

Mignani, G.; Chevalier, C.; Grass, F.; Allmang, G.; Morel, D. *Tetrahedron Lett.*, ***1990***, *31*, 5161.

Me

n-C_6H_{13}—C(OCO$_2$Me)—C≡C−Bu

OTHP; $Pd(PPh_3)_4$, LiCl; CuI , $NHEt_2$, THF; RT , 30 min

n-C_6H_{13}, Me, =C=, Bu, C≡C, OTHP

60%

Mandai, T.; Nakata, T.; Murayama, H.; Yamaoki, H.; Ogawa, M.; Kawada, M.; Tsuji, J. *Tetrahedron Lett.*, ***1990***, *31*, 7179.

Me, =C=C=, Me, Br

Bu_3SnH , AIBN , PhH; 80°C

Me, Me, C≡C, Me

43%

Ziegler, C.B. Jr. *J. Org. Chem.*, ***1990***, *55*, 2983.

Ph_2MeSi−C≡C−H

$RhCl(PPh_3)_2$, RT; PhMe , 30 h

Ph_2MeSi, C≡C−$SiMePh_2$

94%

Ohshita, J.; Furumori, K.; Matsuguchi, A.; Ishikawa, M. *J. Org. Chem.*, ***1990***, *55*, 3277.

Me, H, =C=, CO_2Me, Me

$PhOCH_2C≡C$-H; 4% $Pd(OAc)_2$, TDMPP

Me, CO_2Me, C≡C, PhO

32%

Trost, B.M.; Kottirsch, G. *J. Am. Chem. Soc.*, ***1990***, *112*, 2816.

n-C_6H_{13}−C≡C−C≡C−H

Br; CuI , K_2CO_3

n-C_6H_{13}−C≡C−C≡C−

78%

Balova, I.A.; Remizova, I..; Favorskaya, I.A. *Zhur. Org. Khim.*, ***1990***, *26*, 729.

n-C_5H_{11}−C≡C−C≡C−n-C_5H_{11}

1. e$^-$, 15% $NiBr_2$•dme; pmdta , 20°C , CO_2; (5 atm) , Mg anode; 2. H_2O

n-C_5H_{11}, C≡C, n-C_5H_{11}, CO_2H + n-C_5H_{11}, C≡C, n-C_5H_{11}, HO_2C

25% 58%

pmdta = pentamethyldiethylene triamine

Dérien, S.; Clinet, J-C.; Duñach, E.; Périchon, J. *J. Chem. Soc., Chem. Commun.*, ***1991***, 549.

Ph–(cyclohexenyl)–OTf , DMSO , RT
$Pd(OAc)_2(PPh_3)_2$, CuI , 24 h
iPr_2NH

C≡C–H, CO_2H → C≡C–(cyclohexenyl)–Ph, CO_2H 78%

Arcadi, A.; Cacchi, S.; Del Mastro, M.; Marinelli, F. *SynLett, **1991***, 409.

Bu–C≡C–H
1. $Siamyl_2BH$
2. Bu-C≡C-Li
→ Bu–CH=CH–C≡C–Bu 60%

Negishi, E.; Yoshida, T.; Abramovitch, A.; Lew, G.; Williams, R.M. *Tetrahedron, **1991**, 47*, 343.

Ph–C≡C–H
1. (dibromo-dimethylcyclopentanedione), Py-HF
2. PhC≡CH , $Pd(OAc)_2$
PPh_3 , NEt_3 , 3 h
reflux
→ Ph(F)C=CH–C≡C–Ph 83% (92:8 E:Z)

reacts with alkenes to give dienes

Eddarir, S.; Mestdagh, H.; Rolando, C. *Tetrahedron Lett., **1991**, 32*, 69.

TfO, OTf
2.4 eq. Me_3Si-C≡C-H
4% $PdCl_2(PPh_3)_2$
15% CuI , THF-iPr_2NH
→ Me_3Si–C≡C, C≡C–$SiMe_3$ 97%

Suffert, J.; Brückner, R. *Tetrahedron Lett., **1991**, 32*, 1453.

n-C_6H_{13}, $SiMe_3$, Me_3Sn, C≡C–H
1. I_2 , THF
2. Bu-C≡C-$SnMe_3$
$(MeCN)PdCl_2$
→ n-C_6H_{13}, $SiMe_3$, Bu–C≡C, C≡C–H 74%

Stracker, E.C.; Zweifel, G. *Tetrahedron Lett., **1991**, 32*, 3329.

n-C_5H_{11}–CH=CH–Cl
n-C_5H_{11}-C≡C-H , piperidine
$PdCl_2(PhCN)_2$, 30 min
→ n-C_5H_{11}–CH=CH–C≡C–n-C_5H_{11} 93%

Alami, M.; Linstrumelle, G. *Tetrahedron Lett., **1991**, 32*, 6109.

1. KH
2. Bu-C≡C-I , 7 h
$Pd(PPh_3)_4$

CO_2Me
CO_2Me

Bu-C≡C
CO_2Me
CO_2Me
71%

Bouyssi, D.; Balme, G.; Gore, J. *Tetrahedron Lett.*, ***1991***, *32*, 6541.

H-C≡C

Br , Aliquat-336
30% aq. NaOH , CuCl
CH_2Cl_2 , 20°C , 24 h

C≡C
60%

Grushin, V.V.; Alper, H. *J. Org. Chem.*, ***1992***, *57*, 2188.

$Pd(OAc)_2$, PPh_3 , CuI
Bu_4NCl , Et_2NH , DMF
80°C , 12 h

TBSO
I n-C_5H_{11} , H-C≡C OH

C≡C OH OTBS n-C_5H_{11} + OTBS C≡C n-C_5H_{11} OH
(8 : 1) 87%

Torii, S.; Okumoto, H.; Kotani, T.; Nakayasu, S.; Ozaki, H. *Tetrahedron Lett.*, ***1992***, *33*, 3503.

Br-C≡C OEt OEt

$SnBu_3$, DMF , 20°C
$PdCl_2(MeCN)_2$, 1 h

C≡C OEt OEt
80%

Beaudet, I.; Parrain, J-L.; Quintard, J-P. *Tetrahedron Lett.*, ***1992***, *33*, 3647.

H-C≡C-CO_2Me

1. LiBr-AcOH , MeCN , reflux
2.Ph-C≡C-H , $PdCl_2(PPh_3)_2$

MeO_2C-C≡C Ph
85%

Lu, X.; Huang, X.; Ma, S. *Tetrahedron Lett.*, ***1992***, *33*, 2535.

H-C≡C SEt

1. PhBr , $PdCl_2$•2 PPh_3
PPh_3 , CuBr , NEt_3
reflux
2. Zn , EtOH , reflux
3. $NaNH_2$, $NH_{3\,(liq)}$
4. H_2O

Ph C≡C H
54%
overall

Fossatelli, M.; van der Kerk, A.C.T.H.M.; Vasilevsky, S.F.; Brandsma, L. *Tetrahedron Lett.*, ***1992***, *33*, 4229.

REVIEWS:

"Molecular Design, Chemical Synthesis, and Biological Action of Enediynes"
Nicolaou, K.C.; Smith, A.L. *Accts. Chem. Res.*, ***1992***, *24*, 497.

SECTION 312: CARBOXYLIC ACID - CARBOXYLIC ACID

$NaBO_3 \cdot 4\ H_2O$, AcOH, 95°C — 39%

Muzart, J.; Ajjou, A.N. *Synth. Commun.*, ***1991***, *21*, 575.

montmorillonite, 30°C, $KMnO_4$, PhH, H_2O, 1 h — 92%

Choudary, B.M.; Valli, V.L.K.; Durga Prasad, A. *Synth. Commun.*, ***1991***, *21*, 2007.

SECTION 313: CARBOXYLIC ACID - ALCOHOL, THIOL

1. HCN, iPr_2O, (S)-oxynitrilase; 2. aq. HCl — 70% (98% ee)

Effenberger, F.; Hörsch, B.; Förster, S.; Ziegler, T. *Tetrahedron Lett.*, ***1990***, *31*, 1249.

1. $LiNTMS_2$, -78°C; 2. 2.2 eq. PhCHO, -78°C; 3. aq. NaOH; 4. H_3O^+ — (98.8 : 1.2) 97%

Longmire, C.F.; Evans, S.A. Jr. *J. Chem. Soc., Chem. Commun.*, ***1990***, 922.

5 eq. immobilized *Rhodococcus sp.*, 24 h — 62%

Klempier, N.; de Raadt, A.; Faber, K.; Griengl, H. *Tetrahedron Lett.*, ***1991***, *32*, 341.

1. PhCHO , Zn, TMSCl , PhH; 2. KOH , MeOH

Ph

HOOC, OH, H, Ph 51%

(44% ee , R)

Basavaiah, D.; Bharathi, T.K. *Tetrahedron Lett.*, ***1991***, *32*, 3417.

Me, $B(OiPr)_2$; NEt_3 , CH_2Cl_2

Me, CO_2H → HO, Me, CO_2H, Me 95%

(99:1 syn:anti)

Wang, Z.; Meng, X-J.; Kabalka, G.W. *Tetrahedron Lett.*, ***1991***, *32*, 5677.

CO_2H — BSLDH , NADH, H^+ , 2.8 d → OH, CO_2H 66%

BSLDH = *Bacillus stearothermophilus* lactate dehydrogenase

(99% ee)

Casy, G.; Lee, T.V.; Lovell, H. *Tetrahedron Lett.*, ***1992***, *33*, 817.

Ph — 1. [NaOH/KCl/NaOCl] , MeOH, RT , 3 h; 2. aq. $NaHCO_3$ → OH, Ph, OH, CO_2H + 24% mandelic acid

57%

Guthrie, J.P.; Cossar, J.; Lu, J. *Can. J. Chem.*, ***1991***, *69*, 1904.

Ph — 5% $Ru_2(COD)_2(O_2CCF_3)_4$, 2 dppb , 10 eq. HCOOH; 10 eq. NEt_3 , dioxane, reflux → OH, Ph, CO_2H 80%

Shimizu, I.; Tekawa, M.; Maruyama, Y.; Yamamoto, A. *Chem. Lett.*, ***1992***, 1365.

SECTION 314: CARBOXYLIC ACID - ALDEHYDE

NO ADDITIONAL EXAMPLES

SECTION 315: CARBOXYLIC ACID - AMIDE

1. $Ni(COD)(TMEDA)_2$ THF, reflux
2. H_3O^+

Me CbzHN CO_2H + CO_2H CbzHN

(9 : 1) 60%

1:1 mixture when $Ni(COD)(bipy)_2$ was used

Castaño, A.M.; Echavarren, A.M. *Tetrahedron Lett.*, ***1990***, *31*, 4783.

CbzHN CO_2H → F N N F N F; Py, CH_2Cl_2, 4 h → O F CbzHN 89%

Bertho, J-N.; Loffet, A.; Pinel, C.; Reuther, F.; Sennyey, G. *Tetrahedron Lett.*, ***1991***, *32*, 1303.

t-Bu H-N N-CO_2Me O

1. $Pd(OAc)_2$, $P(o\text{-}Tol)_3$ 100°C, NET_3, DMF, MeO–C6H4–I, 24 h
2. $NaBH_4$; H_3O^+
3. 3N HCl
4. Ac_2O, HCOOH

HOOC NHCHO OMe 66%

Konopelski, J.P.; Chu, K.S.; Negrete, G.R. *J. Org. Chem.*, ***1991***, *56*, 1355.

BocHN CO_2H

1. 3 eq. LDA
2. I
3. H_3O^+

BocHN CO_2H 48%

De Nicola, A.; Einhorn, J.; Luche, J-L. *Tetrahedron Lett.*, ***1992***, *33*, 6461.

Me CO_2H NH_2 → Ac_2O, H_2O,)))))) → Me CO_2H NHAc 96%

Veera Reddy, A.; Ravindranath, B. *Synth. Commun.*, ***1992***, *22*, 257.

SECTION 316: CARBOXYLIC ACID - AMINE

1. $NaN(TMS)_2$
2. (1-chloro-1-nitrosocyclohexane)
3. 1N aq. HCl, RT
4. Zn/aq. HCl, AcOH
5. aq. LiOH, THF
6. ion exchange

HO_2C–CH(NH_2)–CH_3 83% (>99%ee, R)

Oppolzer, W.; Tamura, O. *Tetrahedron Lett.*, ***1990***, *31*, 991.

1. acryloyl chloride
2. NH_3, CH_2Cl_2
3. $ClCO_2Bn$, NEt_3
4. $Hg(O_2CCF_3)_2$
5. $NaBH_4$, MeCN
6. HCl, MeOH, reflux

HOOC–CH(Me)–NH_3•HCl

Amoroso, R.; Cardillo, G.; Tomasini, C. *Tetrahedron Lett.*, ***1990***, *31*, 6413.

Ph–CH_2CH_2–NHBoc

1. *t*-BuLi, THF, -75°C
2. CO_2
3. 20% HCl, reflux 8 h

Ph–CH(CO_2H)–CH_2–NHBoc 55%

Simig, G.; Schlosser, M. *Tetrahedron Lett.*, ***1991***, *32*, 1963, 1965.

Ph–CH(Br)–CO_2Et

1. F_3C–C(=O)–NH_2, TEBA, K_2CO_3, MeCN, 80°C
2. KOH, MeOH, RT, 2 h

Ph–CH(NH_2)–CO_2H 64%

Landini, D.; Penso, M. *J. Org. Chem.*, ***1991***, *56*, 420.

1. LDA
2. BnBr
3. 6N HCl, heat
4. Dowex 50W-X8

H_2N–CH_2–CH(CH_2Ph)–COOH 50%

Juaristi, E.; Quintana, D.; Lamatsch, B.; Seebach, D. *J. Org. Chem.*, ***1991***, *56*, 2553.

REVIEWS:

"New Approaches to the Use of Amino Acids as Chiral Building Blocks in Organic Synthesis" Reetz, M.T. *Angew. Chem. Int. Ed. Engl.*, ***1991***, *30*, 1531.

Related Methods: Section 315 (Carboxylic Acid - Amide).
Section 344 (Amide - Ester). Section 351 (Amine - Ester).

SECTION 317: CARBOXYLIC ACID - ESTER

porcine liver acetone powder
H_2O , pH 7.2 , RT , 7 h

96%

Ahmar, M.; Bloch, R.; Bortolussi, M. *Synth. Commun.*, ***1991***, *21*, 1071.

SECTION 318: CARBOXYLIC ACID - ETHER, EPOXIDE, THIOETHER

1. BuLi , THF , -78°C
2. CO_2
3. H^+

(95%ee)

99%

Chan, P.C-M.; Chong, J.M. *Tetrahedron Lett.*, ***1990***, *31*, 1985.

NaSPh , aq. iPrOH
0°C → RT

95%

Breitschuh, R.; Seebach, D. *Synthesis*, ***1992***, 83.

SECTION 319: CARBOXYLIC ACID - HALIDE, SULFONATE

1. 3 eq. LDA
2. TMSCl
3. H_2O

88%

Welch, J.T.; Plummer, J.S.; Chou, T.S. *J. Org. Chem.*, ***1991***, *56*, 353.

BI_3•NEt_2Ph , PhH
RT , 7 h

87%

Narayana, C.; Reddy, N.K.; Kabalka, G.W. *Tetrahedron Lett.*, ***1991***, *32*, 6855.

1. $CuCl_2$, H_2O_2 aq. MeOH
2. H_2SO_4

85%

Starostin, E.K.; Mazurchik, A.A.; Ignatenko, A.V.; Nikishin, G.I. *Synthesis*, ***1992***, 917.

SECTION 320: CARBOXYLIC ACID - KETONE

1. AcCl, N-methylmorpholine
2. $AlCl_3$

67%

(>20 : 1)

Jefford, C.W.; Tang, Q.; Boukouvalas, J. *Tetrahedron Lett.*, ***1990***, *31*, 995.

1. TFAA
2. $=\!\!<^{OSiMe_3}_{Ph}$, -40°C, 3h

65%

(>95% ee, S)

Michelon, F.; Pouilhès, A.; Bac, N.V.; Langlois, N. *Tetrahedron Lett.*, ***1992***, *33*, 1743.

1. Me–CH(Br)–CO_2Et, Zn, DMF, 70°C, 2 h
2. HCl, 100°C, 1 h

41%

Schick, H.; Ludwig, R. *Synthesis*, ***1992***, 369.

Also via: Section 360 (Ketone - Ester).

SECTION 321: CARBOXYLIC ACID - NITRILE

NO ADDITIONAL EXAMPLES

Also via: Section 361 (Nitrile - Ester).

SECTION 322: CARBOXYLIC ACID - ALKENE

1. 2 eq. $LiNEt_2$
2. 0.5 eq. I_2
3. H_3O^+

62%

Aurell, M.J.; Gil, S.; Parra, M.; Tortajada, A.; Mestres, R. *Tetrahedron*, ***1991***, *47*, 1997.

1. Mg, THF
2. CO_2

(99 : <1) 99%

with 1. Ba, THF; 2. CO_2 (2 : 98) 82%

Yanagisawa, A.; Yasue, K.; Yamamoto, H. *SynLett*, ***1992***, 593.

Also via: Section 313 (Alcohol - Carboxylic Acids). Section 349 (Amide - Alkene).
Section 362 (Ester - Alkene). Section 376 (Nitrile - Alkene).

SECTION 323: ALCOHOL, THIOL - ALCOHOL, THIOL

Pseudomonas putida 39D

Deluca, M.E.; Hudlicky, T. *Tetrahedron Lett.*, ***1990***, *31*, 13.

$Pd(PPh_3)_4$, $(Me_3Sn)_2$
dioxane, sealed tube
100-105°C, 24h

80%

Kelly, T.R.; Li, Q.; Bhushan, V. *Tetrahedron Lett.*, ***1990***, *31*, 161.

air, cat. AIBN
ether

89% (≈2:1 cis:trans)

Kánai, K.; Tömösközi, I *Tetrahedron Lett.*, ***1990***, *31*, 403.

MeO + CuCl$_2$, t-BuNH$_2$ MeOH , 50°C 110 min → CO$_2$Me OH OH MeO 92%

CO$_2$Me OH OH

Hovorka, M.; Gunterová, J.; Závada, J. *Tetrahedron Lett.*, ***1990***, *31*, 413.

Me$_3$Si O OH → 5 eq. NaN$_3$, RT NH$_4$Cl , aq. MeOH → Me$_3$Si OH OH N$_3$ 90%

NaN$_3$ in hot DMF gave azido-allylic alcohols

Chakraborty, T.K.; Reddy, G.V. *Tetrahedron Lett.*, ***1990***, *31*, 1335.

Ph → OsO$_4$ → HO OH Ph 86% (47%ee, S)

Ph Ph N N Ph Ph

Tomioka, K.; Nakajima, M.;Koga, K. *Tetrahedron Lett.*, ***1990***, *31*, 1741.

Ph → dihydroquinidine p-chlorobenzoate; OsO$_4$, K$_3$Fe(CN)$_6$ K$_2$CO$_3$, aq. t-BuOH 0°C → Ph OH OH 73%

improved ee over NMO

Kwong, H-L.; Sorato, C.; Ogino, Y.; Chen, H.; Sharpless, K.B. *Tetrahedron Lett.*, ***1990***, *31*, 2999.

Ph Ph → dihydroquinidine p-chlorobenzoate polymer; 1% OsO$_4$, NMO , 10°C 2-3 d → Ph OH Ph OH

81-87% (85-93% ee)

Kim, B.M.; Sharpless, K.B. *Tetrahedron Lett.*, ***1990***, *31*, 3003.

O OH → MgI$_2$, PhH , AIBN Bu$_3$SnH (added by syringe pump) → OH OH Me 53%

Bonini, C.; Di Fabio, R.; Mecozzi, S.; Right, G. *Tetrahedron Lett.*, ***1990***, *31*, 5369.

OSEM Me → 1. t-BuLi, ether, -78°C; 2. PhHCO → OSEM, OH, Ph, Me 80%

Sengupta, S.; Snieckus, V. *Tetrahedron Lett.*, ***1990***, *31*, 4267.

OAc, $SiMe_2Ph$ → 1. OSEM, O, cat. hydroquinone, xylene, reflux; 2. OsO_4, NMO, aq. THF; 3. KBr, 15% AcOOH, AcOH, NaOAc, RT, 1 h → OAc, HO, CO_2SEM, HO, OH 60% overall

Koreeda, M.; Teng, K.; Murat, T. *Tetrahedron Lett.*, ***1990***, *31*, 5997.

Me Me OH → 1. $(HMe_2Si)_2NH$, NH_3; 2. $[RhCl(ethene)]_2$, R-BINAP, DCE, 30°C, 6.5 d; 3. 30% H_2O_2, KF, $KHCO_3$, MeOH, THF → Me Me OH OH 60% (95/5 syn/anti; 80% ee)

Tamao, K.; Tohma, T.; Inui, N.; Nakayama, O.; Ito, Y. *Tetrahedron Lett.*, ***1990***, *31*, 7333.

Cl, Cp, Mo, (NO), OH OH H Ph → 1. PhCHO, MeOH, CH_2Cl_2; 2. MeOH → OH OH OH, Ph, H, Ph 51%

Lin, S-H.; Vong, W-J.; Cheng, C-Y.; Wang, S-L.; Liu, R-S *Tetrahedron Lett.*, ***1990***, *31*, 7645.

→ OsO_4, $K_3Fe(CN)_6$, t-BuOH, H_2O, RT, K_2CO_3, 24 h → OH OH 94%

Minato, M.; Yamamoto, K.; Tsuji, J. *J. Org. Chem.*, ***1990***, *55*, 766.

O Me → PhCHO, $CrCl_2$, THF, LiI, 0°C, 1 h → Me Ph OH OH + Me Ph OH OH (98 : 2) 95%

Fujimura, O.; Takai, K.; Utimoto, K. *J. Org. Chem.*, ***1990***, *55*, 1705.

OH O

1. catecholborane, THF
-10°C, 5 h
2. aq. Na K tartrate, MeOH

OH OH

82%

(10:1 syn:anti)

Evans, D.A.; Hoveyda, A.H. *J. Org. Chem.*, ***1990***, *55*, 5190.

O H OH

3 eq. BuLi, 0°C
ether-pentane

OH OH Bu

88%

Lautens, M.; Abd-El-Aziz, A.S.; Lough, A. *J. Org. Chem.*, ***1990***, *55*, 5305.

BnO OTBDMS TBDMS-O CO_2Et

5% OsO_4, 25°C
2 eq. NMO
aq. acetone

BnO OTBDMS OH TBDMS-O CO_2Et HO

+

BnO OTBDMS OH TBDMS-O CO_2Et HO

(9 : 1) 96%

Saito, S.; Morikawa, Y.; Moriwake, T. *J. Org. Chem.*, ***1990***, *55*, 5429.

O

aq. Dowex-50W
reflux, 30 min

OH OH

71%

Ranu, B.C.; Chakraborty, R. *Synth. Commun.*, ***1990***, *20*, 1751.

O OH Ph Ph

$MeTi(OiPr)_3$, ether
LiI, -78°C → -20°C

HO Me OH Ph Ph

+

Me OH OH Ph Ph

(99 : 1) 69%

Ukaji, Y.; Kanda, H.; Yamamoto, K.; Fujisawa, T. *Chem. Lett.*, ***1990***, 597.

Ph Ph O

O Me, Yb, RT, 18h
THF, HMPA

Ph OH OH Ph Me

46%
(90% ee)

+

Ph OH Me Ph OH

37%

Takaki, K.; Tanaka, S.; Beppu, F.; Tsubaki, Y.; Fujiwara, Y. *Chem. Lett.*, ***1990***, 1427.

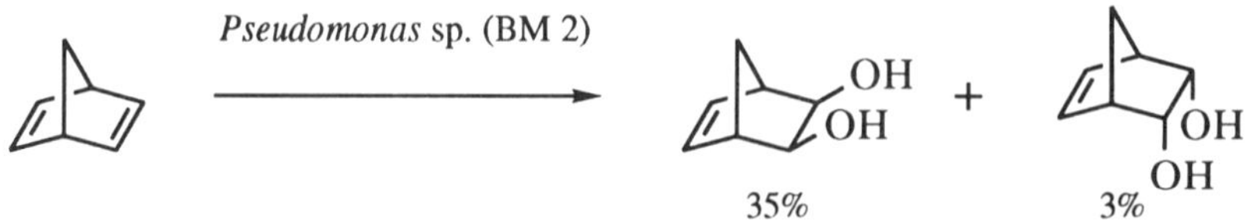

Pseudomonas sp. (BM 2)

OH OH + OH OH

35% 3%

Geary, P.J.; Pryce, R.J.; Roberts, S.M.; Ryback, G.; Winders, J.A. *J. Chem. Soc., Chem. Commun.*, ***1990***, 204.

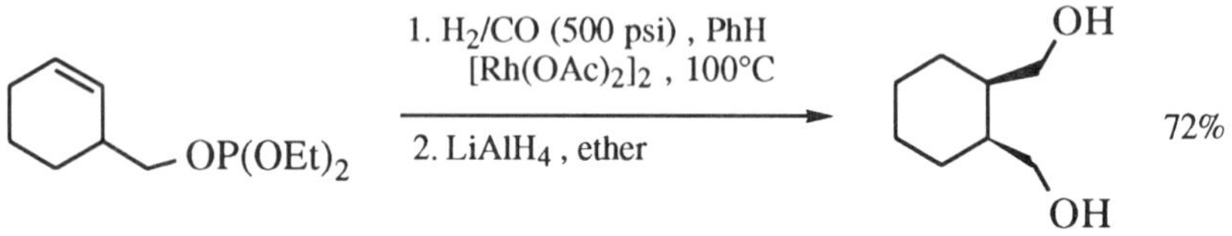

OP(OEt)$_2$

1. H_2/CO (500 psi), PhH [Rh(OAc)$_2$]$_2$, 100°C

2. $LiAlH_4$, ether

OH OH 72%

Jackson, W.R.; Perlmutter, P.; Tasdelen, E.E. *J. Chem. Soc., Chem. Commun.*, ***1990***, 763.

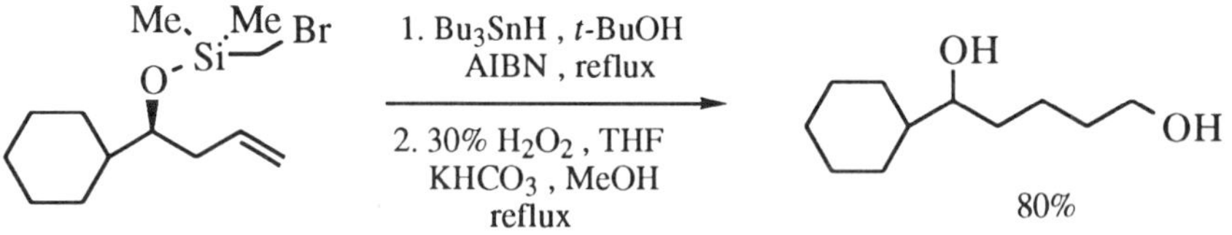

Me, Me Br Si O

1. Bu_3SnH, *t*-BuOH AIBN, reflux

2. 30% H_2O_2, THF $KHCO_3$, MeOH reflux

OH OH 80%

Koreeda, M.; Hamann, L.G. *J. Am. Chem. Soc.*, ***1990***, *112*, 8175.

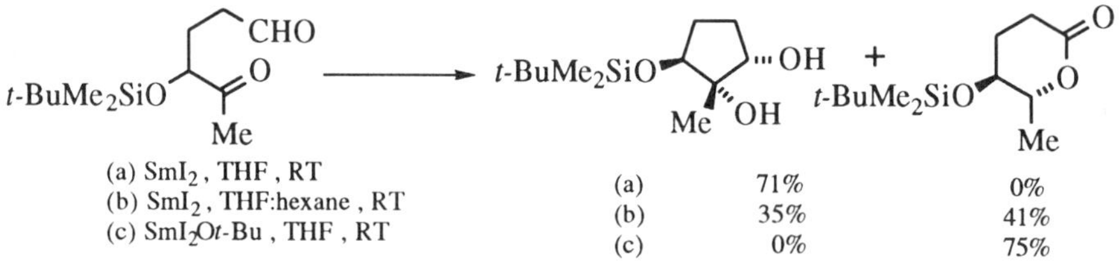

CHO O *t*-BuMe$_2$SiO Me

t-BuMe$_2$SiO OH Me OH + *t*-BuMe$_2$SiO O O Me

Conditions		Diol	Lactone
(a) SmI_2, THF, RT	(a)	71%	0%
(b) SmI_2, THF:hexane, RT	(b)	35%	41%
(c) SmI_2O*t*-Bu, THF, RT	(c)	0%	75%

Uenishi, J.; Masuda, S.; Wakabayashi, S. *Tetrahedron Lett.*, ***1991***, *32*, 5097.

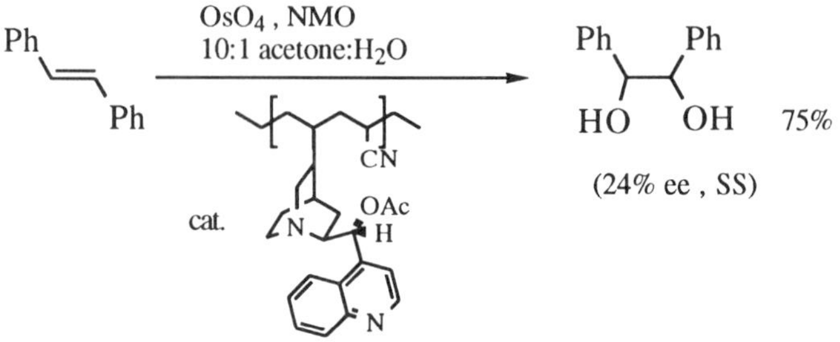

Ph Ph

OsO_4, NMO 10:1 acetone:H_2O

cat. CN N OAc H N

Ph Ph HO OH 75%

(24% ee, SS)

Pini, D.; Petri, A.; Nardi, A.; Rosini, C.; Salvadori, P. *Tetrahedron Lett.*, ***1991***, *32*, 5175.

I, Si, Ph

1. t-BuOK, THF
-78°C

2. H_2O_2, KF
$KHCO_3$

OH, OH

66%

Matsumoto, K.; Miura, K.; Oshima, K.; Utimoto, K. *Tetrahedron Lett.*, ***1991***, *32*, 6383.

Me, O

PhCHO, $SnCl_2$
Bu_4NBr, DMI

25°C, 16 h

Me, Ph, OH, OH + Me, Ph, OH, OH

DMI = dimethylimidazolidin-2-one (93 : 7) 89%

Masuyama, Y.; Nakata, J.; Kurusu, Y. *J. Chem. Soc., Perkin Trans. I*, ***1991***, 2598.

OH

1. Li°, naphthalene, THF
1 h, -78°C

2. iPrCHO, -78°C → RT
3. H_2O

OH OH

68%

Yus, M.; Ramón, D.J. *J. Chem. Soc., Chem. Commun.*, ***1991***, 398.

n-C_8H_{17}

Re_2O_7, H_2O_2, 16 h
90°C, dioxane

n-C_8H_{17}, OH, OH

62%

Warwel, S.; Klass, M.Rg.; Sojka, M. *J. Chem. Soc., Chem. Commun.*, ***1991***, 1578.

Ph, Ph

1. OsO_4, NMO
aq. acetone, 25°C

2. $Na_2S_2O_5$

Ph, OH, OH, HO, HO, Ph + Ph, OH, OH, HO, HO, Ph

(16 : 1) 87%

Park, C.Y.; Kim, B.M.; Sharpless, K.B. *Tetrahedron Lett.*, ***1991***, *32*, 1003.

O, Si, O, Br

1. Bu_3SnH, AIBN
PhMe

2. TBAF

59% HO, HO

Hutchinson, J.H.; Daynard, T.S.; Gillard, J.W. *Tetrahedron Lett.*, ***1991***, *32*, 573.

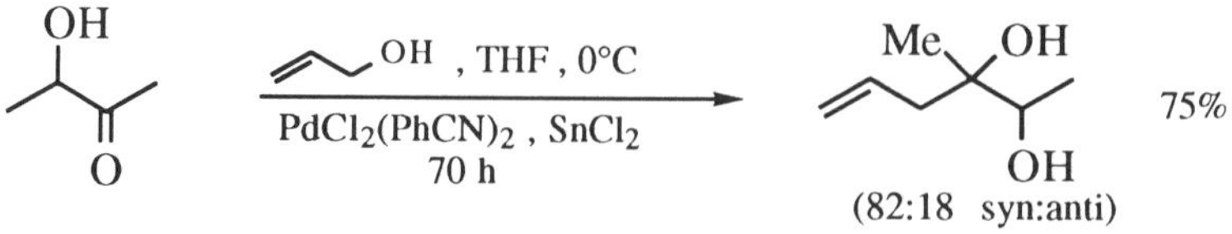

OH, THF, 0°C
$PdCl_2(PhCN)_2$, $SnCl_2$
70 h

Me OH
OH
(82:18 syn:anti) 75%

Takahara, J.P.; Masuyama, Y.; Kurusu, Y. *Chem. Lett.*, ***1991***, 879.

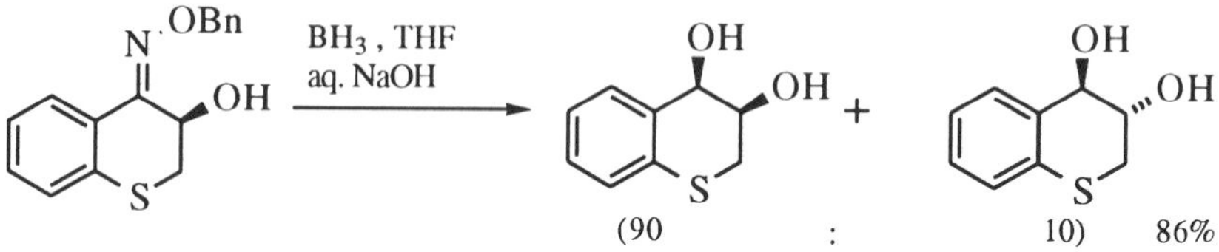

N .OBn
OH
BH_3, THF
aq. NaOH

OH OH + OH OH
S S S
(90 : 10) 86%

Ghosh, A.K.; Mckee, S.P.; Sanders, W.M. *Tetrahedron Lett.*, ***1991***, *32*, 711.

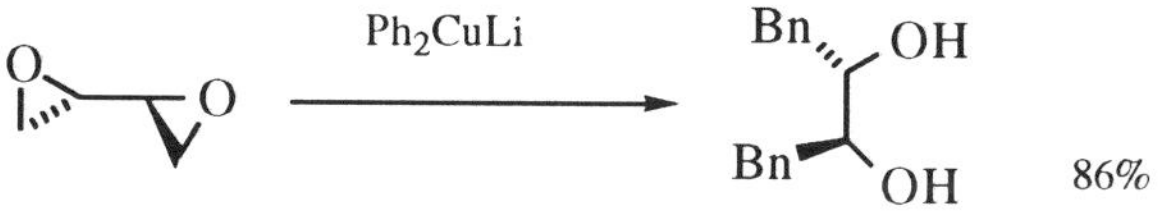

Ph_2CuLi

Bn OH
Bn OH 86%

similar results with RMgX, R_2CuLi

Devine, P.N.; Oh, T. *Tetrahedron Lett.*, ***1991***, *32*, 883.

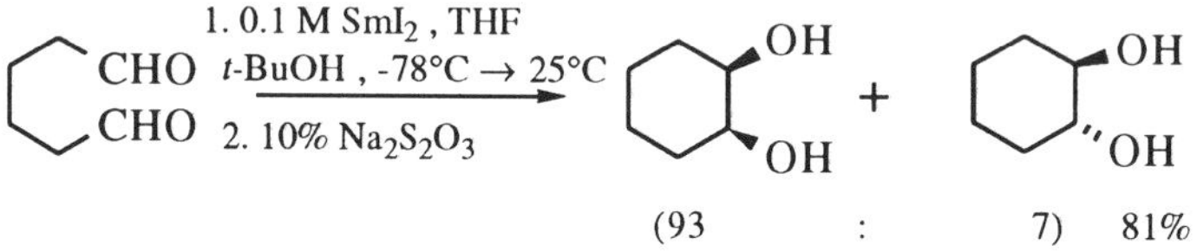

CHO
CHO
1. 0.1 M SmI_2, THF
t-BuOH, -78°C → 25°C
2. 10% $Na_2S_2O_3$

OH OH + OH OH
(93 : 7) 81%

Chiara, J.L.; Cabri, W.; Hanessian, S. *Tetrahedron Lett.*, ***1991***, *32*, 1125.

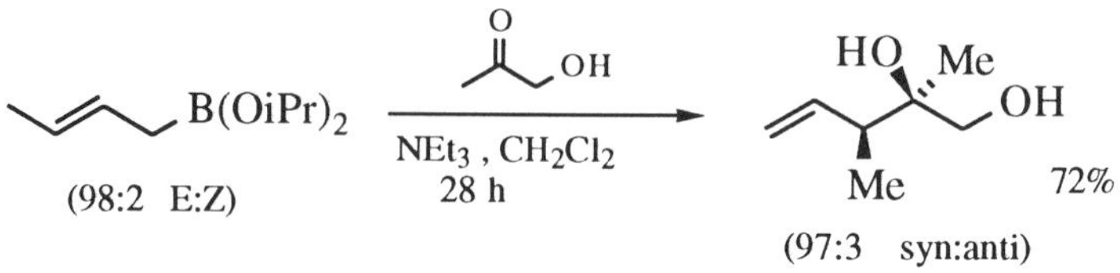

$B(OiPr)_2$
(98:2 E:Z)

O
OH
NEt_3, CH_2Cl_2
28 h

HO Me
OH
Me 72%
(97:3 syn:anti)

Wang, Z.; Meng, X-J.; Kabalka, G.W. *Tetrahedron Lett.*, ***1991***, *32*, 1945.

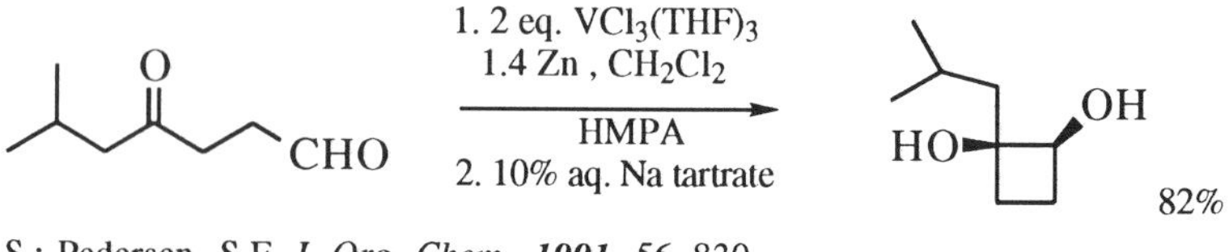

O
CHO
1. 2 eq. $VCl_3(THF)_3$
1.4 Zn, CH_2Cl_2
HMPA
2. 10% aq. Na tartrate

OH
HO
82%

Raw, A.S.; Pedersen, S.F. *J. Org. Chem.*, ***1991***, *56*, 830.
Park, J.; Pedersen, S.F. *Tetrahedron*, ***1992***, *48*, 2069.

Ph-CH=CH-CH=CH$_2$ → 1. catecholborane (B-H), BINAP (PPh$_2$) [Rh(COD)$_2$]$^+$ ClO_4^- 2. NaOH , H_2O_2 → Ph diol (OH OH) + Ph diol (OH OH)

(3 : 1) 62%

40% ee 44% ee

Matsumoto, Y.; Hayashi, T. *Tetrahedron Lett.*, ***1991***, *32*, 3387.

OSiMe$_2$*t*-Bu, Bu, Me → 1. catecholborane , PPh_3 1% $[Rh(COD)Cl]_2$ 2. H_2O_2 , NaOH 3. TBAF → OH OH, Bu, Me + OH OH, Bu, Me

(24 : 1)

When step 1 is 9-BBN/THF/-78°C-25°C (1 : 9)

Burgess, K.; Cassidy, J.; Ohlmeyer, M.J. *J. Org. Chem.*, ***1991***, *56*, 1020.

1. LDBB , THF , -78°C 2. Et_3Al , Me_2O-THF , -70°C 3. iPrCHO → OH, OH 52% + OH 16%

LDBB = 4,4'-di-*t*-butyl biphenylide

Mudryk, B.; Cohen, T. *J. Org. Chem.*, ***1991***, *56*, 5760.

Ph epoxide (O) → 0.2% CAN , H_2O , MeCN 25 min → Ph-CH(OH)-CH$_2$OH 80%

Iranpoor, N.; Baltork. I.M.; Zardaloo, F.S. *Tetrahedron*, ***1991***, *47*, 9861.

O → 1. 2.2 eq. SmI_2 , THF HMPA 2. cyclohexanone (=O) → HO, Me, OH 80%

Molander, G.A.; McKie, J.A. *J. Org. Chem.*, ***1992***, *57*, 3132.

Me, OH → 1. CH$_2$=CH-SiPh$_2$Cl , NEt_3 2. 190°C 3. KF , H_2O_2 → OH, Me, OH + OH, Me, OH

(1 : 1) 71%

Sieburth, S.M.; Fensterbank, L. *J. Org. Chem.*, ***1992***, *57*, 5279.

1. Li , BF_3•OEt_2 , naphthalene
2. iPrCHO
3. aq, HCl

O OH OH 76%

Ramón, D.J.; Yus, M. *Tetrahedron*, ***1992***, *48*, 3585.

1% OsO_4 , $K_3Fe(CN)_6$, H_2O
K_2CO_3 , *t*-BuOH , RT

5% N N $OSiPh_2$*t*-Bu $OSiPh_2$*t*-Bu

OH OH 83%
(41% ee , SS)

Oishi, T.; Hirama, M. *Tetrahedron Lett.*, ***1992***, *33*, 639.

2 $OAlCl_2$ OMe

$FeCl_3$, $MeNO_2$, RT
5 h

OH OH OMe OMe 78%

Sartori, G.; Maggi, R.; Bigi, F.; Arienti, A.; Casnati, G. *Tetrahedron Lett.*, ***1992***, *33*, 2207.

Ph Ph

OsO_4 , aq. MeCN , NMO
polymer-bound
dihydroquinidine , 24 h

OH Ph Ph OH 86%
(82% ee , RR)

Lohray, B.B.; Thomas, A.; Chittari, P.; Ahuja, J.; Dhal, P.K. *Tetrahedron Lett.*, ***1992***, *33*, 5453.

O

1. $BnNEt_3$ BH_4 , Me_3SiCl
CH_2Cl_2 , 0°C
2. 10% K_2CO_3

HO HO 69%

Baskaran, S.; Chidambaram, N.; Narasimhan, N.; Chandrasekaran, S. *Tetrahedron Lett.*, ***1992***, *33*, 6371.

Ph Ph

1. OsO_4 , Pri·N N N N-iPr Ph Ph Ph Ph
2. $NaHSO_3$

HO Ph H H Ph OH 81%
(90% ee)

Fuji, K.; Tanaka, K.; Miyamoto, H. *Tetrahedron Lett.*, ***1992***, *33*, 4021.

Ph–CH=CH–Ph → (cat. OsO_4 , $K_3Fe(CN)_6$, K_2CO_3 , t-BuOH , H_2O; Me, Me, Me, Q, N, S, O_2, N, O) → Ph–CH(OH)–CH(OH)–Ph quant.
(73% ee , RR)

Imada, Y.; Saito, T.; Kawakami, T.; Murahashi, S-I. *Tetrahedron Lett.*, ***1992***, *33*, 5081.

Ph–CH=CH–Me → (DHQD-IND , cat. OsO_4; $K_3Fe(CN)_6$, K_2Co_2 , 0°C, t-BuOH , H_2O) → HO, OH, Ph, Me
72% ee (1R,2S)

DHQD-IND = (N, O, O, Et, N, N, OMe)

Wang, L.; Sharpless, K.B. *J. Am. Chem. Soc.*, ***1992***, *114*, 7568.

Ph–CH=CH–CH=CH–Ph → (1% OsO_4 , $K_3Fe(CN)_6$; $DHQD_2$-PHAL) → Ph–CH(OH)–CH(OH)–CH=CH–Ph 84%
(>99% ee)

$DHQD_2$-PHAL = 1,4-bis-(9-O-dihydroquinidinyl)$_3$
phthalazine - see *J. Org. Chem.*, ***1992***, *57*, 2768.

Xu, D.; Crispino, G.A.; Sharpless, K.B. *J. Am. Chem. Soc.*, ***1992***, *114*, 7570.

(epoxide) O, OBz → (3% $Ni(mac)_2$, O_2; DCE , MS 4Å m 8 h; 100°C) → HO, OBz, OH 90%
(85:15 1,2-diol:1,3-diol)

$Ni(mac)_2$ = bis-(3-methyl-2,4-pentanedionato) nickel (II)

Mukaiyama, T.; Imagawa, K.; Yamada, T.; Takai, T. *Chem. Lett.*, ***1992***, 231.

(γ-butyrolactone) O, O → ($Ru(acac)_3$, tetraglyme; $P(n-C_8H_{17})_3$, H_2; 200°C , H_3PO_4 , 3 h) → OH, OH 27% conversion 95% selectivity

Hara, Y.; Inagaki, H.; Nishimura, S.; Wada, K. *Chem. Lett.*, ***1992***, 1983.

O
Ph Me
Me_3SiCl , NaI , Sm
MeCN , 15 min
Me OH
Ph
Ph
OH Me
84%

Akane, N.; Kanagawa, Y.; Nishiyama, Y.; Ishii, Y. *Chem. Lett.*, ***1992***, 2431.

OMe
1. O_3 , CH_2Cl_2 , 0°C
2. H_2 , Pd-C , EtOH
RT
OH
OH
+
OH
OH

(3 : 1) 41%

Casey, M.; Culshaw, A.J. *SynLett*, ***1992***, 214.

O
Me OH
Et_3Al , PhH-hexane
[PhBr/2 *sec*-BuLi;Al]
0°C → RT
OH
Me OH
Ph
+
Ph
Me OH
OH

(20 : 1) 55%

Ahn, K.H.; Kim, J.S.; Jin, C.S.; Kang, D.H.; Han, D.S.; Shin, Y.S.; Kim, D.H. *SynLett*, ***1992***, 306.

O
Ph N OMe
O Me
$NaBH_4$, EtOH , RT
Ph OH
OH
92%

Sibi, M.P.; Sharma, R. *SynLett*, ***1992***, 497.

REVIEWS:

"Recent Advances in the Asymmetric Dihydroxylation of Alkenes"
Lohray, B.B. *Tetrahedron Asymmetry*, ***1992***, *3*, 1317.

"New Aspects of Stereoselective Syntheses of 1,3-Polyols"
Oishi, T.; Nakata, T. *Synthesis*, ***1990***, 635.

Also via: Section 327 (Alcohol - Ester); Section 357 (Ester - Ester).

SECTION 324: ALCOHOL, THIOL - ALDEHYDE

n-C_4H_9
n-C_5H_{11} OH
OH
NaOCl , 1% TEMPO
$NaHCO_3$, CH_2Cl_2
KBr , Bu_4NCl
n-C_4H_9
n-C_5H_{11} H
OH O
76%

Siedlecka, R.; Skarzewski, J.; Młochowski, J. *Tetrahedron Lett.*, ***1990***, *31*, 2177.

1. PPh_3 , TMSOTf
2. BuLi , THF, -78°C
3. PhCHO , TMSOTf
4. KOH , MeOH
5. H^+

OMe OMe HO Ph CHO 77%

Kim, S.; Kim, Y.C. *Tetrahedron Lett.*, ***1990***, *31*, 2901.

O + 2.1 S S Me$_3$SiO Li

THF , -78°C

S S S S OSiMe$_3$ 86%

Kang, J.; Kim, D.H.; Lee, J.H.; Rim, J.G.; Yoon, Y.B.; Kim, K. *J. Org. Chem.*, ***1990***, *55*, 5555.

1. $MeTi(OiPr)_3$, ether
1 h (>99:1 dr)
2. HCl/H_2O - ether

O Ph N O Me OH Ph CHO 70%

Ukaji, Y.; Yamamoto, K.; Kukui, M.; Fujisawa, T. *Tetrahedron Lett.*, ***1991***, *32*, 2919.

1. Bu_2BOTf , iPr_2NEt
CH_2Cl_2 , 0°C
2. Et_2AlCl , CH_2Cl_2
hexanes , -78°C
(add enolate to Et_2AlCl)
3. iPrCHO

O O N O OH O O Me N O + OH O O Me N O

(95 : 5) 63%

Walker, M.A.; Heathcock, C.H. *J. Org. Chem.*, ***1991***, *56*, 5747.

1. Ph Ph $TolO_2S$-N B N-SO_2Tol C≡C-H , -78°C 2.5 h
2. t-$BuMe_2SiCl$, imidazole
3. O_3

n-$C_5H_{11}CHO$ → n-C_5H_{11} $OSiMe_2t$-Bu CHO 98%

Corey, E.J.; Jones, G.B. *Tetrahedron Lett.*, ***1991***, *32*, 5713.

$Me_3SiO(CN)C(Ph)Ph$ → 1. Dibal, PhMe, 0°C; 2. satd. NH_4Cl; 3. 1.6N H_2SO_4, 18 h; 4. 9N HCl, 15 h → $HO(CHO)C(Ph)Ph$ 69%

Hayashi, M.; Yoshiga, T.; Oguni, N. *SynLett*, ***1991***, 479.

Ph-epoxide-$OSiMe_2t$-Bu → MABR, -78°C, 40 min → Ph-CH(CHO)-CH_2OSiMe_2t-Bu 87% (98% ee)

MABR = Br-(t-Bu)(t-Bu)aryl-O-Al(Me)-O-aryl(t-Bu)(t-Bu)-Br

Maruoka, K.; Ooi, T.; Nagahara, S.; Yamamoto, H. *Tetrahedron*, ***1991***, *47*, 6983.

Related Methods: Section 330 (Alcohol - Ketone).

SECTION 325: ALCOHOL, THIOL - AMIDE

iPrCHO, THF; LHMDS, -100°C → 96% (>99% de 3S,4R)

Ojima, I.; Pei, Y. *Tetrahedron Lett.*, ***1990***, *31*, 977.

HO, H, Ph, O, NHOH → 1. Et_4NIO_4, MeOH, -78°C; 2. cyclopentadiene; 3. Na/Hg, MeOH, NaH_2PO_4 → OH, H, N, O, OH, H, Ph 77%

Miller, A.; Procter, G. *Tetrahedron Lett.*, ***1990***, *31*, 1043.

N-Boc piperidine → 1. *sec*-BuLi, TMEDA, ether; 2. PhCHO → (OH, Ph, N-Boc) 39% + (N, O, O, Ph) 37%

Beak, P.; Lee, W.K. *J. Org. Chem.*, ***1990***, *55*, 2578.

1. LDA, THF, -78°C; 2. MeCHO, -78°C → RT — 80%

PMP = *p*-methoxyphenyl

Durst, T.; Sharma, M.K.; Gabe, E.J.; Lee, F.L. *J. Org. Chem.*, ***1990***, *55*, 5525.

$PhCO_2Me$ → ($HOCH_2CH_2NH_2$, e^-) → BzHN–CH_2CH_2–OH 73%

Arai, K.; Tamura, S.; Masumizu, T.; Kawai, K.; Nakajima, S.; Ueda, A. *Can. J. Chem.*, ***1990***, *68*, 903.

Cl–C_6H_4–NHBoc → (1. 2.2 eq. BuLi; 2. 10°C; 3. H_2O) 82%

Smith, K.; Pritchard, G.J. *Angew. Chem. Int. Ed. Engl.*, ***1990***, *29*, 282.

Et_2NC(O)TeBu → (1. BuLi, THF, -105°C; 2. PhCHO; 3. H^+) Et_2NC(O)CH(OH)Ph 65%

Hiiro, T.; Mogami, T.; Kambe, N.; Fujiwara, S.; Sonoda, N. *Synth. Commun.*, ***1990***, *20*, 703.

Al(Hg), aq. THF,)))))) — 70%

Luzzio, F.A.; O'Hara, L.C. *Synth. Commun.*, ***1990***, *20*, 3223.

BuMnCl — (>99 : <1) 98%

Taniguchi, M.; Fujii, H.; Oshima, K.; Utimoto, K. *Tetrahedron Lett.*, ***1992***, *33*, 4353.

piperidine, $AlCl_3$, 1 h; DCE, 25°C — 91%

Lesimple, P.; Bigg, D.C.H. *Synthesis*, ***1991***, 306.

BnCHO, CH_2Cl_2
RT, 15 h
$[V_2Cl_3(THF)_6]_2[Zn_2Cl_6]$

OMe O N CHO → OMe O N OH HO Ph 73%

(80:20 syn:anti)
(77%ee)

Annunziata, R.; Cinquini, M.; Cozzi, F.; Giaroni, P. *Tetrahedron Asymmetry*, ***1990***, *1*, 355.

O_2N N O O

$PhCO_2H$, PhMe
reflux

O_2N N O BzO OH 92%

Schmidhauser, J.C.; Longley, K.L. *Tetrahedron Lett.*, ***1991***, *32*, 7155.

CHO NHCbz

1. 1.1 eq. $VCl_3(THF)_3$
0.6 Zn
2. 1 M HCl

HO NHCbz CbzHN OH 89%

Konradi, A.W.; Pedersen, S.F. *J. Org. Chem.*, ***1992***, *57*, 28.

Cl N O O CF_3

1. 2 eq. SmI_2, TF
2. cyclohexanone (=O)
3. H_3O^+

N OH O CF_3 53%

Collin, J.; Namy, J.L.; Jones, G.; Kagan, H.B. *Tetrahedron Lett.*, ***1992***, *33*, 2973.

O O H NMe_2 Me

$MeAlCl_2$, 0°C
hexane/CH_2Cl_2
2 h

Me OH O H NMe_2 Me + HO Me O H NMe_2 Me

(97 : 3) 66%

Fujii, H.; Taniguchi, M.; Oshima, K.; Utimoto, K. *Tetrahedron Lett.*, ***1992***, *33*, 4579.

Ph Ph

1. 2 eq. Me_2NCHO
Na, 80°C, 5 h
2. 0°C, ice/H_2O

Ph Ph O NMe_2 OH 93%

Botteghi, C.; Gotta, S.; Marchetti, M.; Melloni, G. *Tetrahedron Lett.*, ***1992***, *33*, 5601.

N-CO_2Ph, OHC, $OSiMe_2t$-Bu —[SmI_2]→ HO, N- CO_2Ph, $OSiMe_2t$-Bu 25%

Baldwin, J.E.; Turner, S.C.M.; Moloney, M.G. *Tetrahedron Lett.*, ***1992***, *33*, 1517.

N, $(CH_2)_8$, O O —[$NaBH_4$, EtOH, 0°C; pH = 8-10]→ N, $(CH_2)_8$, O OH + N, $(CH_2)_8$, OH O

36% 30%

Moeller, K.D.; Hanau, C.E. *Tetrahedron Lett.*, ***1992***, *33*, 6041.

OH, CO_2Et —[1. n-$C_3H_7NH_2$ 2. MeOH, reflux]→ OH, N, O, n-C_3H_7 76%

Amri, H.; El Gaied, M.M.; Ben Ayed, T.; Villeras, J. *Tetrahedron Lett.*, ***1992***, *33*, 6159.

O O NHPh —[ZnBr, THF, 25°C]→ OH O NHPh 91%

Taniguchi, M.; Oshima, K.; Utimoto, K. *Chem. Lett.*, ***1992***, 2135.

SECTION 326: ALCOHOL, THIOL - AMINE

OH, OH, BnO, BnO, OMe —[$PhCH_2NH_2$, Zn, 2h; $NaBH_3CN$, n-PrOH; H_2O, reflux]→ OH, BnO, BnO, NHBn 82%

Bernoas, R.C. *Tetrahedron Lett.*, ***1990***, *31*, 469.

HO, MeHN, Bu, H ←[1. BuCu 2. $LiAlH_4$ 3. H_2, Pd-C]— O, OH, Ph, N, SPh, Me —[1. BuZnI 2. $LiAlH_4$ 3. H_2, Pd-C]→ HO, MeHN, H, Bu

Agami, C.; Couty, F.; Daran, J-C.; Prince, B.; Puchot, C. *Tetrahedron Lett.*, ***1990***, *31*, 2889.

O
Me O
Me O N CO_2H
O Me

e$^-$, NaOMe, MeOH
Bu_4NHSO_4, 15°C
graphite electrodes
500 mA, 80 h
3.9 V

O
Me O
Me O N OH
O Me
75%

Fasseur, D.; Rigo, B.; Cauliez, P.; Debacker, M.; Couturier, D. *Tetrahedron Lett.*, ***1990***, *31*, 1713.

Bn
Boc N CHO
H

$SiMe_3$
$SnCl_4$

Bn
Boc N
H OH
+
Bn
Boc N
H OH
(1 : 6) 65%

Vara Prasad, J.V.N.; Rich, H. *Tetrahedron Lett.*, ***1990***, *31*, 1803.

Me N
t-BuN

1. *t*-BuLi, THF
-78°C
2. PhCHO, -5°C

Ph
HO N
t-BuN
60%

Solladié-Cavallo, A.; Bencheqroun, M. *Tetrahedron Lett.*, ***1990***, *31*, 2157.

OEt
EtO
O
O
N Me

1. BuLi, ether
2. MeOMs
3. H_2O ; aq. HCl

OH
O
O
N Me
Me
67%

Simig, G.; Schlosser, M. *Tetrahedron Lett.*, ***1990***, *31*, 3125.

Ph
Me N O n-C_4H_9

$Mo(CO)_6$, reflux
H_2O/MeCN
2-3 h

Ph n-C_4H_9
MeHN OH
78%

Cicchi, S.; Goti, A.; Brandi, A.; Guarna, A.; De Sarlo, F. *Tetrahedron Lett.*, ***1990***, *31*, 3351.

CHO
NBn_2

$SnCl_4$, CH_2Cl_2
-78°C

HO
NBn_2
+
HO
NBn_2
(73 : 27) 59%

Mikami, K.; Kaneko, M.; Loh, T-P.; Terada, M.; Nakai, T. *Tetrahedron Lett.*, ***1990***, *31*, 3909.

Me, O → Me, OH, NHBn

$BnNH_2$, $LiClO_4$

MeCN, 64 h

80%

Chini, M.; Crotti, P.; Macchia, F. *Tetrahedron Lett.*, ***1990***, *31*, 4661.

Ph, Ph, N, Ph, CO_2Me → H_2N, Ph, Ph, OH

1. Dibal, CH_2Cl_2, -78°C
2. PhMgBr, ether
3. aq. $NaHCO_3$
4. separate
5. H_3O^+

69%

Polt, R.; Peterson, M.A. *Tetrahedron Lett.*, ***1990***, *31*, 4985.

O, n-C_6H_{13} → N_3, OH, n-C_6H_{13} + HO, n-C_6H_{13}, N_3

NaN_3, $LiClO_4$

MeCN, 80°C

24 h

(7.5 : 92.5) 96%

Chini, M.; Crotti, P.; Macchia, F. *Tetrahedron Lett.*, ***1990***, *31*, 5641.

N_2, O, Me, N, O, Me → Me, O, OH, N, Me

$Rh_2(OAc)_4$, PhH

reflux, 5 h

77%

Etkin, N.; Babu, S.D.; Fooks, C.J.; Durst, T. *J. Org. Chem.*, ***1990***, *55*, 1093.

N, NMe_2 → Ph, N, H, OH

1. BuLi
2. PhCHO
3. $NaBH_4$, THF/EtOH

47% overall

Katritzky, A.R.; Lue, P.; Chen, Y-X. *J. Org. Chem.*, ***1990***, *55*, 3688.

N, Me, O → N, Me, $OSiMe_2Ph$

$HSiMe_2Ph$, CH_2Cl_2, 25°C

CsF, 18-crown-6, 9 h

67%

Goldberg, Yu.; Rubina, K.; Shymanska, M.; Lukevics, E. *Synth. Commun.*, ***1990***, *20*, 2439.

2 eq. SmI_2 , ethylene glycol
HMPA , RT
N-OBn

Ph OH NHOBn 83%

Hanamoto, T.; Inanaga, J. *Tetrahedron Lett.*, ***1991***, *32*, 3555.

Ph , MeCN
10 kbar , 24 h

Ph Ph 56%

15% at atmospheric preessure

Kotsuki, H.; Nishiuchi, M.; Kobayashi, S.; Nishizawa, H. *J. Org. Chem.*, ***1990***, *55*, 2969.

1. BuLi , $ClSiMe_2H$, ether
2. $Pt[(CH_2=CH_2)Me_2Si]_2O)_2$ 0.2% , RT , 30 min
3. EDTA•2 Ma , RT
4. 30% H_2O_2 , KF , $KHCO_3$ MeOH , THF , RT

Ph NH_2 → Ph OH NH_2 77% (80:20 syn:anti)

Tamao, K.; Nakagawa, Y.; Ito, Y. *J. Org. Chem.*, ***1990***, *55*, 3438.

CHO + OHC H NHBoc

1. 0.5 $[V_2Cl_3(THF)_6]_2$ $[Zn_2Cl_6]$
1 h addition of amino-aldehyde
2. 10% aq. Na tartrate

OH NHBoc OH 70% (>20:1 dr)

Konradi, A.W.; Pedersen, S.F. *J. Org. Chem.*, ***1990***, *55*, 4506.

1. $SOCl_2$, CCl_4
2. LiN_3 , DMF , 120°C
3. H_2 , Pd-C

OH Ph Ph OH → OH Ph Ph NH_2 81% (>96% ee , RS)

Lohray, B.B.; Ahuja, J.R. *J. Chem. Soc., Chem. Commun.*, ***1991***, 95.

porcine pancreatic lipase
EtOAc

N H OH → N Ac OH + N H OH

(-) , 92% ee , S (9%) (+) , 90% ee , R (90%)

Asensio, G.; Andreu, C.; Marco, J.A. *Tetrahedron Lett.*, ***1991***, *32*, 4197.

PhCHO → 1. $TolSO_2CH_2CN$, 1% AgOTf, 25°C, CH_2Cl_2 (Me, Me, N, N, Fe, PPh_2, PPh_2); 2. $LiAlH_4$, THF → OH, Ph, NHMe

96%

(83% ee)

Sawamura, M.; Hamashima, H.; Ito, Y. *J. Org. Chem.*, ***1990***, *55*, 5935.

O, ⊕ NMe_3 → MeCN, hν (near UV) → HO, ⊕ Me, N-Me quant.

quantum efficiency = 0.56

Wagner, P.J.; Cao, Q. *Tetrahedron Lett.*, ***1991***, *32*, 3915.

O, OH → n-$C_6H_{13}NH_2$, $Ti(OiPr)_4$, CH_2Cl_2 → NHn-C_6H_{13}, OH, OH + OH, OH, NHn-C_6H_{13}

(93 : 7) 96%

Canas, M.; Poch, M.; Verdageur, X.; Moyano, A.; Pericàs, M.A.; Riera, A. *Tetrahedron Lett.*, ***1991***, *32*, 6931.

OH, Me, OH, H, N, $CHPh_2$ → 1.1 MsCl, 3 NEt_3, CH_2Cl_2, 0°C (8 h); RT (36 h) → HO, Me, N, $CHPh_2$ 77%

Poch, M.; Verdaguer, X.; Moyano, A.; Pericàs, M.A.; Riera, A. *Tetrahedron Lett.*, ***1991***, *32*, 6935.

MeO_2C, N, O → 1. O, Ph, N=O; 2. OsO_4, NMO; 3. H_2, Pd-C → HO, OH, N, H

Defoin, A.; Pires, J.; Streith, J. *SynLett*, ***1991***, 417.

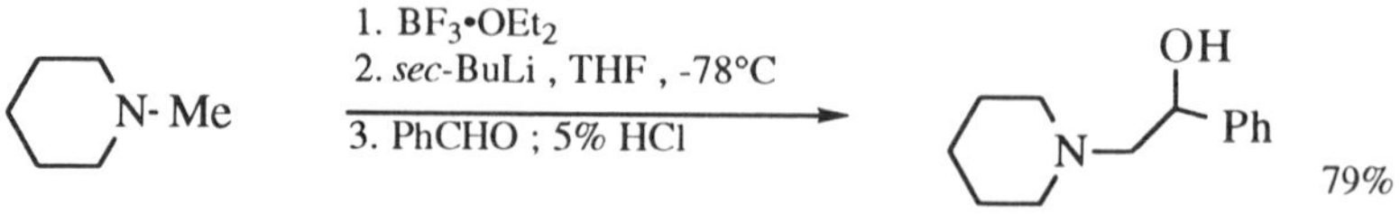

1. BF_3•OEt_2
2. *sec*-BuLi , THF , -78°C
3. PhCHO ; 5% HCl

N-Me → OH, Ph, N 79%

Kessar, S.V.; Singh, P.; Vohra, R.; Kaur, N.P.; Singh, K.N. *J. Chem. Soc., Chem. Commun.*, ***1991***, 568.

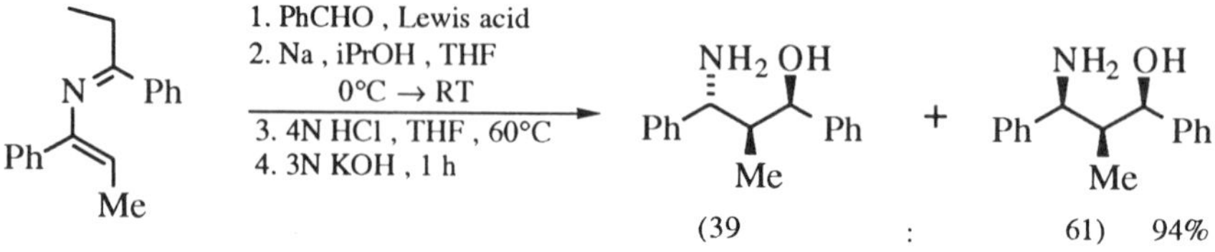

1. PhCHO , Lewis acid
2. Na , iPrOH , THF 0°C → RT
3. 4N HCl , THF , 60°C
4. 3N KOH , 1 h

N, Ph, Ph, Me → NH_2 OH, Ph, Ph, Me + NH_2 OH, Ph, Ph, Me

(39 : 61) 94%

Barluenga, J.; Joglar, J.; González, F.J.; Fustero, S.; Krüger, C.; Tsay, Y-H. *Synthesis,* ***1991***, 387.

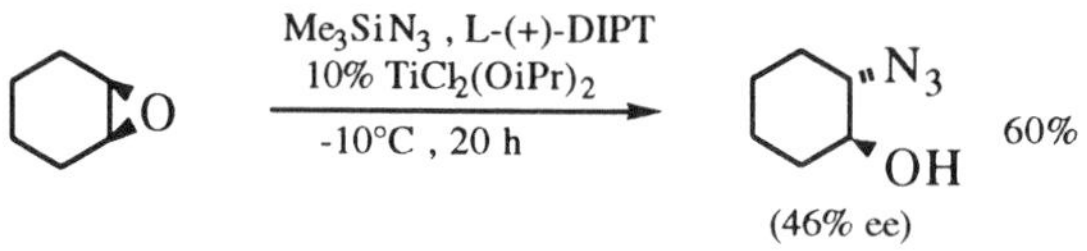

Me_3SiN_3 , L-(+)-DIPT
10% $TiCl_2(OiPr)_2$
-10°C , 20 h

O → N_3, OH 60%

(46% ee)

Hayashi, M.; Kohmura, K.; Oguni, N. *SynLett,* ***1991***, 774.

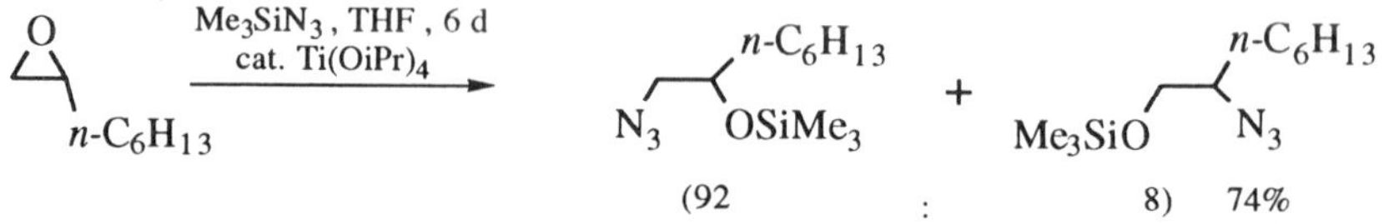

Me_3SiN_3 , THF , 6 d
cat. $Ti(OiPr)_4$

O, n-C_6H_{13} → n-C_6H_{13}, N_3, $OSiMe_3$ + n-C_6H_{13}, Me_3SiO, N_3

(92 : 8) 74%

Sutowardoyo, K.I.; Emziane, M.; Lhoste, P.; Sinou, D. *Tetrahedron,* ***1991***, *47*, 1435.

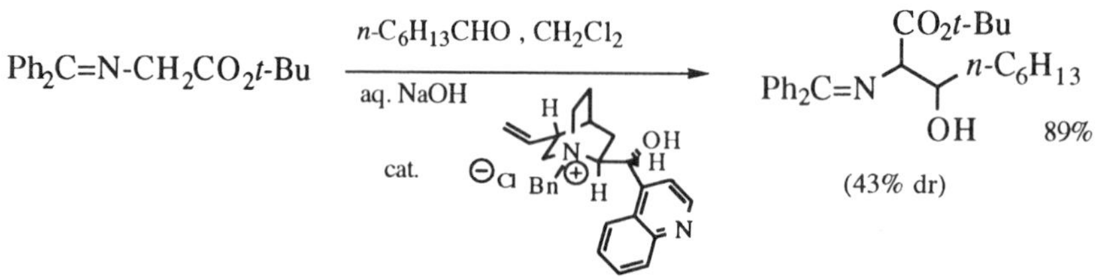

Ph_2C=N-CH_2CO_2t-Bu

n-C_6H_{13}CHO , CH_2Cl_2
aq. NaOH
cat. Cl⁻ Bn N⁺ H H OH H N

→ CO_2t-Bu, Ph_2C=N, n-C_6H_{13}, OH 89%

(43% dr)

Gasparski, C.M.; Miller, M.J. *Tetrahedron,* ***1991***, *47*, 5367.

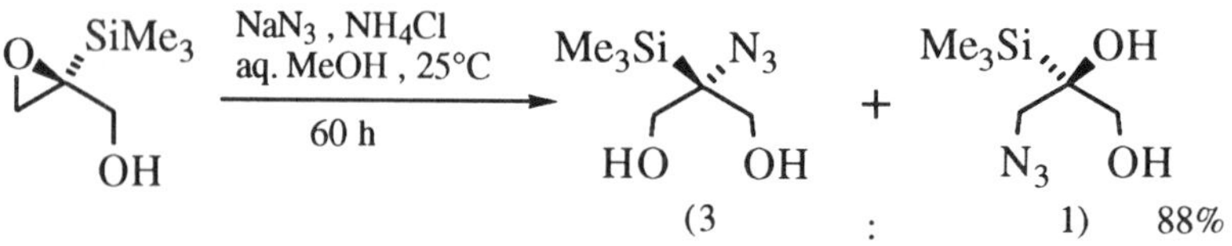

NaN_3 , NH_4Cl
aq. MeOH , 25°C
60 h

O, $SiMe_3$, OH → Me_3Si, N_3, HO, OH + Me_3Si, OH, N_3, OH

(3 : 1) 88%

Chakraborty, T.K.; Reddy, G.V. *Tetrahedron Lett.,* ***1991***, *32*, 679.

e^- , Sn cathode , C anode
5% C , iPrOH , Et_4OTs

NHOMe
HO
88%

Shono, T.; Kise, N.; Fujimoto, T. *Tetrahedron Lett.*, ***1991***, *32*, 525.
Shono, T.; Kise, N.; Kunimi, N.; Nomura, R. *Chem. Lett.*, ***1991***, 2191.

BocHN CO_2H Ph

1. NMM
2. iBuOCOCl , DME
-15°C , 1 min
3. $NaBH_4$, H_2O
-15°C , 30 sec

BocHN OH Ph 81%

Rodriguez, M.; Llinares, M.; Doulut, S.; Heitz, A.; Martinez, J. *Tetrahedron Lett.*, ***1991***, *32*, 923.

N SPh

1. LiDBB , THF , -95°C
2. MeO CHO
-95°C , 90 sec

N OMe OH 79%

LiDBB = lithium di-*t*-butylbiphenyl radical anion

Tsunoda, T.; Fujiwara, K.; Yamamoto, Y.; Itô, S. *Tetrahedron Lett.*, ***1991***, *32*, 1975.

N· Me ⊕ Ph $Br^⊖$ O

1. NaOMe , PhH , reflux
2. $NaBH_4$
3. MeO_2CCl , PhH
reflux

Cl H Ph H H Me–N CO_2Me OCO_2Me 58%

Coates, A.J.; Malone, J.F.; McCarney, M.T.; Stevenson, P.J. *Tetrahedron Lett.*, ***1991***, *32*, 2827.

MeHN PPh_3^+ Br^-

1. Br Ph O OTHP , CH_2Cl_2 NEt_3
2. NaH , THF , heat
3. H^+

OH Ph N Me 76%

Kinder, F.R. Jr.; Jarosinski, M.A.; Anderson, W.K. *J. Org. Chem.*, ***1991***, *56*, 6475.

OSiMe$_2$t-Bu OHC Me → 1. LiHDMS 2. allyl MgCl → OSiMe$_2$t-Bu Me NH$_2$ + OSiMe$_2$t-Bu Me NH$_2$

(4 : 96) 66%

Cainelli, G.; Giacomini, D.; Mezzina, E.; Panunzio, M.; Zarantonello, P. *Tetrahedron Lett.*, ***1991***, *32*, 2967.

S Me N → TsNCO THF 20°C → Ts-N Me-N → 1. LDA, THF -78°C → 0°C 2. PhCHO → Ts-N OH Me N Me + Ts-N OH Me N Me

(82 : 18) 88%

Magnus, P.; Moursounidis, T. *J. Org. Chem.*, ***1991***, *56*, 1529.

N I → 3 eq. SmI$_2$, Me t-Bu (O) (THP)^3 - HMPA → N t-Bu Me OH I

87%

Murakami, M.; Hayashi, M.; Ito, Y. *J. Org. Chem.*, ***1992***, *57*, 793.

Ph Me O Ph NHBz → 1. Dibal, ZnCl$_2$, THF 2. H$_2$O, OH$^-$ 3. LiAlH$_4$, THF → Ph Me OH Ph NHBz + Ph Me OH Ph NHBz

(97 : 3) 92%

Barluenga, J.; Aguilar, E.; Fustero, S.; Olano, B.; Viado, A.L. *J. Org. Chem.*, ***1992***, *57*, 1219.
Barluenga, J.; Aguilar, E.; Olano, B.; Fustero, S. *SynLett*, ***1990***, 463.

+ O Ot-Bu O N=O → Ot-Bu O O N O 94% (96% de) → 1. Na(Hg), aq. MeOH 2. NaOH, aq. EtOH → H$_2$N HO

Martin, S.F.; Hartmann, M.; Josey, J.A. *Tetrahedron Lett.*, ***1992***, *33*, 3583.

1. Dibal
2. NH_4Br
3. $MeNH_2$
4. $NaBH_4$

79%

Zandbergen, P.; van den Nieuwendijk, A.M.C.H.; Brussee, J.; van der Gen, A.; Kruse, C.G. *Tetrahedron*, ***1992***, *48*, 3977.

1. Bn_2NH , RT
2. $NaBH_4$, EtOH , RT

85%

Bégué, J-P.; Bonnet-Deppon, D.; Sdassi, H. *Tetrahedron Lett.*, ***1992***, *33*, 1879.

rat liver microsomes
EtOH , phosphate buffer
36°C

(86% ee) 40%

Kamal, A.; Rao, A.B.; Rao, M.V. *Tetrahedron Lett.*, ***1992***, *33*, 4077.

1. BuLi , THF , -80°C
2. $AlMe_3$, PhMe , -80°C
3. -80°C → 0°C
4. NaF , H_2O

69%

Najime, R.; Pilard, S.; Vaultier, M. *Tetrahedron Lett.*, ***1992***, *33*, 5351.

1. BuLi , -50°C
2. *t*-BuLi , -30°C → 20°C
ether , PMDETA
3. PhCHO
4. H_2O

75%

PMDETA = N,N,N',N'',N''-pentamethyldiethylenediamine

Barluenga, J.; González, R.; Fañanás, F.J. *Tetrahedron Lett.*, ***1992***, *33*, 7573.

$iPrMe_2SiN_3$, TMSOTf
, 0°C , 48 h

86%

(93% ee)

Nugent, W.A. *J. Am. Chem. Soc.*, ***1992***, *114*, 2768.

REVIEWS:

"The Synthetic Utility of α-Amino Alkoxides"
Comins, D.L. *SynLett*, ***1992***, 615.

SECTION 327: ALCOHOL, THIOL - ESTER

baker's yeast, 20% sucrose, H_2O, dialysis tube; 55%

Spiliotis, V.; Papahaatjis, D.; Ragoussis, N. *Tetrahedron Lett.*, ***1990***, *31*, 1615.

n-C_3H_7CHO, 7d; $SO_2N(iPr)_2$; 70% (42% de)

Baylis-Hillman coupling reaction

Basavaiah, D.; Gowriswari, V.V.L.; Sarma, P.K.S.; Dharma Rao, P. *Tetrahedron Lett.*, ***1990***, *31*, 1621.

$OSiMe_3$; Me; O; CO_2iPr; LDA, THF, -100°C; $SiMe_3$ OH; CO_2iPr + $SiMe_3$ OH; CO_2iPr; Me; (35 : 1) 90%

Ishikawa, A.; Uchiyama, H.; Katsuki, T.; Yamaguchi, M. *Tetrahedron Lett.*, ***1990***, *31*, 2415.

CO_2Me; cat. OsO_4, NMO, aq. acetone, RT, 12 h; HO, OH, OH, OH, CO_2Me; 85%

Backenstrass, F.; Streith, J.; Tschamber, T. *Tetrahedron Lett.*, ***1990***, *31*, 2139.

HO, Ph, OH; 1. Bu_2SnO, PhMe reflux, 6 h; 2. BzCl; 3. Me_3SiCl; Ph, Me_3SiO, OBz + Ph, BzO, $OSiMe_3$; (98 : 2) 83%

Reginato, G.; Ricci, A.; Roelens, S.; Scapecchi, S. *J. Org. Chem.*, ***1990***. *55*, 5132.

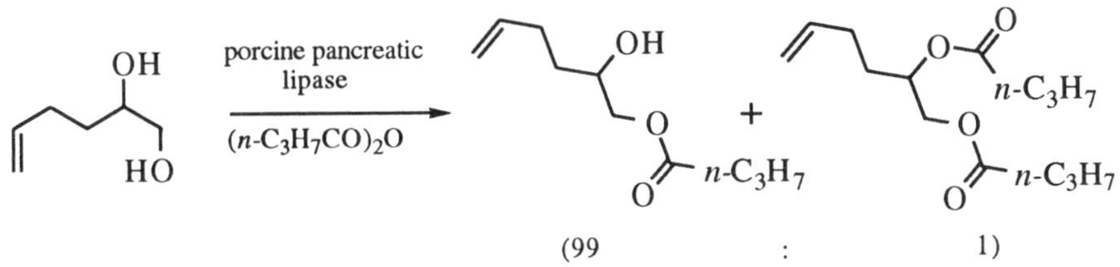

porcine pancreatic lipase

$(n\text{-}C_3H_7CO)_2O$

(99 : 1)

Ramaswamy, S.; Morgan, B.; Oehlschlager, A. *Tetrahedron Lett.*, ***1990***, *31*, 3405.

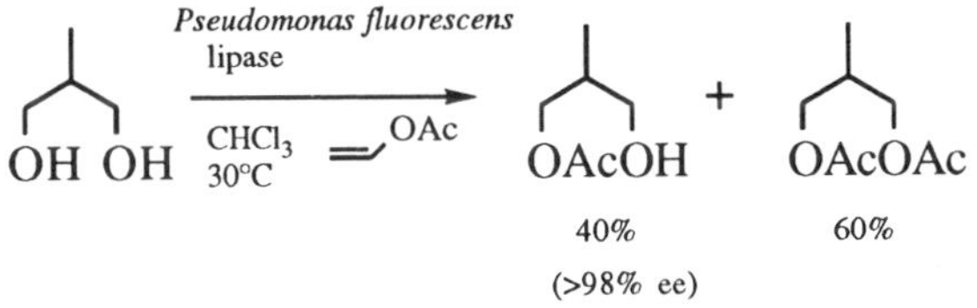

Pseudomonas fluorescens lipase

$CHCl_3$ 30°C

OAcOH 40% (>98% ee) + OAcOAc 60%

Santaniello, E.; Ferrabuschi, P.; Grisenti, P. *Tetrahedron Lett.*, ***1990***, *31*, 5657.

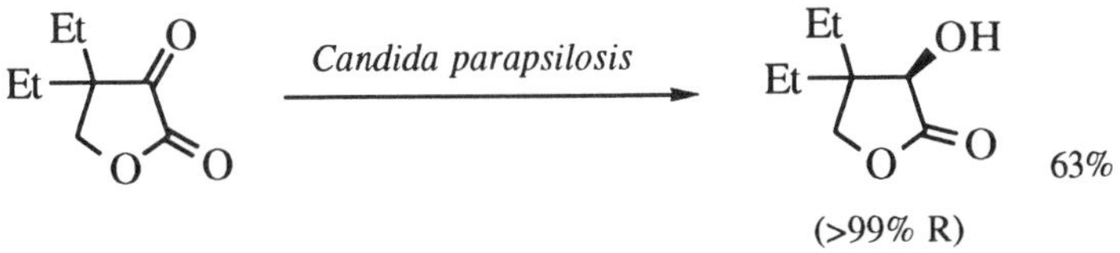

Candida parapsilosis

63% (>99% R)

Hata, H.; Shimizu, S.; Hattori, S.; Yamada, H. *J. Org. Chem.*, ***1990***, *55*, 4377.

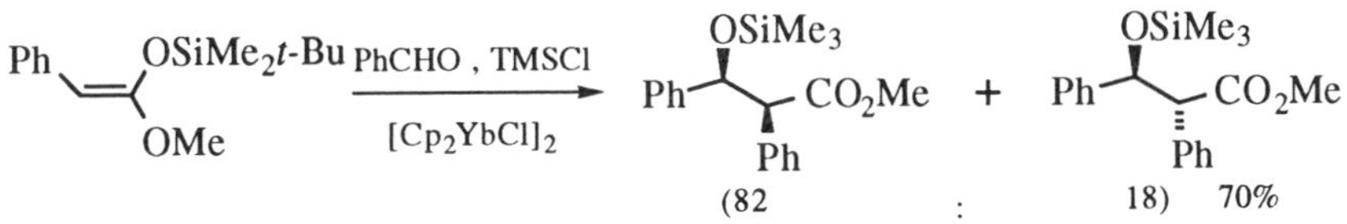

PhCHO , TMSCl

$[Cp_2YbCl]_2$

(82 : 18) 70%

Gong, L.; Streitwieser, A. *J. Org. Chem.*, ***1990***, *55*, 6235.

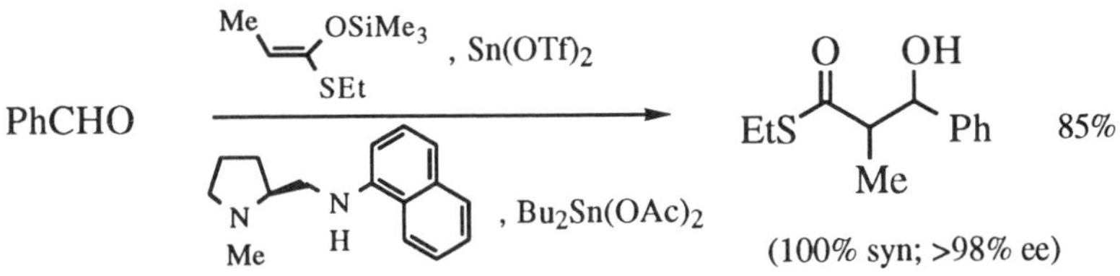

PhCHO

$Sn(OTf)_2$

$Bu_2Sn(OAc)_2$

85% (100% syn; >98% ee)

Mukaiyama, T.; Kobayashi, S.; Uchiro, H.; Shiina, I. *Chem. Lett.*, ***1990***, 129.
Mukaiyama, T.; Uchiro, H.; Shiina, I.; Kobayashi, S. *Chem. Lett.*, ***1990***, 1019.
Mukaiyama, T.; Shiina, I.; Kobayashi, S. *Chem. Lett.*, ***1991***, 1901.
Kobayashi, S.; Uchiro, H.; Fujishita, Y.; Shiina, I.; Mukaiyama, T. *J. Am. Chem. Soc.*, ***1991***, *113*, 4247.

immobilized bakers yeast
MeOH

OH
CO_2Et CO_2Et 56%

(97% ee , S)

Naoshima, Y.; Maeda, J.; Munakata, Y.; Nishiyama, T.; Kamezawa, M.; Tachibana, H. *J. Chem. Soc., Chem. Commun.*, ***1990***, 964.

O , CH_2Cl_2
PhCHO
$BF_3•OEt_2$, $(CH_2OH)_2$
0°C (5 h) → RT (20 h)

Ph O OH
O
61%

Nagumo, S.; Matsukuma, A.; Inoue, F.; Yamamoto, T.; Suemune, H.; Sakai, K. *J. Chem. Soc., Chem. Commun.*, ***1990***, 1538.

OH O
iPrCHO , 15% SmI_2
O
O OH
95%

>99:1 anti:syn)

Evans, D.A.; Hoveyda, A.H. *J. Am. Chem. Soc.*, ***1990***, *112*, 6447.

O
1. Li , NH_3 , THF
-78°C , 30 sec
2. aq. NH_4Cl
$CO_2n\text{-}C_8H_{17}$
HO
$CO_2n\text{-}C_8H_{17}$
72%

van der Baan, J.L.; Barnick, J.W.F.K.; Bickelhaupt, F. *Synthesis*, ***1990***, 897.

PhCHO + $OSiMe_2t\text{-}Bu$ / OEt
20% $ZnEt_2$ / Bn(NHTf)$CO_2SiMe_2t\text{-}Bu$
PhMe , -78°C
EtO_2C $OSiMe_2t\text{-}Bu$ Ph
(60% ee)
79%

Mukaiyama, T.; Takashima, T.; Kusaka, H.; Shimpuku, T. *Chem. Lett.*, ***1990***, 1777.

CO_2Bn
cat. $Mn(dpm)_2$, O_2 , $PhSiH_3$
0°C , iPrOH
OH
CO_2Bn
91%

$Mn(dpm)_2$ = bis(dipivaloylmethanato) manganese (II)

Inoki, S.; Kato, K.; Isayama, S.; Mukaiyama, T. *Chem. Lett.*, ***1990***, 1869.

OH
CO_2Me
1. LDA
2. PhCHO
OH OH
Ph
CO_2Me
+
OH OH
Ph
CO_2Me
(30 : 70) 86%

1. Et_3SiCl 2. PhCHO , $TiCl_4$ (93 : 7) 75%

Shirai, F.; Gu, J-H.; Nakai, T. *Chem. Lett.*, ***1990***, 1931.

OEt
$OSiMe_3$
OEt
$OSiMe_3$
Me O Me O , $TiCl_4$
TiO_2 , CH_2Cl_2 , -80°C
EtO_2C CO_2Et
HO OH
Me Me
50%

Ali, S.M.; Rousseau, G. *SynLett*, ***1990***, 397.

O
CO_2Me
Ph_3SiClO_4 , 1.5 h
CH_2Cl_2 , -78°C → -30°C
OH
CO_2Me
95%

Inokuchi, T.; Kusomoto, M.; Matsumoto, S.; Okada, H.; Torii, S. *Chem. Lett.*, ***1991***, 2009.

O
CO_2Me
SmI_2 , THF-HMPA
t-BuOH
MeO_2C
H OH
70%

Hon, Y-S.; Lu, L.; Chu, k-P. *Synth. Commun.*, ***1991***, *21*, 1981.

PhCHO + CO_2Me
LDA , $CdCl_2$
OH
Ph
CO_2Me
+
OH
Ph
CO_2Me
(15 : 85) 75%

Lei, B.; Fallis, A.G. *Can. J. Chem.*, ***1991***, *69*, 1450.

O
$Cl_2CeCH_2CO_2Et$, THF
-78°C
HO
CO_2Et
96%

Liu, H-J.; Zhu, B-Y. *Can. J. Chem.*, ***1991***, *69*, 2008.

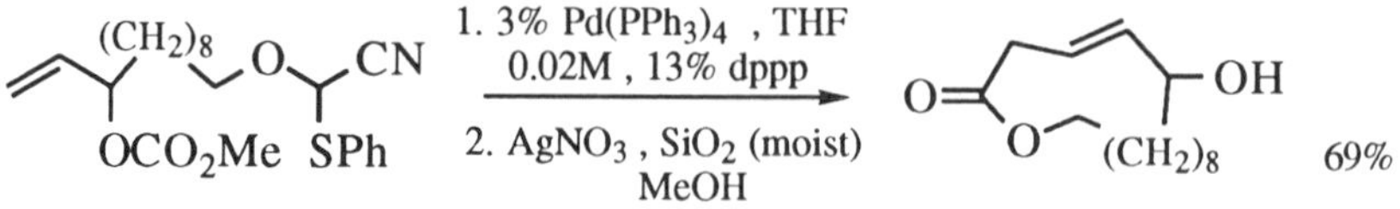

$(CH_2)_8$ O CN
OCO_2Me SPh

1. 3% $Pd(PPh_3)_4$, THF
0.02M , 13% dppp
2. $AgNO_3$, SiO_2 (moist)
MeOH

O= O OH $(CH_2)_8$ 69%

Trost, B.M.; Granja, J.R. *J. Am. Chem. Soc.*, ***1991***, *113*, 1044.

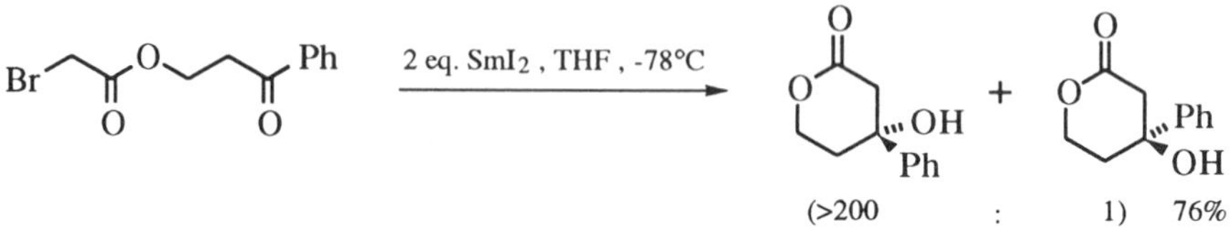

Br O Ph
O O

2 eq. SmI_2 , THF , -78°C

O O OH Ph + O O Ph OH

(>200 : 1) 76%

Molander, G.A.; Etter, J.B.; Harring, L.S.; Thorel, P-J. *J. Am. Chem. Soc.*, ***1991***, *113*, 8036.

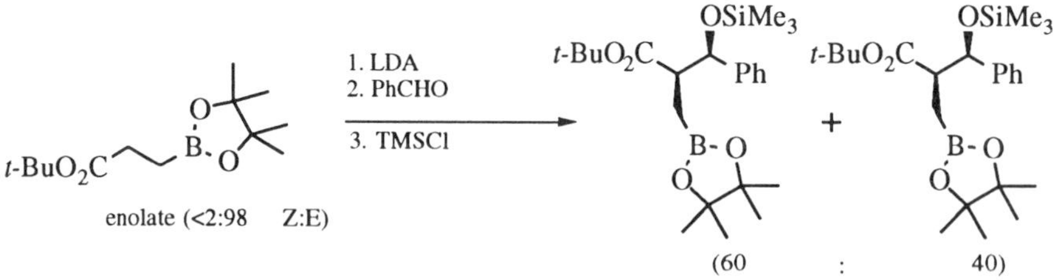

t-BuO_2C B O O

enolate (<2:98 Z:E)

1. LDA
2. PhCHO
3. TMSCl

$OSiMe_3$ *t*-BuO_2C Ph B-O O + $OSiMe_3$ *t*-BuO_2C Ph B-O O

(60 : 40)

Curtis, A.D.M.; Whiting, A. *Tetrahedron Lett.*, ***1991***, *32*, 1507
Whiting, A. *Tetrahedron Lett.*, ***1991***, *32*, 1503.

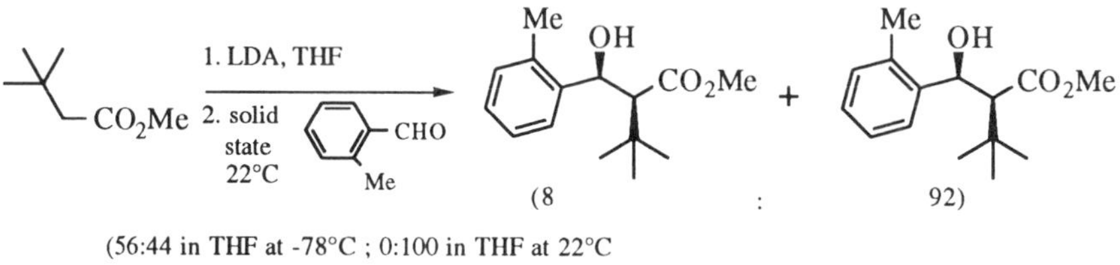

CO_2Me

1. LDA, THF
2. solid state 22°C CHO Me

Me OH CO_2Me + Me OH CO_2Me

(8 : 92)

(56:44 in THF at -78°C ; 0:100 in THF at 22°C

Wei, Y.; Bakthavatchalam, R. *Tetrahedron Lett.*, ***1991***, *32*, 1535.

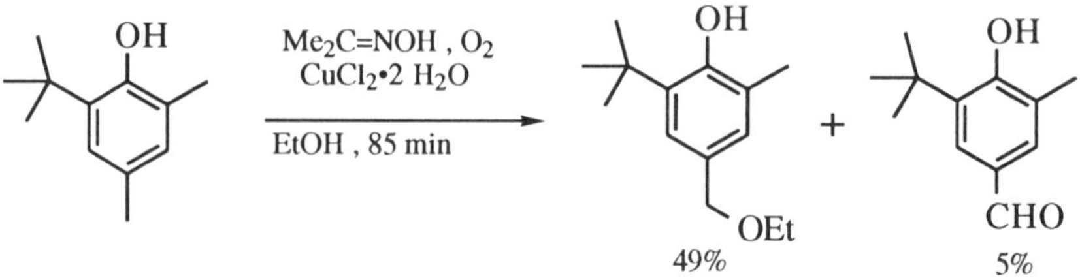

OH

$Me_2C{=}NOH$, O_2
$CuCl_2 \cdot 2\ H_2O$
EtOH , 85 min

OH OEt + OH CHO

49% 5%

Shimizu, M.; Watanabe, Y.; Orita, H.; Hayakawa, T.; Takehira, K. *Tetrahedron Lett.*, ***1991***, *32*, 2053.

1. EtCOCl
2. $AlCl_3$

75%

Sartori, G.; Casnati, G.; Bigi, F.; Baraldi, D. *Tetrahedron Lett.*, ***1991***, *32*, 2153.

1. R^*_2BBr, NEt_3, -78°C
2. PhCHO, CH_2Cl_2

86% (98:2 anti:syn), 96% ee

Bu_3SnH, PhH, AIBN, reflux

74%

Corey, E.J.; Cho, S. *Tetrahedron Lett.*, ***1991***, *32*, 2857.

bakers yeast
50°C, + MVK
30 min

(97 : 3) 94%

Nakamura, K.; Kawai, Y.; Ohno, A. *Tetrahedron Lett.*, ***1991***, *32*, 2927.
Nakamura, K.; Kondo, S.; Kawai, Y.; Ohno, A. *Tetrahedron Lett.*, ***1991***, *32*, 7075.

1. $SOCl_2$
2. LiBr, acetone, 50°C
3. 2.5 Zn, H_2O (cat.)
50°C, 2 h

70%

Gao, Y.; Zepp, C.M. *Tetrahedron Lett.*, ***1991***, *32*, 3155.

$Ti(OiPr)_3OAc$, 1 h
$CHCl_3$

(>98 : <2) 95%

Raifeld, Y.E.; Nikitenko, A.A.; Arshava, B.M. *Tetrahedron Asymmetry*, ***1991***, *2*, 1083.

H_2 (100 atm), 25°C
0.5 $[RuCl_2Ph]_2$
(R)-BINAP, 40 h

97%

(99% ee, R)

Kitamura, M.; Tokunaga, M.; Ohkuma, T.; Noyori, R. *Tetrahedron Lett.*, ***1991***, *32*, 4163.

CO_2Et, Me → (3-pentanone), 2 SmI_2, THF → OH, CO_2Et, Me, HO, Me; 79% (31:1 % dr)

Molander, G.A.; Kenny, C. *J. Org. Chem.*, ***1991***, *56*, 1439.

CO_2Me, N_3 → bakers yeast, 16 h "fermentation conditions" → CO_2Me, N_3 19% (70% ee) + CO_2Me, N_3 49% (70% ee)

Hudlicky, T.; Tsunoda, T.; Gadamasetti, K.G.; Murry, J.A.; Keck, G.E. *J. Org. Chem.*, ***1991***, *56*, 3619.

CO_2Me → $RuCl_2(COD)_n$, NEt_3, PhMe, 140°C, MeOH; cat. PPh_2 PPh_2 (S); H_2 (50 psi), 80°C, 6 h → CO_2Me, HO (97% ee) 90%

Taber, D.F.; Silverberg, L.J. *Tetrahedron Lett.*, ***1991***, *32*, 4227.

Bn_2N, Me, OEt → 1. mCPBA 2. H_2, $Pd(OH)_2$ → Me, OH, OEt (>95% ee) 90%

Reetz, M.T.; Lauterbach, E.H. *Tetrahedron Lett.*, ***1991***, *32*, 4481.

Ph, CO_2Et, Me → $NaBH_4$, $MnCl_2$ → OH, Ph, CO_2Et, Me + OH, Ph, CO_2Et, Me (90 : 10) 90%

Fujii, H.; Oshima, K.; Utimoto, K. *Tetrahedron Lett.*, ***1991***, *32*, 6147.

Ph–C(=O)–CO_2H → (NEt_3, CH_2Cl_2) → HO, Ph, CO_2H 94%

Wang, Z.; Meng, X-J.; Kabalka, G.W. *Tetrahedron Lett.*, ***1991***, *32*, 4619.

MeO_2C–O, Me → (Bu_2BOTf, iPr_2NEt, -78°C → 25°C) → HO, CO_2Me, Me + HO, CO_2Me, Me (8 : 92) 55%

Oh, T.; Wrobel, Z.; Rubenstein, S.M. *Tetrahedron Lett.*, ***1991***, *32*, 4647.

Ph–Br → (1. $Sn[Co(CO)_4]_4$, H_2O, $Ca(OH)_2$, *t*-BuOH, CO (65 bar); 2. H_3O^+) → Ph, Ph, O, O, OH, CO_2H + HO, O, Ph, O, Ph, OH, CO_2H

67%

Monflier, E.; Pellegrini, S.; Mortreux, A.; Petit, F. *Tetrahedron Lett.*, ***1991***, *32*, 4703.

CHO → (CCl_3CF_3, $ClSiMe_3$, e^- (Pt cathode/C anode), DMF, Et_4NOTs) → OH, CF_2CO_2Me

89%

Shono, T.; Kise, N.; Oka, H. *Tetrahedron Lett.*, ***1991***, *32*, 6567.

SH, N, O, O → (1. Me–CH=CH–CO_2Me, NaOMe, MeOH, 48 h; 2. Ph(C=O)SH, $TiCl_4$) → Me, HS, CO_2Me (39% ee) 67%

Shono, T.; Matsumura, Y.; Fujita, T. *Tetrahedron Lett.*, ***1991***, *32*, 6723.

$MeCO_2Me$ → (1. Dibal, -78°C, BF_3•OEt_2; 2. Me_2C=C($OSiMe_3$)(OEt)) → OH, Me, CO_2Et, Me, Me 70%

Kiyooka, S.; Shirouchi, M. *J. Org. Chem.*, ***1992***, *57*, 1.

menthyl–S(O)–Tol → 1. MeO_2C–C(O)– dianion, 4 eq. LDA, THF, -78°C; 2. $ZnCl_2$, Dibal, THF, -78°C; 3. Ni(R) → MeO_2C–CH₂–CH(OH)–CH₃ 81% (97% ee)

Solladié, G.; Almario, A. *Tetrahedron Lett.*, ***1992***, *33*, 2477.

PDC, *t*-BuOOH, CH_2Cl_2, 8 h, 0°C → 25°C

82%

Chidambaram, N.; Bhat, S.; Chandrasekaran, S. *J. Org. Chem.*, ***1992***, *57*, 5013.

$TolO_2S$, Cl_n, Ti, Me, Ph, O, O, Me → PhCHO, -40°C, CH_2Cl_2 → X_c, O, OH, Ph, Me 70% + X_c, O, OH, Ph, Me 18%

Xiang, Y.; Olivier, E.; Ouimet, N. *Tetrahedron Lett.*, ***1992***, *33*, 457.

PHCHO → 1. Me–CH=C(OSiMe$_3$)(OEt), -78°C, 20% BH_3•THF, (catalyst: Me, HO_2C, NHTs); 2. TBAF, THF → Ph, OH, O, SEt, Me (87 (80% ee) + Ph, OH, O, SEt, Me : 13) 89% (94% e)

Parmee, E.R.; Hong, Y.; Tempkin, O.; Masamune, S. *Tetrahedron Lett.*, ***1992***, *33*, 1729.

BzO, O, O, OBz → (ketone, O), 4 eq. SmI_2, THF-HMPA, -78°C → 23°C → O, O, H, HO 46% (35:1)

Enholm, E.J.; Jiang, S. *Tetrahedron Lett.*, ***1992***, *33*, 6069, 313.

PhCHO → $BrCH_2CO_2Et$, Cp_2TiCl_2-Zn, THF, RT, 5 min → Ph–CH(OH)–CH₂–CO_2Et 90%

Ding, Y.; Zhao, G. *J. Chem. Soc., Chem. Commun.*, ***1992***, 941.

3 eq. $LiN(TMS)_2$, -78°C, 1 h — 77%

Brandänge, S.; Leijonmarck, H. *Tetrahedron Lett.*, ***1992***, *33*, 3025.

$iPrCH_2CHO$, $TiCl_4$ — 75%

(59:41)

Vasconcellos, M.L.; Desmaële, D.; Costa, P.R.R.; d'Angelo, J. *Tetrahedron Lett.*, ***1992***, *33*, 4921.

PhCHO — $MeNO_2$, -78°C; 20% — 92%

(90% ee, R)

Kiyooka, S.; Kaneko, Y.; Kume, K. *Tetrahedron Lett.*, ***1992***, *33*, 4927.

BnO⁀CHO, -40°C; 5% $Pr(fod)_3$, CH_2Cl_2 — (1 : 97) 71%

Gu, J-H.; Terada, M.; Mikami, K.; Nakai, T. *Tetrahedron Lett.*, ***1992***, *33*, 1465.

PhCHO — In - ICH_2CO_2Et; THF - pentane, cinchonine — 63%

71% ee, R)

Johar, P.S.; Araki, S.; Butsugan, Y. *J. Chem. Soc., Perkin Trans. I*, ***1992***, 711.

$RuCl_2$•n H_2O, PhH, RT; *t*-BuOOH, 4 h — 79%

Murahashi, S.; Oda, Y.; Naota, T. *Chem. Lett.*, ***1992***, 2237.

PhCHO → ($=C(OSiMe_2t\text{-}Bu)(OMe)$, 20 h; 2% Fe catalyst PF_6^-) → Ph–CH(OSiMe_2t-Bu)–CH_2–CO_2Me 77%

Bach, T.; Fox, D.N.A.; Reetz, M.T. *J. Chem. Soc., Chem. Commun.*, ***1992***, 1634.

Zn , Me_3SiCl → 98%

Escudier, J-M.; Baltas, M.; Gorrichon, L. *Tetrahedron Lett.*, ***1992***, *33*, 1439.

$Ti(OAc)_3$, aq. AcOH; RT , 3 h → 50%

Ferraz, H.M.C.; Ribeiro, C.M.R. *Synth. Commun.*, ***1992***, *22*, 399.

Also via: Section 313 (Alcohol - Carboxylic Acid).

SECTION 328: ALCOHOL, THIOL - ETHER, EPOXIDE, THIOETHER

1. PhSeH, $SnCl_4$ 2. Me_3SiCl → ca. 70%; 1. *n*-BuLi , ether -78°C 2. cylcopentanone → 58% diastereomers (85:15)

Hoffmann, R.W.; Bewersdorf, M. *Tetrahedron Lett.*, ***1990***, *31*, 67.

n-C_3H_7OH , 0.2% DDQ; 90 min → 92%

Iranpoor, N.; Baltork, I.M. *Tetrahedron Lett.*, ***1990***, *31*, 735.
Iranpoor, N.; Baltork, I.M. *Synth. Commun.*, ***1990***, *20*, 2789.

1. $AlMe_3$
2. (cyclohexene oxide)
3. $BF_3 \cdot OEt_2$

(91 : 9) 55%

Alexakis, A.; Hanaïzi, J.; Jachiet, D.; Normant, J-F. *Tetrahedron Lett.*, ***1990***, *31*, 1271.

1. TBAF , THF
2. $Pd(Ph_3)_4$, THF 11.5 h

(92 : 8) 82%

Suzuki, T.; Sato, O.; Hirama, M. *Tetrahedron Lett.*, ***1990***, *31*, 4747.

$ZnCl_2$

(1 : 17) 67%

Angle, S.R.; Arnaiz, D.O. *J. Org. Chem.*, ***1990***, *55*, 3708.

BF_3

71%

(5:1 trans:cis)

Echavarren, A.M. *J. Org. Chem.*, ***1990***, *55*, 4255.

2.2 eq. SmI_2 , 25°C
HMPA/THF

69%

Molander, G.A.; Harring, L.S. *J. Org. Chem.*, ***1990***, *55*, 6171.

1. TMSOTf
2. TMSOTMS, DMAP

60%

Lee, T.V.; Cregg, C. *SynLett*, ***1990***, 317.

V-PILC, PhH, *t*-BuOOH
RT

96%

V-PILC = vanadium-pillared montmorillonite catalyst

Choudary, B.M.; Valli, V.L.K.; Prasad, A.D. *J. Chem. Soc., Chem. Commun.*, ***1990***, 721.

pillared-montmorillonite, 35 h
$Ti(OiPr)_4$, (+)-DIPT

86%

(94% ee)

Choudary, B.M.; Valli, V.L.K.; Prasad, A.D. *J. Chem. Soc., Chem. Commun.*, ***1990***, 1186.

LTA, PhH, RT, 2 h

88%

Yamamoto, M.; Irie, S.; Arase, T.; Kohmoto, S.; Yamada, K. *J. Chem. Soc., Chem. Commun.*, ***1990***, 1492.

1. BuLi
2. Bu_3PbBr
3. $TiCl_4$
$n-C_7H_{15}CHO$

(99 : 1) 84%

(39:61 in 95% yield with $BF_3•OEt_2$ rather than $TiCl_4$

Yamada, J.; Abe, H.; Yamamoto, Y. *J. Am. Chem. Soc.*, ***1990***, *112*, 6118.

1. Me_2BBr, -78°C
2. $BH_3•THF$

(14 : 1) 83%

Guindon, Y.; Girard, Y.; Berthiaume, S.; Gorys, V.; Lemieux, R.; Yoakim, C. *Can. J. Chem.*, ***1990***, *68*, 897.

20% Co(modp)$_2$, iPrOH

t-BuOOH , MS 4Å , O_2

50°C , 30 min

99% trans selective

73%

modp =

Inoki, S.; Mukaiyama, T. *Chem. Lett.*, ***1990***, 67.

$TiCl_4$, Mg , THF

71%

Pons, J-M.; Santelli, M. *Tetrahedron,* ***1990***, *46*, 513.

$ZrCl_4$-$LiAlH_4$, ether , RT

30 min

quant.

Shaozu, W.; Tianhui, R.; Yulan, Z. *Bull. Soc. Chim. Belg.*, ***1991***, *100*, 357.
Shaozu, W.; Yin, C.; Yulan, Z. *Bull. Soc. Chim. Belg.*, ***1991***, *100*, 421 (with $CpTiCl_3$•$LiAlH_4$).

1. SmI_2 , THF , HMPA
RT , 10 min

2. aq. NH_4Cl

55%

Shibuya, K.; Nagaoka, H.; Yamada, Y. *J. Chem. Soc., Chem. Commun.*, ***1991***, 1545.

1. Ph CHO, RT
$TiCl_4$, CH_2Cl_2

2. MeOH , aq. HCl

(1 : 2) 72%

Shimada, S.; Saigo, K.; Makamura, H.; Hagegawa, M. *Chem. Lett.*, ***1991***, 1149.

PhCHO

Me_3Si S $SiMe_3$

TBAF , THF , RT , 20 h

98%

Hosomi, A.; Ogata, K.; Ohkuma, M.; Hojo, M. *SynLett*, ***1991***, 557.

O SnBu$_3$ Ph

1. BuLi , TMEDA , THF
-90°C
2. PhCHO

Ph O OH Ph

87%

Lohse, P.; Loner, H.; Acklin, P.; Sternfeld, F.; Pfaltz, A. *Tetrahedron Lett.*, ***1991***, *32*, 615.

O O Ph Ph

1. TMSOTf , CH_2Cl_2
BH_3•SMe_2 , -78°C
2. $NaHCO_3$

O OH Ph Ph

95%

Hunter, R.; Bartels, B.; Michael, J.P. *Tetrahedron Lett.*, ***1991***, *32*, 1095.

$MeCO_2$ O_2CMe O

porcine pancreatic
lipase
pH 6.5 , -5°C , 26 h

HO O_2CMe O

54% conversion
(90% ee)

Grandjean, D.; Pale, P.; Chuche, J. *Tetrahedron Lett.*, ***1991***, *32*, 3043.

OAc OBz

1. $Zn(BH_4)_2$, DME
24 h)))))))))
2. H_2O

OH OBz

quant.

Ranu, B.C.; Basu, M.K. *Tetrahedron Lett.*, ***1991***, *32*, 3243.

Ph O O

$ZrCl_4$-$LiAlH_4$, ether
30°C , 12 h

Ph O OH 93% + Ph OH 7%

Shaozu, W.; Tianhui, R.; Ning, C.; Yulan, Z. *Org. Prep. Proceed. Int.*, ***1991***, *23*, 427.

MeO_2C CO_2Me O O c-C_6H_{11}

1. Me_2BBr , 20°C
2. $Me_2Cu(CN)Li_2$
-30°C

MeO_2C CO_2Me O OH c-C_6H_{11} Me + MeO_2C CO_2Me O OH c-C_6H_{11} Me

(17 : 1) 83%

Guidon, Y.; Simoneau, B.; Yoakim, C.; Gorys, V.; Lemieux, R.; Ogilvie, W. *Tetrahedron Lett.*, ***1991***, *32*, 5453.

1. 2.2 eq. BuLi , DME , 0°C
2. 2.0 eq. $B(OMe)_3$, 0°C
3. PhBr , $Pd(PPh_3)_4$, reflux
H_2O , DME

OH SiMe$_2$t-Bu → Ph OH SiMe$_2$t-Bu 68%

Cristofoli, W.A.; Keay, B.A. *Tetrahedron Lett.*, ***1991***, *32*, 5881.

CHO O → CH_2Cl_2 , MS 4Å , 0°C, 1 d, 20% Ti(ClO_4)$_2$ → OH O + OH O (80 : 20) 50%

Mikami, K.; Terada, M.; Sawa, E.; Nakai, T. *Tetrahedron Lett.*, ***1991***, *32*, 6571.

OH → mCPBA , aq. $NaHCO_3$, 25°C , 7 h → O OH 83%

Fringuelli, F.; Germani, R.; Pizzo, F.; Santinelli, F.; Savelli, G. *J. Org. Chem.*, ***1992***, *57*, 1198.

H Ph O → $SnCl_4$, MeOH , -30°C, 30 min → Ph H OH MeO (95% ee) 88%

Moberg, C.; Rákos, L.; Tottie, L. *Tetrahedron Lett.*, ***1992***, *33*, 2191.

Me O O Ph → hν (350 nm), PhH , 4 h → Me O Ph OH 90%

Sumathi, T.; Balasubramanian, K.K. *Tetrahedron Lett.*, ***1992***, *33*, 2213.

OH → t-BuOOH , $CHCl_3$, MS 4Å , 45°C , 72 h → O OH 73%

Antonioletti, R.; Bonadies, F.; Locati, L.; Screttri, A. *Tetrahedron Lett.*, ***1992***, *33*, 3205.

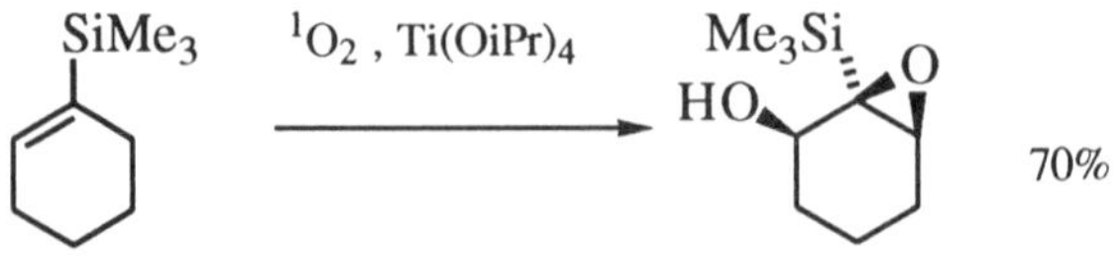
SiMe3
$^1O_2$, Ti(OiPr)4
Me3Si
HO
O
70%

Adam, W.; Richter, M. *Tetrahedron Lett.*, ***1992***, *33*, 3461.

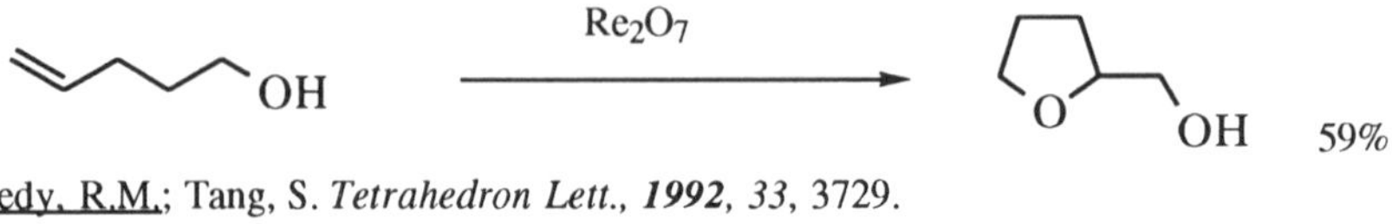
OH
Re2O7
O
OH
59%

Kennedy, R.M.; Tang, S. *Tetrahedron Lett.*, ***1992***, *33*, 3729.

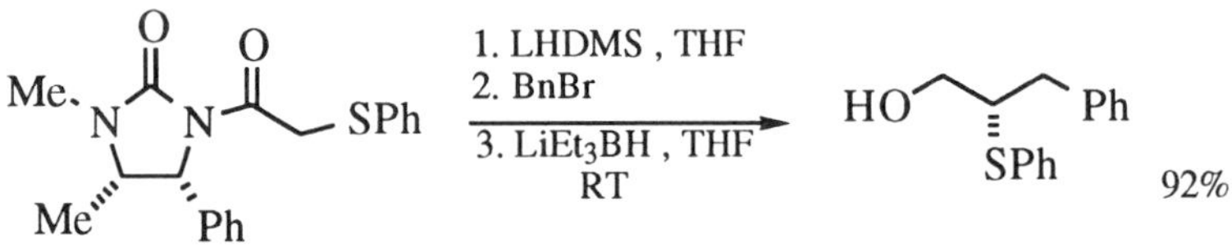
O
O
Me
N
N
SPh
Me
Ph
1. LHDMS , THF
2. BnBr
3. LiEt3BH , THF
RT
HO
Ph
SPh
92%

Orena, M.; Porzi, G.; Sandri, S. *Tetrahedron Lett.*, ***1992***, *33*, 3797.

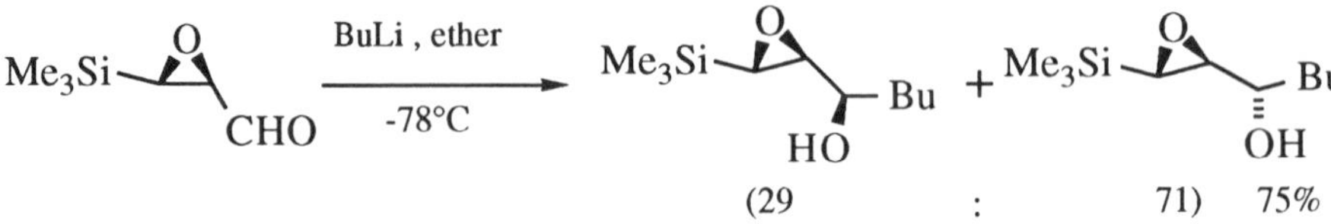
Me3Si
O
CHO
BuLi , ether
-78°C
Me3Si
O
Bu
HO
+
Me3Si
O
Bu
OH
(29 : 71) 75%

Urabe, H.; Matsuka, T.; Sato, F. *Tetrahedron Lett.*, ***1992***, *33*, 4179.

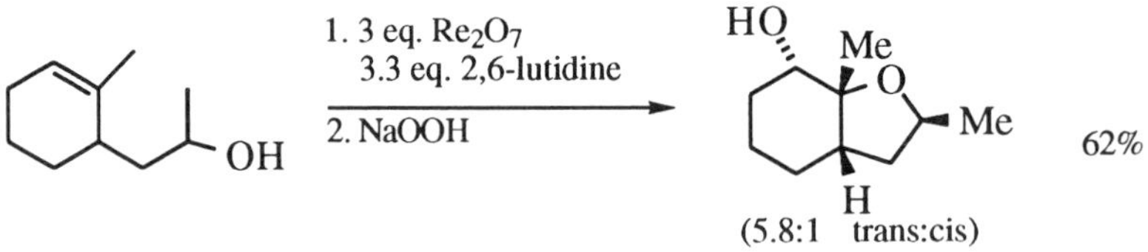
OH
1. 3 eq. Re2O7
3.3 eq. 2,6-lutidine
2. NaOOH
HO
Me
O
Me
H
62%
(5.8:1 trans:cis)

Tang, S.; Kennedy, R.M. *Tetrahedron Lett.*, ***1992***, *33*, 5299, 5303.

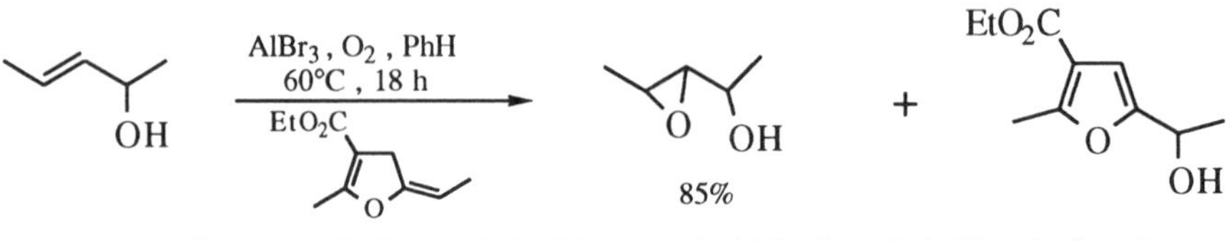
OH
AlBr3 , O2 , PhH
60°C , 18 h
EtO2C
O
O
OH
85%
+
EtO2C
O
OH

Antonioletti, R.; Bonadies, F.; Lattanzi, A.; Monteagudo, E.S.; Scettri, A. *Tetrahedron Lett.*, ***1992***, *33*, 5433.

NaBH$_4$-CeCl$_3$ / MeOH

(9 : 91) 82%

Li, K.; Hamann, L.G.; Koreeda, M. *Tetrahedron Lett.*, ***1992***, *33*, 6569.

1. SmI$_2$
2. Me(C=O)n-C$_6$H$_{13}$

80%

Curran, D.P.; Totleben, M.J. *J. Am. Chem. Soc.*, ***1992***, *114*, 6050.

NaBH$_4$-CaCl$_2$ / THF , 0°C

(85 : 15) 85%

Fujii, H.; Oshima, K.; Utimoto, K. *Chem. Lett.*, ***1992***, 967.

1. PhCH$_2$CH$_2$NHC(O)S$^-$ $H_3N^+CH_2CH_2Ph$, MeOH , RT , 5 h
2. aq. HCl , CH$_2$Cl$_2$
3. DMSO , PhCH(Me)NH$_2$ 50°C , 30 min

80%

Karikomi, M.; Narabu, S.; Yoshida, M.; Toda, T. *Chem. Lett.*, ***1992***, 1655.

REVIEWS:

"Rearrangement of Epoxyalcohols and Related Compounds"
Magnusson, G. *Org. Prep. Proceed. Int.*, ***1990***, *22*, 547.

"Transformation of β-Hydroxyalkylselenides to Aldehydes and Ketones. A Review"
Krief, A.; Laboureur, J.L.; Dumont, W.; Labar, D. *Bull. Soc. Chim. Fr.*, ***1990***, *127*, 681.

"Stereocontrolled Manipulations of Chromatographically Resolved Pyranosides"
Mash, E.A. *SynLett*, ***1991***, 529.

SECTION 329: ALCOHOL, THIOL - HALIDE, SULFONATE

Bu_2SnH_2 , 0°C

(0 : 100) 82%

with Bu_2SnH_2/DNB (90 : 10) 75%

DNB = *p*-dinitrobenzene (radical inhibitor)

Shibata, I.; Nakamura, K.; Baba, A.; Matsuda, H. *Tetrahedron Lett.*, ***1990***, *31*, 6381.

PhCHO, $CrCl_2$, RT

85% 13% 2%

Wender, P.A.; Grissom, J.W.; Hoffmann, U.; Ma, R. *Tetrahedron Lett.*, ***1990***, *31*, 6605.

LDA , PhCHO, -90°C , 2 h

80%

also undergoes conjugate addition and alkylation reactions

Qian, C-P.; Nakai, T. *Tetrahedron Lett.*, ***1990***, *31*, 7043.

$Bu_4N^+ H_2F_3^-$, 120°C, KHF_2 , 8 h

(39 : 61) 74%

Landini, D.; Penso, M. *Tetrahedron Lett.*, ***1990***, *31*, 7209.

I_2-Ti(OiPr)$_4$, CH_2Cl_2 , 0°C

obtain a 1:1 mixture at 25°C (7 : 1) 89%

Alvarez, E.; Nuñez, M.T.; Martín, V.S. *J. Org. Chem.*, ***1990***, *55*, 3429.

1. allyl-$SiMe_3$, $TiCl_4$/CH_2Cl_2, -78°C , 1 h
2. NH_4F , H_2O

85%

Imai, T.; Mishida, S. *J. Org. Chem.*, ***1990***, *55*, 4849.

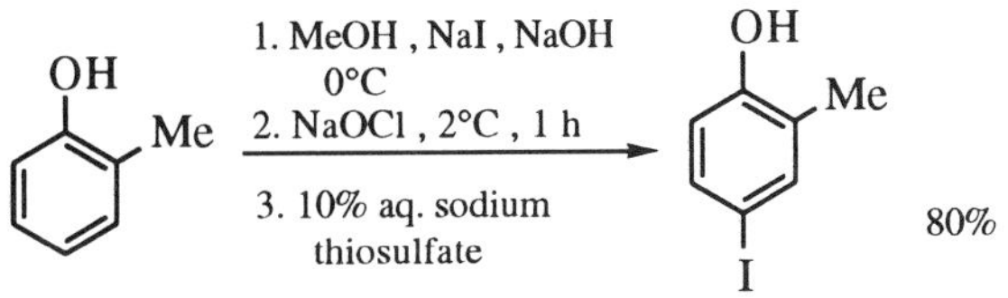

1. MeOH, NaI, NaOH, 0°C
2. NaOCl, 2°C, 1 h
3. 10% aq. sodium thiosulfate

OH, Me → OH, Me, I 80%

Edgar, K.J.; Falling, S.N. *J. Org. Chem.*, ***1990***, *55*, 5287.

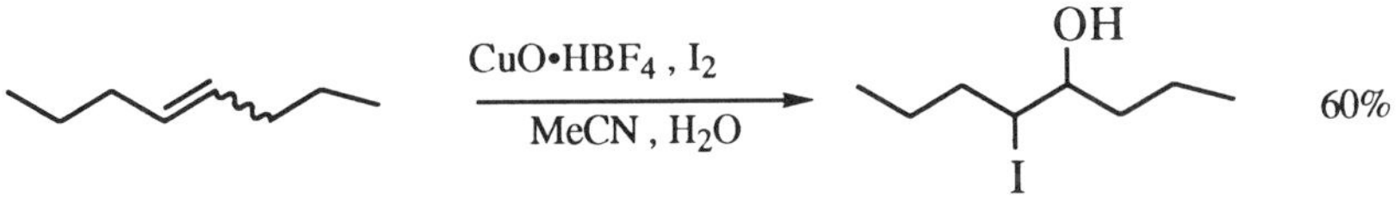

$CuO•HBF_4$, I_2
MeCN, H_2O

OH, I 60%

Barluenga, J.; Rodríguez, M.A.; Campos, P.J. *J. Chem. Soc., Perkin Trans. I*, ***1990***, 2807.

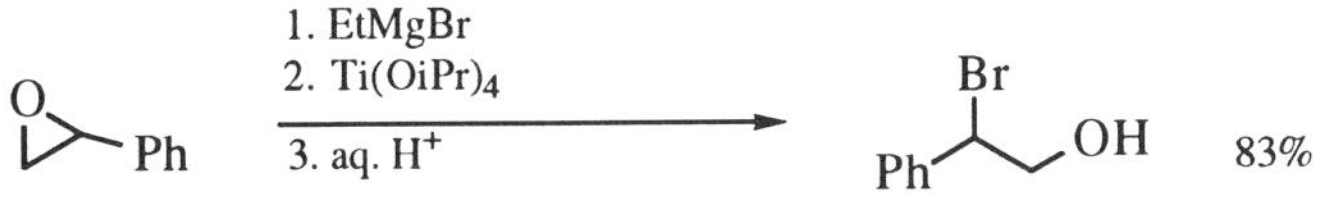

O, Ph
1. EtMgBr
2. $Ti(OiPr)_4$
3. aq. H^+

Br, Ph, OH 83%

Sosnovskii, G.M.; Astapovich, I.V. *Zhur. Org. Khim.*, ***1990***, *26*, 911 (Engl., p. 781).

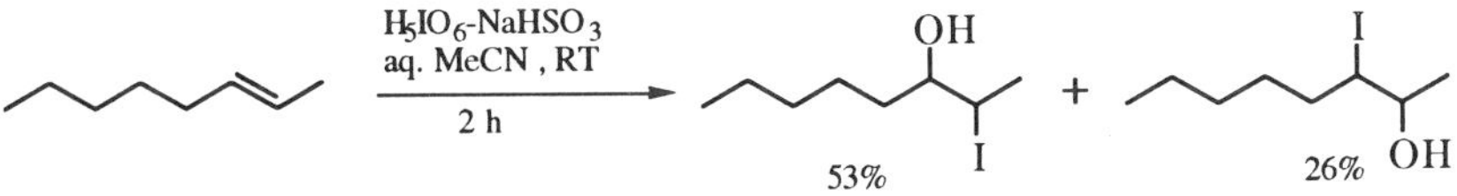

H_5IO_6-$NaHSO_3$
aq. MeCN, RT
2 h

OH, I 53% + I, OH 26%

Ohta, H.; Sakata, Y.; Takeuchi, T.; Ishii, Y. *Chem. Lett.*, ***1990***, 733.

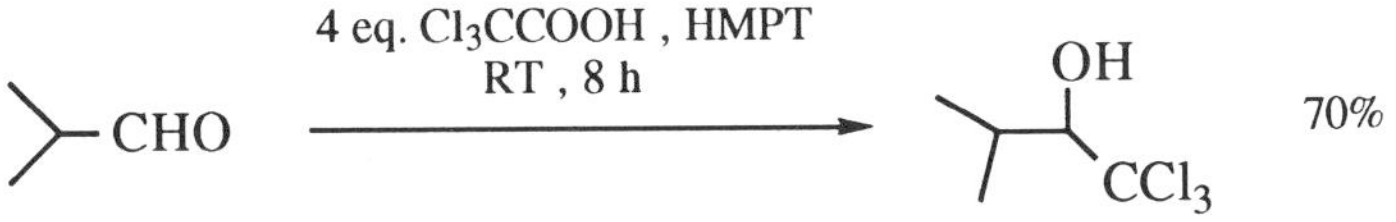

CHO
4 eq. Cl_3CCOOH, HMPT
RT, 8 h

OH, CCl_3 70%

Ferraccioli, R.; Gallina, C.; Giordano, C. *Synthesis*, ***1990***, 327.

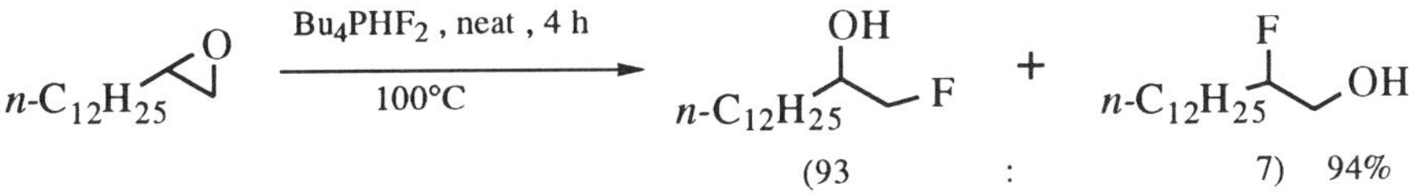

n-$C_{12}H_{25}$, O
Bu_4PHF_2, neat, 4 h
100°C

OH, n-$C_{12}H_{25}$, F + F, n-$C_{12}H_{25}$, OH
(93 : 7) 94%

Seto, H.; Qian, Z.; Yoshioka, H.; Uchibori, Y.; Umeno, M. *Chem. Lett.*, ***1991***, 1185.

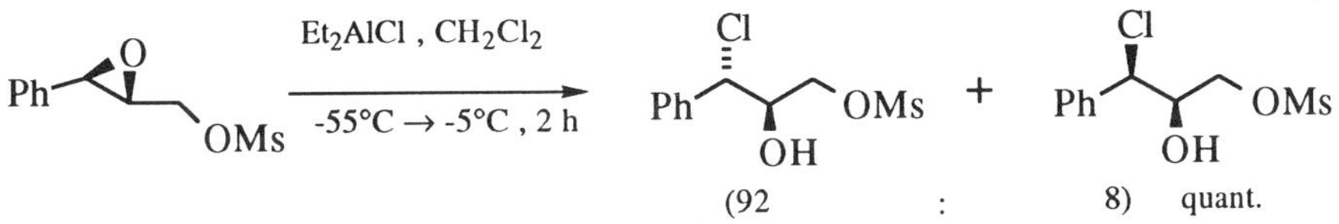

Ph, O, OMs
Et_2AlCl, CH_2Cl_2
-55°C → -5°C, 2 h

Cl, Ph, OMs, OH + Cl, Ph, OMs, OH
(92 : 8) quant.

Gao, L.; Saitoh, H.; Feng, F.; Murai, A. *Chem. Lett.*, ***1991***, 1787.

Li_2CuBr_4 , THF , 4 h

OH

Br

96%

Ciaccio, J.A.; Heller, E.; Talbot, A. *SynLett, 1991*, 248.

O

n-C_6H_{13} Me

CH_2I_2 , Sm , THF

OH I

n-C_6H_{13} Me

93%

diols available from chloromethyl ethers

Imamoto, T.; Hatajima, T.; Takiyama, N.; Takeyama, T.; Kamiya, Y.; Yoshizawa, T. *J. Chem. Soc., Perkin Trans. I, 1991*, 3127.

O

Ph Cl

bakers yeast , EtOH , sucrose
(ketone added to yeast slurry)

OH

Ph Cl

(44% ee)

84%

de Carvalho, M.; Okamoto, M.T.; Moran, P.J.S.; Rodrigues, J.A.R. *Tetrahedron, 1991, 47*, 2073.

O

LiI , AcOH , THF

RT , 20 min

OH

I

97%

Bajwa, J.S.; Anderson, R.C. *Tetrahedron Lett., 1991, 32*, 3020.

O

Ph Ph

6 eq. HMDS , DMF , CHF_3
e^- , Pt electrodes

[CHF_3/pyrrolidinone/DMF/ Bu_4NX]

CF_3

Ph OH

Ph

84%

Shono, T.; Ishifune, M.; Okada, T.; Kashimura, S. *J. Org. Chem., 1991, 56*, 2.

O

1. Me_3SiCF_3 , THF
TBAF

2. TBAF , THF

OH

CF_3

77%

Krishnamurti, R.; Bellew, D.R.; Prakash, G.K.S. *J. Org. Chem., 1991, 56*, 984.

Ph-CHO

$BrCF_2CH{=}CH_2$, Zn

THF , 0 → 25°C

OH

Ph C

F_2

67%

Yang, Z-Y.; Burton, D.J. *J. Org. Chem., 1991, 56*, 1037.

PhCHO → 1. $IZn(CN)Cu(CH_2)_4$-ZnI, THF, -60°C → -25°C; 2. I_2, -70°C → RT → OH, Ph, I 59%

Achyutha Rao, S.; Knochel, P. *J. Org. Chem.*, ***1991***, *56*, 4593.

CF_3, O → Me_3Al, -78°C → HO CF_3 80%

Abouabdellah, A.; Bégué, J-P.; Connet-Delpon, D.; Lequeux, T. *J. Org. Chem.*, ***1991***, *56*, 5800.

O, OH → MgI_2, PhMe, -60°C → I, OH, OH 85%

Bonini, C.; Righi, G.; Sotgiu, G. *J. Org. Chem.*, ***1991***, *56*, 6206.

I, O → $(Bu_3Sn)_2$, PhH, hν (sunlamp) → I, Me, OH 66%

Rawal, V.H.; Iwasa, S. *Tetrahedron Lett.*, ***1992***, *33*, 4687.

H, O, Ph, OH → $BH_2Cl{\bullet}SMe_2$, -60°C → Cl, OH, Ph, OH (>98:2)

(1,2:1,3-diol)

Bovicelli, P.; Lupattelli, P.; Bersani, M.T.; Mincione, E. *Tetrahedron Lett.*, ***1992***, *33*, 6181.
Bovicelli, P.; Mincione, E.; Ortaggi, G. *Tetrahedron Lett.*, ***1992***, *32*, 3719.

Ph, O, OH → $TiCl_4$ - LiCl, THF, -20°C, 15 min → Ph, Cl, HO, OH + Ph, OH, Cl, OH

(89 : 1) 96%

Shimizu, M.; Yoshida, A.; Fujisawa, T. *SynLett*, ***1992***, 204.

0.5 Br_2 , CH_2Cl_2

n-C_6H_{13} → OH, Br, n-C_6H_{13} + Br, OH, n-C_6H_{13}

(0.7→3.0 : 1) 54%

Konaklieva, M.I.; Dahl, M.L.; Turos, E. *Tetrahedron Lett.*, ***1992***, *33*, 7093.

SECTION 330: ALCOHOL, THIOL - KETONE

CHO

1. dihydroxy acetone phosphate
fructose-1,6-diphosphate aldolase (EC 4.1.2.13)
2. acid phophatase pH 4.8

OH O OH OH

in equilibrium with (+)-*exo*-brevicomin 40%

Schultz, M.; Waldmann, H.; Vogt, W.; Kunz, H. *Tetrahedron Lett.*, ***1990***, *31*, 867.

OHC SiMe$_3$ → OSiMe$_3$ / TBAF , SiO_2 → Me_3Si O OH 60%

Lee, T.V.; Roden, F.S. *Tetrahedron Lett.*, ***1990***, *31*, 2067.

Cl O O

1. $LiC_{10}H_8$, -78°C
2. PhCHO , -78°C → +20°C
3. H_2O

OH O O Ph 60%

Ramón, D.J.; Yus, M. *Tetrahedron Lett.*, ***1990***, *31*, 3763, 3767.

O OSiEt$_3$ H Me → OM / -78°C → Et$_3$SiO HO Me O + Et$_3$SiO HO Me O

M = Li , THF 18 : 82)
M = SiMe$_2t$-Bu
BF_3•OEt_2 , CH_2Cl_2 98 : 2)

Evans, D.A.; Gage, J.R. *Tetrahedron Lett.*, ***1990***, *31*, 6129.

PhCHO , Et_2MeSiH
PPh_2Me , Tol
-5°C , 2 h

O OSiMeEt$_2$ Ph 99%

(83:17 syn:anti)

Matsuda, I.; Takahashi, K.; Sato, S. *Tetrahedron Lett.*, ***1990***, *31*, 5331.

1. LDA
2. PhCHO

OSiMe$_3$ → Me$_3$SiO Me OH Ph + Me$_3$SiO Me OH Ph

(10 : 90)

with ClTi(OiPr)$_3$ (99 : 1)

Panyachotipun, C.; Thornton, E.R. *Tetrahedron Lett.*, ***1990***, *31*, 6001.

$(PhSe)_2$, $CH_2(CO_2Me)_2$
e$^-$, Pt electrodes , 3 V
MeOH , $NaClO_4$

Ph OH O 70%

Inokuchi, T.; Kusumoto, M.; Torii, S. *J. Org. Chem.*, ***1990***, *55*, 1548.

OSiMe$_3$ EtCHO , $ZnCl_2$ / CH_2Cl_2 → O OSiMe$_3$ Et 61%

Moiseenkov, A.M.; Czeskis, B.A.; Ivanova, N.M.; Nefedov, O.M. *Org. Prep. Proceed. Int.*, ***1990***, *22*, 215.

OSiMe$_3$ OBz iPrCHO , CH_2Cl_2 / $TiCl_4$, 3.5 h → OH O OBz 88%

(92:8 dr)

Trost, B.M.; Urabe, H. *J. Org. Chem.*, ***1990***, *55*, 3982.

Ph Ph O MoOPD → OH Ph Ph O 47%

a safer alternative to MoOPH

MoOPD = oxodiperoxymolybdenum (pyridine)-1,3-dimethyl-3,4,5,6-tetrahydro-2(1H)pyrimidinone = MoO_5•Py•DMPU

Anderson, J.C.; Smith, S.C. *SynLett*, ***1990***, 107.

OH OMe TFA, aq. THF O OH 76%

Stone, G.B.; Liebeskind, L.S. *J. Org. Chem.*, ***1990***, *55*, 4614.

PhCHO 1. (bromoacetone), $Zr(Ot\text{-}Bu)_4$, THF, -30°C; 2. MsCl, NEt_3, CH_2Cl_2, -78°C → -30°C → Ph, Br, O 56%

Sasai, H.; Kirio, Y.; Shibasaki, M. *J. Org. Chem.*, ***1990***, *55*, 5306.

MeO_2C (O) 1. ethylene glycol, TsOH, PhH, reflux; 2. $LiAlH_4$, THF, 0°C; 3. wet silica gel → HO (O) 81%

Hitchcock, S.R.; Perron, F.; Martin, V.A.; Albizati, K.F. *Synthesis*, ***1990***, 1059.

PhCHO t-BuS(C=CH$_2$)OSiMe$_2t$-Bu, PhMe, 20% (binaphthol)Ti=O → t-BuS(O)…(OSiMe$_2t$-Bu)Ph 91% (60% ee)

Mukaiyama, T.; Inubushi, A.; Suda, S.; Hara, R.; Kobayashi, S. *Chem. Lett.*, ***1990***, 1015.

Ph…(O)Ph 1. Yb, RT; 2. D_2O → Ph, O, Ph, OH, Ph, Ph, D 73%

Takaki, K.; Beppu, F.; Tanaka, S.; Tsubaki, Y.; Jintoku, T.; Fujiwara, Y. *J. Chem. Soc., Chem. Commun.*, ***1990***, 516.

Ph(O)…(epoxide)Ph Bu_3SnH, PhH, hν, 30 min → Ph(O)…(OH)Ph 93%

Hasegawa, E.; Ishiyama, K.; Horaguchi, T.; Shimizu, T. *J. Chem. Soc., Chem. Commun.*, ***1990***, 550.

PhCHO → [OSiMe3 / Me–CH=C–SEt, CH_2Cl_2, -78°C; SnO + Me_3SiOTf; N-methylpyrrolidinylmethyl-N-(1-naphthyl)amine] → EtS–C(=O)–CH(Me)–CH(OSiMe3)–Ph + EtS–C(=O)–CH(Me)–CH(OSiMe3)–Ph

(94 : 6) 82%

(91% ee)

Mukaiyama, T.; Uchiro, H.; Kobayashi, S. *Chem. Lett.*, ***1990***, 1147.

Ph–C(=O)–TeBu → [1. BuLi; 2. Me–C(=O)–*t*-Bu; 3. H^+] → Ph–C(=O)–C(Me)(OH)–*t*-Bu 85%

Hiiro, T.; Morita, Y.; Inoue, T.; Kambe, N.; Ogawa, A.; Ryu, I.; Sonoda, N. *J. Am. Chem. Soc.*, ***1990***, *112*, 455.

[Oxazolidinone (Bn) N-acyl ketone] → iPrCHO → [aldol products, Me, OH] + [aldol product, Me, OH]

$Sn(OTf)_2$, NEt_3	(95	:	5) 83%
$TiCl_4$, iPr_2NEt	(<1	:	99) 86%

Evans, D.A.; Clark, J.S.; Metternich, R.; Novack, V.J.; Sheppard, G.S. *J. Am. Chem. Soc.*, ***1990***, *112*, 866.
Evans, D.A.; Rieger, D.L.; Bilodeau, M.T.; Urpí, F. *J. Am. Chem. Soc.*, ***1991***, *113*, 1047.

Cyclohexanone → [25°C, 10 h; 2 eq. $[Bn–C_6H_4]_3N^{+•}$ $SbCl_6^-$] → 2-hydroxycyclohexanone 88%

Schulz, M.; Kluge, R.; Siviai, L.; Kamm, B. *Tetrahedron*, ***1990***, *46*, 2371.

Ph–C(=CH2)–OSiMe3 → [1. PhCHO, $BiCl_3$, CH_2Cl_2, RT, 25 min; 2. MeOH, 1N HCl] → Ph–C(=O)–CH2–CH(OH)–Ph 94%

Wada, M.; Takeichi, E.; Matsumoto, T. *Bull. Chem. Soc. Jpn.*, ***1991***, *64*, 990.

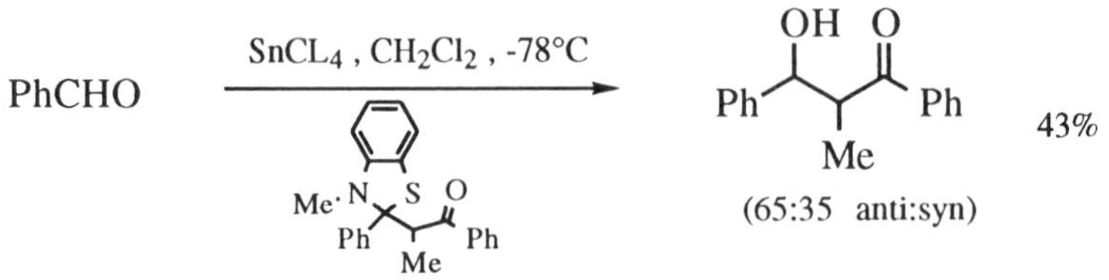

PhCHO → ($SnCL_4$, CH_2Cl_2, -78°C) → 43% (65:35 anti:syn)

Chikashita, H.; Tame, S.; Yamada, S.; Itoh, K. *Bull. Chem. Soc. Jpn.*, ***1990***, *63*, 497.

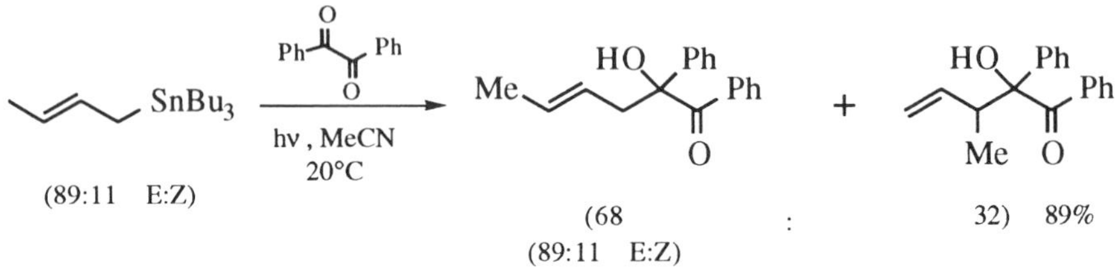

(89:11 E:Z) → (hν, MeCN, 20°C) → (68 : 32) 89% (89:11 E:Z)

Takuwa, A.; Nishigaichi, Y.; Yamaoka, T.; Iichama, K. *J. Chem. Soc., Chem. Commun.*, ***1991***, 1359.

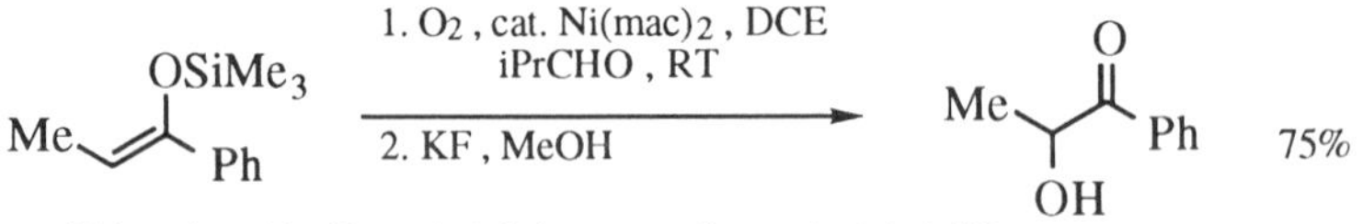

1. O_2, cat. $Ni(mac)_2$, DCE, iPrCHO, RT
2. KF, MeOH

75%

$Ni(mac)_2$ = bis-(3-methyl-2,4-pentanedionato) nickel (II)

Takai, T.; Yamada, T.; Rhode, O.; Mukaiyama, T. *Chem. Lett.*, ***1991***, 281.
Takai, T.; Yamada, T.; Mukaiyama, T. *Chem. Lett.*, ***1991***, 1499.

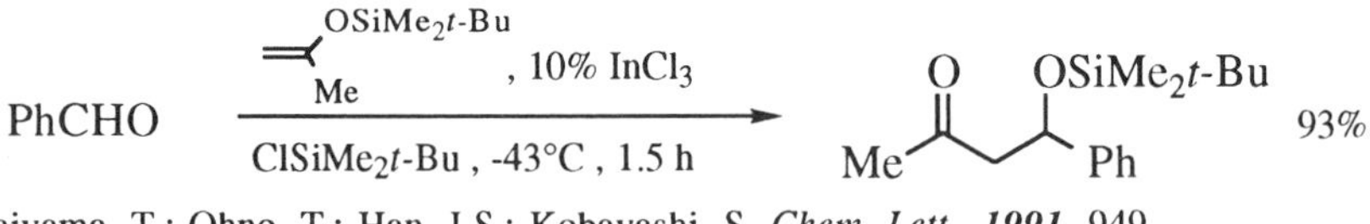

PhCHO → (10% $InCl_3$; $ClSiMe_2t$-Bu, -43°C, 1.5 h) → 93%

Mukaiyama, T.; Ohno, T.; Han, J.S.; Kobayashi, S. *Chem. Lett.*, ***1991***, 949.

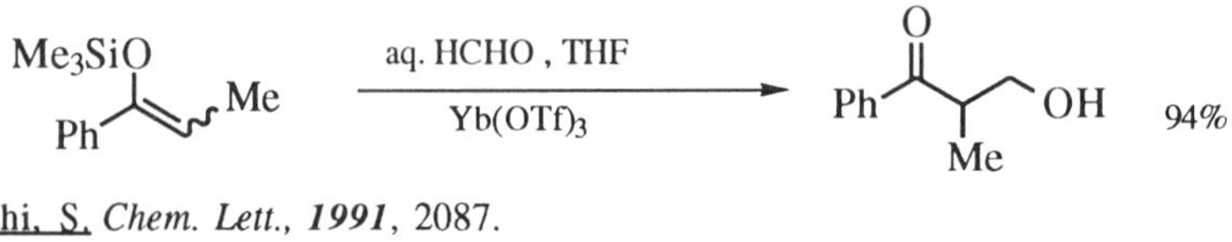

aq. HCHO, THF; $Yb(OTf)_3$ → 94%

Kobayashi, S. *Chem. Lett.*, ***1991***, 2087.

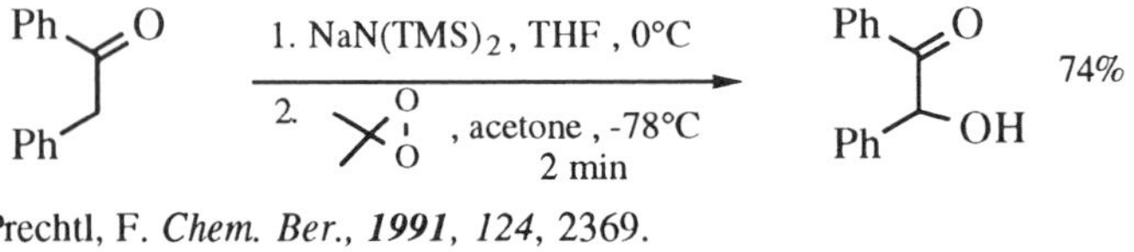

1. $NaN(TMS)_2$, THF, 0°C
2. dimethyldioxirane, acetone, -78°C, 2 min

74%

Adam, W.; Prechtl, F. *Chem. Ber.*, ***1991***, *124*, 2369.

1. LDA , THF , -78°C
2. (dimethyldioxirane) , acetone
-78°C , 30 min
(inverse addition)

77%

Guertin, K.R.; Chan, T-H. *Tetrahedron Lett.*, ***1991***, *32*, 715.

1. 2-ICr_2BH , $CDCl_3$
1,5 h , 0°C
2. PhCHO
3. H_2O

2-ICr_2BH = di-(2-isocaranyl)borane

Boldrini, G.P.; Bortolitti, M.; Tagliavini, E.; Trombini, C.; Umani-Ronchi, A. *Tetrahedron Lett.*, ***1991***, *32*, 1229.

1. LDA , THF , 0°C
2. (oxaziridine reagent, OMe, OMe, N, O, S, O_2)

OMe O ... OH
OMe 66%
(94% ee , R)

Davis, F.A.; Kumar, A.; Chen, B-C. *J. Org. Chem.*, ***1991***, *56*, 1143.
Davis, F.A.; Weismiller, M.C. *J. Org. Chem.*, ***1990***, *55*, 3715. [dimethoxy derivative]

n-$C_8H_{17}COCl$ → 1. 2 eq. SmI_2 2. H_3O^+ → n-$C_8H_{17}COCH(OH)$-n-C_8H_{17} 75%

Aryl acid chlorides lead to 1,2-diketones

Collin, J.; Namy, J-L.; Dallemer, F.; Kagan, H.B. *J. Org. Chem.*, ***1991***, *56*, 3118.

LDA , THF
$MgBr_2•OEt_2$
-70°C→ RT

(98 : 2) 73%

Swiss, K.A.; Choi, W-B.; Liotta, D.C.; Abdel-Magid, A.F.; Maryanoff, C.A. *J. Org. Chem.*, ***1991***, *56*, 5978.

PCWP , 35% H_2O_2
CHCl$_3$, reflux , 16 h

OH OH → O OH 90%

PCWP = peroxytungsto phosphate

Sakata, Y.; Ishii, Y. *J. Org. Chem.*, ***1991***, *56*, 6233.

e$^-$, Sn cathode
iPrOH , 25°C

O Me CN → HO O Me 60% + O Me 5%

Shono, T.; Kise, N.; Fujimoto, T.; Tominaga, N.; Morita, H. *J. Org. Chem.*, ***1992***, *57*, 7175.
Shono, T.; Kise, N. *Tetrahedron Lett.*, ***1990***, *31*, 1303.

$Bu_3SnSiMe_3$
$BnEt_3NCl$

O Me O Br → Me O HO 74%

Mori, M.; Kaneta, N.; Isono, N.; Shibasaki, M. *Tetrahedron Lett.*, ***1991***, *32*, 6139.

1. CH_2N_2
2. acetone , (dimethyldioxirane)

O Cl → O OH OH 65%

Ihmels, H.; Maggini, M.; Prato, M.; Scorrano, G. *Tetrahedron Lett.*, ***1991***, *32*, 6215.

1. iPrCHO , CH_2Cl_2
$BiCl_3$, 3 NaI
2. MeOH , HCl

O OSiMe$_3$ → O O Me OH 90%

Le Roux, C.; Maraval, M.; Borredon, M.E.; Gaspard-Iloughmane, H.; Dubac, J. *Tetrahedron Lett.*, ***1992***, *33*, 1053.

PhCHO , Gd(OTf)$_3$
1:4 (H_2O:THF) , RT
20 h

OSiMe$_3$ → O OH Ph + O OH Ph

(78 : 22) 88%

Kobayashi, S.; Hachiya, I. *Tetrahedron Lett.*, ***1992***, *33*, 1625.

Ph OSiMe$_3$, <6 min; [Ru(salen)(NO)(H$_2$O)]$^+$

CHO → Ph(CO)CH$_2$CH(OSiMe$_3$)Ph 90%

Odenkirk, W.; Whelan, J.; Bosnich, B. *Tetrahedron Lett.*, ***1992***, *33*, 5729.

PhI(O$_2$CCF$_3$)$_2$, TFA , 3 h; MeCN , H$_2$O

PhC(O)Me → PhC(O)CH$_2$OH 69%

Moriarty, R.M.; Berglund, B.A.; Penmasta, R. *Tetrahedron Lett.*, ***1992***, *33*, 6065.

PhC(O)Cl + PhC(O)Me; 1. 2.2 eq. SmI$_2$, MeCN; 2. H$_3$O$^+$ → PhC(O)C(OH)(Me)Ph 78%

Ruder, S.M. *Tetrahedron Lett.*, ***1992***, *33*, 2621.

1. HSiMeEt$_2$, CO (50 atm), PhH , 140°C, cat. [RhCl(CO)$_2$]$_2$; 2. aq. TsOH

OSiEt$_2$Me 62%

Ikeda, S.; Chatani, N.; Kajikawa, Y.; Ohe, K.; Murai, S. *J. Org. Chem.*, ***1992***, *57*, 2.

OH; 1. BuLi; 2. 10% (PCy$_3$)$_2$NiCl$_2$; 3. PhCHO , -78°C

O H OH Ph + O H OH Ph

(4.2 : 1) 64%

Motherwell, W.B.; Sandham, D.A. *Tetrahedron Lett.*, ***1992***, *33*, 6187.

Ph OSiMe$_3$; 0.5% TiCp$_2$(CF$_3$CsO$_3$)$_2$, PhCHO , CD$_2$Cl$_2$, <5 min

Ph(CO)CH$_2$CH(OSiMe$_3$)Ph 90%

Hollis, T.K.; Robinson, N.P.; Bosnich, B. *Tetrahedron Lett.*, ***1992***, *33*, 6423.

OSiMe$_3$; cat. Mn (salen) derivative; PhIO , MeCN , 25°C, 70 min

O OH + O OSiMe$_3$

(2.6 : 1) 70%
(30% ee)

Reddy, D.R.; Thornton, E.R. *J. Chem. Soc., Chem. Commun.*, ***1992***, 172.

35% H_2O_2 , 1.6% PCWP
$CHCl_3$, H_2SO_4 , reflux
24 h

O
OH

PCWP = $[\pi C_5H_5N^+(CH_2)_{15}Me]_3 \{PO_4[W(O)(O_2)_2]_4\}^{-3}$

Sakata, Y.; Katayama, Y.; Ishii, Y. *Chem. Lett.*, ***1992***, 671.

O NO_2 Ph Me
aq. Dowex-50 , reflux
Ph O Me OH 83%

Chakraborty, R.; Das, A.R.; Ranu, B.C. *Synth. Commun.*, ***1992***, *22*, 1523.

CHO
1. OSiMe3 Ph , EtCN
H O O N N B H Ts Bu
-78°C , 14 h
2. aq. H^+

HO H O Ph 67%
(93% ee)

Corey, E.J.; Cywin, C.L.; Roper, T.D. *Tetrahedron Lett.*, ***1992***, *33*, 6907.

REVIEWS:

"Uses of the Fries Rearrangement for the Preparation of Hydroxyarylketones"
Martin, R. *Org. Prep. Proceed. Int.*, ***1992***, *24*, 392.

"Selective Oxidation of α,β-Unsaturated Ketones at the α-Position"
Demir, A.S.; Jeganathan, A. *Synthesis*, ***1992***, 235.

SECTION 331: ALCOHOL, THIOL - NITRILE

PhCHO
$BrZnCH_2C\equiv N$
Me_3SiCl , THF
reflux, 30 min

$OSiMe_3$
Ph C≡N 86%

Palomo, C.; Aizpurua, J.M.; López, M.C.; Aurrekoetxea, N. *Tetrahedron Lett.*, ***1990***, *31*, 2205.

n-C_6H_{13} O
$[YbCl_3/BuLi]$, THF
Me_3SiCN

n-C_6H_{13} CN OTMS >99%

Matsubura, S.; Onishi, H.; Utimoto, K. *Tetrahedron Lett.*, ***1990***, *31*, 6209.

PhCHO → (2% cyclo(Phe-His) catalyst; HCN, PhMe, -20°C, 8 h) → Ph–CH(OH)–CN 97% (97% ee)

Tanaka, K.; Mori, A.; Inoue, S. *J. Org. Chem.*, ***1990***, *55*, 181.

PhCHO → (Me_3SiCN, $Ti(OiPr)_4$, (+)-DIPT, CH_2Cl_2; 0°C, 18 h) → Ph–CH($OSiMe_3$)–CN 95% (78% ee, R)

Hayashi, M.; Matsuda, T.; Oguni, N. *J. Chem. Soc., Chem. Commun.*, ***1990***, 1364.

PhCHO → ($PhCH_2CN$, e^-, -40°C, DMF-H_2O, 80 min) → HO–CH(Ph)–CH(Ph)–CN 75% + Ph–CH=C(CN)–Ph 8%

Torii, S.; Kawafuchi, H.; Inokuchi, T. *Bull. Chem. Soc. Jpn.*, ***1990***, *63*, 2430.

epoxide OAc → (Me_3SiCN - CaO, 60°C, heptane) → CN / $OSiMe_3$ / OAc + $OSiMe_3$ / CN / OAc (96 : 4) 75%

Sugita, K.; Ohta, A.; Onaka, M.; Izumi, Y. *Chem. Lett.*, ***1990***, 481.

3-PhO-C6H4-CHO → (HCN, neat, 1 h; cyclo(His-Phe) catalyst) → 3-PhO-C6H4-CH(OH)CN 90% (37% ee, S)

Danda, H. *Bull Chem. Soc. Jpn.*, ***1991***, *64*, 3743.

H–C(O)–CH2CH2–CH(OBn)–CH(CH3)2 → (TMSCN, CH_2Cl_2, cat. TMSOTf, -78°C) → N≡C–CH(OH)–CH2CH2–CH(OBn)–CH(CH3)2 50% (7:1 dr)

Molander, G.A.; Haar, J.P. Jr. *J. Am. Chem. Soc.*, ***1991***, *113*, 3608.

PhCHO → (Me_3SiCN, CH_2Cl_2; t-Bu, OH, N, OH ligand • 20% $Ti(OiPr)_4$, -30°C, 44 h) → Ph–CH(OSiMe$_3$)CN 90% (67% ee, R)

Hayashi, M.; Miyamoto, Y.; Inoue, T.; Oguni, N. *J. Chem. Soc., Chem. Commun.*, ***1991***, 1752.

MeO–C$_6$H$_4$–CHO → (HO, CN, $Ti(OiPr)_4$, RT; OH, O, H, N, N ligand, CH_2Cl_2, 3 h) → MeO–C$_6$H$_4$–CH(OH)CN 87%

Mori, A.; Inoue, S. *Chem. Lett.*, ***1991***, 145.
Mori, A.; Kinoshita, K.; Osaka, M.; Inoue, S. *Chem. Lett.*, ***1990***, 1171.

Ph–CH$_2$CH$_2$–CHO → (Me_3SiCN, CH_2Cl_2; 10% $EtN(iPr)_2$, 0°C) → Ph–CH$_2$CH$_2$–CH(OSiMe$_3$)CN 99%

Kobayashi, S.; Tsuchiya, Y.; Mukaiyama, T. *Chem. Lett.*, ***1991***, 537, 541.

cyclopentanone → (Me_3SiCN, $Yb(CN)_3$, THF; 0°C, 1.5 h) → 1-OSiMe$_3$-1-CN-cyclopentane 93%

Matsubara, S.; Takai, T.; Utimoto, K. *Chem. Lett.*, ***1991***, 1447.

PhCHO → (HCN; Me_3Al • OH, O, N, Bn, N, H, CO_2Me ligand) → Ph–CH(OH)CN

-78°C, 102 h (41%, 61% ee, R)
-78°C, 72 h (87%, 36% ee, R)

Mori, A.; Ohno, H.; Nitta, H.; Tanaka, K.; Inoue, S. *SynLett*, ***1991***, 567.

Me_3Si–CH$_2$CN → (PhCHO, KF, 3 min; microwaves (440 W)) → Ph–CH(OSiMe$_3$)–CH$_2$CN 62%

Latouche, R.; Texier-Boullet, F.; Hamelin, J. *Tetrahedron Lett.*, ***1991***, *32*, 1179.

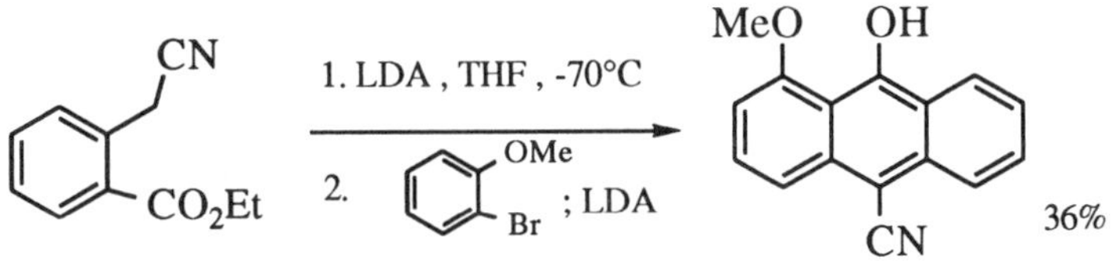

1. LDA, THF, -70°C
2. (2-bromoanisole); LDA

MeO OH CN 36%

Bhawal, B.M.; Khanapure, S.P.; Zhang, H.; Biehl, E.R. *J. Org. Chem.*, ***1991***, *56*, 2846.

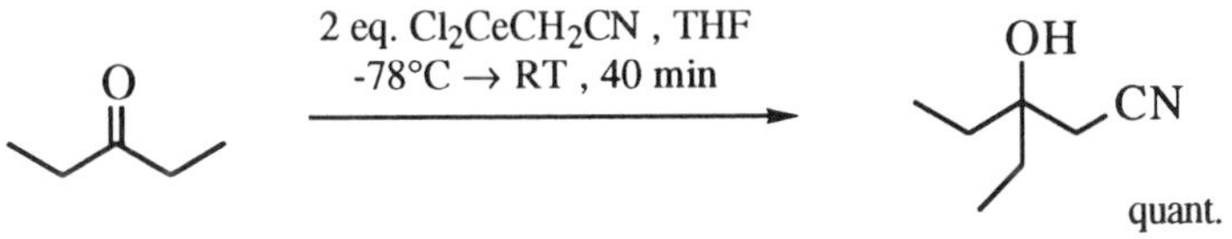

2 eq. Cl_2CeCH_2CN, THF
-78°C → RT, 40 min

OH CN quant.

Liu, H-J.; Al-Said, N.H. *Tetrahedron Lett.*, ***1991***, *32*, 5473.

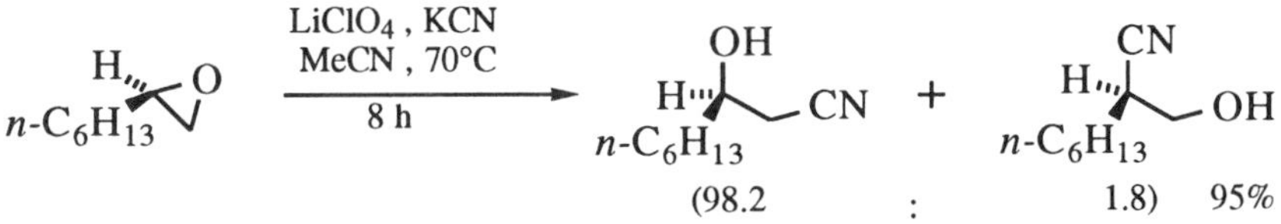

$LiClO_4$, KCN
MeCN, 70°C
8 h

n-C_6H_{13} (98.2 : 1.8) 95%

Chini, M.; Crotti, P.; Favero, L.; Macchia, F. *Tetrahedron Lett.*, ***1991***, *32*, 4775.

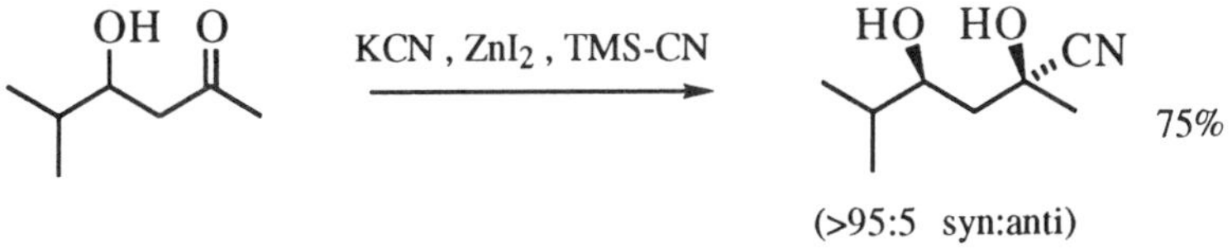

KCN, ZnI_2, TMS-CN

HO HO CN 75%

(>95:5 syn:anti)

Brunet, E.; Batra, M.S.; Aguilar, F.J.; Ruano, J.L.G. *Tetrahedron Lett.*, ***1991***, *32*, 5423.

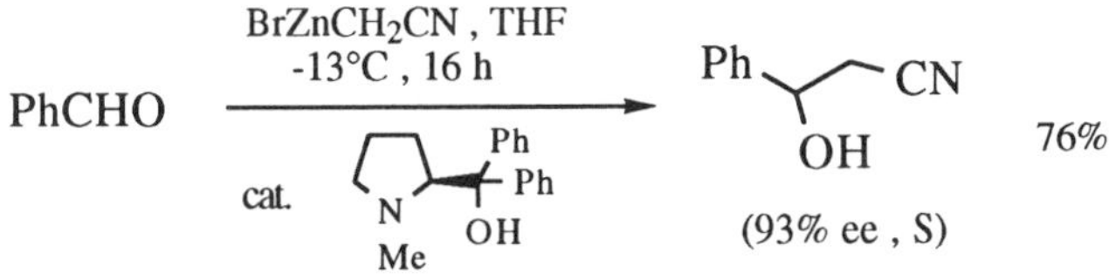

PhCHO

$BrZnCH_2CN$, THF
-13°C, 16 h
cat. (N-Me pyrrolidine, Ph, Ph, OH)

Ph CN OH 76%

(93% ee, S)

Soai, K.; Hirose, Y.; Sakata, S. *Tetrahedron Asymmetry*, ***1992***, *3*, 677.

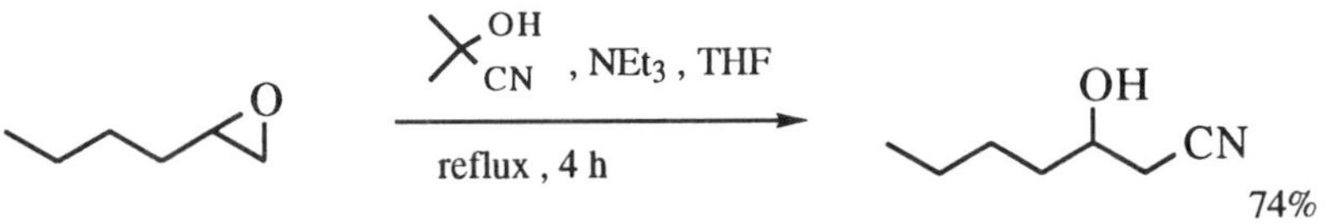

(acetone cyanohydrin: OH, CN), NEt_3, THF
reflux, 4 h

OH CN 74%

Mitchell, D.; Koenig, T.M. *Tetrahedron Lett.*, ***1992***, *33*, 3281.

3.5 eq. LiCN , THF; reflux , 4 h — 71%

Ciaccio, J.A.; Stanescu, C.; Bontemps, J. *Tetrahedron Lett.*, ***1992***, *33*, 1431.

TMSCN , CH_2Cl_2 , 20°C, 18 h; t-Bu, OH • $Ti(OiPr)_4$, N OH — CN, $OSiMe_3$ — 80%

Hayashi, M.; Tamura, M.; Oguni, N. *SynLett*, ***1992***, 663.

SECTION 332: ALCOHOL, THIOL - ALKENE

$B(IPC)_2$ — 1. CH_3CHO, -78°C , 1h; 2. NaOH , H_2O_2 — OH — 65% (90% ee)

Brown, H.C.; Randad, R.S. *Tetrahedron Lett.*, ***1990***, *31*,455.

Me, SPh, OTBS — PhCHO , Me_2AlCl, CH_2Cl_2 , -23°C, 20 h — OH SPh, Ph, OTBS, Me + OH SPh, Ph, OTBS, Me (94 : 6) 86%

Tanino, K.; Nakamura, T.; Kuwajima, I. *Tetrahedron Lett.*, ***1990***, *31*, 2165.

Ph, Ph, P, O — 1. OsO_4 , NMO, aq. t-BuOH , RT, 24 h; 2. KOH , DMSO, 50°C — OH

Harmat, N.J.S.; Warren, S. *Tetrahedron Lett.*, ***1990***, *31*, 2743.

OH, Bu — [iPrMgCl , Cp_2TiCl_2], ether , Bu_3SnCl; THF , reflux, 6 h — OH, $SnBu_3$, Bu — 75%

Lautens, M.; Huboux, A.H. *Tetrahedron Lett.*, ***1990***, *31*, 3105.

montmorillonite clay
RT , 9 h

OH

53%

clay catalyzed Claisen rearrangement

Dauben, W.G.; Cogen, J.M.; Behar, V. *Tetrahedron Lett.*, ***1990***, *31*, 3241.

$SiMe_3$ O BnO $OSiPh_3$

2 $SiMe_3$

2.2 $TiCl_4$, -78°C
CH_2Cl_2

Me_3Si BnO OH $OSiPh_3$

88%

(2,3 anti/syn = 33/1)

Shimazaki, M.; Morimoto, M.; Suzuki, K. *Tetrahedron Lett.*, ***1990***, *31*, 3335.

Zn/Ag-graphite
THF , RT , 10 min

OH

90%

Fürstner, A. *Tetrahedron Lett.*, ***1990***, *31*, 3735.

C≡C- H O Me

SmI_2 , THF , HMPA

t-BuOH , 0°C

OH Me 55%

Shim, S.C.; Hwang, J-T.; Kang, H-Y.; Chang, M.H. *Tetrahedron Lett.*, ***1990***, *31*, 4765.

TsO O H H *n*-C_5H_{11}

Te , $NaBH_4$

DMF

OH *n*-C_5H_{11}

91%

Discordia, R.P.; Murphy, C.K.; Dittmer, D.C. *Tetrahedron Lett.*, ***1990***, *31*, 5603.

$SnBu_3$ OBn

1. $SnCl_4$, -78°C , 5 min

2. PhCHO , -78°C , 1 h

OH OBn Ph 90%

McNeill, A.H.; Thomas, E.J. *Tetrahedron Lett.*, ***1990***, *31*, 6239.

1. NBS , CCl_4 , K_2CO_3
2. KO_2 , 18-crown-6

OH 45%

Brimble, M.A.; Edmonds, M.K.; Williams, G.M. *Tetrahedron Lett.*, ***1990***, *31*, 7509.

t-BuMe$_2$SiO CHO

1. PhMe , MS 4Å
$(C_6H_{11}O)Me_2Si$ B CO$_2$iPr CO$_2$iPr
2. H_2O_2 , $KHCO_3$, MeOH THF , 55°C

t-BuMe$_2$SiO OH OH 67%

Roush, W.R.; Grover, P.T.; Lin, X. *Tetrahedron Lett.*, ***1990***, *31*, 7563.
Roush, W.R.; Grover, P.T. *Tetrahedron Lett.*, ***1990***, *31*, 7567.

SiMe$_3$

1. BBr_3 , CH_2Cl_2 -78°C → 25°C
2. 30% H_2O_2 3N NaOH , reflux

HO 97%

Ryu, I; Hirai, A.; Suzuki, H.; Sonoda, N.; Murai, S. *J. Org. Chem.*, ***1990***, *55*, 1409.

SiMe$_3$ HO

1. KH , I_2 , THF 65°C
2. TBAF , THF , 65°C

H HO 74%

tandem anionic oxy-Cope•allyl silane cyclization

Jisheng, L.; Gallardo, T.; White, J.B. *J. Org. Chem.*, ***1990***, *55*, 5426.

O Ph-P CHO Ph

1. iPrCHO $[V_2Cl_3(THF)_6]_2[Zn_2Cl_6]$
2. > 2 eq. NaH

OH 71%

Park, J.; Pedersen, S.F. *J. Org. Chem.*, ***1990***, *55*, 5924.

3 O

PhI , 10% $Pd(OAc)_2$, Bu_4NCl
3 HCO_2Li , 3 $EtN(iPr)_2$, LiCl
DMF , 50°C , 1 d

Ph OH

(79:21 E:Z) 78%

Larock, R.C.; Leung, W-Y. *J. Org. Chem.*, ***1990***, *55*, 6244.

1. catechol (OH, OH), NEt_3, 5 h, 45°C
2. Me–C₆H₄–CHO, NEt_3, $CHCl_3$, reflux
3. 1M HCl ; 1M NaOH

$SiMe_3$ → OH, Me — 84%

Hosomi, A.; Kohra, S.; Ogata, K.; Yanagi, T.; Tominaga, Y. *J. Org. Chem.*, ***1990***, *55*, 2415.

CHO → Ph, Br, RT / Zn, DMF → OH, Ph — 87%

(75:15 erythro:threo)

Shono, T.; Ishifune, M.; Kashimura, S. *Chem. Lett.*, ***1990***, 449.

Me, CHO, Me → Me_2AlCl, CH_2Cl_2, -78°C → Me, OH + Me, OH — (9 : 1) 65%

MABR, CH_2Cl_2, -78°C → (1 : 32) 85%

MABR = Br, t-Bu, t-Bu, O, Al, O, Br, t-Bu, Me, t-Bu

Maruoka, K.; Ooi, T.; Yamamoto, H. *J. Am. Chem. Soc.*, ***1990***, *112*, 9011.

SO_2Ph, O, C, $SiMe_3$ → Et_3Al, CH_2Cl_2, RT → PhO_2S, $SiMe_3$, HO, Et — 80%

Trost, B.M.; Urabe, H. *J. Am. Chem. Soc.*, ***1990***, *112*, 4982.

Cu → 1. ICH_2ZnI 2. n-$C_5H_{11}CHO$ → OH, n-C_5H_{11} — 86%

Knochel, P.; Rao, S.A. *J. Am. Chem. Soc.*, ***1990***, *112*, 6146.

Bu_3Sn (77:23 E:Z) $TiCl_4$, -78°C → HO ... + HO ... (74 : 26) quant.

Yoshitake, M.; Yamamoto, M.; Kohmoto, S.; Yamada, K. *J. Chem. Soc., Perkin Trans. I*, ***1990***, 1226.

n-C_6H_{13} epoxide; Ti(OPh)$_3$, THF, -78°C → RT → n-C_6H_{13} ... OH + n-C_6H_{13} ... OH (22 : 1) 77%

Tanaka, T.; Inoue, T.; Kamei, K.; Murakami, K.; Iwata, C. *J. Chem. Soc., Chem. Commun.*, ***1990***, 906.

Me_3SiO, SePh, C=C=C–H; Bu_3SnH, AIBN, PhH → OH 54% + HO ... C=C–H 22%

Keck, G.E.; Tafesh, A.M. *SynLett*, ***1990***, 257.

O; LDA, t-BuOK, 0°C, 1 h → OH 88%

Mordini, A.; Ben Rayana, E.; Margot, C.; Schlosser, M. *Tetrahedron*, ***1990***, *46*, 2401.

O, F_3C, CO_2Et; 1. $TiCl_4$, 0°C, 3.5 h 2. SiO_2 → CO_2Et, F_3C OH 75%

Abouabdellah, A.; Aubert, C.; Bégué, J-P.; Bonnet-Delpon, D.; Guilhem, J. *J. Chem. Soc., Perkin Trans. I*, ***1991***, 1397.

Ph, O, Br; Zn(Cu), EtOH, H_2O,))))))) → OH, Ph 95%

Sarandeses, L.A.; Mouriño, A.; Luche, J-L. *J. Chem. Soc., Chem. Commun.*, ***1991***, 818.

Bu_3SnH, AIBN
PhH, reflux
syringe pump addition
over 8 h

46%

(55:45 EE:ZE)

Clive, D.L.J.; Cole, D.C. *J. Chem. Soc., Perkin Trans. I*, ***1991***, 3263.

$c\text{-}C_6H_{11}CHO$
$BF_3\bullet OEt_2$

(95 : 5) 74%

Marshall, J.A.; Welmaker, G.S.; Gung, B.W. *J. Am. Chem. Soc.*, ***1991***, *113*, 647.
Marshall, J.A.; Luke, G.P. *J. Org. Chem.*, ***1991***, *56*, 484.

1. PhCHO, $BF_3\bullet OEt_2$, -78°C
2. $Me_3N\text{-}O$, CH_2Cl_2, 23°C

68%

Lin, S-H.; Yang, Y-J.; Liu, R-S. *J. Chem. Soc., Chem. Commun.*, ***1991***, 1004.

1. $n\text{-}C_6H_{13}MgBr$, ether
$CuBr\bullet SMe_2$, -70°C
2. H_2O

49%

Barber, C.; Bury, P.; Kocieński, P.; O'Shea, M. *J. Chem. Soc., Chem. Commun.*, ***1991***, 1595.

1. $Co(dmgH)_2Py$
2. hν (sunlamp)

76%

Harrowven, D.C.; Pattenden, G. *Tetrahedron Lett.*, ***1991***, *32*, 243.

2 eq. Riecke Zn
$THF/MeOH/H_2O$
30 min

60%
(95:5 cis:trans)

Chou, W-N.; Clark, D.L.; White, J.B. *Tetrahedron Lett.*, ***1991***, *32*, 299.

10% Bu_4NReO_4, 5 eq. $TsOH \cdot H_2O$, CH_2Cl_2, RT — 40%

Narasaka, K.; Kusama, H.; Hayashi, Y. *Chem. Lett.*, ***1991***, 1413.

1. Me_2CuLi 2. $LiAlH_4$, THF 3. KH, THF — Me_3Si ... n-C_5H_{11} → OH, Me, n-C_5H_{11} 69%

Ito, T.; Yamakawa, I.; Okamoto, S.; Kobayashi, Y.; Sato, F. *Tetrahedron Lett.*, ***1991***, *32*, 371.

1. PhCHO, 24 h 2. MeOH — OH, Ph, Me 99% (8% de)

If quenched after 1 h: 45% yield, 98% de

Faller, J.W.; DiVerdi, M.J.; John, J.A. *Tetrahedron Lett.*, ***1991***, *32*, 1271.

Ph_3SnH, AIBN, PhH (3 h addition), reflux — (62 : 38) 80%

Rochigneux, I.; Fontanel, M-L.; Malanda, J-C.; Doutheau, A. *Tetrahedron Lett.*, ***1991***, *32*, 2017.

1. $CeCl_3$, -20°C, 30 min 2. BnCOBn, -78°C — 9% + 19%

Imamoto, T.; Hatajima, T.; Ogata, K. *Tetrahedron Lett.*, ***1991***, *32*, 2787.

$Br^{\ominus}$ $Ph_3P^{\oplus}$ — 1. BuLi 2. BuLi-TMEDA, THF 3. PhCHO, 2 h — 53%

Shen, Y.; Wang, T. *Tetrahedron Lett.*, ***1991***, *32*, 4353.

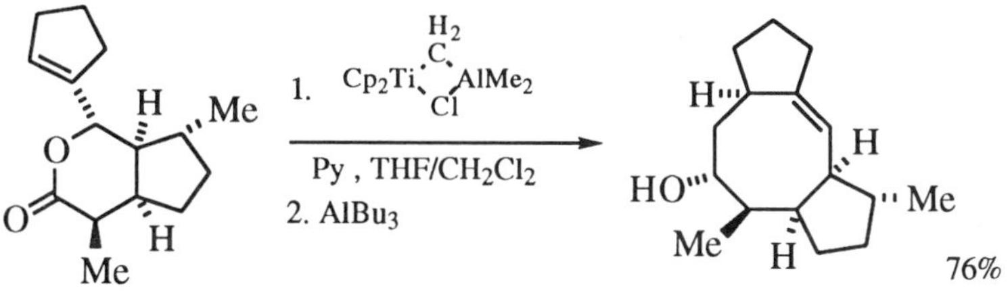

1. $Cp_2Ti(\mu\text{-}CH_2)(\mu\text{-}Cl)AlMe_2$, Py, THF/CH_2Cl_2
2. $AlBu_3$

76%

Paquette, L.A.; Friedrich, D.; Rogers, R.D. *J. Org. Chem.*, ***1991***, *56*, 3841.
Paquette, L.A.; Philippo, C.M.G.; Vo, N.H. *Can. J. Chem.*, ***1992***, *70*, 1356.

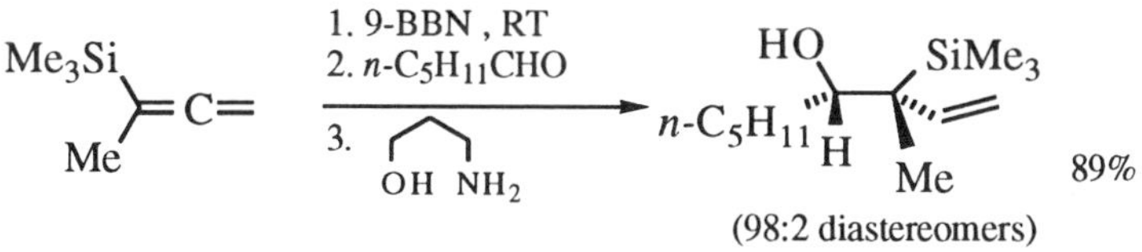

Me_3Si, Me

1. 9-BBN, RT
2. $n\text{-}C_5H_{11}CHO$
3. OH NH_2

HO, $SiMe_3$, $n\text{-}C_5H_{11}$, H, Me

(98:2 diastereomers)

89%

Gu, Y.G.; Wang, K.K. *Tetrahedron Lett.*, ***1991***, *32*, 3029.

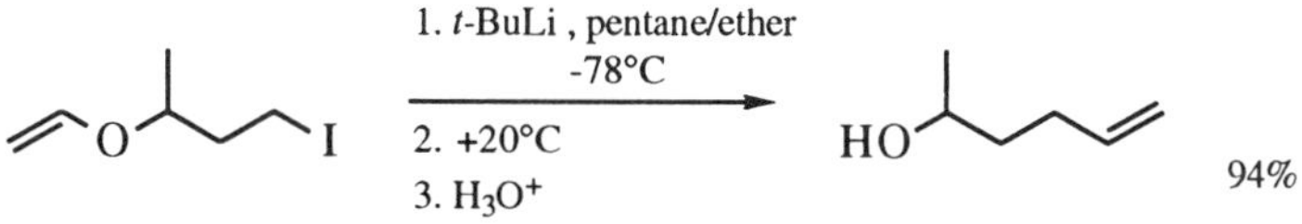

1. *t*-BuLi, pentane/ether -78°C
2. +20°C
3. H_3O^+

HO

94%

Bailey, W.F.; Zarcone, L.M.J. *Tetrahedron Lett.*, ***1991***, *32*, 4425.

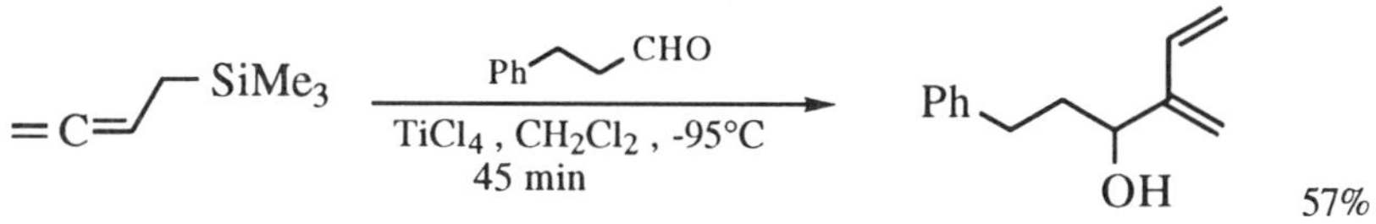

$SiMe_3$

Ph CHO

$TiCl_4$, CH_2Cl_2, -95°C
45 min

Ph, OH

57%

Hatakeyama, S.; Sugawara, K.; Kawamura, M.; Takano, S. *Tetrahedron Lett.*, ***1991***, *32*, 4509.

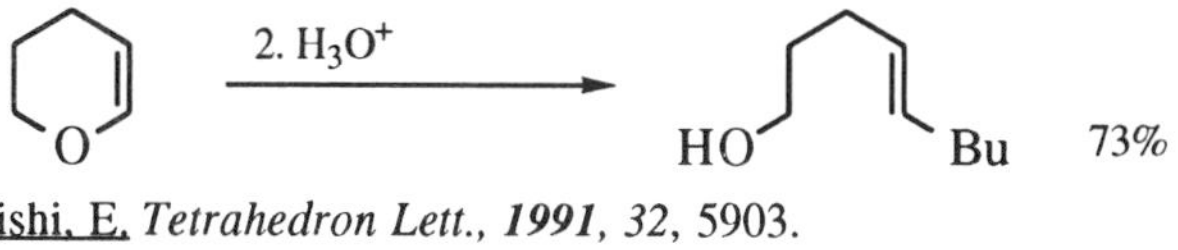

1. BuLi
2. H_3O^+

HO, Bu 73%

Nguyen, T.; Negishi, E. *Tetrahedron Lett.*, ***1991***, *32*, 5903.

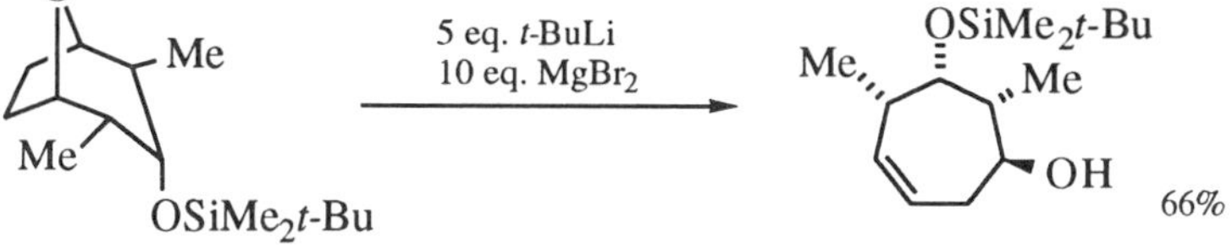

O, Me, Me, $OSiMe_2t$-Bu

5 eq. *t*-BuLi
10 eq. $MgBr_2$

$OSiMe_2t$-Bu, Me, Me, OH

66%

Lautens, M.; Chiu, P. *Tetrahedron Lett.*, ***1991***, *32*, 4827.

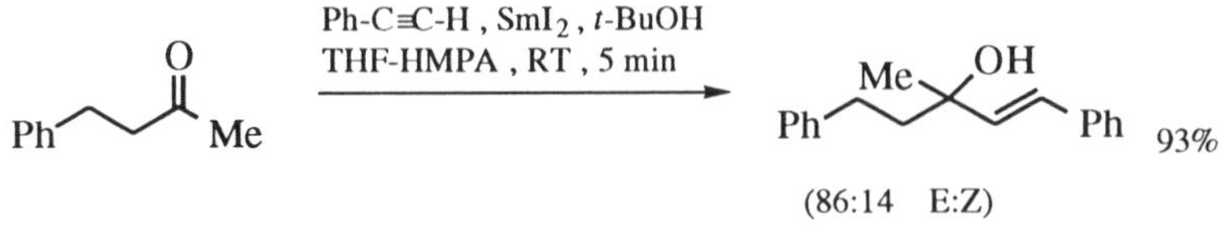
O
Ph
Me
Ph-C≡C-H , SmI2 , t-BuOH
THF-HMPA , RT , 5 min
Me
OH
Ph
Ph
93%
(86:14 E:Z)

Inanaga, J.; Katsuki, J.; Ujikama, O.; Yamaguchi, M. *Tetrahedron Lett.*, ***1991***, *32*, 4921.

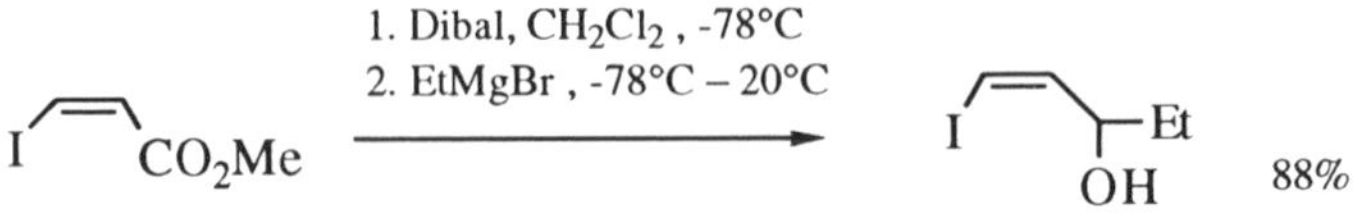
I
CO2Me
1. Dibal, CH2Cl2 , -78°C
2. EtMgBr , -78°C – 20°C
I
Et
OH
88%

Marek, I.; Alexakis, A.; Normant, J-F. *Tetrahedron Lett.*, ***1991***, *32*, 5329.

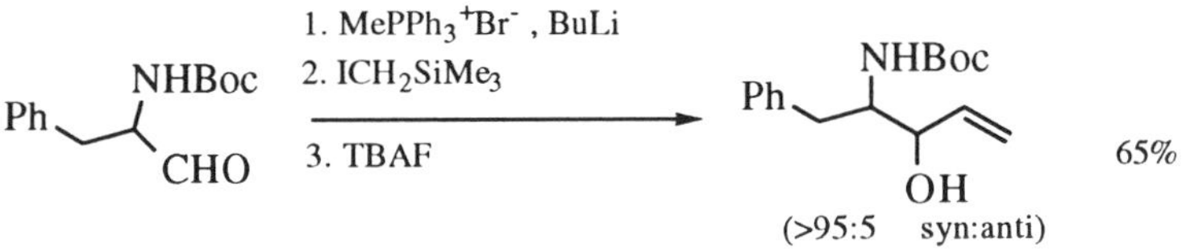
NHBoc
Ph
CHO
1. MePPh3+Br- , BuLi
2. ICH2SiMe3
3. TBAF
NHBoc
Ph
OH
(>95:5 syn:anti)
65%

Franciotti, M.; Mann, A.; Taddei, M. *Tetrahedron Lett.*, ***1991***, *32*, 6783.

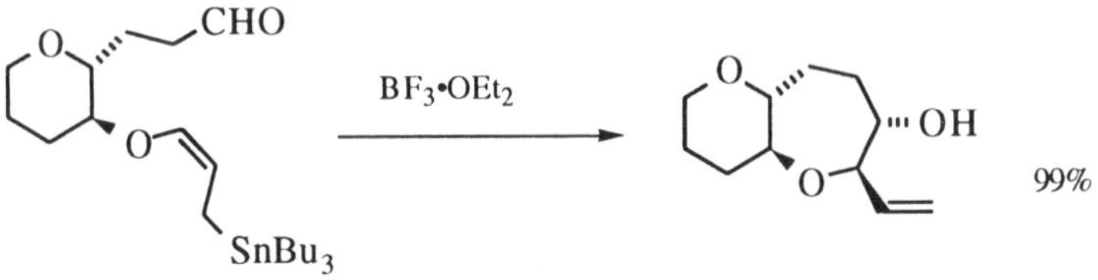
O
CHO
O
SnBu3
BF3•OEt2
O
OH
O
99%

Yamamoto, Y.; Yamada, J.; Kadota, I. *Tetrahedron Lett.*, ***1991***, *32*, 7069.

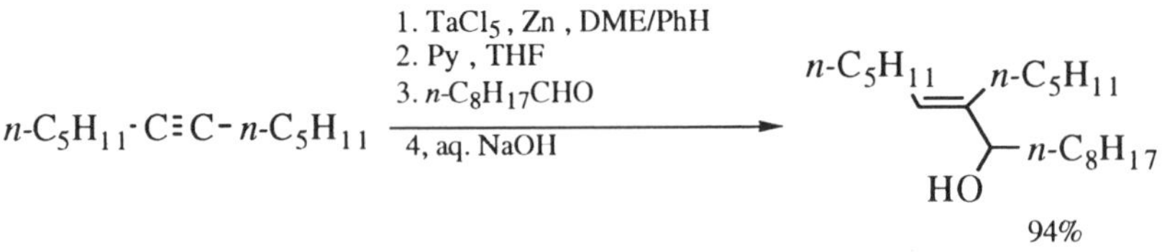
n-C5H11-C≡C-n-C5H11
1. TaCl5 , Zn , DME/PhH
2. Py , THF
3. n-C8H17CHO
4, aq. NaOH
n-C5H11
n-C5H11
n-C8H17
HO
94%

Kataoka, Y.; Miyai, J.; Oshima, K.; Takai, K.; Utimoto, K. *J. Org. Chem.*, ***1992***, *57*, 1973.
Takai, K.; Kataoka, Y.; Utimoto, K. *J. Org. Chem.*, ***1990***, *55*, 1707.

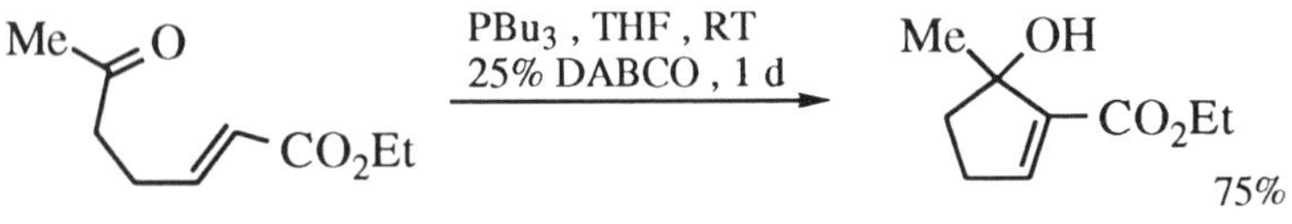
Me
O
CO2Et
PBu3 , THF , RT
25% DABCO , 1 d
Me
OH
CO2Et
75%

Roth, F.; Gygax, P.; Fráter, G. *Tetrahedron Lett.*, ***1992***, *33*, 1045.

OH

Ph

1% Bu_4N ReO_4 , CH_2Cl_2
5% *p*-TsOH•H_2O , 5 min

Ph

HO

47%

(2.5:1 cis:trans)

dienes formed with longer reaction times

Narasaka, K.; Kusama, H.; Hayashi, Y. *Tetrahedron*, ***1992***, *48*, 2059.

1. $^1O_2$, 25,C , CCl_4
2. PPh_3

HO + HO

10-20% conversion (92 : 8)

Dussault, P.H.; Hayden, M.R. *Tetrahedron Lett.*, ***1992***, *33*, 451.

1. B , 25°C
2. $NaBO_3$•4 H_2O

Ph

OH

Ph

97%

(>95:5 *p:m*)

Singleton, D.A.; Martinez, J.P.; Watson, J.V. *Tetrahedron Lett.*, ***1992***, *33*, 1017.
Singleton, D.A.; Martinez, J.P. *Tetrahedron Lett.*, ***1991***, *32*, 7381.

OAc

O

hν (Hg lamp) , AIBN
PhSSPh

OAc H

OH

60%

(4:1 dr)

Rawal, V.H.; Krishnamurthy, V. *Tetrahedron Lett.*, ***1992***, *33*, 3439.

Ph OMe

PhCHO
[$(Cp_2ZrCl_2)_2$/BuLi]

OH

Ph

Ph

79%

(10:1 threo:erythro)

Ito, H.; Taguchi, T.; Hanzawa, Y. *Tetrahedron Lett.*, ***1992***, *33*, 1295.

Me_3Si ... Br → (Me_3Si, Br, PhCHO; $CrCl_2$, THF, RT, 3 h) → Ph, OH, $SiMe_3$ 69%

Hodgson, D.M.; Wells, C. *Tetrahedron Lett.*, ***1992***, *33*, 4761.

Bu_3Sn ... Me, OBn → (1. $SnCl_4$, -78°C, 5 min; 2. Me, BnO ... CHO, -78°C) → BnO, Me, OH, BnO 76%

(<4% other diastereomers)

McNeill, A.H.; Thomas, E.J. *Tetrahedron Lett.*, ***1992***, *33*, 1369.

Me, $SnBu_3$, OBn O → (BuLi) → Me, OBn OH + Me, OBn OH

(91 : 9) 94%

Tomooka, K.; Igarashi, T.; Watanabe, M.; Nakai, T. *Tetrahedron Lett.*, ***1992***, *33*, 5795.

H-C≡C-Bu → (1. $Cp_2Zr(H)Cl$, CH_2Cl_2, RT, 10 min; 2. PhCHO, cat. $AgClO_4$) → Ph, OH, Bu 92%

Maeta, H.; Hashimoto, T.; Hasegawa, T.; Suzuki, K. *Tetrahedron Lett.*, ***1992***, *33*, 5965.

S → (1. LDA, THF, -78°C; 2. Me_3SiCl; 3. MeOH, 20°C, 1h) → SH

Le Nocher, A-M.; Metzner, P. *Tetrahedron Lett.*, ***1992***, *33*, 6151.

Me, Ph, Si, Ph, O, O, Me, Me, MeO_2C → (1. PhMe (93 mM), 172°C, 24 h; 2. aq. HF, MeCN, $NaHCO_3$, 1 h, RT) → Me, OH, Me, MeO_2C, OH, Me + Me, OH, Me, MeO_2C, OH, Me

(1 : 1) 75%

Craig, D.; Reader, J.C. *Tetrahedron Lett.*, ***1992***, *33*, 6165.

1. Cu*
2. PhCHO
3. allyl chloride

83%

Cu* = zero-valent copper (from CuCN•2 LiCl)

Stack, D.E.; Rieke, R.D. *Tetrahedron Lett.*, ***1992***, *33*, 6575.

1. Cp_2Bu_2 , PhMe -78°C → RT
2. PhCHO , "X" , 3 h -78°C
3. 1N HCl

X = no Lewis acid (35 : 65) 70%
X = BF_3•OEt_2 (71 : 29) 82%

Ito, H.; Taguchi, T.; Hanzawa, Y. *Tetrahedron Lett.*, ***1992***, *3*, 7873.

Cp_2TiCl , THF

(E+Z) 91%

Yadav, J.S.; Shekharam, T.; Srinivas, D. *Tetrahedron Lett.*, ***1992***, *33*, 7973.
Yadav, J.S.; Shekharam, T.; Gadgil, V.R. *J. Chem. Soc., Chem. Commun.*, ***1990***, 843.

$MeNO_2$ CH_3SO_3H , 16 h
75°C

75% (E+Z)
33% Z

Blanchot-Courtois, V.; Hanna, I. *Tetrahedron Lett.*, ***1992***, *33*, 8087.

PhCHO

allyl bromide , Al-$InCl_3$, RT
THF-H_2O , 50 h

88%

Araki, S.; Jin, S-J.; Idou, Y.; Butsugan, Y. *Bull. Chem. Soc. Jpn.*, ***1992***, *65*, 1736.

PhCHO

BF_3•OEt_2
CH_2Cl_2 , -78°C

83%

Narasaka, K.; Shibata, T.; Hayashi, Y. *Bull. Chem. Soc. Jpn.*, ***1992***, *65*, 2825.

$[CH_2=CHCH_2]_4Sn$, THF; BuLi, $SmCl_3$; 0°C → RT

(85 : 15) 97%

Fukuzawa, S.; Sakai, S. *Bull. Chem. Soc. Jpn.*, ***1992***, *65*, 3308.

1. PhCHO, $SnCl_2$, $PdCl_2(PhCN)_2$; 15°C, 4 h,))))))))); 2. RT, 48 h

(98 : 2) 57%

(67:33 E:Z) (70:30 syn:anti)

Masuyama, Y.; Hayakawa, A.; Kurusu, Y. *J. Chem. Soc., Chem. Commun.*, ***1992***, 1102.

1. THF, 160°C, 1 h; 2. $PdCl_2(PhCN)_2$; $SnCl_2$, DMI, H_2O; RT, 5 d

35%

(32:68 syn:anti)

Masuyama, Y.; Nimura, Y.; Kurusu, Y. *Tetrahedron Lett.*, ***1992***, *33*, 6477.

1. LDA, *t*-BuOK, -50°C; 2. H_2O

72%

Degl'Innocenti, A.; Mordini, A.; Pecchi, S.; Pinzani, D.; Reginato, G.; Ricci, A. *SynLett*, ***1992***, 753, 803.

Allylic and benzylic hydroxylation (C=C-C-H → C=C-C-OH, etc.) is listed in Section 41 (Alcohols from Hydrides). Also via: Section 302 (Alkyne - Alcohol).

SECTION 333: ALDEHYDE - ALDEHYDE

t-BuOH, H_2O_2, $WO_3{\cdot}H_2O$

CHO CHO 80%

Jingfa, D.; Xinhua, X.; Haiying, C.; Anren, J. *Tetrahedron*, ***1992***, *48*, 3503.

e^-, Pt anode, $LiClO_4$; 10% MeOH/MeCN; 2,6-lutidine — 68%

Moeller, K.D.; Tinao, L.V. *J. Am. Chem. Soc.*, ***1992***, *114*, 1033.

REVIEWS:

"Oxidative Rearrangement of Furyl Carbinols to 6-Hydroxyl 2H-pyran-3(6H)ones, a Useful Synthon for the Preparation of a Variety of Heterocyclic Compounds"
Georgiadia, M.P.; Albizati, KF.; Georgiadia, T.M. *Org. Prep. Proceed. Int.*, ***1992***, *24*, 97.

SECTION 334: ALDEHYDE - AMIDE

1. CH_2Cl_2; 2. silica gel

Wasserman, H.H.; Cook, J.D.; Vu, C.B. *J. Org. Chem.*, ***1990***, *55*, 1701.

1. PhMe, NEt_3, RT, 5 h; 2. 5% aq. HCl, RT — 73%

Ar = 4-(MeO)-C_6H_4

Alcaide, B.; Martín-Cantalejo, Y.; Plumet, J.; Rodríquez-López, J.; Sierra, M.A. *Tetrahedron Lett.*, ***1991***, *32*, 803.

NBS, DME, H_2O, 0°C, 2 h

(>95:5 dr)

Jäger, V.; Hümer, W. *Angew. Chem. Int. Ed. Engl.*, ***1990***, *29*, 1171.

1% TEMPO ($N-O^{\cdot}$), NaOCl; NaBr, aq. $NaHCO_3$, 0°C

N-Boc-prolinol (OH) → N-Boc-prolinal (CHO) 82% (>98% ee)

Leanna, M.R.; Sowin, T.J.; Morton, H.E. *Tetrahedron Lett.*, ***1992***, *33*, 5029.

SECTION 335: ALDEHYDE - AMINE

H_2/CO, PPh_3, $[Rh(OAc)_2]_2$; EtOAc, 55°C, 22h

PPh_2 → CHO, PPh_2 64% + CHO, PPh_2 <4%

Jackson, W.R.; Perlmutter, P.; Tasdelen, E.E. *Tetrahedron Lett.*, ***1990***, *31*, 2461.

$[Rh(OAc)_2]_2$, PPh_3, H_2, CO (400 psi); 80°C, 20 h

PhHN → OHC, Ph, N, Me + CHO, Ph, N, Me (67 : 33) 76%

Anastasiou, D.; Campi, E.M.; Chaouk, H.; Jackson, W.R.; McCubbin, Q.J. *Tetrahedron Lett.*, ***1991***, *33*, 2211.

1. MeOH, reflux, 23 h
2. 6N HCl, H_2O/CH_2Cl_2
reflux, 26 h

N, H, Cl → CHO, N-H 58%

De Kimpe, N.; Boeykens, M.; Boelens, M.; De Buck, K.; Cornelis, J. *Org. Prep. Proceed. Int.*, ***1992***, *24*, 679.

1. BuLi, -78°C
2. $MgBr_2$, 0°C
3. PhCHO, -65°C

N, O, N → N, O, N, Ph, OH 50% (80% ds)

Zhang, P.; Gawley, R.E. *Tetrahedron Lett.*, ***1992***, *33*, 2945.

CHO → PhNHMe , cat. DBU; THF , -15°C , 40 min → PhMeN CHO >95%

Markó, I.E.; Chesney, A. *SynLett*, ***1992***, 275.
Chesney, A.; Markó, I.E. *Synth. Commun.*, ***1990***, *20*, 3167.

REVIEWS:

"Synthesis of α-Amino Aldehydes and α-Amino Ketones"
Fisher, L.E.; Muchowski, J.M. *Org. Prep. Proceed. Int.*, ***1990***, *22*, 399.

SECTION 336: ALDEHYDE - ESTER

Me_2N-N= CHO → 1. EtO-P(O)(OEt) CO_2Et, K_2CO_3 , reflux; 2. H(O)C-C(O)H , H^+ → H(O)C-CH=CH-CO_2Et 85%

Schmitt, M.; Bourguignon, J.J.; Wermuth, C.G. *Tetrahedron Lett.*, ***1990***, *31*, 2145.

NMe_2 O O OMe → 1. EtOTf , CH_2Cl_2 RT; 2. 2 eq. L-Selectride THF , -78°C → H O O OMe 69%

Tsay, S-C.; Robl, J.A.; Hwu, J.R. *J. Chem. Soc., Perkin Trans. I*, ***1990***, 757.

Ph CHO → 1. NaI , Me_3SiCl , DMF NEt$_3$; 2. N_2CHCO_2Me , heat $Cu(acac)_2$; 3. NEt_3•HF , CH_2Cl_2 → H O Ph CO_2Me 35%

Kunz, T.; Janowitz, A.; Reißig, H-U. *Synthesis*, ***1990***, 43.

CO_2Me CO_2Me → 1. e^- , NaI , MeOH 2.2.F/mol; 2. H^+ → H O Me CO_2Me CO_2Me 80%

also for preparation of ketone-esters

Nikishin, G.I.; Elinson, M.N.; Feducovich, S.K. *Tetrahedron Lett.*, ***1991***, *32*, 799.

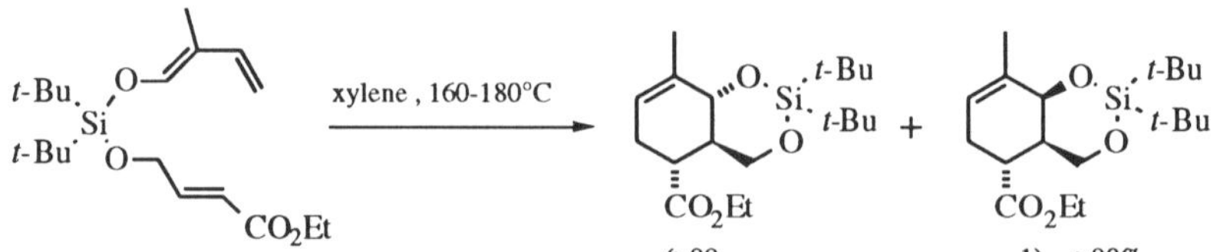

xylene , 160-180°C

(>99 : 1) >90%

Gillard, J.W.; Fortin, R.; Grimm, E.L.; Maillard, M.; Tjepkema, M.; Bernstein, M.A.; Glaser, R. *Tetrahedron Lett.*, ***1991***, *32*, 1145.

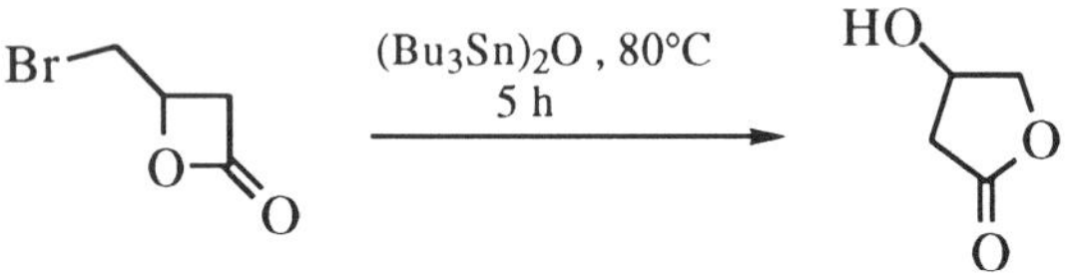

$(Bu_3Sn)_2O$, 80°C, 5 h

80%

Shibata, I.; Matsuo, F.; Baba, A.; Matsuda, H. *J. Org. Chem.*, ***1991***, *56*, 475.

SECTION 337: ALDEHYDE - ETHER, EPOXIDE, THIOETHER

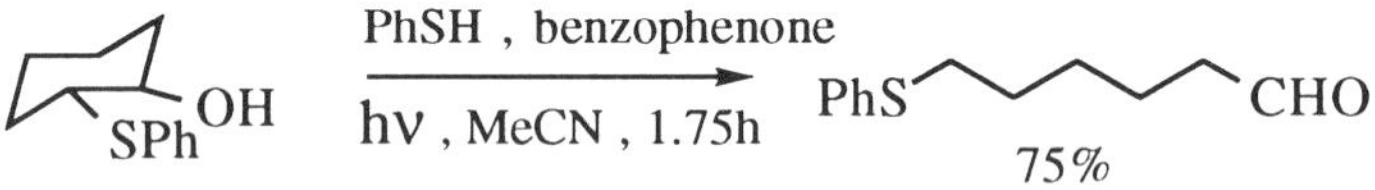

PhSH , benzophenone; hν , MeCN , 1.75h

75%

Gravel, D.; Farmer, L.; Ayotte, C. *Tetrahedron Lett.*, ***1990***, *31*, 63.

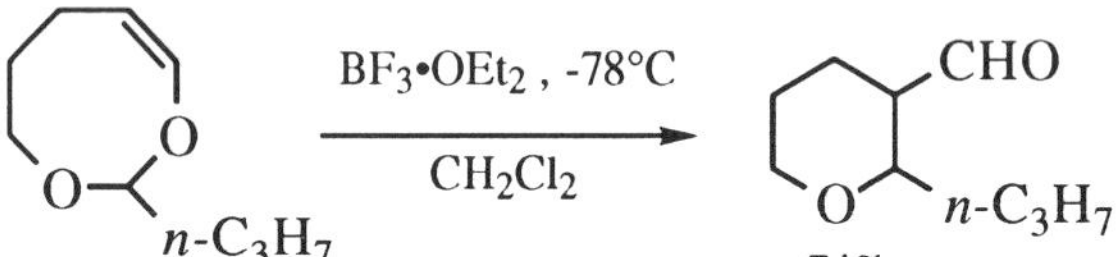

BF_3•OEt_2 , -78°C; CH_2Cl_2

74%

Frauenrath, H.; Sawicki, M. *Tetrahedron Lett.*, ***1990***, *31*, 649.

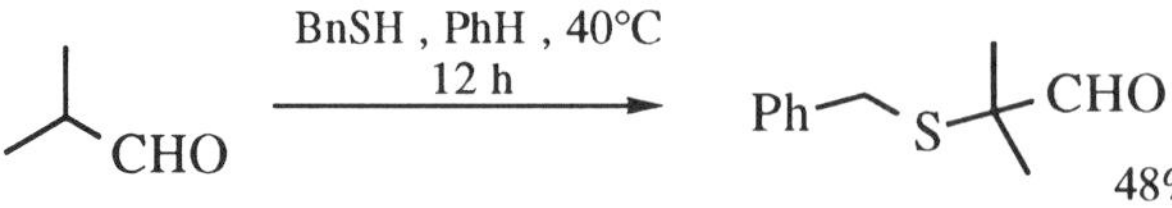

BnSH , PhH , 40°C, 12 h

48%

Jones, G.S. Jr.; Elmaleh, D.R. *Org. Prep. Proceed. Int.*, ***1990***, *22*, 112.

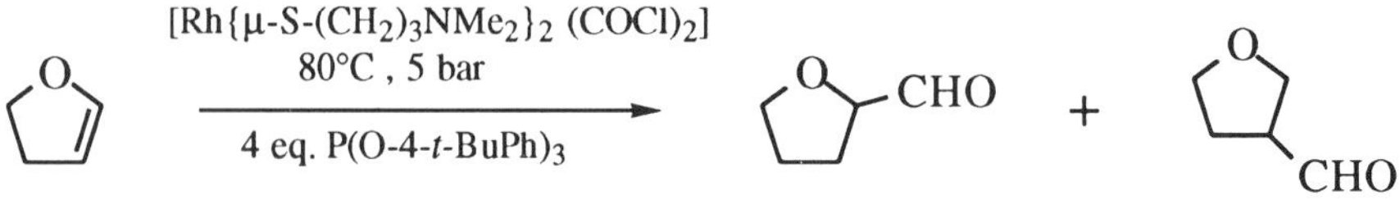

[Rh{μ-S-$(CH_2)_3NMe_2$}$_2$ $(COCl)_2$]; 80°C , 5 bar; 4 eq. P(O-4-*t*-BuPh)$_3$

(77 : 23) 98%

Polo, A.; Real, J.; Claver, C.; Castillón, S.; Bayón, J.C. *J. Chem. Soc., Chem. Commun.*, ***1990***, 600.

Ph, O, O, Ph, Ph → $POCl_3$, DMF; 80°C , 5 h → Ph, O, Ph, OHC, CHO, Ph 52%

Venugopal, M.; Umarani, R.; Perumal, P.T.; Rajadurai, S. *Tetrahedron Lett.*, ***1991***, *32*, 3235.

Ph, S, Ph, O → 1. NaOAc , Ac_2O , Tf_2O; 2. $NaHCO_3$; 3. 5 eq. K_2CO_3 , MeOH; NH_4Cl → SPh, Ph, CHO 55%

Craig, D.; Daniels, K.; MacKenzie, A.R. *Tetrahedron Lett.*, ***1991***, *32*, 6973.

H- C≡ C- n-C_8H_{17} → n-$C_6H_{13}SH$, PhH; AIBN , CO (80 atm); 100°C → n-$C_6H_{13}S$, n-C_8H_{17}, CHO 70% + n-$C_6H_{13}S$, n-C_8H_{17} 9%

Nakatani, S.; Yoshida, J.; Isoe, S. *J. Chem. Soc., Chem. Commun.*, ***1992***, 880.

SECTION 338: ALDEHYDE - HALIDE, SULFONATE

O, N, F → 1. n-BuLi , -78°C; 2. $PhCH_2Br$ → O, N, F, Ph 76% → 1. $NaBH_4$; 2. H_3O^+ → O, H, F, Ph 51%

Patrick, T.B.; Hosseini, S.; Bains, S. *Tetrahedron Lett.*, ***1990***, *31*, 179.

CHO → 1. $iPrNH_2$; 2. NCS , CCl_4; 3. $AlCl_3$, CS_2; 4. oxalic acid; aq. CH_2Cl_2 → CHO, Cl 32%

Stevens, C.; De Kimpe, N. *Org. Prep. Proceed. Int.*, ***1990***, *22*, 589.

O → DMF , $POCl_3$; water bath → CHO, Cl, CHO

Evans, S.L.; Lloyd, H.A.; LaBeau, D.; Sokoloski, E.B. *Org. Prep. Proceed. Int.*, ***1990***, *22*, 764.

1. Br_2, 35°C
2. H_2O

Br CHO + CHO Br + CHO Br

(84 : 13 : 3) 74%

Duhamel, L.; Guillemont, J.; Poirier, J-M.; Chabardes, P. *Tetrahedron Lett.*, ***1992***, *33*, 5051.

N *t*-Bu H

1. LDA, THF, 0°C
2. Me_3SiCl
3. NCS, CCl_4, RT
4. 10 eq. MeOH, CCl_4, RT
5. $(CO_2H)_2$, H_2O-CH_2Cl_2

CHO Cl 63%

De Kimpe, N.; Coppens, W.; Krauze, A.; De Corte, B.; Stevens, C.; Welch, J. *Bull. Soc. Chim. Fr.*, ***1992***, *101*, 237.

SECTION 339: ALDEHYDE - KETONE

O Me Me SBu

1. [$PhSCH_2OMe$, THF / BuLi, -78°C] THF, -78°C, 2 h
2. $HgCl_2$, H^+

O Me Me CHO 54%

Guerrero, A.; Parrilla, A.; Camps, F. *Tetrahedron Lett.*, ***1990***, *31*, 1873.

O N-H OEt

1. BuLi, THF, -78°C
2. *t*-BuO_2CCl
3. *n*-$C_6H_{13}MgBr$
4. 6N NH_3

O *n*-C_6H_{13} CHO 74%

Savoia, D.; Concialini, V.; Roffia, S.; Tarsi, L. *J. Org. Chem.*, ***1991***, *56*, 1822.

1. PhSeBr, $HgCl_2$, $AgNO_2$
2. DMAP (38%)
OEt
3. 10% HCl (95%)

O CHO

Bäckvall, J-E.; Karlsson, U.; Chinchilla, R. *Tetrahedron Lett.*, ***1991***, *32*, 5607.

Me O

O O, acetone
RT

Me O CHO >95%

Adger, B.M.; Barrett, C.; Brennan, J.; McKervey, M.A.; Murray, R.W. *J. Chem. Soc., Chem. Commun.*, ***1991***, 1553.

SH SH

Al_2O_3, 4 h

C≡C-H

70%

Ranu, B.C.; Bhar, S.; Chakraborti, R. *J. Org. Chem.*, ***1992***, *57*, 7349.

OSiMe$_2$*t*-Bu

$SnBu_3$, CAN

MeCN, MS 4Å, -23°C

87%

Narasaka, K.; Okauchi, T.; Arai, N. *Chem. Lett.*, ***1992***, 1229.

SECTION 340: ALDEHYDE - NITRILE

NO ADDITIONAL EXAMPLES

SECTION 341: ALDEHYDE - ALKENE

For the oxidation of allylic alcohols to alkene aldehydes, see Section 48 (Aldehydes from Alcohols).

Me_3Si / $SiMe_3$

PhCHO, cat. $ZnBr_2$, THF

$ZnCl_2$, aq. ether

Ph—CHO 86% E only

Gaudemar, M.; Bellassoued, M. *Tetrahedron Lett.*, ***1990***, *31*, 349.

EtO-P(O)(OEt)

1. *t*-BuOK, THF
2. PhCHO
3. H_3O^+

Ph—CHO 86%

Duhamel, L.; Guillemont, J.; LeGallic, Y.; Plé, G.; Poirier, J-M.; Ramondenc, Y.; Chabardes, P. *Tetrahedron Lett.*, ***1990***, *31*, 3129.

C≡C-H

O_2, 5% $PdBr_2$, THF, 65°C, 1.5 h

CHO 65%

Kataoka, H.; Watanabe, K.; Goto, K. *Tetrahedron Lett.*, ***1990***, *31*, 4181.

n-C_6H_{13}, I → (CH_2=CHCH$_2$OH, MeCN; cat. $Pd(OAc)_2$, Ag_2CO_3; Bu_4N $NHSO_4$) → n-C_6H_{13}, CHO 60%

Jeffery, T. *Tetrahedron Lett.*, ***1990***, *31*, 6641.

OH → V-PILC, *m*-xylene; reflux, 1.5 h → CHO 84%

V-PILC = vanadium-pillared montmorillonite clay

Choudary, B.M.; Prasad, A.D.; Valli, V.L.K. *Tetrahedron Lett.*, ***1990***, *31*, 7521.

Ph_3P, O → 1. BuLi, THF, -23°C; 2. PhCHO; 3. H_2O; 4. KH, 18-crown-6, THF → Ph, CHO 36% overall

Enholm, E.J.; Satici, H.; Prasad, G. *J. Org. Chem.*, ***1990***, *55*, 324.

n-C_5H_{11}CHO → 1. [LDA, THF, -78°C; O, EtO, P, OEt, N]; THF, 2 h; 2. SiO_2 → n-C_5H_{11}, CHO 78% (84:14:2 all E:2E,4Z:2Z,4E)

Kann, N.; Rein, T.; Åkermark, B.; Helquist, P. *J. Org. Chem.*, ***1990***, *55*, 5312.

N, iPr, H, Cl → 1. NaOMe, MeOH, heat; 2. $(CO_2H)_2$•2 H_2O, heat; H_2O, CH_2Cl_2 → CHO 85-95%

De Kimpe, N.; Stevens, C. *Bull. Soc. Chim. Belg.*, ***1990***, *99*, 41.

O → (Br, *t*-Bu, *t*-Bu, O, Al, O, Me, *t*-Bu, *t*-Bu, Br) → CHO 64% (7:93 E:Z)

Nonoshita, K.; Banno, H.; Maruoka, K.; Yamamoto, H. *J. Am. Chem. Soc.*, ***1990***, *112*, 316.

2 CHO → Ti(OiPr)$_4$, hexane, -68°C, 4 h → CHO 76%

Mahrwald, R.; Schick, H. *Synthesis*, ***1990***, 592.

PhCHO → 1. (Li, OSiMe$_3$) 2. 1N HCl → Ph CHO

Contreras, B.; Duhamel, L.; Ple, G. *Synth. Commun.*, ***1990***, *20*, 2983.

CHO + → CH_2Cl_2, -78°C; 20% (H, B, O, N, S, O_2, iPr, iPr, iPr, O) → CHO 73% (74% ee)

Takasu, M.; Yamamoto, H. *SynLett*, ***1990***, 194.

CHO + → (H, B, O, N, S, O_2, Me, O) → CHO 95% (<1:99 endo:exo) (64% ee)

Sartor, D.; Saffrich, J.; Helmchen, G. *SynLett*, ***1990***, 197.

O → 3M $LiClO_4$, ether → CHO 94%

Grieco, P.A.; Clark, J.D.; Jagoe, C.T. *J. Am. Chem. Soc.*, ***1991***, *113*, 5488.

Me CHO → , (N, Me, Ph, Ph, OH) → CHO Me 84% (>99:1 exo:endo ; 91% ee)

Kobayashi, S.; Murakami, M.; Harada, T.; Mukaiyama, T. *Chem. Lett.*, ***1991***, 1341.

PhH; (P)-(C$_6$H$_4$-CH$_2$N$^+$Me$_3$)$_2$ $Cr_2O_7^{2-}$ — OH → CHO 81%

no oxidation of non-allylic alcohols

Yang, H.; Li, B. *Synth. Commun.*, ***1991***, *21*, 1521.

MeO, MeO, OMe; DMF, $POCl_3$; 0°C → RT, 4 h; MeO, MeO, CHO, OMe 60%

Ramesh, N.G.; Balasubramanian, K.K. *Tetrahedron Lett.*, ***1991***, *32*, 3875.

MeO_2C CO_2Me, HO, Br; 5% $Pd(OAc)_2$, 15% PPh_3, 80°C, 15 h; MeO_2C CO_2Me, CHO 60%

Gaudin, J-M. *Tetrahedron Lett.*, ***1991***, *32*, 6113.

PhCHO; CH$_2$=CH-CN, $ISiCl_3$; DCE, H_2O, 25°C, 4 h; Ph-CH=N-CH=CH-CHO 57%

Elmorsy, S.S.; Badawy, D.S.; Nour, M.A.; Pelter, A. *Tetrahedron Lett.*, ***1991***, *32*, 5421.

NO_2, O; 1. Bu-CH=CH-$Al(iBu)_2$; 2. 3N HCl; Bu, CHO 71%

Menicagli, R.; Malanga, C.; Guagnano, V. *Tetrahedron Lett.*, ***1992***, *33*, 2867.

OEt, OEt; $SnCl_2 \cdot 2H_2O$, CH_2Cl_2; CHO 94%

Ford, K.L.; Roskamp, E.J. *Tetrahedron Lett.*, ***1992***, *33*, 1135.

H-C≡C-CH$_2$-NH_2; 2 eq. Me_2PhSiH, CO, PhH; $Rh_4(CO)_{12}$, 100°C, 2 h; $PhMe_2Si$, OHC 94%

Matsuda, I.; Sakakibara, J.; Inoue, H.; Nagashima, H. *Tetrahedron Lett.*, ***1992***, *33*, 5799.

CHO , RT , 4 h

10% $(MeO)_3P$–Fe(Cp)(CO)(THF)

CHO 64%

Olson, A.S.; Seitz, W.J.; Hossain, M.M. *Tetrahedron Lett.*, ***1991***, *32*, 5299.

t-Bu

Cl

$Cl_2C{=}CHCl$, 55°C , 48 h

t-Bu CHO 80%

Katritzky, A.R.; Scherbakova, I.V.; Taek, R.D.; Steel, P.J. *Can. J. Chem.*, ***1992***, *70*, 2040.

Also via β-Hydroxy aldehydes: Section 324 (Alcohols - Aldehyde).

SECTION 342: AMIDE - AMIDE

NH_2 NH_2

$Pb(OAc)_4$, MeOH

95%

Beckwith, A.L.J.; Dyall, L.K. *Aust. J. Chem.*, ***1990***, *43*, 451.

Ph N Me N Me Ph

1. $LiNTMS_2$, THF
2. aq. H^+

Ph NHMe Ph NHMe

70%

(70:30 threo:erythro)

Magedov, I.V.; Smushkevich, Y.I. *J. Chem. Soc., Chem. Commun.*, ***1990***, 1686.

H-N N-H S CO_2H

Ph NH_2

DMF , aq. dioxane , 10 min

iPr_2NEt Me_2N Me_2N $BF_4^{\ominus}$

H-N N-H Ph N-H S

78%

Bannworth, W.; Knorr, R. *Tetrahedron Lett.*, ***1991***, *32*, 1157.

1. 2.5 eq. LDA, THF, 0°C → 20°C
2. H_3O^+

(4 : 1) 50%

Endo, Y.; Shudo, K. *Tetrahedron Lett.*, ***1991***, *32*, 4517.

Also via Dicarboxylic Acids: Section 312 (Carboxylic Acid - Carboxylic Acid)
Diamines Section 350 (Amines - Amines)

SECTION 343: AMIDE - AMINE

1. LDA
2. Ph–CH=N–C₆H₄–OMe

63%

Alcaide, B.; Plumet, J.; Rodríguez-Lopez, J.; Sánche-Cantelejo, Y.M. *Tetrahedron Lett.*, ***1990***, *31*, 2493.

1. OsO_4, $NaIO_4$, NMMO
2. Et_3SiI, TFA

78%

Holladay, M.W.; Nadzan, A.M. *J. Org. Chem.*, ***1991***, *56*, 3900.

CO, ether, hν

44%

Söderberg, B.C.; Hegedus, L.S. *J. Org. Chem.*, ***1991***, *56*, 2209.

N=C=O, RT

84%

Rigby, J.H.; Qabar, M. *J. Am. Chem. Soc.*, ***1991***, *113*, 8975.

N - CO_2Me
Ph
O
, $CHCl_3$
NH_2
RT , 70 min
NH-$NHCO_2Me$
80%

Vidal, J.; Drouin, J.; Collet, A. *J. Chem. Soc., Chem. Commun.*, ***1991***, 435.

O
Me
CO_2Et
H_2N NH_2 , Al_2O_3
microwaves (125 W)
15 min
H Me
N
N
O
90%

Pilard, J-F.; Klein, B.; Texier-Boullet, F.; Hamelin, J. *SynLett*, ***1992***, 219.

REVIEWS:

"The Chemistry of Pyrazolidinones"
Claramunt, R.M.; Elguero, J. *Org. Prep. Proceed. Int.*, ***1991***, *23*, 275.

SECTION 344: AMIDE - ESTER

O
O
O
+
O
O
N
PMB
1. $AlMe_3$, CH_2Cl_2
-78°C → 0°C
2. CH_2N_2
CO_2Me
O
O
N
PMB
O
>60%

PMB = *p*-methoxybenzyl
91:9 (◄ CO_2Me : ···· CO_2Me)

Clark, R.D.; Souchet, M. *Tetrahedron Lett.*, ***1990***, *31*, 193.

Ph
Ph
OAc
Na
MeO_2C CO_2Me
NHAc
$[(\pi\text{-}C_3H_5)PdCl]_2$
THF , S-BINAP
120 h , 25°C
Ph Ph
MeO_2C CO_2Me
NHAc
92%
+
Ph Ph
MeO_2C CO_2Me
NHAc

(95% ee , S)

Yamaguchi, M.; Shima, T.; Yamagishi, T.; Hida, M. *Tetrahedron Lett.*, ***1990***, *31*, 5049.

O
N
Boc
CO_2t-Bu
OH , KCN
THF , RT , 24 h
CO_2t-Bu
t-BuO_2C
NHBoc
77%

Schoenfelder, A.; Mann, A. *Synth. Commun.*, ***1990***, *20*, 2585.

OH, CO_2t-Bu

1. 2 LDA , THF, -78°C → -50°C

2. SPh, N. CO_2Me, 2 eq. LiCl

OH $NHCO_2Me$, SPh, CO_2t-Bu + OH $NHCO_2Me$, SPh, CO_2t-Bu

(96 : 4) 85%

Hatanaka, M.; Park, O-S.; Ueda, I. *Tetrahedron Lett.*, ***1990***, *31*, 7631.

Me, Me, MeO, MeO, CO_2Et

1. Ph N-Ts , $TiCl_4$, CH_2Cl_2 , RT

2. EtOH

Me, Me, O, CO_2Et, N, Ts, Ph + Me, Me, O, CO_2Et, N, Ts, Ph

(98 : 2) 81%

Saigo, K.; Shimada, S.; Hasegawa, M. *Chem. Lett.*, ***1990***, 905.

O, N, CO_2H, CO_2Bn

mCPBA , DCC , DMF, 25°C , 1 h

0% ester with free lactam

O, N, O, CO_2Bn, O, Cl

42%

Shiozaki, M. *Synthesis*, ***1990***, 691.

O, NHBoc

K $CH(CO_2Me)_2$, 110°C

$Mn(OAc)_3$, AcOH, 10 min

O, O, O, NHBoc

47%

del Rosario-Chow, M.; Ungwitayatorn, J.; Currie, B.L. *Tetrahedron Lett.*, ***1991***, *32*, 1011.

O, N, H

1. NaOAc , $OsCl_3$, 30% $MeCO_3H$, AcOH , EtOAc, RT

2. 5% Na_2SO_3

OAc, O, N, H

78%

Murahashi, S.; Saito, T.; Naota, T.; Kumobayashi, H.; Akutagawa, S. *Tetrahedron Lett.*, ***1991***, *32*, 2145.

Me, OH, CbzHN, CO_2Me

1. $Pb(OAc)_4$, EtOAc , MS 4Å

2. DABCO

3. EtSH

SEt, CbzHN, CO_2Me

86%

Apitz, G.; Steglich, W. *Tetrahedron Lett.*, ***1991***, *32*, 3163.

1. $RuCl_3 \cdot 3H_2O$, NaOAc, AcOH, MeCHO, EtOAc, 40°C, 3 h
2. 10% aq. Na_2SO_3

Me, N, H, O → Me, OAc, N, H, O 72%

Murahashi, S-I.; Saito, T.; Naota, T.; Kumobayashi, H.; Akutagawa, S. *Tetrahedron Lett.*, ***1991***, *32*, 5991.

Boc-Trp-COOH + H_2N-Trp-OMe → Boc-Trp-Trp-OMe 96%

DMF, iPr_2NEt, RT, 1.5 h

new peptide coupling reagent

O, Ph-P, Ph, O, F, F, F, F, F

Chen, S.; Xu, J. *Tetrahedron Lett.*, ***1991***, *32*, 6711.

$PdCl_2$, $CuCl_2$, NaOAc, CO, MeOH-AcOH, 30°C, 87 h

O, NHTs, O → CO_2Me, O, N-Ts, O 89%

(1:7 cis:trans)

Tamaru, Y.; Tanigawa, H.; Itoh, S.; Kimura, M.; Tanaka, S.; Fugami, K.; Sekiyama, T.; Yoshida, Z. *Tetrahedron Lett.*, ***1992***, *33*, 631.

[acetophenone/LDA] THF

O, N, Boc, CO_2Et → OH, O, Ph, CO_2Et, NHBoc 78%

Ezquerra, J.; de Mendoza, J.; Pedregal, C.; Ramírez, C. *Tetrahedron Lett.*, ***1992***, *33*, 5589.

N·Boc; $Rh_2(Ooctyl)_4$, pentane

O, O, CO_2Et, N_2 → Boc, N, O, O, CO_2Et 82%

(66% dr, 1R)

Davies, H.M.L.; Huby, N.J.S. *Tetrahedron Lett.*, ***1992***, *33*, 6935.

1. "$NaPhSeBH_3$", heat
2. H^+, CH_2N_2

O, NHBz, O, $OSiPh_2t$-Bu → BzHN, MeO_2C, $OSiPh_2t$-Bu, SePh 85%

Pedersen, M.L.; Berkowitz, D.B. *Tetrahedron Lett.*, ***1992***, *33*, 7315.

1. Tf_2O, 2,6-di-t-Bu-4-Me-pyridine, CH_2Cl_2, -10°C
2. RT, cyclohexene
3. H_2O, CCl_4, RT
4. mCPBA, $NaHCO_3$, CH_2Cl_2, 0°C

Ts(Me)N ... → N(Me)Ts 75% (93% ee)

Genicot, C.; Ghosez, L. *Tetrahedron Lett.*, ***1992***, *33*, 7357.

1. $OSiMe_3$ / $OSiMe_3$, -78°C, $TiBr_4$, CH_2Cl_2
2. H_2O
3. CH_2N_2

Ph, N. Ts → Ph, Me, CO_2Me, NHTs + Ph, Me, CO_2Me, NHTs (92 : 8) 88%

Shimada, S.; Saigo, K.; Abe, M.; Sudo, A.; Hasegawa, A. *Chem. Lett.*, ***1992***, 1445.
Shiozaki, M. *Synthesis*, ***1990***, 691.

CHO → MeO·P(O)(MeO)–CH(CO_2Me)(NHCbz), DBU, CH_2Cl_2, -20°C → RT → CO_2Me, NHCbz quant. (3:97 E:Z)

Schmidt, U.; Griesser, H.; Leitenberger, V.; Lieberknecht, A.; Mangold, R.; Meyer, R.; Riedl, B. *Synthesis*, ***1992***, 487.

CO_2Me, Cl–C–NHCHO, Me → 2 eq. cyclopentenyl-$SiMe_3$, TMSOTf, 4 h → Me, CO_2Me, NHCHO 84%

Roos, E.C.; Hiemstra, H.; Speckamp, W.N.; Kaptein, B.; Kamphuis, J.; Schoemaker, H.E. *SynLett*, ***1992***, 451.

Related Methods: Section 315 (Carboxylic Acid - Amide). Section 316 (Carboxylic Acid - Amine). Section 351 (Amine - Ester)

SECTION 345: AMIDE - ETHER, EPOXIDE, THIOETHER

NHBoc → 4% $RhH(PPh_3)_4$, MeOH, sealed tube, 100°C → NHBoc, OMe 84%

Nemoto, H.; Jimenez, H.N.; Yamamoto, Y. *J. Chem. Soc., Chem. Commun.*, ***1990***, 1304.

1. NBS, hv, 15 min, CCl_4, CH_2Cl_2; 2. EtOH, 2,6-lutidine

41%

Easton, C.J.; Pitt, M.J. *Tetrahedron Lett.*, ***1990***, *31*, 3471.

HBF_4, H_2NCO_2Et, 45°C, 10 h

N-SePh

PhSe ... $NHCO_2Et$ + EtO_2CHN ... SePh

n-$C_{10}H_{21}$

61% 25%

Francisco, C.G.; Hernández, R.; León, E.I.; Salazar, J.A.; Suárez, E. *J. Chem. Soc., Perkin Trans. I*, ***1990***, 2417.

PhSH, hv

54% (trans) 12% (2:1 cis:trans)

Naito, T.; Honda, Y.; Miyata, O.; Nonomiya, I. *Heterocycles*, ***1991***, *32*, 2319.

2% $VO(dpn)_2$, O_2, DCE, iBuCHO, RT, 12 h

87%

$VO(dpm)_2$ = bis-(dipivaloylmethanato)oxovanadium (IV)

Inoki, S.; Takai, T.; Yamada, T.; Mukaiyama, T. *Chem. Lett.*, ***1991***, 893.

$TolSo_2NH_2$, PhSeCl, DCE; cat. $ZnCl_2$/ether, 24 h

SePh, $NHSO_2Tol$ 85%

Toshimitsu, A.; Kusumoto, T.; Oida, T.; Tanimoto, S. *Bull. Chem. Soc. Jpn.*, ***1991***, *64*, 2148.

$OSiMe_3$ → Me_2NCHO, N_3CO_2Et, heat

CHO, N, Me 38%

Mitani, M.; Watanabe, K.; Tachizawa, O.; Koyama, K. *Chem. Lett.*, ***1992***, 813.

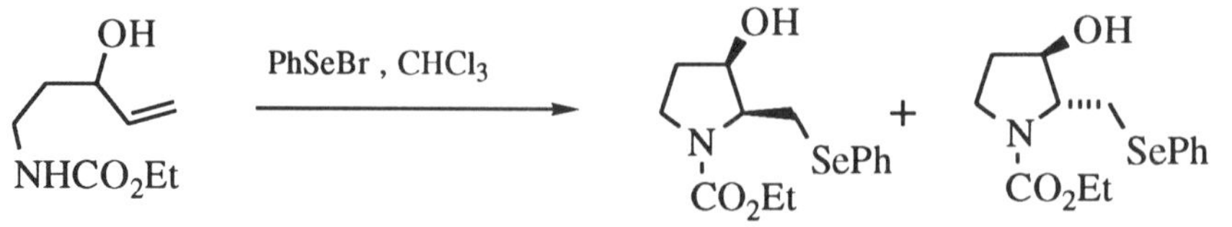

OH

NHCO$_2$Et

PhSeBr , CHCl$_3$

OH N CO$_2$Et SePh + OH N CO$_2$Et SePh

(87 : 13) 95%

Cooper, M.A.; Ward, A.D. *Tetrahedron Lett.*, ***1992***, *33*, 5999.

SECTION 346: AMIDE - HALIDE, SULFONATE

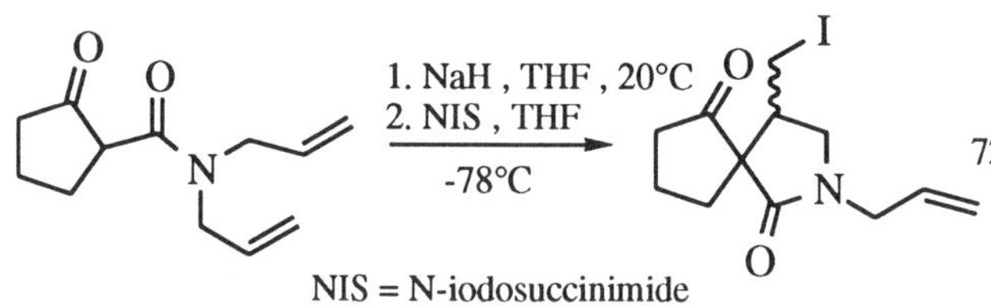

1. NaH , THF , 20°C
2. NIS , THF
-78°C

72%

NIS = N-iodosuccinimide

Cossy, J.; Thellend, A. *Tetrahedron Lett.*, ***1990***, *31*, 1427.

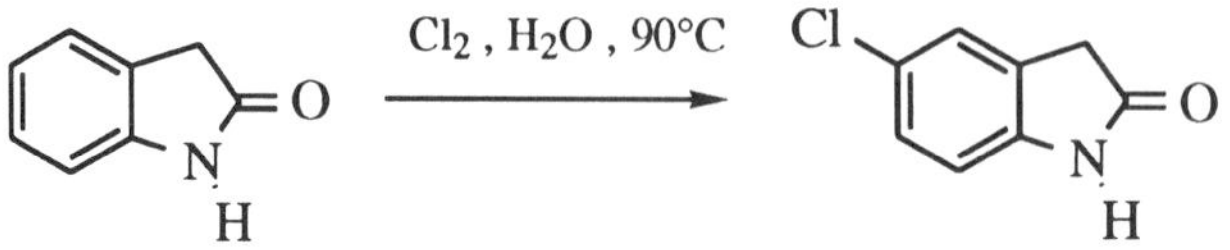

Cl$_2$, H$_2$O , 90°C

39%

Pranc, P. *Org. Prep. Proceed. Int.*, ***1990***, *22*, 104.

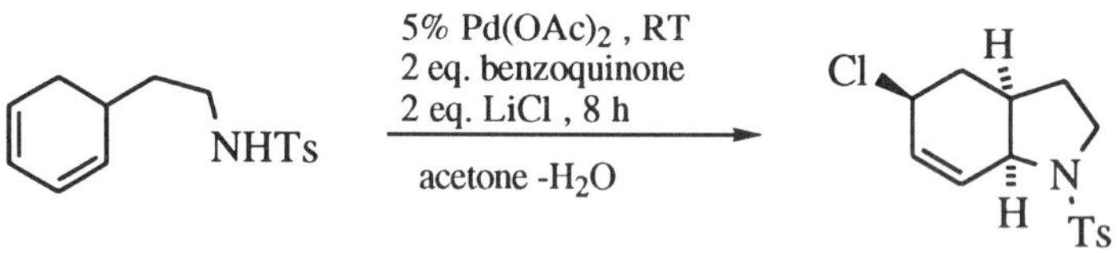

NHTs

5% Pd(OAc)$_2$, RT
2 eq. benzoquinone
2 eq. LiCl , 8 h
acetone -H$_2$O

90%

(>98% cis)

Bäckvall, J-E.; Andersson, P.G. *J. Am. Chem. Soc.*, ***1990***, *112*, 3683.

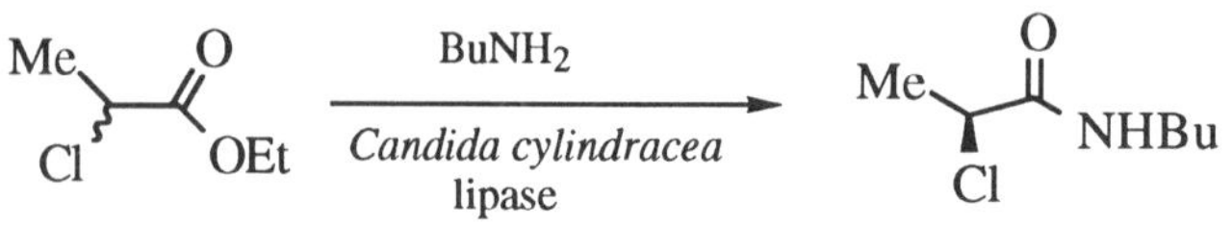

Me, Cl, OEt

BuNH$_2$
Candida cylindracea
lipase

Me, Cl, NHBu

62%

(95% ee)

Gotor, V.; Brieva, R.; González, C.; Rebolledo, F. *Tetrahedron*, ***1991***, *47*, 9207.
Brieva, R.; Rebolledo, F.; Gotor, V. *J. Chem. Soc., Chem. Commun.*, ***1990***, 1386 [with subtilisin].

O
F
Me
OTMP

1. $LiN(SiMe_3)_2$
2. Ph-HC=N-(C6H4)-OMe

F Ph
Me
O N
+
F Ph
Me
O N
OMe OMe

(1 : 3.5) 68%

Araki, K.; Wichtowski, J.A.; Welch, J.T. *Tetrahedron Lett.*, ***1991***, *32*, 5461.

O
O N
Ph Me
O
Bu

1. LDA
2. O_2 S N · F S O_2

O O
O N
Bu
F
Ph Me
(97% de)

88%

Davis, F.A.; Han, W. *Tetrahedron Lett.*, ***1992***, *33*, 1153.

Ph
O
NEt_2

I_2, s-collidine, RT
CH_2Cl_2, 18 h

Ph
O
NEt_2
I
84%

Kitagawa, O.; Hanano, T.; Hirata, T.; Inoue, T.; Taguchi, T. *Tetrahedron Lett.*, ***1992***, *33*, 1299.

Me
MeS
N
Bn

I_2, THF

Me
O N
Bn I
56%
+
Me
O N
Bn I
4%

Takahata, H.; Wang, E-C.; Ikuro, K.; Yamazaki, T.; Momose, T. *Heterocycles*, ***1992***, *34*, 435.

SECTION 347: AMIDE - KETONE

OMe
$Si(iPr)_3$
N

1. ROCOCl
2. MeMgCl, -78°C
PhMe/THF
3. H_3O^+

O
Cl
Me N
CO_2R
major
+
O
Cl
Me N
CO_2R
minor

R = 8-phenylmenthol

92% (91% de)

Comins, D.L.; Guehring, R.R.; Joseph, S.P.; O'Connor, S. *J. Org. Chem.*, ***1990***, *55*, 2574.

1. $CH_2=CH-C(O)-C(O)-CO_2t\text{-Bu}$
MeOH , CH_2Cl_2
2. pyridinium *p*-toluenesulfonate
CH_2Cl_2, reflux , 1 h

Wasserman, H.H.; Henke, S.L.; Luce, P.; Nakanishi, E.; Schulte, G. *J. Org. Chem.*, ***1990***, *55*, 5821.

1. Bn-N=C=O , NaH ,THF
2. H_2O

75%

Ognyanov, V.I.; Hesse, M. *Helv. Chim. Acta*, ***1990***, *73*, 272.

1. Ac•BF_4•MeCN
2. aq. NaOH

30-40%

Gridnev, I.D.; Balenkova, E.S. *Zhur. Org. Khim.*, ***1990***, *26*, 46 (Engl., p. 37).

MeO–CH_2–N(Cbz)(*t*-Bu)
TMSOTf

85%

Earle, M.J.; Fairhurst, R.A.; Heaney, H.; Papageorgiou, G. *SynLett*, ***1990***, 621.

1. 20% HCl
2. MsCl , $Bu_4N\ HSO_4$
$H_2O/CHCl_3$

54%

Buchholz, R.; Hoffmann, H.M.R. *Helv. Chim. Acta*, ***1991***, *74*, 1213.

$PhI(O_2CCF_3)_2$, MeOH
RT , 10 min

76%

Kita, Y.; Tohma, H.; Kikuchi, K.; Inagaki, M.; Yakura, T. *J. Org. Chem.*, ***1991***, *56*, 435.

1. cyclohexyl-CHO, LDA, THF, 0°C, 1 h
2. , Py, CH_2Cl_2, RT, 2 h
I(OAc)$_3$
SPh
70%

Linde, R.G. II; Deroncic, L.O.; Danishefsky, S.J. *J. Org. Chem.*, ***1991***, *56*, 2534.

$Rh_2(OAc)_4$, $CHCl_3$
25°C, 45 min
Me
CHN_2
N Ph
N-Ph
79%

Padwa, A.; Chinn, R.L.; Hornbuckle, S.F.; Zhang, Z.J. *J. Org. Chem.*, ***1991***, *56*, 3271.

1. $TiCl_4$, CH_2Cl_2
iPr$_2$NEt, 0°C, 1 h
2.
-25°C, 0.75 h
Bn
Me
Bn
88%
(>99:1)

Evans, D.A.; Bilodeau, M.T.; Somers, T.C.; Clardy, J.; Cherry, D.; Kato, Y. *J. Org. Chem.*, ***1991***, *56*, 5750.

hν (λ = 313 nm), MeCN
(10^{-2} M), 75 min
Bn
N
Bz
Ph
N-Bz
70%

Le Blanc, S.; Pete, J-P.; Piva, O. *Tetrahedron Lett.*,***1992***, *33*, 1993.

230°C, 5 h
Me
Me
(85 : 15) 95%

Cossy, J.; Bouzide, A. *Tetrahedron Lett.*, ***1992***, *33*, 2505.

1. NEt_3, Me_3SiCl NaI, MeCN
2. $MeNHSO_2Cl$, NEt_3 MeCN

OSiMe$_3$ → SO_2NHMe 65%

Vega, J.A.; Molina, A.; Alajarín, R.; Vaquero, J.J.; García Navío, J.; Alvarez-Builla, J. *Tetrahedron Lett.*, ***1992***, *33*, 3677.

Bz_2O, DMF, 55°C, 3 h

Ph–CO–CH_2NH_2•HCl → Ph–CO–CH_2NHBz 82%

Shipton, M.R. *SynLett*, ***1992***, 491.

SECTION 348: AMIDE - NITRILE

80°C, 17 h

NC, N, Ac, Ph → Ph, Ph, NC, N, Me, Ac + Ph, Ph, NC, N, Me, Ac

(5 : 1) 80%

Teng, M.; Fowler, F.W. *J. Org. Chem.*, ***1990***, *55*, 5646.

1. LiNHBn, THF, -70°C
2. EtOH

NC–CH_2–CO_2Et → NC–CH_2–CO–NHBn 75%

Bhawal, B.M.; Khanapure, S.P.; Biehl, E.R. *Synth. Commun.*, ***1990***, *20*, 3235.

SECTION 349: AMIDE - ALKENE

3 eq. PhO–CH_2–COCl
NEt_3, CH_2Cl_2 25°C, 12h

Ph–CH=N–CH=CH_2 → PhO, Ph, O, N (93%) + PhO, Ph, O, N (3%)

Georg, G.I.; He, P.; Kant, J.; Mudd, J. *Tetrahedron Lett.*, ***1990***, *31*, 451.

(epoxide), PPh_3
HCO_2Ac

N-STol → NHCHO 76%

Baldwin, J.E.; Aldous, D.J.; O'Neil, I.A. *Tetrahedron Lett.*, ***1990***, *31*, 2051.

1. LDA , THF -78°C; 2. xylene , 4h 135°C

NH*n*-Bu + NH*n*-Bu

(99.5 : 0.5) 92%

Tsunoda, T.; Sasaki, O.; Itô, S *Tetrahedron Lett*, ***1990***, *31*, 727.

2 H-C≡C-C(Me)(Me)OH → CO_2 , $BuNH_2$; PBu_3 , 110°C → 56%

Fournier, J.; Bruneau, C.; Dixneuf, P.H. *Tetrahedron Lett.*, ***1990***, *31*, 1721.

1. Ac_2O *; 2. LDA; 3. 85% H_3PO_4 PhMe , 60°C

n-C_6H_{13} 85%

* see *Tetrahedron Lett.*, ***1988***, *29*, 4855

Barluenga, J.; Tomás, M.; Suárez-Sobrino, A.; Rubio, E. *Tetrahedron Lett.*, ***1990***, *31*, 2189.

PhCHO , PPh_3 iPrOH , reflux 48 h

71% (2:1 E:Z)

Dellaria, J.F. Jr.; Sallin, K.J. *Tetrahedron Lett.*, ***1990***, *31*, 2661.

1. $LiN(TMS)_2$; TMSCl THF , 0°C; 2. ; 3. xylene, reflux , 20 h

57%

Uyehara, T.; Suzuki, I.; Yamamoto, Y. *Tetrahedron Lett.*, ***1990***, *31*, 3753.

p-Tol-N=C=O dioxane 110°C

76%

Liebeskind, L.S.; Johnson, S.A.; McCallum, J.S. *Tetrahedron Lett.*, ***1990***, *31*, 4397.

NHBoc
BnO OBz

1. t-$BuMe_2SiOTf$
2,6-lutidine
2. TBAF, THF, RT
10% $Pd(PPh_3)_4$, 1 h

H. N O O
BnO
78%
(9:1 syn:anti)

Spears, G.W.; Nakanishi, K.; Ohfune, Y. *Tetrahedron Lett.*, ***1990***, *31*, 5339.

O N I

Bu_3SnH, PhH
AIBN, reflux
8 h

O N
84%

Knapp, S.; Gibson, F.S.; Choe, Y.H. *Tetrahedron Lett.*, ***1990***, *31*, 5397.

N OMe
CO_2t-Bu

BuMgBr

N Bu
CO_2t-Bu
+
Bu
N
CO_2t-Bu

+ BF_3•OEt_2	(100	:	0)	63%
+ CuBr	(0	:	100)	77%

Shono, T.; Terauchi, J.; Ohki, Y.; Matsumura, Y. *Tetrahedron Lett.*, ***1990***, *31*, 6385.

BocHN CHO

1. $(CH_2=CH)_2ZnCl$
2. Ac_2O
3. $H_2C(CO_2Et)_2$, NaH
$Pd(PPh_3)_4$, THF

BocHN CO_2Et CO_2Et
87%

Thompson, W.J.; Tucker, T.J.; Schwering, J.E.; Barnes, J.L. *Tetrahedron Lett.*, ***1990***, *31*, 6819.

H N CCl_3
Ph O

1. THF, KH, -78°C
Br
2. MeCN, 105°C
12 h

O N CCl_3
Ph
70% (4 diastereomers)
51:31:12:7 syn ax.:anti ax.:anti:eq.:syn eq.

Guy, A.; Graillot, Y. *Tetrahedron Lett.*, ***1990***, *31*, 7315.

1. $RuCl_2(PPh_3)_3$ PhH, *t*-BuOOH
2. Na bisulfite
3. $TiCl_4$, CH_2Cl_2 -78°C 4 h, allyl-$SiMe_3$

N-CO_2Me ... 89%

Naota, T.; Nakato, T.; Murahashi, S. *Tetrahedron Lett.*, ***1990***, *31*, 7475.

PhI, NEt_3, 3% $Pd(OAc)_2$

N-CO_2Me → Ph, N-CO_2Me

Nilsson, K,; Hallberg, A. *J. Org. Chem.*, ***1990***, *55*, 2464.

OH, N-CO_2Et; allyl-$SiMe_3$, $SnCl_4$; CH_2Cl_2, -78°C, 30 min

N-CO_2Et + N-CO_2Et (50 : 1) 90%

Kozikowski, A.P.; Park, P. *J. Org. Chem.*, ***1990***, *55*, 4668.

NHTs, I; vinylcyclopropane, DMF; KOAc, 80°C, 4 d, Bu_4Cl

Ts-N ... 77%

Larock, R.C.; Yum, E.K. *SynLett*, ***1990***, 529.

PhN_3
1. PPh_3, PhMe, RT
2. adipoyl chloride, PhMe
reflux, 3 h

O, N−Ph, Cl 45%

Aubert, T.; Farnier, M.; Guilard, R. *Synthesis*, ***1990***, 149.

1. Ac•BF_4•MeCN
2. MeCN
3. aq. NaOH

NHAc 60-80%

Gridnev, I.D.; Balenkova, E.S. *Zhur. Org. Khim.*, ***1990***, *26*, 50 (Engl., p. 40).

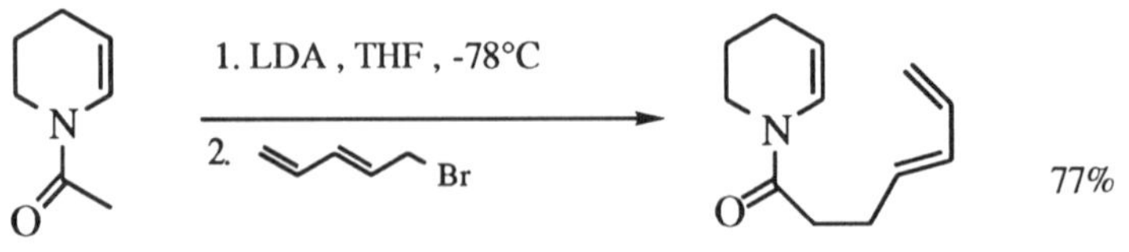

1. LDA, THF, -78°C
2. Br

77%

Diels-Alder reported previously

Chockalingam, K.; Harirchian, B.; Bauld, N.L. *Synth. Commun.*, ***1990***, *20*, 189.

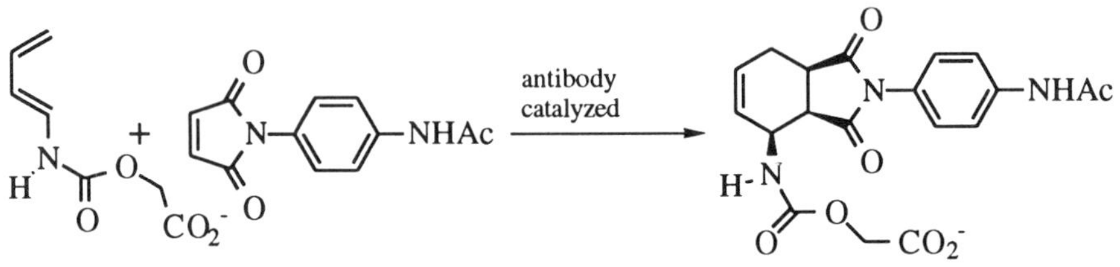

antibody catalyzed

NHAc

Braisted, A.C.; Schultz, P.G. *J. Am. Chem. Soc.*, ***1990***, *112*, 7430.

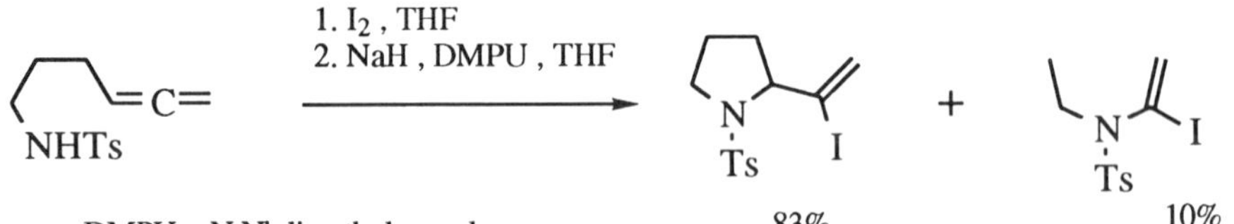

1. I_2, THF
2. NaH, DMPU, THF

83% 10%

DMPU = N,N'-dimethylpropylene urea

Shaw, R.; Anderson, M.; Gallagher, T. *SynLett*, ***1990***, 584.

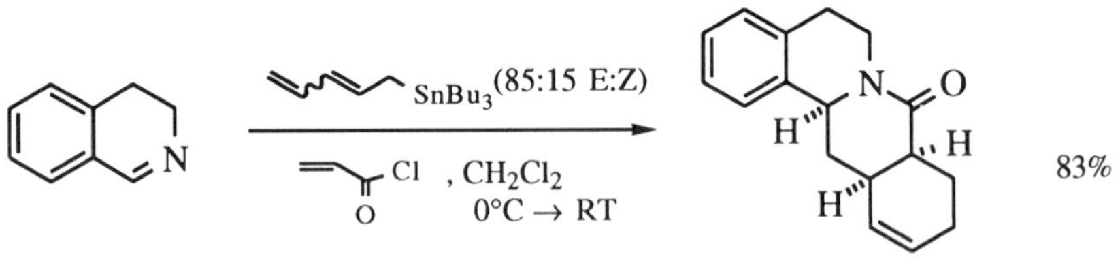

$SnBu_3$ (85:15 E:Z)

, CH_2Cl_2
0°C → RT

83%

Yamaguchi, R.; Hamasaki, T.; Utimoto, K.; Kozima, S.; Takaya, H. *Chem. Lett.*, ***1990***, 2161.

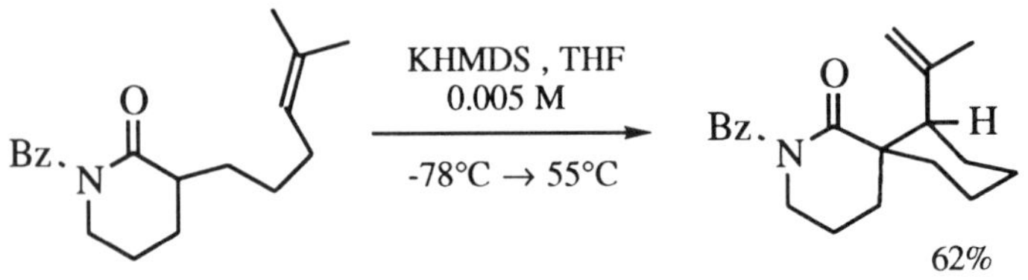

KHMDS, THF
0.005 M
-78°C → 55°C

62%

Kim, D.; Kim, H.S.; Yoo, J.Y. *Tetrahedron Lett.*, ***1991***, *32*, 1577.

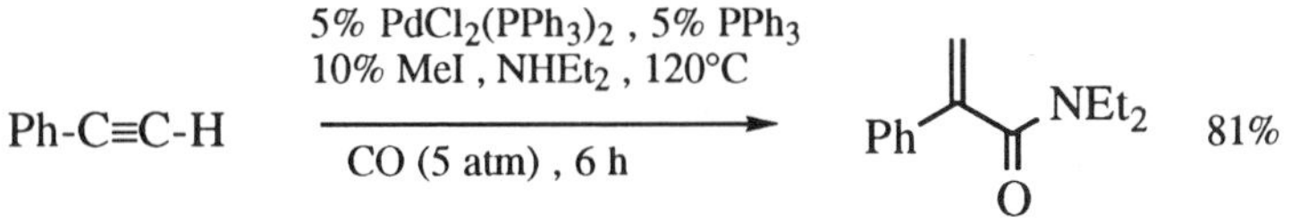

Ph-C≡C-H → 5% $PdCl_2(PPh_3)_2$, 5% PPh_3, 10% MeI , $NHEt_2$, 120°C, CO (5 atm) , 6 h → Ph, NEt_2, O 81%

Torii, S.; Okumoto, H.; Sadakane, M.; Xu, L.H. *Chem. Lett.*, ***1991***, 1673.

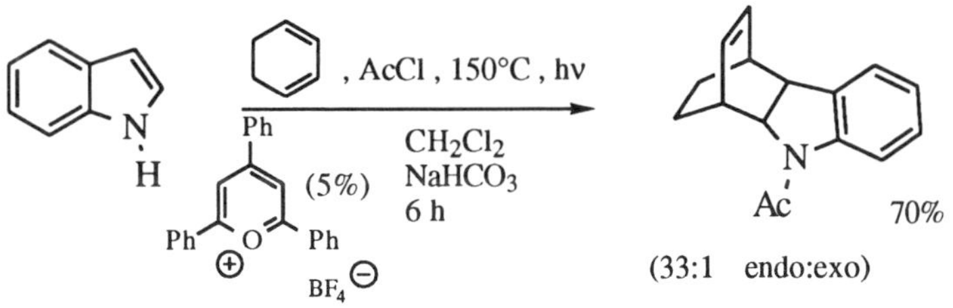

N, H; , AcCl , 150°C , hν; Ph, Ph, O, Ph (5%); CH_2Cl_2, $NaHCO_3$, 6 h; $BF_4^{\ominus}$ → N, Ac 70% (33:1 endo:exo)

Giesler, A.; Steckhan, E.; Wiest, O.; Knoch, F. *J. Org. Chem.*, ***1991***, *56*, 1405.

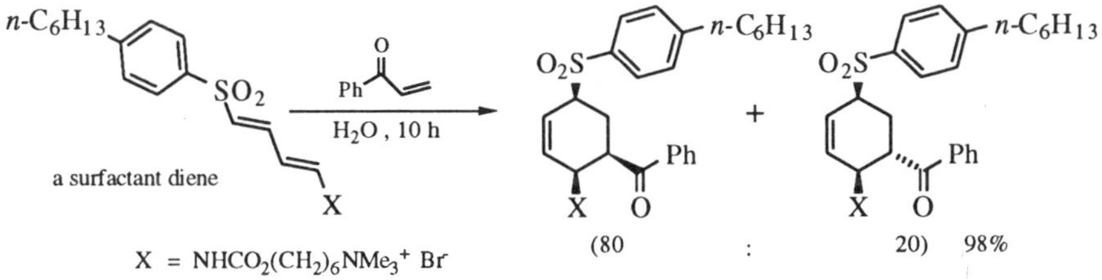

n-C_6H_{13}, SO_2, a surfactant diene, X; Ph, O; H_2O , 10 h → O_2S, n-C_6H_{13}, Ph, X, O + O_2S, n-C_6H_{13}, Ph, X, O (80 : 20) 98%

X = $NHCO_2(CH_2)_6NMe_3^+$ Br^-

Jaeger, D.A.; Shinozaki, H.; Goodson, P.A. *J. Org. Chem.*, ***1991***, *56*, 2482.

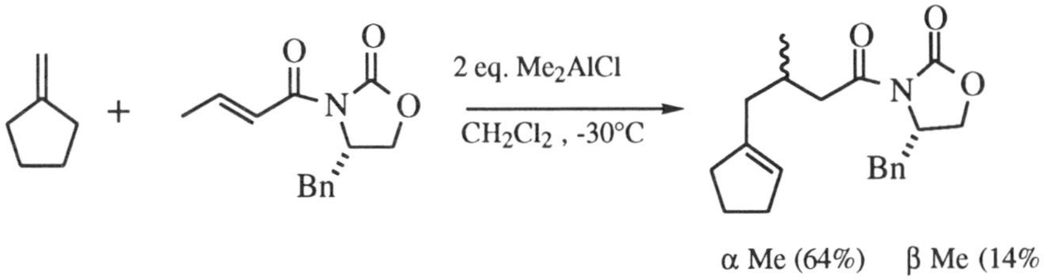

+ O, O, N, O, Bn; 2 eq. Me_2AlCl, CH_2Cl_2 , -30°C → O, O, N, O, Bn

α Me (64%) β Me (14%

Snider, B.B.; Zhang, Q. *J. Org. Chem.*, ***1991***, *56*, 4908.

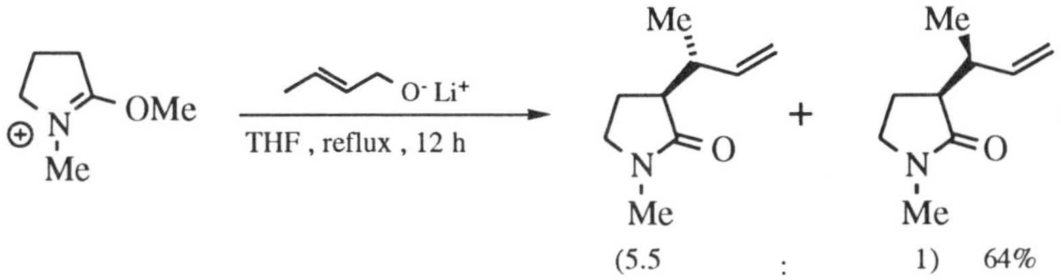

N, OMe, Me; O^- Li^+; THF , reflux , 12 h → Me, N, O, Me + Me, N, O, Me (5.5 : 1) 64%

Coates, B.; Montgomery, D.; Stevenson, P.J. *Tetrahedron Lett.*, ***1991***, *32*, 4199.

H_2N–CH(Me)Ph, xylene, reflux — 76% (>95:5 E:Z)

Janecki, T.; Bodalski, R. *Tetrahedron Lett.*, ***1991***, *32*, 6231.

5 eq. Ac_2O — 60%

Roos, E.C.; Bernabé, P.; Hiemstra, H.; Speckamp, W.N. *Tetrahedron Lett.*, ***1991***, *32*, 6633.

H_2N C≡C-H; *t*-BuMe$_2$Si-H, CO, $Rh_4(CO)_{12}$, DBU, PhH, 100°C; *t*-BuMe$_2$Si — 64%

Matsuda, I.; Sakakibara, J.; Nagashima, H. *Tetrahedron Lett.*, ***1991***, *32*, 7431.

CN; $RuH_2(PPh_3)_4$, DME, H_2O, 120°C, 24 h — 94%

Murahashi, S.; Sasao, S.; Saito, E.; Naota, T. *J. Org. Chem.*, ***1992***, *57*, 2521.

C=CH$_2$, NHTs; $PdCl_2(PhCN)_2$, THF, NEt_3, RT, allyl Cl; N-Ts — 53%

Kimura, M.; Fugami, K.; Tanaka, S.; Tamaru, Y. *J. Org. Chem.*, ***1992***, *57*. 6377.
Kimura, M.; Kure, S.; Yoshida, Z.; Tanaka, S.; Fugami, K.; Tamaru, Y. *Tetrahedron Lett.*, ***1990***, *31*, 4887 [with CuCl, NEt_3].
Kimura, M.; Fugami, K.; Tanaka, S.; Tamaru, Y. *Tetrahedron Lett.*, ***1991***, *32*, 6359 [with AgNCO].

O
NH_2
Me

PhCHO , PPE , 24 h
35°C

O
N·H
Ph
Me

63%

PPE - see *J. Org. Chem.*, ***1969***, *34*, 2665

Marson, C.M.; Grabowska, U.; Walsgrove, T. *J. Org. Chem.*, ***1992***, *57*, 5045.

Et
O
N·Ph
Me

5% $PdCl_2(PhCN)_2$
THF , 20°C , 16 h

Et
O
N-Ph
Me

79%

Metz, P.; Mues, C.; Schoop, A. *Tetrahedron*, ***1992***, *48*, 1071.

CO_2H
NHCbz

OTf , DMF , K_2CO_3
10% $Pd(OAc)_2$, Bu_4NCl
20% $P(o\text{-}Tol)_3$

CO_2H
NHCbz

77%

Criso, G.T.; Glink, P.T. *Tetrahedron*, ***1992***, *48*, 3541.

O O
EtO-P
EtO
N·OMe
Me

n-$C_6H_{13}CHO$, ether
80% NaH , 0°C → RT
1 h

n-C_6H_{13}
N·OMe
O Me

94%

(95:5 E:Z)

Netz, D.F.; Seidel, J.L. *Tetrahedron Lett.*, ***1992***, *33*, 1957.

O
OAc
$(CH_2)_8NHBoc$

1. TFA
2. NMI
3. hν (λ = 340 nm)
CH_2Cl_2

$(CH_2)_6$
OAc
N
H O

52%

Quinkert, G.; Nestler, H.P.; Schumacher, B.; del Grosso, M.; Dürner, G.; Bats, J.W. *Tetrahedron Lett.*, ***1992***, *33*, 1977.

O
N·H

CO_2Me , DME , O_2
5% $PdCl_2(MeCN)_2$
5% CuCl , 60°C

O
N
CO_2Me

85%

Hosokawa, T.; Takano, M.; Kuroki, Y.; Murahashi, S-I. *Tetrahedron Lett.*, ***1992***, *33*, 6643.

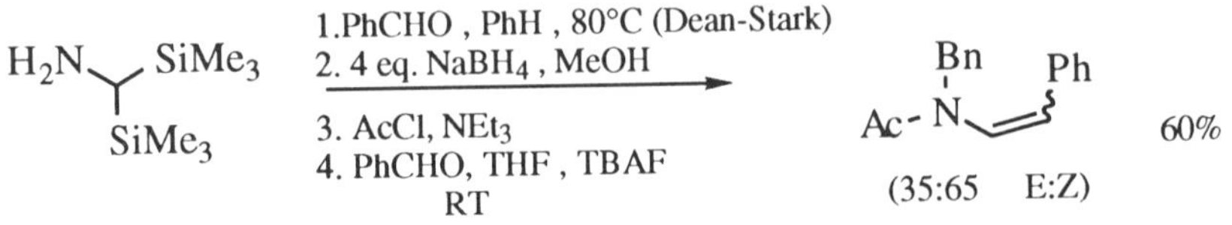

H_2N $SiMe_3$ / $SiMe_3$ → 1. PhCHO, PhH, 80°C (Dean-Stark); 2. 4 eq. $NaBH_4$, MeOH; 3. AcCl, NEt_3; 4. PhCHO, THF, TBAF, RT → Bn, Ph, Ac-N (35:65 E:Z) 60%

Palomo, C.; Aizpurua, J.M.; Legido, M.; Picard, J.P.; Dunogues, J.; Constantieux, T. *Tetrahedron Lett.*, ***1992***, *33*, 3903.

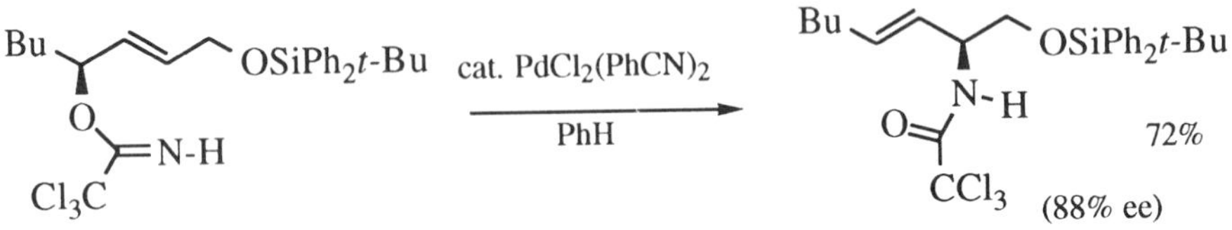

Bu, $OSiPh_2t$-Bu, O, N-H, Cl_3C → cat. $PdCl_2(PhCN)_2$, PhH → Bu, $OSiPh_2t$-Bu, N-H, O, CCl_3 72% (88% ee)

Mehmandoust, M.; Petit, Y.; Larchevêque, M. *Tetrahedron Lett.*, ***1992***, *33*, 4313.

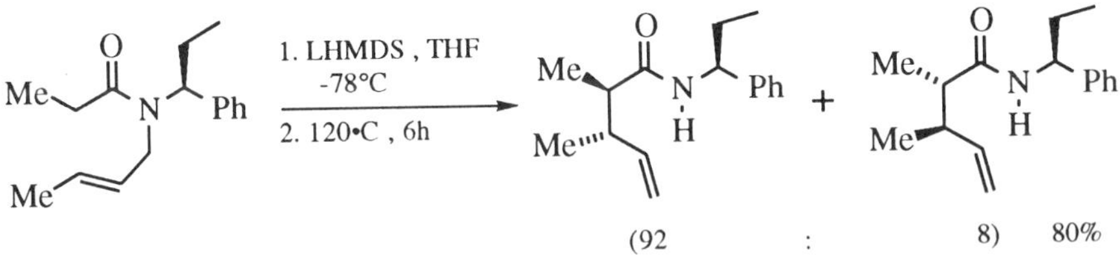

O, Me, N, Ph, Me → 1. LHMDS, THF, -78°C; 2. 120•C, 6h → O, Me, N, Ph, H, Me + O, Me, N, Ph, H, Me (92 : 8) 80%

Tsunoda, T.; Sakai, M.; Sasaki, O.; Sako, Y.; Hondo, Y.; Itô, S. *Tetrahedron Lett.*, ***1992***, *33*, 1651.

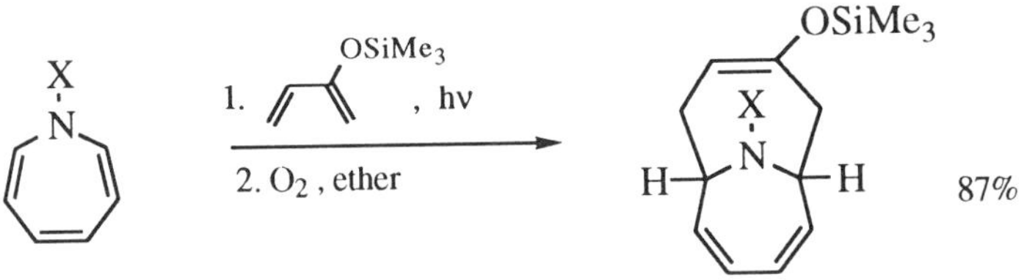

X, N → 1. $OSiMe_3$, hν; 2. O_2, ether → $OSiMe_3$, X, N, H, H 87%

X = CO_2Me

Rigby, J.H.; Ateeq, H.S.; Krueger, A.C. *Tetrahedron Lett.*, ***1992***, *33*, 5873.

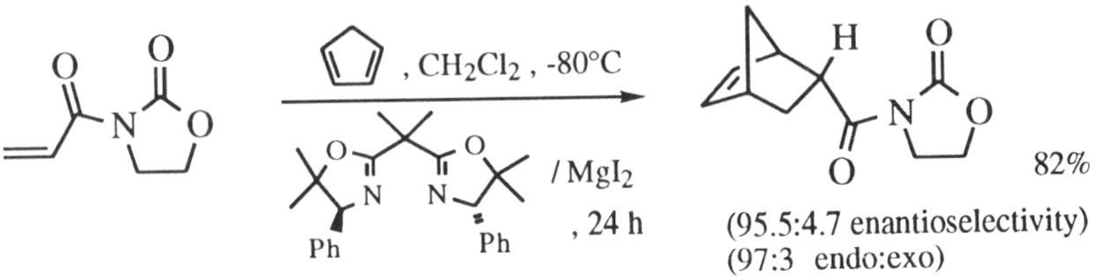

O, O, N, O → , CH_2Cl_2, -80°C; O, N, N, O, Ph, Ph / MgI_2, 24 h → H, O, N, O, O 82% (95.5:4.7 enantioselectivity) (97:3 endo:exo)

Corey, E.J.; Ishihara, K. *Tetrahedron Lett.*, ***1992***, *33*, 6807.

H-C≡C-CH2CH2-NHTs → 1. BuLi, THF 2. 5% $Pd(OAc)_2$, THF 10% PPh_3 3. 3 eq. PhI, 60°C → N-Ts 2-benzylidenepyrrolidine 57%

Luo, F-T.; Wang, R-T. *Tetrahedron Lett.*, ***1992***, *33*, 6835.

n-C_5H_{11}-C≡C-n-C_5H_{11} → 1. $TaCl_5$, Zn DME, PhH 2. THF 3. PhN=C=O 4. aq. NaOH → 80%

Kataoka, Y.; Oguchi, Y.; Yoshizumi, K.; Miwatashi, S.; Takai, K.; Utimoto, K. *Bull. Chem. Soc. Jpn.*, ***1992***, *65*, 1543.
Takai, K.; Kataoka, Y.; Yoshizumi, K.; Oguchi, Y.; Utimoto, K. *Chem. Lett.*, ***1991***, 1479.

NH_2 → 10% $Pd(OAc)_2$, THF 2 eq. $CuCl_2$, O_2, 60°C 24 h → 90%

Andersson, P.G.; Bäckvall, J-E. *J. Am. Chem. Soc.*, ***1992***, *114*, 8696.

PhCHO → Br-CH2-C(O)-N(piperidine), Bu_3As 25% $Pd(PPh_3)_4$ → 30%

Shen, Y.; Zhou, Y. *Synth. Commun.*, ***1992***, *22*, 657.
Zheng, J.; Wang, Z.; Shen, Y. *Synth. Commun.*, ***1992***, *22*, 1611 [with Zn, Bu_3P, 70%].

CHO → $Ph_3\overset{\oplus}{P}CH_2C_3H_7$ $Br^{\ominus}$ BuLi, THF → 52%

Smith, M.B.; Kwon, T.W. *Synth. Commun.*, ***1992***, *22*, 2865.

REVIEWS:

"N-Dienyl Amides and Lactams: Preparation and Diels-Alder Reactivity"
Smith, M.B. *Org. Prep. Proceed. Int.*, ***1990***, *22*, 315.

Also via Alkenyl Acids: Section 322 (Carboxylic Acid -Alkene)

SECTION 350: AMINE - AMINE

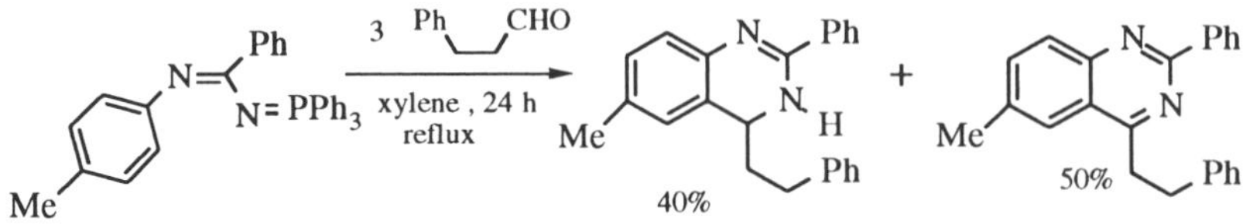

Ph; N=; N= PPh_3; Me; 3 Ph CHO; xylene , 24 h reflux; Me; N; Ph; N H; Ph; 40%; +; Me; N; Ph; N; 50%; Ph

Rossi, E.; Celentano, G.; Stradi, R.; Strada, A. *Tetrahedron Lett.*, ***1990***, *31*, 903.

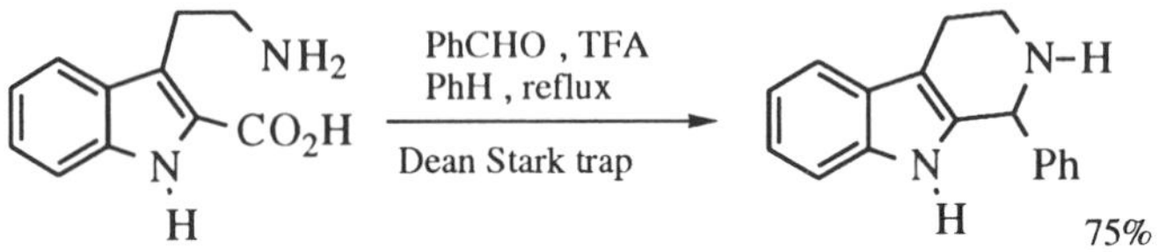

NH_2; CO_2H; N H; PhCHO , TFA; PhH , reflux; Dean Stark trap; N-H; N H; Ph; 75%

Narayanan, K.; Cook, J.M. *Tetrahedron Lett.*, ***1990***, *31*, 3389.

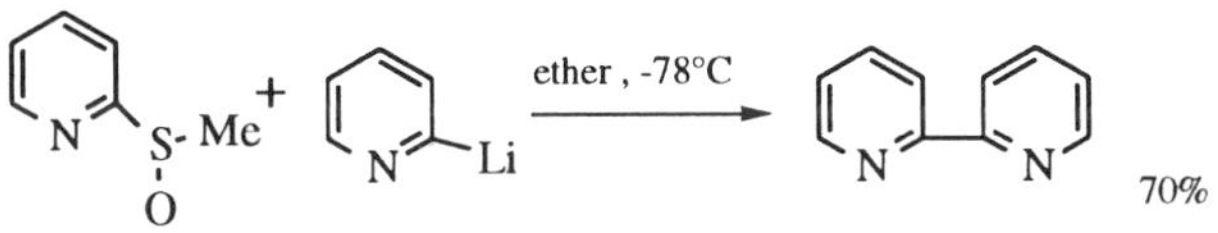

N; S- Me; O; +; N; Li; ether , -78°C; N; N; 70%

Uenishi, J.; Tanaka, T.; Wakabayashi, S.; Oae, S. *Tetrahedron Lett.*, ***1990***, *31*, 4625.

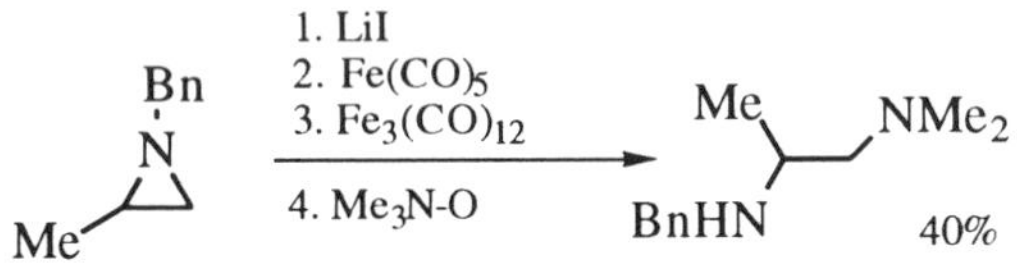

Bn; N; Me; 1. LiI; 2. $Fe(CO)_5$; 3. $Fe_3(CO)_{12}$; 4. Me_3N-O; Me; NMe_2; BnHN; 40%

Chamchaang, W.; Pinhas, A.R. *J. Org. Chem.*, ***1990***, *55*, 2531.

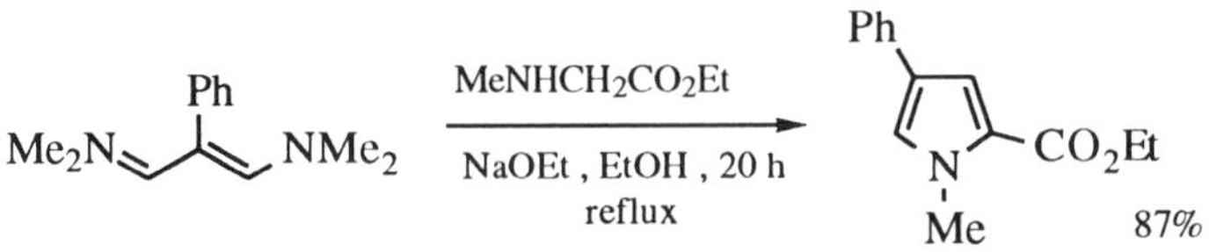

Ph; Me_2N; NMe_2; $MeNHCH_2CO_2Et$; NaOEt , EtOH , 20 h reflux; Ph; CO_2Et; N Me; 87%

Gupton, J.T.; Krolikowski, D.A.; Yu, R.H.; Riesinger, S.W.; Sikorski, J.A. *J. Org. Chem.*, ***1990***, *55*, 4735.

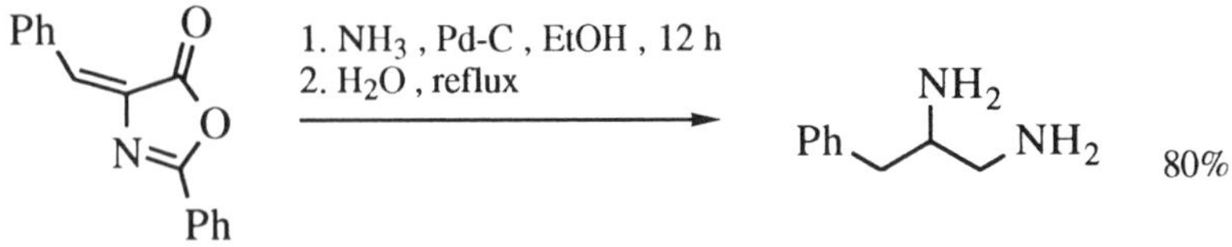

Ph; O; O; N; Ph; 1. NH_3 , Pd-C , EtOH , 12 h; 2. H_2O , reflux; NH_2; Ph; NH_2; 80%

Kidwai, M. *J. Indian Chem. Soc.*, ***1990***, *67*, 470.

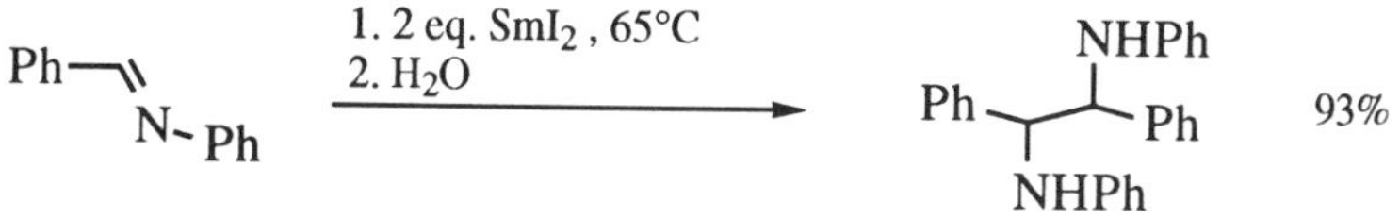

1. 2 eq. SmI_2, 65°C
2. H_2O

93%

Imamoto, T.; Nishimura, S. *Chem. Lett.*, ***1990***, 1141.

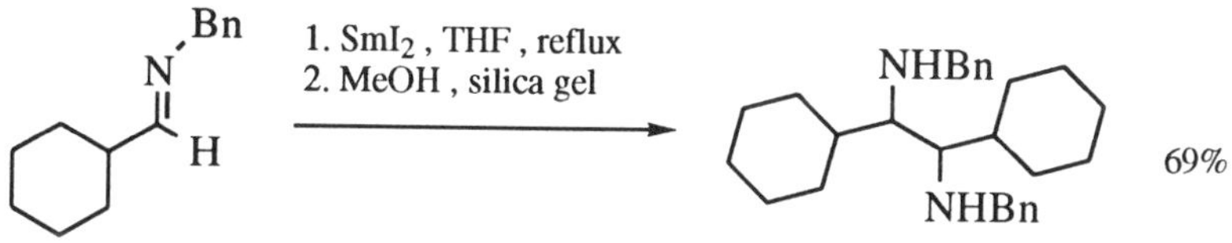

1. SmI_2, THF, reflux
2. MeOH, silica gel

69%

Enholm, E.J.; Forbes, D.C.; Holub, D.P. *Synth. Commun.*, ***1990***, *20*, 981.

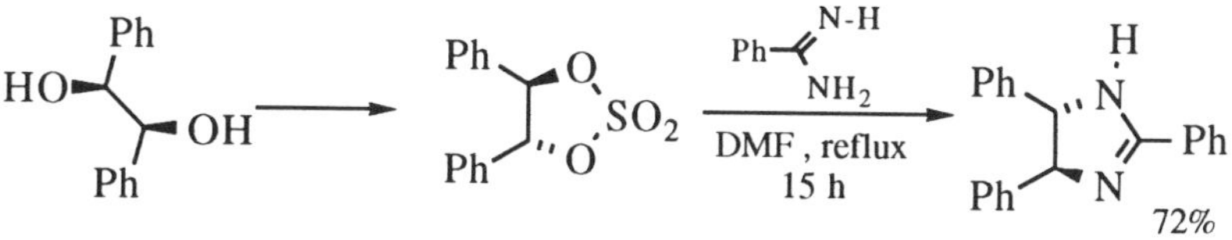

DMF, reflux
15 h

72%

Oi, R.; Sharpless, K.B. *Tetrahedron Lett.*, ***1991***, *32*, 999.

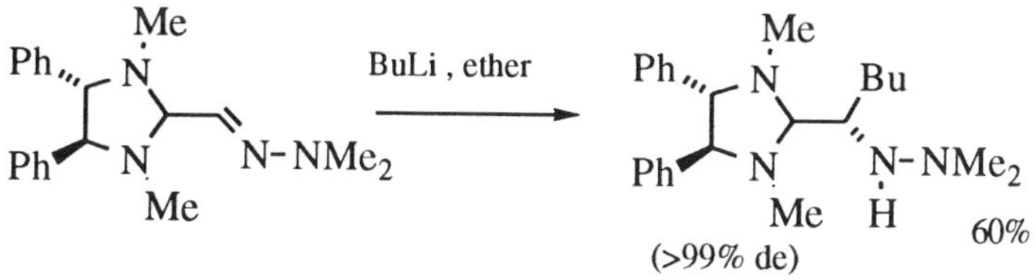

BuLi, ether

(>99% de)

60%

Alexakis, A.; Lensen, N.; Mangeney, P. *Tetrahedron Lett.*, ***1991***, *32*, 1171.

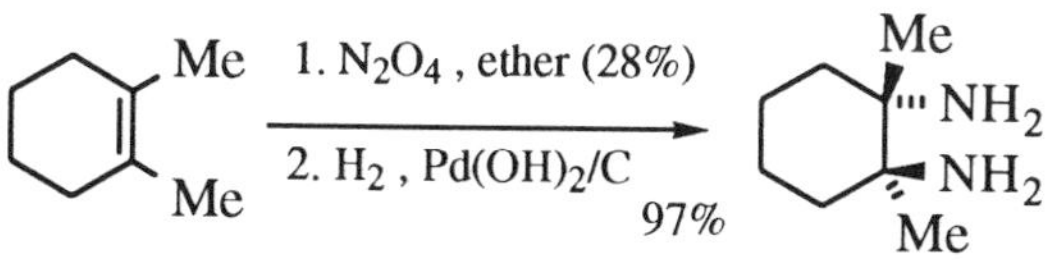

1. N_2O_4, ether (28%)
2. H_2, $Pd(OH)_2/C$

97%

Zhang, W.; Jacobsen, E.N. *Tetrahedron Lett.*, ***1991***, *32*, 1711.

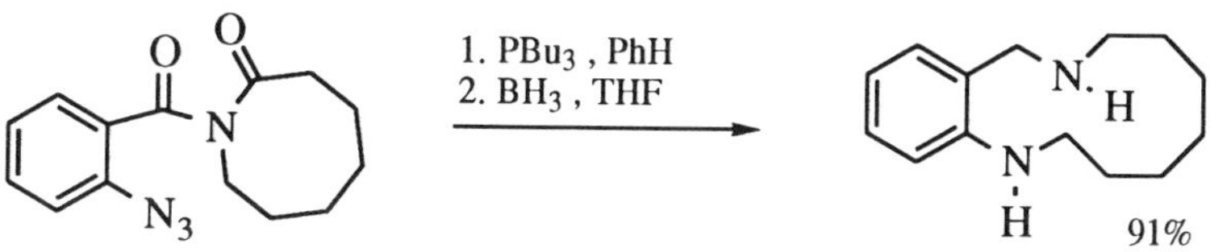

1. PBu_3, PhH
2. BH_3, THF

91%

Takeuchi, H.; Matsushita, Y.; Eguchi, S. *J. Org. Chem.*, ***1991***, *56*, 1535.

1. Sn, 12 N HCl
2. NaOH, CH_2Cl_2

NO_2 NO_2 → NH_2 NH_2 82%

Asaro, M.F.; Nakayama, I.; Wilson, R.B. Jr. *J. Org. Chem.*, ***1992***, *57*, 778.

O N=PPh_3

1. PhNCO, PhMe, reflux 30 min
2. N (enamine), PhMe, reflux, 20 h

O N NHPh 45%

Noguchi, M.; Onimura, K.; Isomura, Y.; Kajigaeshi, S. *J. Het. Chem.*, ***1991***, *28*, 885.

NH_2

1. oxalyl chloride, THF reflux
2. $NaBH_4$, TFA 25°C

PhHN NHPh 73%

Nutaitis, C.F. *Synth. Commun.*, ***1992***, *22*, 1081.

REVIEWS:

"The Chemistry of 1,3-Dioximes"
Kotali, A.; Papageorgiou, V.P. *Org. Prep. Proceed. Int.*, ***1991***, *23*, 595.

SECTION 351: AMINE - ESTER

SPh N CO_2Me CO_2Me

Bu_3SnH, AIBN
PhMe, 90°C, 8h

N CO_2Me CO_2Me

92% (35:65 cis:trans)

Esch, P.M.; Hiemstra, H.; Speckamp, W.N. *Tetrahedron Lett.*, ***1990***, *31*, 759.

MeO N CO_2Me CO_2Me

BuCu
BF_3

Bu N CO_2Me CO_2Me + Bu N CO_2Me CO_2Me

(96 : 4) 84%

Skrinjar, M.; Wistrand, L-G. *Tetrahedron Lett.*, ***1990***, *31*, 1775.

EtO_2C, CO_2Et aziridine, N-Ts → ($LiMe_2Cu$, ether, -78°C) → Me, CO_2Et, EtO_2C, NHTs 68%

Tanner, D.; Birgersson, C.; Dhaliwal, H.K. *Tetrahedron Lett.*, ***1990**, 31*, 1903.

CO_2Me, ONs → (Boc-$NHNH_2$, MeCN, reflux) → CO_2Me, NHNHBoc 80%

ONs = 2-nosyloxy - 2-(((4-nitrobenzene)sulfonyl)oxy)

Hoffman, R.V.; Kim, H-O. *Tetrahedron Lett.*, ***1990**, 31*, 2953.

SO_2NCy_2, O, N, Ph, Ph → (1. LDA, THF, -78°C; 2. DIOP (-), -50°C, 4h; Cl–allyl–Pd) → Cl, SO_2NCy_2, O, N, Ph, Ph 80% (99% de)

Genet, J.P.; Kopola, N.; Juge, S.; Ruiz-Montes, J.; Antunes, O.A.C.; Tanier, S. *Tetrahedron Lett.*, ***1990**, 31*, 3133.

N·SPh → (1. methyl acrylate; 2. Bu_3SnH) → N, CO_2Me 83% + CO_2Me, N, CO_2Me 12%

Boivin, J.; Fouquet, E.; Zard, S.Z. *Tetrahedron Lett.*, ***1990**, 31*, 3545.

Ph_2C=N–, CO_2Me, OAc → ($NaCH(CO_2Me)_2$, MeCN, 25°C, 2 h; cat. $Pd(PPh_3)_4$) → Ph_2C=N–, CO_2Me, MeO_2C, CO_2Me 79%

O'Donnell, M.J.; Yang, X.; Li, M. *Tetrahedron Lett.*, ***1990**, 31*, 5135.

MOMO, N, CO_2t-Bu → (1. 3 LDA, THF, -78°C; 2. BnBr) → MOMO, N, CO_2t-Bu, Ph 66%

Williams, D.R.; Benbow, J.W. *Tetrahedron Lett.*, ***1990**, 31*, 5881.

CO_2Et CO_2Et N_3 — $P(OEt)_3$, PhMe, RT → CO_2Et CO_2Et N 73%

Luheshi, A-B.N.; Salem, S.M.; Smalley, R.K. *Tetrahedron Lett.*, ***1990***, *31*, 6561.

N CO_2Me Me — 1. LiBr, DBU, THF; 2. CO_2Me → Me N CO_2Me Me CO_2Me 61%

Kanemasa, S.; Uchida, O.; Wada, E. *J. Org. Chem.*, ***1990***, *55*, 4411.

CO_2Me — $LDA•Bn_2NCH_2CO_2Et$, THF, -78°C → Me NBn_2 MeO_2C CO_2Et 56%

threo selective

Yamaguchi, M.; Torisu, K.; Minami, T. *Chem. Lett.*, ***1990***, 377.

Ph N-Ph — $OSiMe_3$ Me OMe, -78°C; 5% $TrSbCl_6$, CH_2Cl_2 15 min → NHPh Ph CO_2Me Me + NHPh Ph CO_2Me Me (92 : 8) 95%

other Lewis acids used, with varying selectivity

Mukaiyama, T.; Akamatsu, H.; Han, J.S. *Chem. Lett.*, ***1990***, 889.
Kato, K.; Mukaiyama, T. *Chem. Lett.*, ***1990***, 1917.

$NHCO_2SiMe_2t$-Bu MeO_2C Cl — 2 eq. AgF, MeCN RT, 24 h → O H N O MeO_2C + O H N O MeO_2C (3 : 2) 63%

Sakaitani, M.; Ohfune, Y. *J. Am. Chem. Soc.*, ***1990***, *112*, 1150.

NBn_2 $(CO)_5Cr$ Me — hν, CO (60 psi), MeOH → CO_2Me Me NBn_2 87%

Hegedus, L.S.; Schwindt, M.A.; DeLombaert, S.; Imwinkelried, R. *J. Am. Chem. Soc.*, ***1990***, *112*, 2264.

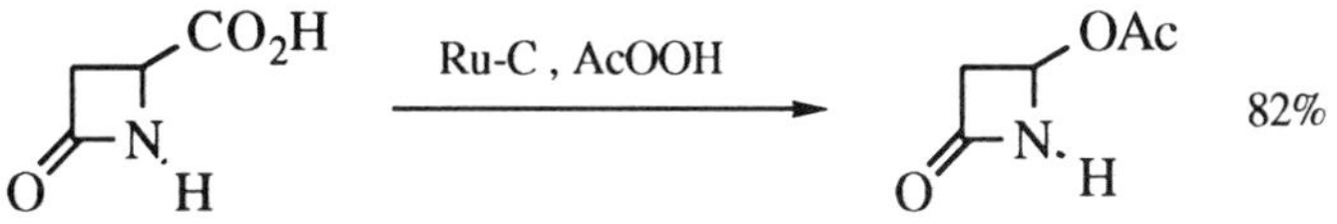

Ru-C , AcOOH

82%

Murahashi, S.; Naota, T.; Kuwabara, T.; Saito, T.; Kumobayashi, H.; Akutagawa, S. *J. Am. Chem. Soc.*, ***1990***, *112*, 7820.

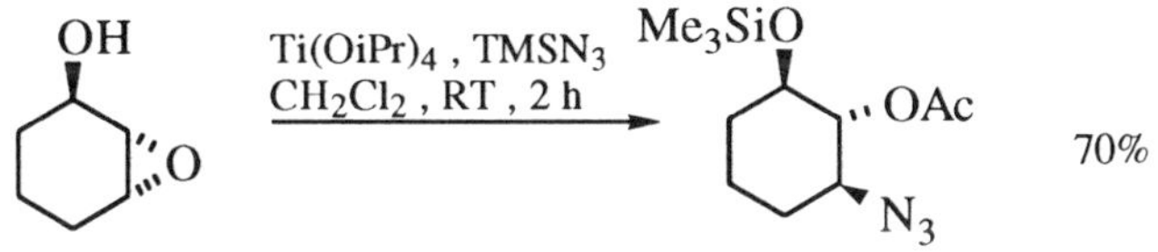

$Ti(OiPr)_4$, $TMSN_3$
CH_2Cl_2 , RT , 2 h

70%

Sutowardoyo, K.I.; Sinou, D. *Bull. Soc. Chim. Fr.*, ***1991***, *128*, 387.

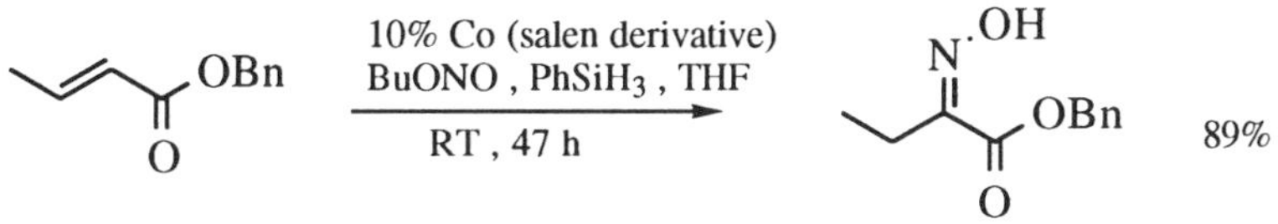

10% Co (salen derivative)
BuONO , $PhSiH_3$, THF
RT , 47 h

89%

Kato, K.; Mukaiyama, T. *Bull. Chem. Soc. Jpn.*, ***1991***, *64*, 2948.
Kato, K.; Mukaiyama, T. *Chem. Lett.*, ***1990***, 1917.

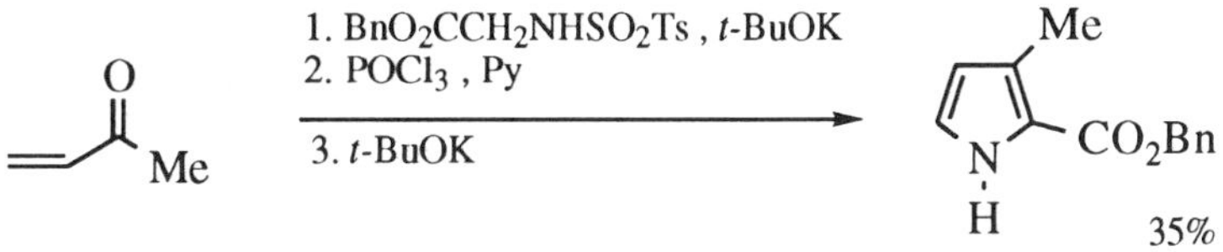

1. $BnO_2CCH_2NHSO_2Ts$, *t*-BuOK
2. $POCl_3$, Py
3. *t*-BuOK

35%

Lash, T.D.; Hoehner, M.C. *J. Het. Chem.*, ***1991***, *28*, 1671.

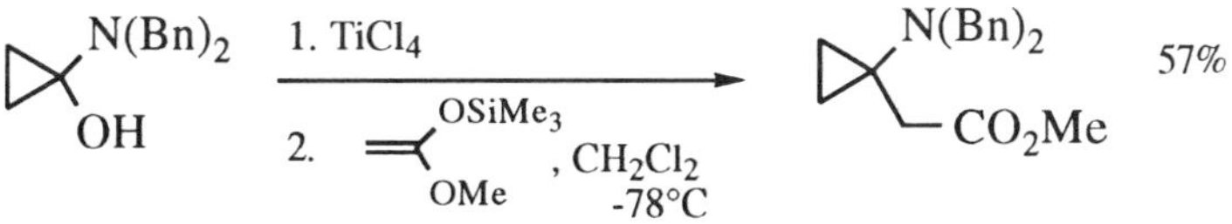

1. $TiCl_4$
2. (OSiMe₃/OMe ketene acetal) , CH_2Cl_2 -78°C

57%

Mertin, A.; Thiemann, T.; Hanss, I.; de Meijere, A. *SynLett*, ***1991***, 87.

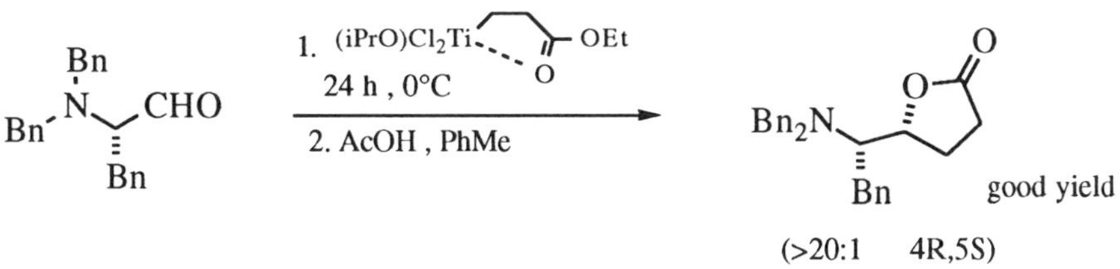

1. $(iPrO)Cl_2Ti$ (OEt homoenolate)
24 h , 0°C
2. AcOH , PhMe

good yield

(>20:1 4R,5S)

De Camp, A.E.; Kawaguchi, A.T.; Volante, R.P.; Shinkai, I. *Tetrahedron Lett.*, ***1991***, *32*, 1867.

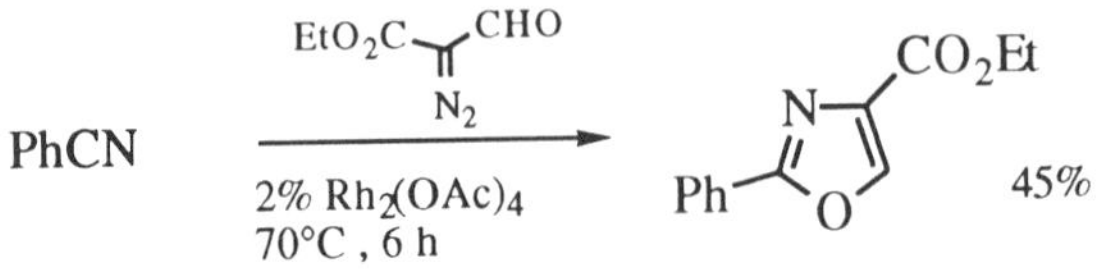

Connell, R.D.; Tebbe, M.; Helquist, P.; Åkermark, B. *Tetrahedron Lett.*, ***1991***, *32*, 17.

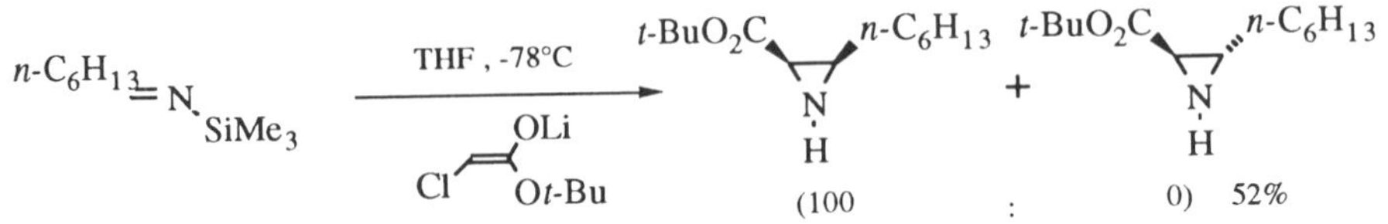

Cainelli, G.; Panunzio, M.; Giacomini, D. *Tetrahedron Lett.*, ***1991***, *32*, 121.

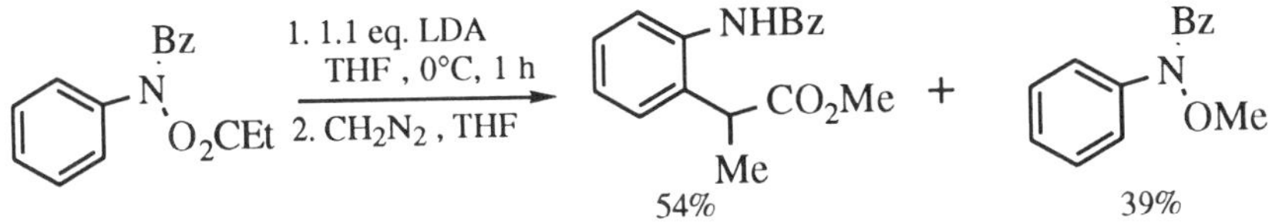

Endo, Y.; Hizatate, S.; Shudo, K. *Tetrahedron Lett.*, ***1991***, *32*, 2803.

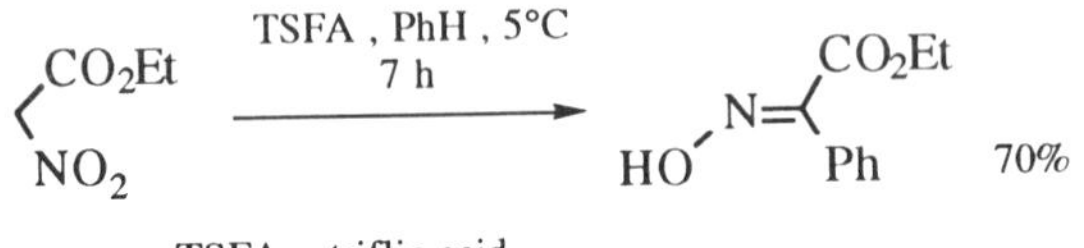

TSFA = triflic acid

Coustard, J-M.; Jacquesy, J-C.; Violeau, B. *Tetrahedron Lett.*, ***1991***, *32*, 3075.

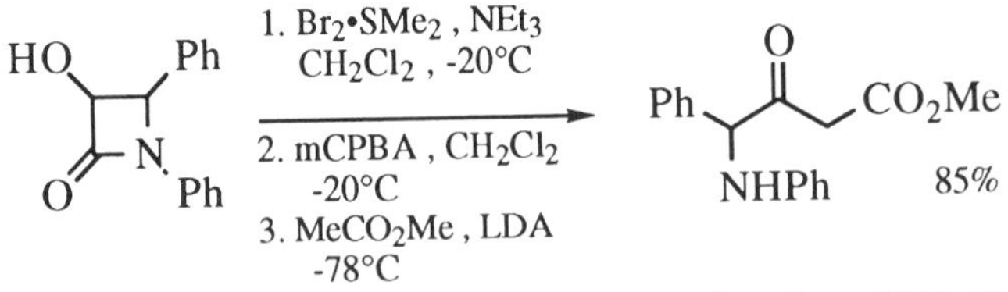

Palomo, C.; Cossío, F.P.; Rubiales, G.; Apáricio, D. *Tetrahedron Lett.*, ***1991***, *32*, 3115.

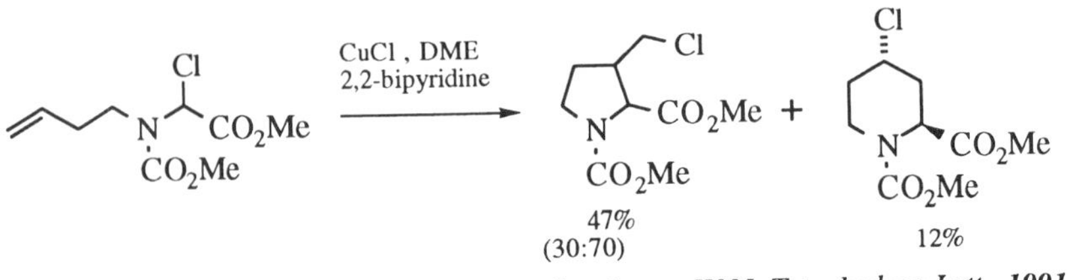

Udding, J.H.; Hiemstra, H.; van Zanden, M.N.A.; Speckamp, W.N. *Tetrahedron Lett.*, ***1991***, *32*, 3123.

LDA , EtONO; THF - pet ether

CH=NOH; CO_2Et 46%

Moorhoff, C.M.; Paquette, L.A. *J. Org. Chem.*, ***1991***, *56*, 703.

MeO_2C, MeO_2C, C≡C-H

1. NaH , THF
2. Ph, N, N, N, SO_2Me
3. reflux

Ph, N, CO_2Me, CO_2Me 41%

Taylor, E.C.; Macor, J.E.; French, L.G. *J. Org. Chem.*, ***1991***, *56*, 1807.

Me_2N CO_2Et

PhCHO , LDA; BEt_3 , -20°C

OH, Ph, CO_2Et, NMe_2 + OH, Ph, CO_2Et, NMe_2

(>95 : <5) 93%

Georg, G.I.; Akgün, E. *Tetrahedron Lett.*, ***1991***, *32*, 5521.

Ph CO_2Et

$BnNH_2$, EtOH; bakers yeast; β-cyclodextrin , 37°C; pH 7.2 , 48 h

Ph, CO_2Et, NHBn (73% ee) 70%

Rama Rao, K.; Nageswar, Y.V.D.; Kumar, H.M.S. *Tetrahedron Lett.*, ***1991***, *32*, 6611.

N

$Na_2S_2O_8$, $AgNO_3$; H_2SO_4 , CH_2Cl_2; reflux , 1.5 h; EtO, O, OH, O

N, CO_2Et 58% + CO_2Et, N 27%

84% conversion

Copa, F.; Fontana, F.; Lazzarini, E.; Minisci, F.; Pianese, G.; Zhao, L. *Tetrahedron Lett.*, ***1992***, *33*, 3057.

NEt_2

CO_2Me , NMP; Ru(COD)(COT) , 70°C; 20 h

CO_2Me, Me, NEt_2 85%

Mitsudo, T.; Zhang, S-W.; Satake, N.; Kondo, T.; Watanabe, Y. *Tetrahedron Lett.*, ***1992***, *33*, 5533.

1. MeLi
2. $NaBH_4$
3. H_2SO_4

51%

(4:1 β:α methyl)

Fray, A.H.; Meyers, A.I. *Tetrahedron Lett.*, ***1992***, *33*, 3575.

1. NaHMDS, EtO_2C–$N=CHPh$
2. 1N HCl, RT

53%

Kotha, S.; Kuki, A. *Tetrahedron Lett.*, ***1992***, *33*, 1565.

PhCH=N-Ph
$ZnCl_2$, THF
RT

(85 : 15) 72%

Beslin, P.; Marion, P. *Tetrahedron Lett.*, ***1992***, *33*, 5339.

$[Bn(TMS)N]_2CuLi$
-78°C

54% 10%

Yamamoto, Y.; Asao, N.; Uyehara, T. *J. Am. Chem. Soc.*, ***1992***, *114*, 5427.

REVIEWS:

"The Asymmetric Michael Addition Reaction Using Chiral Imines"
d'Angelo, J; Desmaële, D.; Dumas, F.; Guingant, A. *Tetrahedron Asymmetry*, ***1992***, *3*, 227.

Related Methods: Section 315 (Carboxylic Acid - Amide). Section 316 (Carboxylic Acid - Amine). Section 344 (Amide - Ester)

SECTION 352: AMINE - ETHER, EPOXIDE, THIOETHER

1. ME_3SiCl, NEt_3
2. TBAF

NO_2

74% (>99:1 trans:cis)

Dehaen, W.; Hassner, A. *Tetrahedron Lett.*, ***1990***, *1990*, 743.

PhCHO

$Au[c\text{-}C_6H_{11}\text{-}N{\equiv}C]_2BF_4$

CH_2Cl_2 , cat.

C≡N

CO_2Me

Ph CO_2Me + Ph CO_2Me

(89 : 12)

92% ee 1% ee

cat. = Fe PPh_2 MeO_2C Me N Me N Me PPh_2

Pastor, S.D.; Togni, A. *Tetrahedron Lett.*, ***1990***, *31*, 839.

baker's yeast
30% EtOH
37°C , 30 h

Ph CO_2Et

pH 7.2 buffer

Cl C≡N-O Cl

Cl N O Cl EtO_2C Ph

lower selectivity with ethyl cinnamate or 2,4,6-trimethylphenyl nitrile oxide

Rao, K.R.; Bhanumathi, N.; Srinivasan, T.N.; Sattur, P.B. *Tetrahedron Lett.*, ***1990***, *31*, 899.

O^- Na^+ N H MeO

DMF , 120°C
16 h

Br

MeO O N H + MeO O N H

(4 : 1) 61%

Grigg, R.; Markandu, J.; Surendrakumar, S. *Tetrahedron Lett.*, ***1990***, *31*, 1191.

NHMe

$Hg(OAc)_2$, AcOH
MeOH , RT

Me N OMe

80%

Majumdar, K.C.; De, R.N.; Saha, S. *Tetrahedron Lett.*, ***1990***, *31*, 1207.

n-Bu
N- H
n-Bu

1. S O N O $Cl^{\ominus}$

0.15M malonic acid
MeCN , t-BuSH

2. 2.6 eq. OEt

n-Bu
N OEt
n-Bu
69%

Newcomb, M.; Kumar, M.U. *Tetrahedron Lett.*, ***1990***, *31*, 1675.

O
N H
O
R^2
R^1

1. 1.5 $Hg(OAc)_2$
MeCN , 25°C
2. aq. NaCl
3. 2 $LiBH_4$
THF , -78°C

O Me
N R^2
O R^1
+
O Me
N R^2
O R^1

R^1 = OBn; R^2 = H (>50 : 1) 89%
R^1 = H; R^2 = OBn (1 : >50) 78%

Takacs, J.M. Helle, M.A.; Sanyal, B.J.; Eberspacher, T.A. *Tetrahedron Lett.*, ***1990***, *31*, 6765.

Me N

$PhCO_3H$, $BnEt_3NCl$
CH_2Cl_2 - H_2O

Me N
O
85-95%

Balachkova, A.I.; Koldobskii, G.I.; Drozdetskii, A.G.; Tereshchenko, G.F. *Zhur. Org. Khim.*, ***1990***, *26*, 216 (Engl., p. 185).

Me Me
O
N OH

n-BuLi-t-BuLi
THF , RT

Me Me
N O
+
Me Me
N O

(85 : 15) 80%

Roussi, G.; Zhang, J. *Tetrahedron Lett.*, ***1991***, *32*, 1443.

OMe
N S Ph
O

OMe
MgBr
THF

OMe
N
MeO
79%

Shibutani, T.; Fujihara, H.; Furukawa, N. *Tetrahedron Lett.*, ***1991***, *32*, 2943.

2 eq. NaH
2 eq. $CuBr \cdot SMe_2$
Py , 18 h
0.004 M

58%

Boger, D.L.; Yohannes, D. *J. Org. Chem.*, ***1991***, *56*, 1763.

$Me_2NCH(OMe)_2$
DMF , PhH , reflux
Amberlite , 16 h

81%

Leonard, W.R.; Romine, J.L.; Meyers, A.I. *J. Org. Chem.*, ***1991***, *56*, 1961.

$TiCl_3(OiPr)$, CH_2Cl_2
-78°C

98%

Newcomb, M.; Ha, C. *Tetrahedron Lett.*, ***1991***, *32*, 6493.

BuLi , THF

(2 : 98) 46%

$BuCuMgBrI \cdot BF_3$
THF , MDS

(99 : 1) 54%

Cainelli, G.; Giacomini, D.; Panunzio, M.; Zarantonello, P. *Tetrahedron Lett.*, ***1992***, *33*, 7783.

, acetone
20°C

93%

Adam, W.; Peters, E-M.; Peters, K.; von Schnering, H.G.; Voerckel, V. *Chem. Ber.*, ***1992***, *125*, 1263.

REVIEWS:

"Preparation of Diaminoethers and Polyamines"
Bradshaw, J.S.; Krakowiak, K.E.; Izatt, R.M. *Tetrahedron*, ***1992***, *48*, 4475.

SECTION 353: AMINE - HALIDE, SULFONATE

OMe, MeO, NH_2 — Ph- C≡C–C(O)CF_3, TFA, reflux, 2 h → OMe Ph, MeO, N, CF_3

48%

Linderman, R.J.; Kirollos, K.S. *Tetrahedron Lett.*, ***1990***, *31*, 2689.

O, NHTs, O, n-C_6H_{11}, =C= — 1. 2 I_2, K_2CO_3, ether, RT; 2. MeOH, NaOH; 3. Ac_2O, Py → OAc I, n-C_6H_{11}, NHTs + OAc I, n-C_6H_{11}, NHTs

(40 : 1) 76%

Friesen, R.W. *Tetrahedron Lett.*, ***1990***, *31*, 4249.

N, Cl — AcCl, NaI, MeCN, reflux, 24 h → N, I

55%

Corcoran, R.C.; Bang, S.H. *Tetrahedron Lett.*, ***1990***, *31*, 6757.

N — AcOF, CH_2Br_2 → N, Br

60%

Hebel, D.; Rozen, S. *J. Org. Chem.*, ***1991***, *56*, 6298.

MeS, CN, N, H, CO_2Me — 1. PhOPh, reflux; 2. $POCl_3$ → Cl, CN, N, SMe

49%

Tominaga, Y.; Michioka, T.; Moriyama, K.; Hosomi, A. *J. Het. Chem.*, ***1990***, *27*, 1217.

Me_3SiN_3, CH_2Cl_2-$MeNO_2$, NBS, -5°C → RT → Br, N_3

67%

Olah, G.A.; Wang, Q.; Li, X-Y.; Surya Prakash, G.K. *SynLett*, ***1990***, 487.

$BH_3 \cdot SMe_2$, CH_2Cl_2
reflux , 7 h

NH*t*-Bu
N *t*-Bu
H
Cl
80%

De Kimpe, N.; Stevens, C. *Tetrahedron*, ***1991***, *47*, 3407.

1. benzyne , dioxane reflux , iAmONO
2. aq. KOH

Ph
N N
Cl
13%

Gonsalves, A.M.d'A.R.; Pinho e Melo, T.M.V.D.; Gilchrist, T.L. *Tetrahedron*, ***1992***, *48*, 6821.

SECTION 354: AMINE - KETONE

OH
Me
Me N
Bn
diglyme , reflux
3 h
Me O
Me
NHBn
65%

Vatele, J-M.; Dumas, D.; Gore, J. *Tetrahedron Lett.*, ***1990***, *31*, 2277.

OH
CO_2t-Bu
N
Br
NaH , PhH
80°C
O
CO_2t-Bu
N
88%

Wasserman, H.H.; Cook, J.D.; Vu, C.B. *Tetrahedron Lett.*, ***1990***, *31*, 4945.

1. $NaN(CHO)_2$, EtOH
RT → 50°C
2. HCl , reflux

O
Ph Br
O
Ph $NH_2 \cdot HCl$
91%

Yinglin, H.; Hongwen, H. *Synthesis*, ***1990***, 615.

1. Ph
2. PhMe , reflux

⊕N
O⊖
O N Ph + O N Ph + Ph O H N
[2 : 1]
(0.67 : 1) 95%

Cordero, F.M.; Brandi, A.; Querci, C.; Goti, A.; DeSarlo, F.; Guarno, A. *J. Org. Chem.*, ***1990***, *55*, 1762.

1. LDA, THF, 0°C
2. $PhCO_2Me$, THF, 0°C

68%

Bartoli, G.; Cimarelli, C.; Palmieri, G.; Bosco, M.; Dalpozzo, R. *Synthesis*, ***1990***, 895.

PhLi, THF
-78°C → 0°C

73%

Kraus, G.A.; Shi, J. *J. Org. Chem.*, ***1990***, *55*, 5423.

t-BuOK, DMSO

52%

Arenz, T.; Frauenrath, H. *Angew. Chem. Int. Ed. Engl.*, ***1990***, *29*, 932.

NEt_3

46%

Tominaga, Y.; Ichihara, Y.; Mori, T.; Kamio, C.; Hosomi, A. *J. Het. Chem.*, ***1990***, *27*, 263.

$CH_2(NMe_2)_2$, CH_2Cl_2

68%

Nolen, E.G.; Allocco, A.; Vitarius, J.; McSorely, K. *J. Chem. Soc., Chem. Commun.*, ***1990***, 1532.

hn (>320 nm), DCA
MeOH - MeCN

90%

DCA = 9,10-dicyanoanthracene

Jeon, Y.T.; Lee, C-P.; Mariano, P.S. *J. Am. Chem. Soc.*, ***1991***, *113*, 8847.

3 eq. [3 THF-Mg_2Cl_2O•TiNCO]
5% $Pd(PPh_3)_4$, NMP , 12 h
100°C

Me

87%

Uozumi, Y.; Mori, M.; Shibasaki, M. *J. Chem. Soc., Chem. Commun.*, ***1991***, 81.

Me

$BnNH_2$, AcOH , PhMe
reflux

Me

NHBn

67%

Otto, A.; Schick, H. *Synthesis*, ***1991***, 115.

SO_2Ph

1. $BnNH_2$, CH_2Cl_2
MeOH
2. NaOMe , MeOH

Bn

65%

Chou, S-S.P.; Yuan, T-M. *Synthesis*, ***1991***, 171.

NHiPr

Ph

1. LiTMP , THF , 0°C
2. MeI

NHiPr

Me

Ph

86%

Bartoli, G.; Bosco, M.; Cimarelli, C.; Dalpozzo, R.; Palmieri, G. *SynLett*, ***1991***, 229.

1. H_2O , MeOH
2. $CH_3COCH_2CO_2H$

85%

Baxter, G.; Melville, J.C.; Robins, D.J. *SynLett*, ***1991***, 359.

$OBBu_2$

Ph

+ Me_2N NMe_2 Ph

CH_2Cl_2

NMe_2 Ph Ph Me + NMe_2 Ph Ph Me

-78°C	33%	(4	: 1)
RT	95%	(1	: 17)

Nolen, E.G.; Allocco, A.; Broody, M.; Zuppa, A. *Tetrahedron Lett.*, ***1991***, *32*, 73.

Ph-C≡CH , CO
5% $PdCl_2(PPh_3)_2$
Et_2NH , 120°C
6 h

90%

Torii, S.; Okumoto, H.; Xu, L.H. *Tetrahedron Lett.*, ***1991***, *32*, 237.

$PhI(O_2CCF_3)_2$
CF_3CH_2OH
RT , 30 min

51%

Kita, Y.; Tohma, H.; Inagaki, M.; Hatanaka, K.; Kikuchi, K.; Yakura, T. *Tetrahedron Lett.*, ***1991***, *32*, 2035.

$(CO)_3Co$ OH Co

1. NHTos
$BF_3{\bullet}OEt_2$, -78°C
CH_2Cl_2
2. $Me_3N\text{-}O$, O_2
CH_2Cl_2 , RT , 1 h

Tos-N ... =O 85%

Jeong, N.; Yoo, S.; Lee, S.J.; Lee, S.H.; Chung, Y.K. *Tetrahedron Lett.*, ***1991*** *32*, 2137.

1. PhCHO , NH_2
conc. HCl
2. 10% aq. $NaHCO_3$

88%

Yi, L.; Zou, J.; Lei, H.; Lin, X.; Zhang, M. *Org. Prep. Proceed. Int.*, ***1991***, *23*, 673.

e^- (graphite electrodes)
MeOH , PhH , NaOH
(-1.78 V) , 30°C , 8 h

65%

Rajanarayanan, A.; Jeyaraman, R. *Tetrahedron Lett.*, ***1991***, *32*, 3873.

4 eq. iPrNH$_2$, ether
RT , 17 h

(98 : 2) 72%

De Kimpe, N.; D'Hondt, L.; Moens, L. *Tetrahedron,* ***1992***, *48*, 3183.

Ph-C≡C-H , $PdCl_2$(dppf)
Et_2NH , autoclave
CO (20 atm) , 120°C , 1 h

70%

Kalinin, V.N.; Shostakovsky, M.V.; Ponomaryov, A.B. *Tetrahedron Lett.,* ***1992***, *33*, 373.

5 eq. NaN_3
5 eq. CAN

72%

Magnus, P.; Barth, L. *Tetrahedron Lett.,* ***1992***, *33*, 2789.

PhMe , 80°C , 1 h

80%

Desmaële, D.; Champion, N. *Tetrahedron Lett.,* ***1992***, *33*, 4447.

N-$SiMe_3$, $CHCl_3$
RT , 1 h

63%

Degl'Innocenti, A.; Capperucci, A.; Reginato, G.; Mordini, A.; Ricci, A. *Tetrahedron Lett.,* ***1992***, *33*, 1507.

1. Bu-CH=CH-BCl_2 , PhMe , 12 h
-78°C – RT
2. BnN_3
3. MeOH

72%

Noiret, N.; Youssofi, A.; Carboni, B.; Vaultier, M. *J. Chem. Soc., Chem. Commun.,* ***1992***, 1105.

REVIEWS:

"Synthesis of α-Amino Aldehydes and α-Amino Ketones"
Fisher, L.E.; Muchowski, J.M. *Org. Prep. Proceed. Int.*, ***1990***, *22*, 399.

"Further Advances in the Chemistry of Mannich Bases"
Tramontini, M.; Angiolini, L. *Tetrahedron*, ***1990***, *46*, 1791.

SECTION 355: AMINE - NITRILE

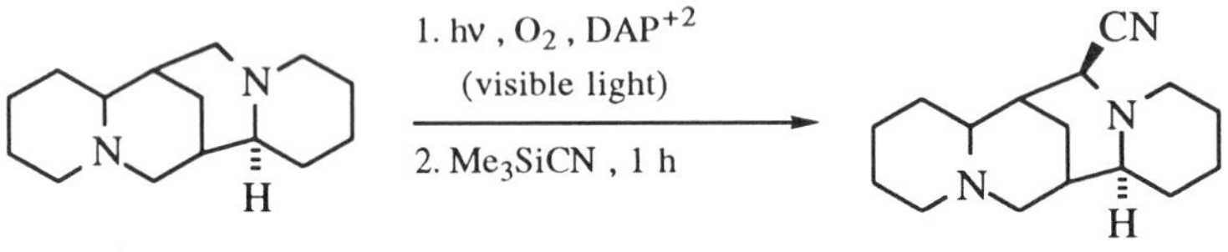
1. hν , O2 , DAP+2
(visible light)
2. Me3SiCN , 1 h
CN
N
H

DAP^{+2} = N,N'-dimethyl-2,7-deazapyrenium bis-tetrafluoroborate 90%

Santamaria, J.; Kaddachi, M.T.; Riglaudy, J. *Tetrahedron Lett.*, ***1990***, *31*, 4735.

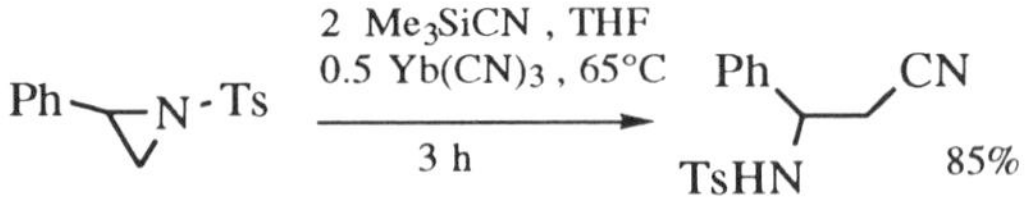
2 Me3SiCN , THF
0.5 Yb(CN)3 , 65°C
3 h
Ph
N-Ts
CN
TsHN
85%

Matsubara, S.; Kodama, T.; Utimoto, K. *Tetrahedron Lett.*, ***1990***, *31*, 6379.

SECTION 356: AMINE - ALKENE

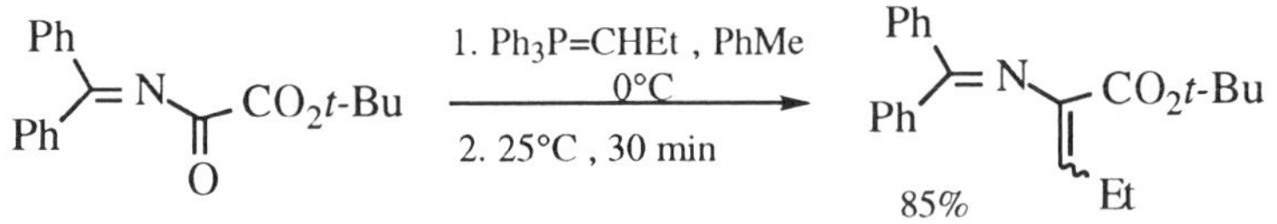
Ph
CO2t-Bu
1. Ph3P=CHEt , PhMe
0°C
2. 25°C , 30 min
85%
Et

85:15 Z:E

O'Donnell, M.J.; Arasappan, A.; Hornback, W.J.; Huffman, J.C.
Tetrahedron Lett., ***1990***, *31*, 157.

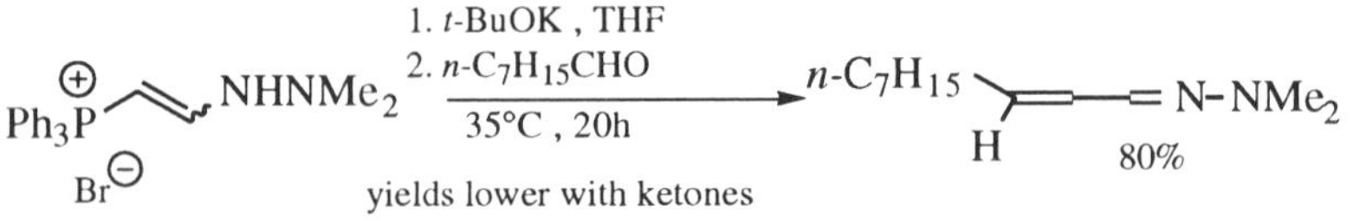
1. t-BuOK , THF
2. n-C7H15CHO
35°C , 20h
Ph3P
NHNMe2
Br
n-C7H15
N-NMe2
H
80%

yields lower with ketones

Cristau, H.-J.; Gasc, M.-B. *Tetrahedron Lett.*, ***1990***, *31*, 341.

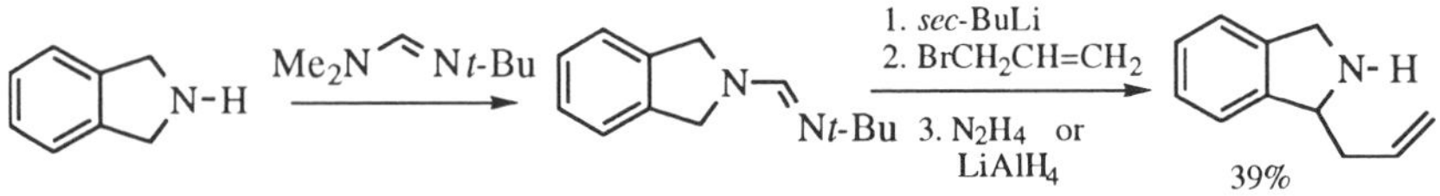
N-H
Me2N Nt-Bu
Nt-Bu
1. sec-BuLi
2. BrCH2CH=CH2
3. N2H4 or
LiAlH4
N- H
39%

Beeley, L.J.; Rockell, C.J.M. *Tetrahedron Lett.*, ***1990***, *31*, 417.

O OMe O → $(PhCH_2)_2NH$ 0°C, 90 h; PPh_2 Fe Me H Ph_2P N Me HO OH → NBu_2 (97 : 3) + NBu_2 87%; 84% ee, S; 91:9 E:Z

Hayashi, T.; Kishi, K.; Yamamoto, A.; Ito, Y. *Tetrahedron Lett.*, ***1990***, *31*, 1743.

Ph OTf → $N(Boc)_2$; Bu_3N, DMF, 80°C; 5% $Pd(OAc)_2(PPh_3)_2$; 5h → Ph $N(Boc)_2$ 36%

Arcadi, A.; Bernocchi, E.; Cacchi, S.; Caglioti, L.; Marinelli, F. *Tetrahedron Lett.*, ***1990***, *31*, 2463.

CHO → 1. $BnNH_2$ 2. NCS 3. 1.3 *t*-BuOK, RT ether, 5h → N Ph 91%

De Kimpe, N.; Yao, Z-P.; Boeykens, M.; Nagy, M. *Tetrahedron Lett.*, ***1990***, *31*, 2771.

Bn N Ph → Br, cat. $TiCl_4$; THF, Al_{foil} → NHBn Ph 83%

Tanaka, H.; Inoue, K.; Pokorski, U.; Taniguichi, M.; Torii, S. *Tetrahedron Lett.*, ***1990***, *31*, 3023.

OBn N CO_2Me C C H → 1. Bu_3SnH, PhH AIBN, 80°C 2. AcOH, MeOH → (BnO)HN CO_2Me 82%

Enholm, E.J.; Burroff, J.A.; Jaramillo, L.M. *Tetrahedron Lett.*, ***1990***, *31*, 3727.

MsO OMs → CH_2Cl_2, RT; NH_2 → N 73%

Ding, Z.; Tufariello, J.J. *Synth. Commun.*, ***1990***, *20*, 227.

Ph NHMe Ph — BuLi, THF; -78°C → -10°C → Ph N Me Ph 85%

LiAlH$_4$ 86%

Tokuda, M.; Fujita, H.; Suginome, H. *Tetrahedron Lett.*, ***1990***, *31*, 5353.

$Ph_3P^{\oplus}$ $^{\ominus}Br$ — 1. N-H; 2. NaN(TMS)$_2$; 3. iPrCHO → N 81%

Barluenga, J.; Merino, I.; Palacios, F. *Tetrahedron Lett.*, ***1990***, *31*, 6713.

MeO$_2$C-O O-CO$_2$Me — MeHN NHMe, THF; Pd$_2$(dba)$_3$•CHCl$_3$, RT; P(OiPr)$_3$, 4 h → Me-N N-Me 74%

Tsuda, T.; Kiyoi, T.; Saegusa, T. *J. Org. Chem.*, ***1990***, *55*, 3388.

NO$_2$ — K$_2$CO$_3$, H$_2$O, CS$_2$; TEBA, CH$_2$Cl$_2$ → N-OH 52%

Albanese, D.; Landini, D.; Penso, M. *Synthesis*, ***1990***, 333.

=C=C(Bn)(SMe) — iPrCH=NBn, AlCl$_3$, PhMe; RT → BnHN SMe Ph 81%

(3:1 cis:trans)

Hayashi, Y.; Shibata, T.; Narasaka, K. *Chem. Lett.*, ***1990***, 1693.

PhHN — PhNH$_2$•HCl; 140°C → NH$_2$ 60% + NH$_2$ 4.5%

Abdrakhmanov, I.B.; Gataullin, R.R.; Mustafin, A.G.; Tolstikov, G.A.; Baikova, I.P.; Fatykhov, A.A.; Panasenko, A.A. *Zhur. Org. Khim.*, ***1990***, *26*, 1527 (Engl., p. 1319).

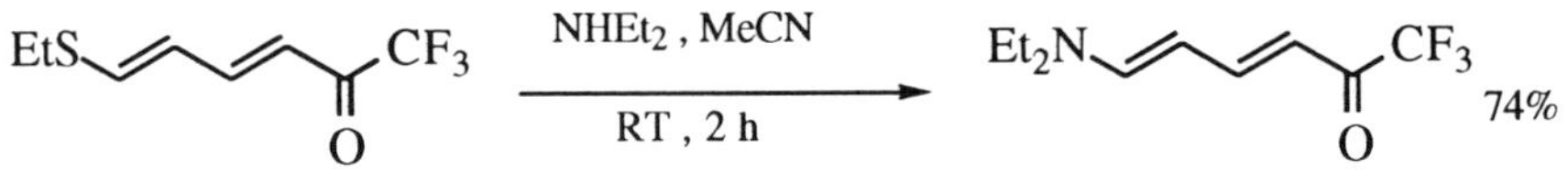
EtS
CF3
O
NHEt2 , MeCN
RT , 2 h
Et2N
CF3
O
74%

Hojo, M.; Masuda, R.; Okada, E. *Chem. Lett.*, ***1990***, 2095.

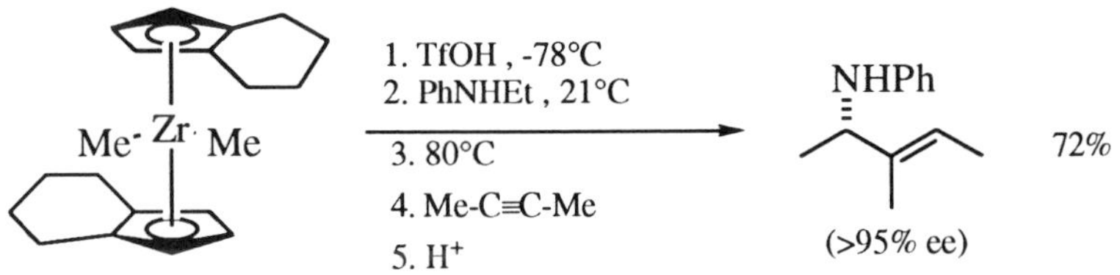
Me Zr Me
1. TfOH , -78°C
2. PhNHEt , 21°C
3. 80°C
4. Me-C≡C-Me
5. H+
NHPh
72%
(>95% ee)

Grossman, R.B.; Davis, W.M.; Buchwald, S.L. *J. Am. Chem. Soc.*, ***1991***, *113*, 2321.

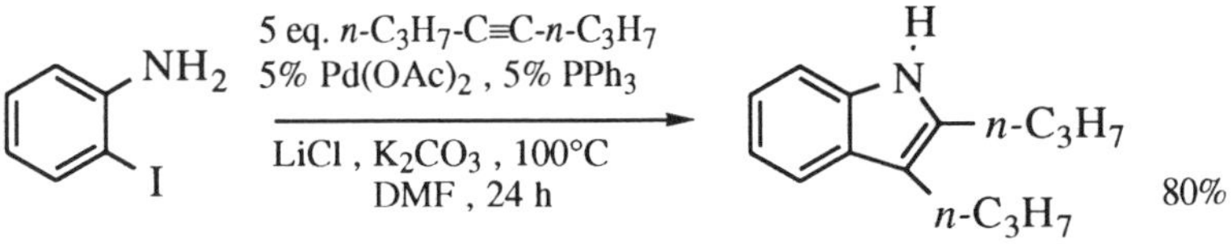
NH2
I
5 eq. n-C3H7-C≡C-n-C3H7
5% Pd(OAc)2 , 5% PPh3
LiCl , K2CO3 , 100°C
DMF , 24 h
H
N
n-C3H7
n-C3H7
80%

Larock, R.C.; Yum, E.K. *J. Am. Chem. Soc.*, ***1991***, *113*, 6689.

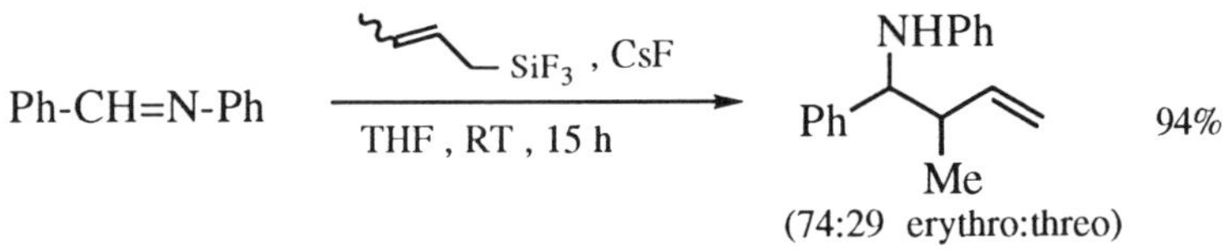
Ph-CH=N-Ph
SiF3 , CsF
THF , RT , 15 h
NHPh
Ph
Me
94%
(74:29 erythro:threo)

Kira, M.; Hino, T.; Sakurai, H. *Chem. Lett.*, ***1991***, 277.

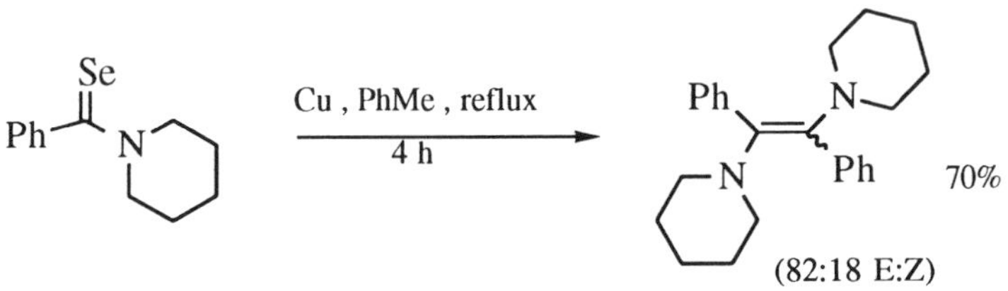
Se
Ph
N
Cu , PhMe , reflux
4 h
Ph
N
N
Ph
70%
(82:18 E:Z)

Sekiguchi, M.; Ogawa, A.; Kambe, N.; Sonoda, N. *Chem. Lett.*, ***1991***, 315.

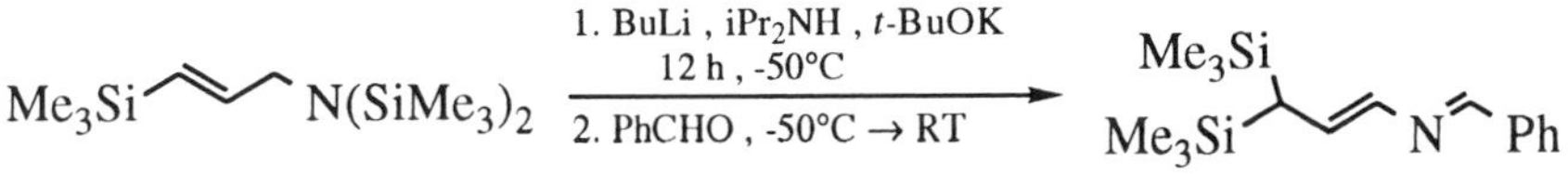
Me3Si
N(SiMe3)2
1. BuLi , iPr2NH , t-BuOK
12 h , -50°C
2. PhCHO , -50°C → RT
Me3Si
Me3Si
N
Ph

Degl'Innocenti, A.; Mordini, A.; Pinzani, D.; Reginato, G.; Ricci, A. *SynLett*, ***1991***, 712.

1. LDA , THF
2. MeCHO , -78°C → RT
3. DCC , cat. CuCl , ether
3 h

32%

Schuster, E.; Hesse, C.; Schumann, D. *SynLett*, ***1991***, 916.

Me Ph N Ph

Br , ether

Me Ph N Ph H

(43:57 diastereomers) quant.

Bocoum, A.; Boga, C.; Savoia, D.; Umani-Ronchi, A. *Tetrahedron Lett.*, ***1991***, *32*, 1367.

I F_3C N

Ph , PhMe , K_2CO_3
$Pd_2(dba)_3$•$CHCl_3$
65°C , 26 h

Ph F_3C N

95%

Uneyama, K.; Watanabe, H. *Tetrahedron Lett.*, ***1991***, *32*, 1459.

Ph N

B O O Bu
CCl_4 , reflux

Ph N H 86%

Wuts, P.G.M.; Jung, Y-W. *J. Org. Chem.*, ***1991***, *56*, 365.

N I

Bu_3SnH , AIBN , ACN
PhH , reflux , 3 h
slow addition of tin hydride

N 65% + N

ACN = azobis(cyclohexane)carbonitrile

Ziegler, F.E.; Jeroncic, L.O. *J. Org. Chem.*, ***1991***, *56*, 3479.

N

0.3 eq. HCl , dioxane
101°C , 20 h

N 74%

Cook, G.R.; Stille, J.R. *J. Org. Chem.*, ***1991***, *56*, 5578.

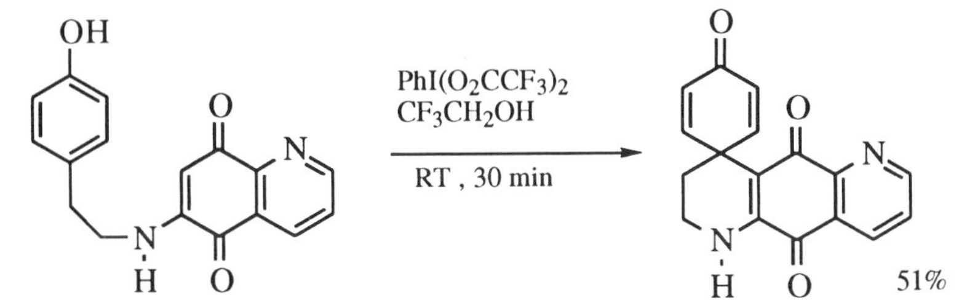

$PhI(O_2CCF_3)_2$
CF_3CH_2OH
RT, 30 min

51%

Kita, Y.; Tohma, H.; Inagaki, M.; Hatanaka, K.; Kikuchi, K.; Yakura, T. *Tetrahedron Lett.*, ***1991***, *32*, 2035.

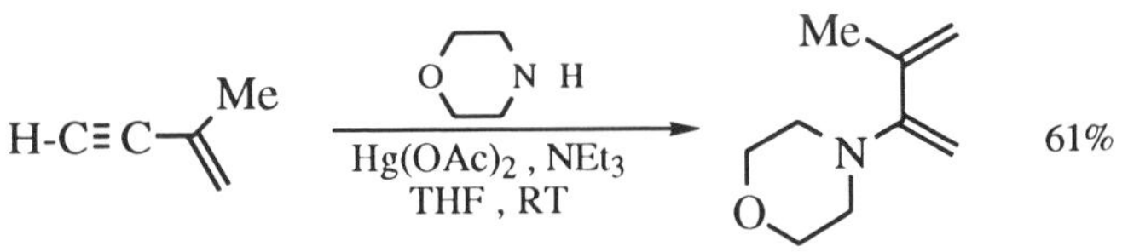

$Hg(OAc)_2$, NEt_3
THF, RT

61%

Barluenga, J.; Aznar, F.; Valdés, C.; Cabal, M-P. *J. Org. Chem.*, ***1991***, *56*, 6166.

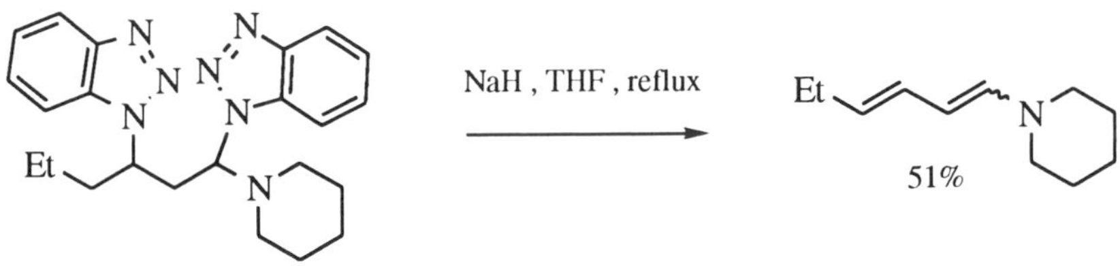

NaH, THF, reflux

51%

Katritzky, A.R.; Long, Q-H.; Lue, P. *Tetrahedron Lett.*, ***1991***, *32*, 3597.

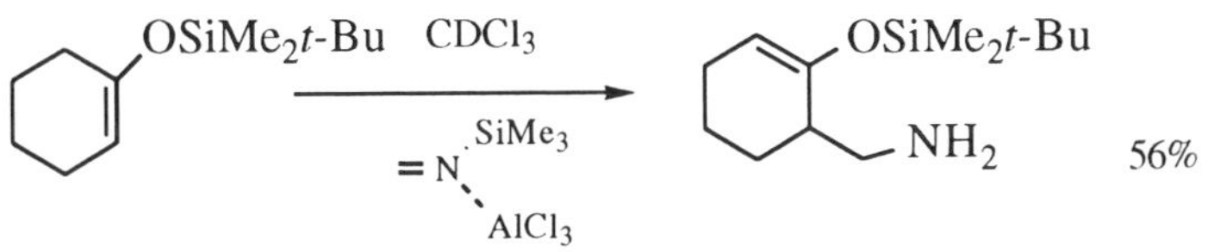

$CDCl_3$

56%

Tanino, K.; Takahashi, M.; Murayama, K.; Kuwajima, I. *J. Org. Chem.* ***1991***, *57*, 7009.

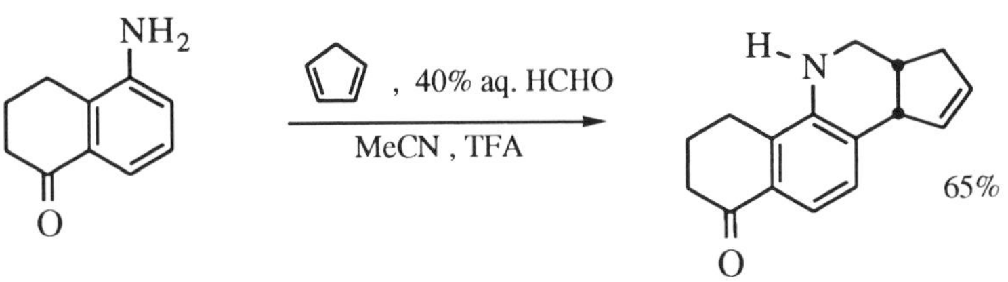

, 40% aq. HCHO
MeCN, TFA

65%

Gregoire, P.J.; Mellor, J.M.; Merriman, G.D. *Tetrahedron Lett.*, ***1991***, *32*, 7099.

$H-C\equiv C-CH_2-N(SiMe_3)_2$ → 1. Bu_3SnH, AIBN; 2. PhBr, PhMe; 2% $Pd(PPh_3)_4$, 72 h → Ph–CH=CH–CH$_2$–NH_2 65%

Corriu, R.J.P.; Bolin, G.; Moreau, J.J.E. *Tetrahedron Lett.*, ***1991***, *32*, 4121.

1. PhNHMe; 2. allyl bromide; $BiCl_3$•Al, aq. THF → NMePh 85%

Katritzky, A.R.; Shoban, N.; Harris, P.A. *Tetrahedron Lett.*, ***1991***, *32*, 4247.

1. allyl MgCl (2 eq.), THF; 2. TFA, Et_3SiH → H_2N NH_2 76%

Neumann, W.L.; Rogic, M.M.; Dunn, T.J. *Tetrahedron Lett.*, ***1991***, *32*, 5865.

10 eq. HCOOH, THF → 49%

Rutjes, F.P.J.T.; Paz, M.M.; Hiemstra, H.; Speckamp, W.N. *Tetrahedron Lett.*, ***1991***, *32*, 6629.

1. xylene/PhMe, -78°C, $TiCl_4$; 2. $LiAlH_4$ → 71%

Cook, G.R.; Barta, N.S.; Stille, J.R. *J. Org. Chem.*, ***1992***, *57*, 461.
Barta, N.S.; Cook, G.R.; Landis, M.S.; Stille, J.R. *J. Org. Chem.*, ***1992***, *57*, 7188.

Br, NHTs → 5% $Pd(OAc)_2$, DMF, $P(o\text{-Tol})_3$, Na_2CO_3; NBu_4Cl, 75°C, 22 h → Me, N-Ts 67%

Harris, G.D. Jr.; Herr, R.J.; Weinreb, S.M. *J. Org. Chem.*, ***1992***, *57*, 2528.

1. LDA, THF, -78°C 85%
2. $(EtO)_2POCl$
3. base
4. hexanal

Me, N, n-C_4H_9 → N, n-C_4H_9, n-C_4H_9

Highet, R.J.; Jones, T.H. *J. Org. Chem.*, ***1992***, *57*, 4038.

$Sn[N(iPr)_2]$, ether

Ph, CHO → Ph, $N(iPr)_2$ 70%

Burnell-Curty, C.; Roskamp, E.J. *J. Org. Chem.*, ***1992***, *57*, 5063.

Bu_3SnH, AIBN
THF, reflux

Phth, Me, N, S, O, N, N → Me, N-Phth + Me, N-Phth

(1 : 1) 16%

Dickinson, J.M.; Murphy, J.A. *Tetrahedron*, ***1992***, *48*, 1317.

TFA, CH_2Cl_2
0°C

OH, N, O, Me, $SiMe_3$ → N, Me 96%

Gelas-Mialhe, Y.; Gramain, J-C.; Louvet, A.; Remuson, R. *Tetrahedron Lett.*, ***1992***, *33*, 73.

MeO-C_6H_4-I, MeCN
$Pd(PPh_3)_4$, K_2CO_3, 2 h

C≡C-Bu, O, N-H, CF_3 → Ph, Bu, N-H 82%

Arcadi, A.; Cacchi, S.; Marinelli, F. *Tetrahedron Lett.*, ***1992***, *33*, 3915.

2 eq. MeCHO, MeCN
5 eq. Bu_4NI, 120°C
2 eq. $MeSO_3H$, 18 h

C≡C-Me, NHBn → I, Me, Me, N-Bn 71%

Overman, L.E.; Sarkar, A.K. *Tetrahedron Lett.*, ***1992***, *33*, 4103.

MeO NHt-Bu → heat → N–t-Bu

De Kimpe, N.; De buck, K.; Booten, K. *Tetrahedron Lett., **1992**, 33*, 393.

OTf → 1. $Pd(PPh_3)_4$, H_2N, CO (40 psi), 55°C, HO; 2. $SOCl_2$ → O, N 83%

Meyers, A.I.; Robichaud, A.J.; McKennon, M.J. *Tetrahedron Lett., **1992**, 33*, 1181.

$OSiMe_3$ + Ph N Ph → 1. $TfOSiMe_2t$-Bu, CH_2Cl_2, 20°C; 2. $NaHCO_3$ → Ph N Ph $OSiMe_3$ + Ph N Ph $OSiMe_3$

better selectivity-lower yield with $AlCl_3$ (54 : 28) 86%

Nogue, D.; Paugam, R.; Wartski, L. *Tetrahedron Lett., **1992**, 33*, 1265.

Me_3Si Me C= → 1. 9-BBN; 2. N CHO; 3. NaOOH , 1 h → N Me 92%

(98:2 Z:E)

Sattsangi, P.D.; Wang, K.K. *Tetrahedron Lett., **1992**, 33*, 5025.

N Et → $SiMe_3$, hν , DCN , MeCN:H_2O, 10 h → N Et 70%

Pandey, G.; Rani, K.S.; Lakshmaiah, G. *Tetrahedron Lett., **1992**, 33*, 5107.

O O N_3 → 1. CF_3SO_3H , 0°C, CH_2Cl_2; 2. NaOH → O O N 71%

Pearson, W.H.; Schkeryantz, J.M. *Tetrahedron Lett., **1992**, 33*, 5291.

1. Cp_2ZrCl_2 , 2 BuLi , THF , -78°C
2. n-C_3H_7 C≡ C- n-C_3H_7 , THF
PhNC , 20°C → 67°C , 3 h
3. MeOH
4. H_2O

H NHPh n-C_3H_7 H n-C_3H_7 56%

Davis, J.M.; Whitby, R.J.; Jaxa-Chamiec, A. *Tetrahedron Lett.*, ***1992***, *33*, 5655.

Ph N Ph — Br, In , THF , 1 h → NHPh Ph 91%

Beuchet, P.; Le Marrec, N.; Mosset, P. *Tetrahedron Lett.*, ***1992***, *33*, 5959.

CN
1. Cp_2TiMe_2 , PhH
60°C , 5 d
2. Na_2SO_4

NH_2 N-H quant.

Barluenga, J.; del Pozo Losada, C.; Olano, B. *Tetrahedron Lett.*, ***1992***, *33*, 7579.

O Ph NEt_2 — Sm , SmI_2 , THF , 4 h, 67°C → Ph NEt_2 Et_2N Ph 72%

(38/62 E/Z)

Ogawa, A.; Takami, N.; Sekiguchi, M.; Ryu, I.; Kambe, N.; Sonoda, N. *J. Am. Chem. Soc.*, ***1992***, *114*, 8729.

PhNHOH , dioxane , 100°C
MoO_2(dipoc)(HMPA) , 8 h

NHPh 52%

dipoc = 2,6-pyridine dicarboxylate

Srivastava, A.; Ma, Y.; Pankayatselvan, R.; Dinges, W.; Nicholas, K.M. *J. Chem. Soc., Chem. Commun.*, ***1992***, 853.

n-C_5H_{11} C≡ C- n-C_5H_{11}

1. $TaCl_5$, Zn , DME
PhH
2. Me_3Al , THF
3. Ph N-NMe_2 H
4. aq. NaOH

n-C_5H_{11} n-C_5H_{11} Ph H- N NMe_2 80%

Takai, K.; Miwatashi, S.; Kataoka, Y.; Utimoto, K. *Chem. Lett.*, ***1992***, 99.

PhHN

1. BuLi
2. $Cp_2ZrMeCl$, 59°C , 6 h
3. n-$C_3H_7C{\equiv}Cn$-C_3H_7
4. MeOH

PhHN

92%

Coles, N.; Whitby, R.J.; Blagg, J. *SynLett*, ***1992***, 143.

NHBn
C≡ C-Me

3 eq. $(CH_2O)_n$, MeCN
6 eq. Bu_4NI , 80°C
1.3 eq. $MeSO_3H$

Bn N Me I 62%

+

Bn. N I 9%

Overman, L.E.; Rodriguez-Campos, I.M. *SynLett*, ***1992***, 995.

REVIEWS:

"Diels-Alder Reactions of Azadienes: Scope and Limitations"
Boger, D.L. *Bull. Soc. Chim. Belg.*, ***1990***, *99*, 599.

"Electronically Neutral 2-Aza-1,3-dienes: Are They Useful Intermediates in Organic Synthesis?"
Barluenga, J.; Joglar, J.; González, F.J.; Fustero, S. *SynLett*, ***1990***, 129.

"The aza-Wittig Reaction in Heterocyclic Synthesis"
Eguchi, S.; Matsushita, Y.; Yamashita, K. *Org. Prep. Proceed. Int.*, ***1992***, *24*, 209.

SECTION 357: ESTER - ESTER

Me O O (O O)

$Pd(OAc)_2$, NEt_3 , MeCN
CO , RT overnight
O Cl (O)

O O (O O) 80%

Barton, D.H.R.; Langlois, P.; Okano, T.; Ozbalik, N. *Tetrahedron Lett.*, ***1990***, *31*, 325.

O O Me N_2

$Rh_2(OAc)_4$, RT
CH_2Cl_2 , 15 h

EtO_2C O Me O 50%

+

EtO_2C O O Me Me 10%

Lee, E.; Jung, K.W.; Kim, Y.S. *Tetrahedron Lett.*, ***1990***, *31*, 1023.

e^-, MeCN, $CH_2(CO_2Me)_2$; Pt foil anode, Hg pool cathode, Bu_4NBr, -2.1 V

(1 : 1) 90%

Bode, H.E.; Sowell, C.G.; Little, R.D. *Tetrahedron Lett.*, ***1990***, *31*, 2525.

1. acetyl bromide, $ZnCl_2$
2. Zn/THF/DMSO
3. CuCN, 2 LiCl, 0°C
4. THF, LiCl, -78°C

CHO → OAc CO_2Et 95%

Br, CO_2Et

Chou, T-S.; Knochel, P. *J. Org. Chem.*, ***1990***, *55*, 4791.

1. deposit on Celite powder
2. FVP (500°C)

10^{-2}-10^{-4} mbar

Ene reaction

57%

Choi, S.S-M.; Kirby, G.W.; Mahajan, M.P. *J. Chem. Soc., Chem. Commun.*, ***1990***, 138.

CN → $PhI(OAc)_2$, HBF_4, 6 h; aq. AcOH → OAc 83%

Tiecco, M.; Testaferri, L.; Tingoli, M.; Bartoli, D. *Tetrahedron*, ***1990***, *46*, 7139.

$OSiMe_3$, Me, OMe → 1. *p*-NBSP 2. NaOAc, DMF, 60°C → Me, CO_2Me, OAc 33%

p-NBSP = *p*-((nitrobenzene)sulfonyl)peroxide

Hoffman, R.V.; Stoll, D. *Synth. Commun.*, ***1991***, *21*, 211.

SPh, CO_2Me → "base", DMF, -78°C → SPh, CO_2Me (>90% de) 88%

Rodríguez, C.M.; Martín, V.S. *Tetrahedron Lett.*, ***1991***, *32*, 2165.

10% $PdCl_2$, $CuCl_2$, 1 atm. CO, RT, 1 d; propylene oxide, MeOH, $MeC(OEt)_3$, CH_2Cl_2 — OH → CO_2Me, O, O — 55%

Tamaru, Y.; Hojo, M.; Yoshida, Z. *J. Org. Chem.*, ***1991***, *56*, 1099.

O, O, CO_2Et — LiI, DMF, 40 min → O, O, CO_2Et — 61%

Kasatkin, A.N.; Kulak, A.N.; Tolstikov, G.A.; Shitikova, O.V. *Tetrahedron Lett.*, ***1991***, *32*, 4591.

CHO, Br, CO_2Me — 1. SmI_2, THF, 0°C 2. Ac_2O, DMAP → OAc, CO_2Me — 68%

Inanaga, J.; Yokoyama, Y.; Handa, Y.; Yamaguchi, M. *Tetrahedron Lett.*, ***1991***, *32*, 6371.

Ph, CO_2Me — SmI_2-THF, HMPA, MeOH, RT, < 1 min → Ph, Ph, CO_2Me, CO_2Me — 70% (1.9:1 dr)

Inanaga, J.; Handa, Y.; Tabuchi, T.; Otsubo, K.; Yamaguchi, M.; Hanamoto, T. *Tetrahedron Lett.*, ***1991***, *32*, 6557.

CO_2n-$C_{12}H_{25}$, OH, CO_2t-Bu — porcine pancreatic lipase, 28 h → t-BuO_2C, O, O — 37% (49 %ee) + CO_2n-$C_{12}H_{25}$, OH, CO_2t-Bu

Sibi, M.P.; Gaboury, J.A. *Tetrahedron Lett.*, ***1992***, *33*, 5681.

CO_2Me, Br, Br, CO_2Me — e^-, Pt cathode/Cu anode, NaI, 0.1 M Et_4NClO_4, DMF → CO_2Me, CO_2Me — 77% (36:64 trans:cis)

Tokuda, M.; Hayashi, A.; Suginome, H. *Bull. Chem. Soc. Jpn.*, ***1991***, *64*, 2590.

Me, CO_2Me, O, EtO, I

10 eq. $=\!\!/^{CO_2Et}$, AIBN
0.1 Bu_3SnCl, *t*-BuOH
2 eq. $NaBH_3Cn$, reflux

Me, CO_2Me, O, CO_2Et, O

65%

Stork, G.; Franklin, P.J. *Aust. J. Chem.*, ***1991***, *44*, 275.

Me

$Bi(OAc)_3$, I_2

Me, OAc, OAc + Me, OAc, OAc

(70 : 30) 71%

Trainor, R.W.; Deacon, G.B.; Jackson, W.R.; Giunta, N. *Aust. J. Chem.*, ***1992***, *45*, 1265.

Also via Dicarboxylic Acids: Section 312 (Carboxylic Acids - Carboxylic Acids).
Hydroxy-esters Section 327 (Alcohol - Ester). Diols Section 323 (Alcohol - Alcohol)

SECTION 358: ESTER - ETHER, EPOXIDE, THIOETHER

O

MeOH, RT
$CuCl_2$, $PdCl_2$
CO, $(MeO)_3CH$
20 h

MeO_2C, O O, CO_2Me

75%

Yadav, J.S.; Rao, E.S.; Rao, V.S.; Choudary, B.M. *Tetrahedron Lett.*, ***1990***, *31*, 2491.

O, Ph

2 iodosobenzene, TMOF
conc. H_2SO_4, overnight

OMe, Ph, CO_2Me

82%

TMOF = trimethylorthoformate

Singh, O.V. *Tetrahedron Lett.*, ***1990***, *31*, 3055.

PhSe, CO_2Et, CO_2Et

$=\!\!/^{n\text{-}C_6H_{13}}$
hv (sunlamp), PhH
reflux

EtO_2C, EtO_2C, SePh, n-C_6H_{13}

95%

Byers, J.H.; Lane, G.H. *Tetrahedron Lett.*, ***1990***, *31*, 5697.

1. poly *L*-leucine , H_2O_2 NaOH , CCl_4
2. mCPBA , CH_2Cl_2

Ph Ph O → Ph O O OPh 56% (>99% ee)

Baures, P.W.; Eggleston, D.S.; Flisak, J.R.; Gombatz, K.; Lantos, I.; Mendelson, W.; Remich, J.J. *Tetrahedron Lett.*, ***1990***, *31*, 6501.

BnO O O Me OTf → K_2CO_3 , MeOH → BnO O Me CO_2Me 80%

Witty, D.R.; Fleet, G.W.J.; Choi, S.; Vogt, K.; Wilson, F.X.; Wang, Y.; Storer, R.; Myers, P.L.; Wallis, C.J. *Tetrahedron Lett.*, ***1990***, *31*, 6927.

SPh Me OH

1. $HOOCCH_2CO_2Me$, CH_2Cl_2 DCC , DMAP , 30°C
2. TsN_3 , MeCN , NEt_3
3. $Rh_2(OAc)_3$, PhH , reflux

Me PhS O EtO_2C 40%

Kido, F.; Kazi, A.B.; Yoshikoshi, A. *Chem. Lett.*, ***1990***, 613.

O OH → *n*-C_7H_{15}Br , MeCN , 100°C KCNO , Bu_4NBr → O O O NH*n*-C_7H_{15} 61%

Prashad, M.; Tomesch, J.C.; Houlihan, W.J. *Synthesis*, ***1990***, 477.

O CO_2H → , H_2SO_4 → O O O*t*-Bu 63%

Jarvis, B.B.; Wells, K.M.; Kaufmann, T. *Synthesis*, ***1990***, 1079.

EtO CO_2Et

1. OH , NBS CH_2Cl_2
2. $Cr(OAc)_2$, aq. THF RT

Me CO_2Et O OEt 67%

(1:3.9 cis Me:trans Me)

Lübbers, T.; Schäfer, H.J. *SynLett*, ***1990***, 44.

MeO O CHN$_2$ O → $Rh_2(MEPY)_4$, CH_2Cl_2 / reflux → O O OMe (91% ee , S) 62%

MEPY = MeO_2C N O

Doyle, M.P.; Van Oeveren, A.; Westrum, L.J.; Protopopova, M.N.; Clayton, T.W. Jr. *J. Am. Chem. Soc.*, ***1991***, *113*, 8982.

N *t*-Bu O_2N → $BnEt_3NCl$, $CHCl_3$ / K_2CO_3 , H_2O , 3 h / oxone , pH 1.5 → *t*-Bu N O O_2N 83%

Bulachkova, A.I.; Koldobskii, G.I.; Drozdetskii, A.G.; Tereshchenko, G.F. *Zhur. Org. Khim.*, ***1991***, *27*, 215 (Engl., p. 189).

OMe Ph OMe → $BF_3\bullet OEt_2$, -78°C → 0°C / $OSiMe_3$ CO_2Et → MeO O Ph CO_2Et 86%

Sugimura, H.; Shigekawa, Y.; Uematsu, M. *SynLett*, ***1991***, 153.

CO_2H HO Me H + H *t*-Bu MeO OMe → PPTS , EtOAc / cyclohexane / Dean-Stark trap / 80°C → *t*-Bu O O H O Me + H O O *t*-Bu O Me

(97 : 3) 59%

Chapel, N.; Greiner, A.; Ortholand, J-Y. *Tetrahedron Lett.*, ***1991***, *32*, 1441.
Greiner, A.; Ortholand, J-Y. *Tetrahedron Lett.*, ***1990***, *31*, 2135.

Et SPh O CO_2Et 80% E → Bu_3SnH , AIBN , PhH / reflux , 6 h → O CO_2Et (35:65 cis:trans) 77%

Lolkema, L.D.M.; Hiemstra, H.; Ghouch, A.A.A.; Speckamp, W.N. *Tetrahedron Lett.*, ***1991***, *32*, 1491.

1. 2 eq. BuLi , THF , -78°C
2. CHO / HO
3. H_3O^+

EtO, EtO P(O) CO$_2$H SMe → MeS ... O 68%

Mikołajczyk, M.; Midura, W.H. *SynLett*, ***1991***, 245.

OSiMe$_2$*t*-Bu

SnCl$_4$, CH$_2$Cl$_2$, -78°C reflux

EtO$_2$C OBu → EtO$_2$C OBu *t*-BuMe$_2$SiO 69%

Komatsu, M.; Suehiro, I.; Horiguchi, Y.; Kuwajima, I. *SynLett*, ***1991***, 771.

1. [TBDMS-OTf/NaMn(CO)$_5$]
2. CO$_2$Me
3. hν
4. CSA

O → O O O 76%

De Shong, P.; Rybczynski, P.J. *J. Org. Chem.*, ***1991***, *56*, 3207.

BnSH , BuLi
THF , 0°C

Me Me CO$_2$Me → Me Me BnS CO$_2$Me + Me Me BnS CO$_2$Me

(98 : 2) 98%

Miyata, O.; Shinada, T.; Ninomiya, I.; Naito, T.; Date, T.; Okamura, K.; Inagaki, S. *J. Org. Chem.*, ***1991***, *56*, 6556.

4 eq. *Candida cylindracea*
hexanes , 0.005 M
OAc , 25°C , 48 h

HO OBn OBn OH OBn → AcO OBn OBn OH OBn 58%

(98% de)

Burgess, K.; Henderson, I. *Tetrahedron Lett.*, ***1991***, *32*, 5701.

1. Zn , AcOH , 60°C
2. NaCl , DMSO/H$_2$O
150°C

OH SPh → O O SPh CO$_2$Et 56%

Kido, F.; Kawada, Y.; Kato, M.; Yoshikoshi, A. *Tetrahedron Lett.*, ***1991***, *32*, 6159.

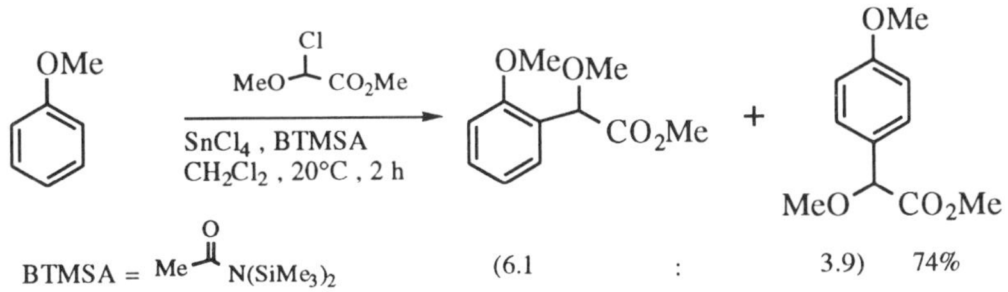

Earle, M.J.; Fairhurst, R.A.; Heaney, H. *Tetrahedron Lett.*, ***1991***, *32*, 6171.

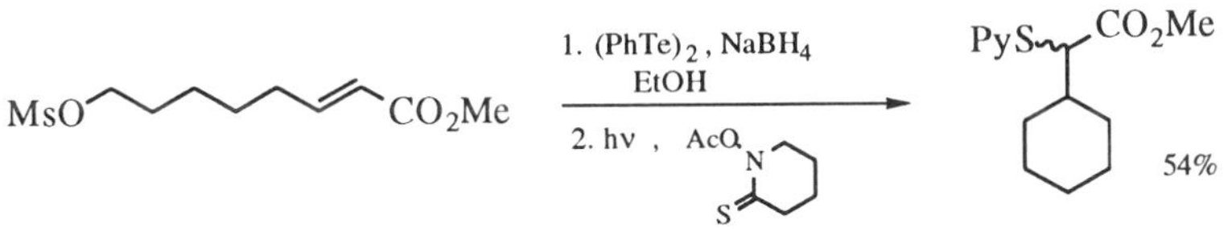

Barton, D.H.R.; Dalko, P.I.; Géro, S.D. *Tetrahedron Lett.*, ***1991***, *32*, 4713.

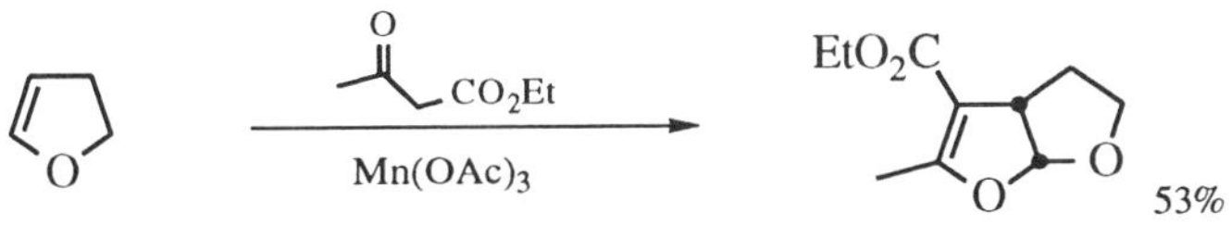

Mellor, J.M.; Mohammed, S. *Tetrahedron Lett.*, ***1991***, *32*, 7107, 7111.

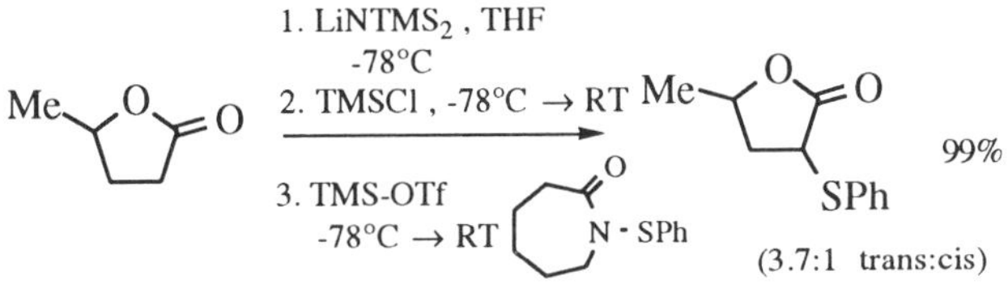

Wilson, L.J.; Liotta, D.C. *J. Org. Chem.*, ***1992***, *57*, 1948.

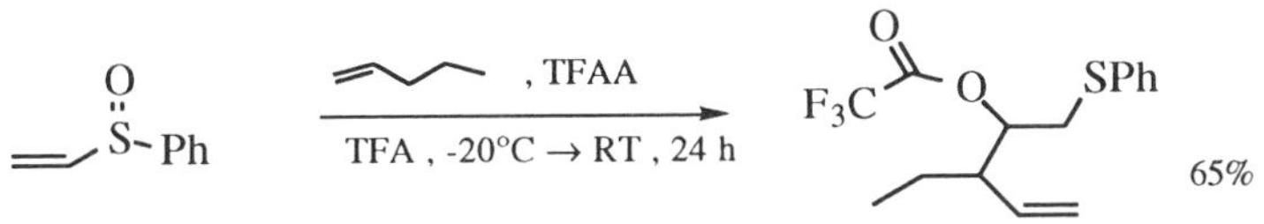

Brichard, M-H.; Janousek, Z.; Merény, R.; Viehe, H.G. *Tetrahedron Lett.*, ***1992***, *33*, 2511.
Brichard, M-H.; Musick, H.; Janousek, Z.; Viehe, H.G. *Synth. Commun.*, ***1990***, *20*, 2379.

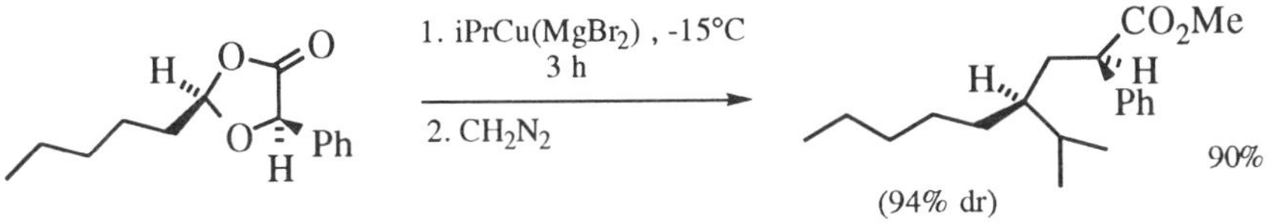

Heckmann, B.; Mioskowski, C.; Yu, J.; Falck, J.R. *Tetrahedron Lett.*, ***1992***, *33*, 5201.

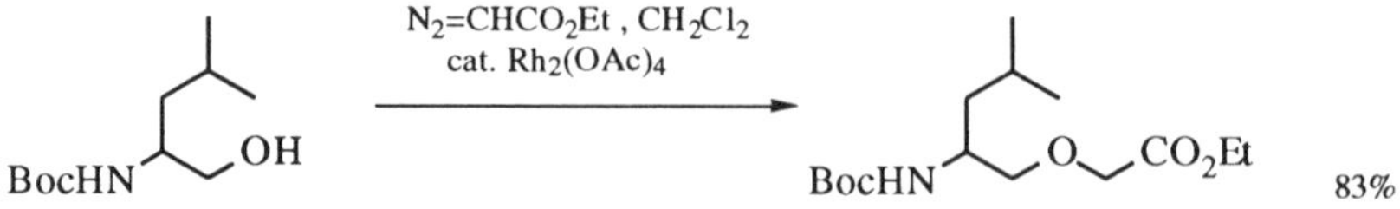

N_2=CHCO$_2$Et , CH$_2$Cl$_2$
cat. Rh$_2$(OAc)$_4$

BocHN OH → BocHN O CO$_2$Et 83%

Hlavacek, J.; Kral, V. *Coll. Czech. Chem. Commun.*, ***1992***, *57*, 525.

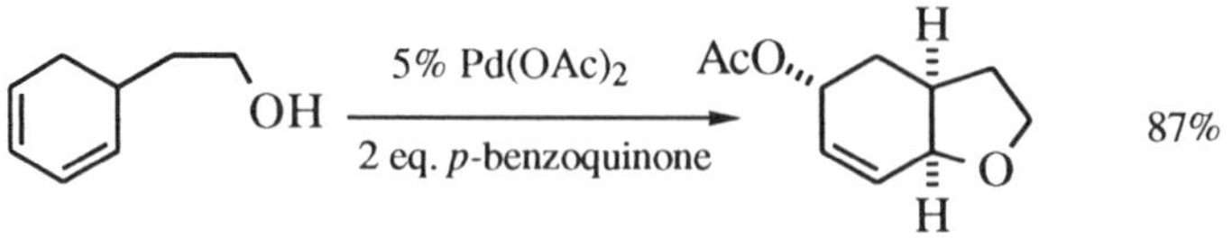

OH 5% Pd(OAc)$_2$ / 2 eq. *p*-benzoquinone → AcO H O H 87%

Bäckvall, J-E.; Andersson, P.G. *J. Am. Chem. Soc.*, ***1992***, *114*, 6374.

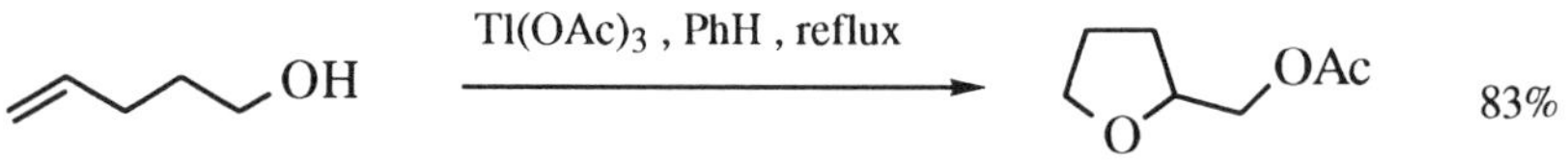

OH Tl(OAc)$_3$, PhH , reflux → O OAc 83%

Mihailović, M.Lj.; Vukićević, R.; Konstantinović, S.; Milosavljević, S.; Schroth, G. *Liebigs Ann. Chem.*, ***1992***, 305.

SECTION 359: ESTER - HALIDE, SULFONATE

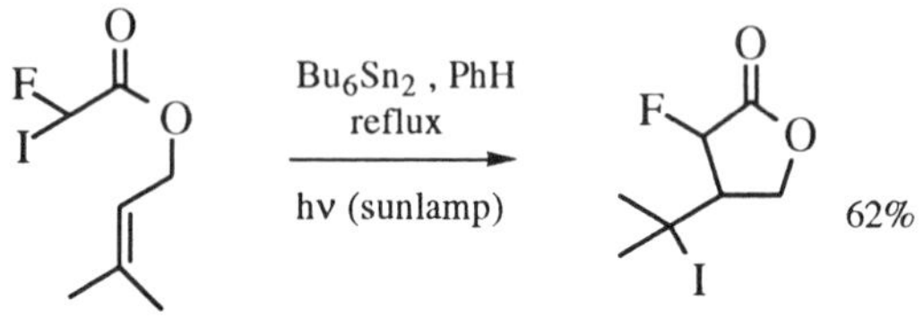

F I O O Bu$_6$Sn$_2$, PhH reflux / hν (sunlamp) → F O O I 62%

Barth, F.; O-Yang, C. *Tetrahedron Lett.*, ***1990***, *31*, 1121.

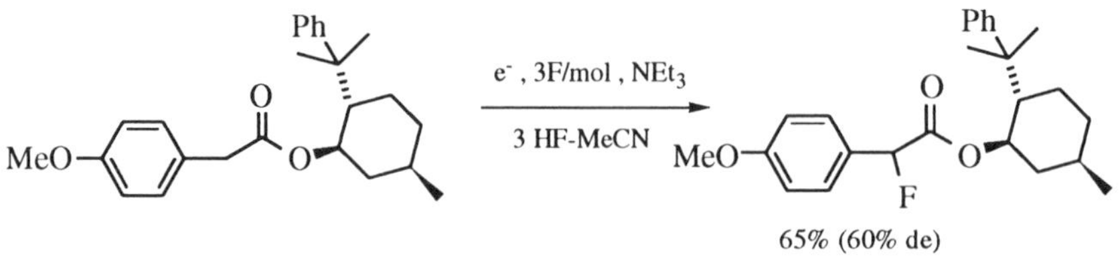

MeO O Ph O e$^-$, 3F/mol , NEt$_3$ / 3 HF-MeCN → MeO O F Ph O

65% (60% de)

Kabore, L.; Chebli, S.; Faure, R.; Laurent, E.; Marquet, B. *Tetrahedron Lett.*, ***1990***, *31*, 3137.

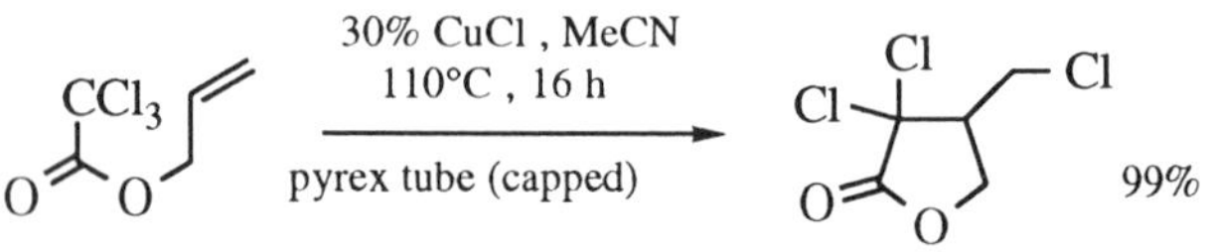

CCl$_3$ O O 30% CuCl , MeCN 110°C , 16 h / pyrex tube (capped) → Cl Cl Cl O O 99%

Nagashima, H.; Seki, K.; Ozaki, N.; Wakamatsu, H.; Itoh, K.; Tomo, Y.; Tsuji, J. *J. Org. Chem.*, ***1990***, *55*, 985.

$n\text{-}C_7H_{15}\text{-}Br$ → 1. $Bu_3P{=}CFCO_2Et$, THF, -78°C; 2. aq. $NaHCO_3$ → $n\text{-}C_7H_{15}CFHCO_2Et$

Thenappan, A.; Burton, D.J. *J. Org. Chem.*, ***1990***, *55*, 2311.

OSiMe$_3$ → 1. $[Rh(OAc)_2]_2$, N_2=C(CF_3)CO_2Et, ether, reflux; 2. TBAF, THF → O, CF_3, CO_2Et 83%

Shi, G.; Xu, Y. *J. Org. Chem.*, ***1990***, *55*, 3383.

CO_2Et → $Ca(OCl)_2$, AcOH, CH_2Cl_2, H_2O → CO_2Et, Cl 80%

Mathew, J.; Alink, B. *J. Org. Chem.*, ***1990***, *55*, 3850.

O, OH, Me, Me → 1. $Tf_2O{\bullet}NEt(iPr)_2$, CH_2Cl_2, -78°C; 2. $PhI(OAc)_2$, hν → O, O, Me, I, Me

Kaino, M.; Naruse, Y.; Ishihara, K.; Yamamoto, H. *J. Org. Chem.*, ***1990***, *55*, 5814.

→ TFA, $CuCl_2$, $Pb(OAc)_4$ → O_2CCF_3, Cl 83%

Serguchev, Yu.A.; Gutsulyak, R.B. *Zhur. Org. Khim.*, ***1990***, *26*, 2066 (Engl., p. 1784).

CO_2H → $CHCl_3$, iPr_2NEt, DMSO-TMSBr, reflux, 12 h → Br, O, O 55%

Iwata, C.; Tanaka, A.; Mizuno, H.; Miyashita, K. *Heterocycles*, ***1990***, *31*, 987.

$OSiMe_2t\text{-}Bu$, $On\text{-}C_8H_{17}$ → $CHBr_3$, BEt_3, 1.5 h → Br, $CO_2n\text{-}C_8H_{17}$ 60%

Sugimoto, J.; Miura, K.; Oshima, K.; Utimoto, K. *Chem. Lett.*, ***1991***, 1319.

n-C_6H_{13} O NMe$_2$ HO

I_2, aq. THF, RT

n-C_6H_{13} O O HO I + n-C_6H_{13} O O HO I

(7 : 1) 60%

Rotella, D.P.; Li, X. *Heterocycles*, ***1990***, *31*, 1205.

ICF_2CO_2Me, DMF
Cu, 6 h

I CF_2CO_2Me 79%

(5:3 trans:cis)

Kitagawa, O.; Miura, A.; Kobayashi, Y.; Taguchi, T. *Chem. Lett.*, ***1990***, 1011.

Me CO_2H OH

1. $Ti(OiPr)_4$, 0°C, 2 h
2. NIS, CH_2Cl_2

OH Me O O I 70%

(23:1 cis:trans)

Kitagawa, O.; Sato, T.; Taguchi, T. *Chem. Lett.*, ***1991***, 177.

Ph CO_2Me

1. LDA, THF, -78°C
2. THF, -78°C

N · F S O_2

Ph CO_2Me F + Ph CO_2Me F F

(95 : 5) 40%

Differding, E.; Rüegg, G.M.; Lang, R.W. *Tetrahedron Lett.*, ***1991***, *32*, 1779.

Me CO_2Et Ph

1. LDA, THF, 0°C
2. $(CF_3SO_2)_2NF$
-80°C

Me CO_2Et Ph F 83%

Resnati, G.; Des Marteau, D.D. *J. Org. Chem.*, ***1991***, *56*, 4925.

ICF_2CO_2Me, Cu
55°C

I CF_2CO_2Me 78%

Yang, Z-Y.; Burton, D.J. *J. Org. Chem.*, ***1991***, *56*, 5125.
Yang, Z-Y.; Burton, D.J. *J. Chem. Soc., Chem. Commun.*, ***1992***, 233.

TMPAP , dioxane
RT

Br
Br
52%

TMPAP = trimethyl(phenyl)ammonium perbromide

Collado, I.G.; Massanet, G.M.; Alonso, M.S. *Tetrahedron Lett.*, ***1991***, *32*, 3217.

$AgNO_3$, Br_2
NaOH , $MeCN/H_2O$

NBS , $KHCO_3$
CH_2Cl_2/H_2O

CO_2H

Br 74%

Br 93%

Dai, W.; Katzenellenbogen, J.A. *J. Org. Chem.*, ***1991***, *56*, 6893.

$VO(OEt)Cl_2$, $CBrCl_3$

$Br\diagdown\diagup\diagdown CO_2Et$

60%

Hirao, T.; Fujii, T.; Miyata, S.; Ohshiro, Y. *J. Org. Chem.*, ***1991***, *56*, 2264.

1. PhLi , ether , -78°C
2. HgO , I_2 , PhH
3. hv , 25°C

Ph I
56%

Kobayashi, K.; Sasaki, A.; Kanno, Y.; Suginome, H. *Tetrahedron*, ***1991***, *47*, 7245.

1. Br_2 , PBr_3 , 120°C , 6 h
2. CH_2Cl_2 , PhSH , RT
16 h

CO_2H

SPh 85%
Br

Liu, H-J.; Luo, W. *Synth. Commun.*, ***1991***, *21*, 2097.

THF , 1% AIBN
1% BPO , reflux
12 h

EtO_2C F F Me

EtO_2C F F Me
73%

BPO = benzoyl peroxide

Bumgardner, C.L.; Burgess, J.P. *Tetrahedron Lett.*, ***1992***, *33*, 1683.

$Ti(Ot\text{-}Bu)_4$, I_2, CH_2Cl_2 , RT; CO_2Me, CO_2Me, CO_2Me, CO_2Me, I — 74%

Kitagawa, O.; Inoue, T.; Taguchi, T. *Tetrahedron Lett.*, ***1992***, *33*, 2167.

PhSe, CO_2Et; e^- , $Et_3N{\bullet}3$ HF , MeCN; PhSe, CO_2Et, F — 87%

Fuchigami, T.; Hayashi, T.; Konno, A. *Tetrahedron Lett.*, ***1992***, *33*, 3161.

O, $CHCl_2$, O; Cu(bpy)Cl , reflux, PhH (0.1 M) , 18 h; Cl, O, Cl, O — 92% (70:30 E:Z)

Pirrung, F.O.H.; Steeman, W.J.M.; Hiemstra, H.; Speckamp, W.N.; Kaptein, B.; Boesten, W.H.J.; Schoemaker, H.E.; Kamphuis, J. *Tetrahedron Lett.*, ***1992***, *33*, 5141.

OBoc; IBr , PhMe , -80°C; O, O, O, I + O, O, O, I — (21.1 1) 89%

Duan, J.J-W.; Sprengeler, P.A.; Smith, A.B. III *Tetrahedron Lett.*, ***1992***, *33*, 6439.

O, Cl, Br; 1. Me_2NEt , THF , -78°C; 2. O, O, HO; O, O, O, O, Cl — 71% (85% dr)

Durst, T.; Koh, K. *Tetrahedron Lett.*, ***1992***, *33*, 6799.

DMF (1M HCl) , RT, oxone , 1 h; Cl, OCHO — 85%

Kim, J.N.; Kim, H.R.; Ryu, E.K. *Synth. Commun.*, ***1992***, *22*, 2521.

BzCl , Bu_2SnCl_2-PPh_3, 70°C , 15 h

$SnCl_2$, PPh_3 , BzCl, 70°C , 11 h

(87 : 13) 80%

(1 : 99) 58%

Shibata, I.; Yoshimura, N.; Baba, A.; Matsuda, H. *Tetrahedron Lett.*, ***1992***, *33*, 7149.

$CH_2(CO_2Et)_2$, NaH , THF , 25°C, 2 h

(2 : 3) 66%

Lawrence, R.M.; Perlmutter, P. *Chem. Lett.*, ***1992***, 305.

SECTION 360: ESTER - KETONE

$Mn(OAc)_3$, $Cu(OAc)_2$, AcOH, 25°C , 3h

61%

Method extended to include formation of bridged molecules

White, J.D.; Somers, T.C.; Yager, K.M. *Tetrahedron Lett.*, ***1990***, *31*, 59.

1. $Me_2C=C(OMe)OSiMe_3$, $TiCl_4$, CH_2Cl_2 2. aq. $NaHCO_3$

60%

Collins, D.J.; Dosen, M.; Jhingran, A.G. *Tetrahedron Lett.*, ***1990***, *31*, 421.

$CH_2=CHCO_2Me$, Ph_2S_2 , AIBN, PhCl , hν (sunlamp)

(1 : 1) 46%

Feldman, K.S.; Vong, A.K.K. *Tetrahedron Lett.*, ***1990***, *31*, 823.

PhI(OH)OTs

CH_2Cl_2 , reflux 15h

Ph, CO_2H, O, 74%

Moriarty, R.M.; Vaid, R.K. Hopkins, T.E.; Vaid, B.K.; Prakash, O. *Tetrahedron Lett.*, ***1990***, *31*, 201.

1. O_3 , CH_2Cl_2 , MeOH -78°C

2. PhH , CH_2Cl_2 , 0°C $SnCl_2$ N_2 CO_2Et O

CO_2Et 63%

Holmquist, C.R.; Roskamp, E.J. *Tetrahedron Lett.*, ***1990***, *31*, 4991.

1. PPh_3 CO_2t-Bu

2. oxone , aq. THF RT

O Cl CO_2t-Bu 86%

Wasserman, H.H.; Vu, C.B. *Tetrahedron Lett.*, ***1990*** , *31*, 5205.

1. LDA , THF , -78°C
2. Ph_2SiCl_2
3. OH O CO_2Me O
4. PhH , reflux , 4 h
5. KH , MeOH

CO_2Me CO_2Me H H

61% overall

intramolecular Diels-Alder using a "disposable tether"

Shea, K.J.; Zandi, K.S.; Staab, A.J.; Carr, R. *Tetrahedron Lett.*, ***1990***, *31*, 5885.

18 nM , DMSO aq. NaOH , 50°C

20 min

O Cl CO_2H O O 72%

Karim, M.R.; Sampson, P. *J. Org. Chem.*, ***1990***, *55*, 598.

HI , Ac_2O , RT

OAc

quant.

Zoeller, J.R.; Ackerman, C.J. *J. Org. Chem.*, ***1990***, *55*, 1354.

OLi

1. TBSO ... OEt, Br

2. TMSI , HMDS
-20°C → RT
3. $HClO_4$, aq. THF

TBSO H O

EtO_2C 76%

Hudlicky, T.; Heard, N.E.; Fleming, A. *J. Org. Chem.*, ***1990***, *55*, 2570.

O Br CO_2Et

e^- (4 F/mol)
Pb cathode
TMS-Cl-DMF

CO_2Et 76%

Shono, T.; Kise, N.; Uematsu, N.; Morimoto, S. *J. Org. Chem.*, ***1990***, *55*, 5037.

n-C_3H_7 C CO_2H

acetone

n-C_3H_7 OH + n-C_3H_7 OH

(2 : 3) 84%

Crandall, J.K.; Rambo, E. *J. Org. Chem.*, ***1990***, *55*, 5929.

Me

EtO_2C CO_2Et

MeO O O O

$Pd(OAc)_2$, PPh_3 , THF
reflux , 4 h

Me O

EtO_2C CO_2Et 84%

Ikeda, I.; Gu, X-P.; Okuhara, T.; Okahara, M. *Synthesis*, ***1990***, 32.

O CO_2Et Bu

1. NaOEt , EtOH
2. O_3
3. Me_2S

Bu CO_2Et 48%

Si, Z-X.; Jiao, X-Y.; Hu, B-F *Synthesis*, ***1990***, 509.

1. 2 eq. MeLi, THF, HMPA
2. BnBr
3. H^+
4. CH_2N_2

57%

Tapia, I.; Alcazar, V.; Moran, J.R.; Caballero, C.; Grande, M. *Chem. Lett.*, ***1990***, 697.

1. O_3
2. O_2

52%

Scholz, G.; Konusch, J.; Tochtermann, W. *Liebigs Ann. Chem.*, ***1990***, 593.

Bu_3SnH, AIBN

63%

Boger, D.L.; Mathvink, R.J. *J. Am. Chem. Soc.*, ***1990***, *112*, 4003, 4008.
Boger, D.L.; Mathvink, R.J. *J. Org. Chem.*, ***1992***, *57*, 1429.

1. LDA, ether
2. Me–CH=CH–CO_2Me, -100°C
3. aq. NH_4Cl
4. $CuSO_4$, H_2O

(93% ee, R) 69%

Enders, D.; Gerdes, P.; Kipphardt, H. *Angew. Chem. Int. Ed. Engl*, ***1990***, *29*, 179.

Bu_3SnH, PhH
0.01 M, AIBN
reflux

(>50 : 1) 82%

Enholm, E.J.; Kinter, K.S. *J. Am. Chem. Soc.*, ***1991***, *113*, 7784.

n-$C_7H_{15}CHO$

1. Zn, THF
2. Ac_2O, Py
3. NBS, AgOAc
4. t-BuOOH, OsO_4, MeOH
epichlorohydrin

60%

Chen, C.; Crich, D. *J. Chem. Soc., Chem. Commun.*, ***1991***, 1289.

Et, MeO, OMe, CO_2Et

1. $PhCO_2Me$, $TiCl_4$, CH_2Cl_2
2. MeOH, -78°C → 0°C
3. NEt_3

Et, OMe, OMe, Ph, MeO_2C, CO_2Et 90% + Et, O, MeO_2C, Ph, CO_2Et 0%

Shimada, S.; Saigo, K.; Hashimoto, Y.; Maekawa, Y.; Yamashita, T.; Yamamoto, T.; Hasegawa, M. *Chem. Lett.*, ***1991***, 1475.

CHO

1. $(CO_2Et)_2$, NaOMe, 2 h, MeOH, 10°C
2. conc. HCl, RT, 1 h

O, O, O 92%

Hata, H.; Morishita, T.; Akutsu, S.; Kawamura, M. *Synthesis*, ***1991***, 289.

$OSiMe_2t$-Bu, OiPr

Ph(C=O)Cl, CH_2Cl_2
$Pd(Bz)Cl(PPh_3)_2$, 2.5 h

Ph, O, OiPr, O 97%

Fujimura, T.; Aoki, S.; Nakamura, E. *J. Org. Chem.*, ***1991***, *56*, 2809.

O, CO_2Et, NMe_2

1. $LiN(SiMe_3)_2$, THF, -78°C, BzCl
2. aq. HCl

O, CO_2Et, Ph, O 63%

[O, CO_2Et, Ph, N, H 58% with $NH_4OAc/AcOH$ at 60°C]

McCombie, S.W.; Metz, W.A.; Nazareno, D.; Shankar, B.B.; Tagat, J. *J. Org. Chem.*, ***1991***, *56*, 4963.

O, Me, MeO

1. LDA
2. CO_2Me

O, Me, CO_2Me, MeO, Me 95%

Schinzer, D.; Kalesse, M. *Tetrahedron Lett.*, ***1991***, *32*, 4691.

O, CO_2Et

O, 3% LiI
DME, reflux, 24 h

O, O, CO_2Et 90%

Antonioletti, R.; Bonadies, F.; Monteagudo, E.S.; Scettri, A. *Tetrahedron Lett.*, ***1991***, *32*, 5373.

$PhI(OAc)_2$, I_2
CH_2Cl_2 , hν (Pyrex)

OH

O

I

54%

Galatsis, P.; Millan, S.D. *Tetrahedron Lett.*, ***1991***, *32*, 7493.

OMe
OSiMe$_3$, TASF , -78°C

O

O

O

O

CO_2Me

64%

TASF = tris(dimethylamino)sulfonium trimethyl difluorosiliconate

Klimko, P.G.; Singleton, D.A. *J. Org. Chem.*, ***1992***, *57*, 1733.

O

CO_2Me

OMe

1. $TosN_3$, Al_2O_3 KF , THF , RT
2. $Rh_2(OAc)_4$, RT

O

CO_2Me

46%

Ceccherelli, P.; Curini, M.; Marcotullio, M.C.; Rosati, O. *Org. Prep. Proceed. Int.*, ***1992***, *24*, 497.

Me

C

C

O

O

NMO , $Co_2(CO)_6$
CH_2Cl_2 , 5 h

O

Me

O

O

66%

Krafft, M.E.; Romero, R.H.; Scott, I.L. *J. Org. Chem.*, ***1992***, *57*, 5277.

O

I

CO_2Me

Bu_3SnH , AIBN
PhH , reflux

O

CO_2Me

79%

Dowd, P.; Choi, S-C. *Tetrahedron*, ***1992***, *48*, 4773.
Dowd, P.; Choi, S-C. *Tetrahedron Lett.*, ***1991***, *32*, 565.

2% $Rh_2(TPA)_4$, CH_2Cl_2, 0°C, 1.5 h

TPA = triphenyl acetate (96 : 4) 75%

Hashimoto, S.; Watanabe, N.; Ikegami, S. *Tetrahedron Lett.*, ***1992***, *33*, 2709.
Hashimoto, S.; Watanabe, N.; Ikegami, S. *Tetrahedron Lett.*, ***1990***, *31*, 5173.

MeO(CH2)COCl; $AlCl_3$, $MeNO_2$, 80°C, 2 h — 68%

Sartori, G.; Bigi, F.; Tao, X.; Casnati, G.; Canali, G. *Tetrahedron Lett.*, ***1992***, *33*, 4771.

1. pyrrolidine
2. LDA, THF
3. MeI, -78°C → RT
4. H_3O^+

80%

Hodgson, A.; Marshall, J.; Hallett, P.; Gallagher, T. *J. Chem. Soc., Perkin Trans. I*, ***1992***, 2169.

$KMnO_4 \cdot CuSO_4 \cdot 5\ H_2O$, cat. H_2O, CH_2Cl_2, 7 h — 47%

Baskaran, S.; Islam, I.; Vankar, P.S.; Chandrasekaran, S. *J. Chem. Soc., Chem. Commun.*, ***1992***, 626.

$PhI(O_2CCF_3)_2$, MeCN, 0°C — 83%

McKillop, A.; McLaren, L.; Taylor, R.J.K.; Watson, R.J.; Lewis, N. *SynLett*, ***1992***, 201.

Also via Ketoacids — Section 320 (Carboxylic Acid - Ketone)
Hydroxyketones — Section 330 (Alcohol - Ketone)

SECTION 361: ESTER - NITRILE

CN

iPrOH , CH_2Cl_2
$Co_2(CO)_8$, 130°C
200 bar

CN CN CO$_2$iPr 89% + NC CO$_2$iPr 8%

Sisak, A.; Ungváry, F.; Markó, L. *J. Org. Chem.*, ***1990***, *55*, 2508.

N S S CO$_2$Me Me

1. Bu_2CuLi , THF -30°C
2. BuBr

CO$_2$Me Me 80%

Calò, V.; Lopez, L.; Pesce, G. *Gazz. Chim. Ital.*, ***1990***, *120*, 203.

PhCOOH → $ClCH_2CN$, DMSO / Na_2CO_3 → Ph O CN, O 97%

Byers, J.H.; Baran, R.C.; Craig, M.E.; Jackson, J.T. *Org. Prep. Proceed. Int.*, ***1991***, *23*, 373.

PhCOOH → $ClCH_2CN$, NEt_3 → $PhCO_2CH_2C{\equiv}N$ 90%

80% ← Na_2S , aq. acetone

Hugel, H.M.; Bhaskar, K.V.; Longmore, R.W. *Synth. Commun.*, ***1992***, *22*, 693.

SECTION 362: ESTER - ALKENE

This section contains syntheses of enol esters and esters of unsaturated acids as well as ester molecules bearing a remote alkenyl unit.

O

1. 1.2 eq. LDA
2. 2 eq. $HgCl_2$

O HgCl

Li_2PdCl_4
THF-HMPA
25°C , 4d

O O 63%

Larock, R.C.; Stinn, D.E.; Kuo, M-Y. *Tetrahedron Lett.*, ***1990***, *31*, 17.

OH SnBu$_3$

1. Ac_2O
2. KHMDS
t-$BuMe_2SiCl$
3. heat

CO$_2$Me SnBu$_3$ 60%

Ritter, K. *Tetrahedron Lett.*, ***1990***, *31*, 869.

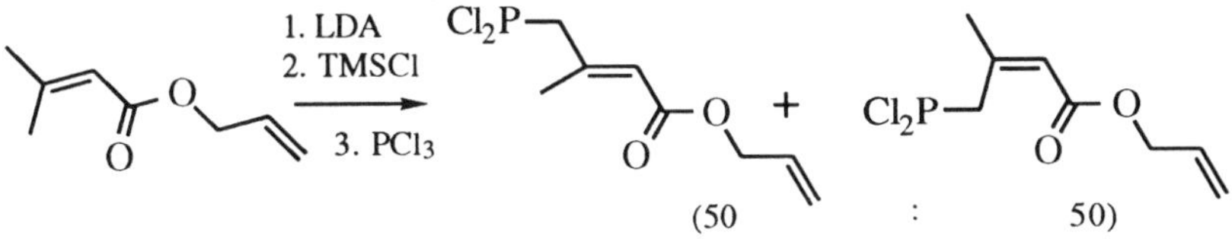

1. LDA
2. TMSCl
3. PCl_3

Cl_2P ... + Cl_2P ...

(50 : 50)

Pellon, P.; Himdi-Kabbab, S.; Rault, I.; Tonnard, F.; Hamelin, J. *Tetrahedron Lett.*, ***1990***, *31*, 1147.

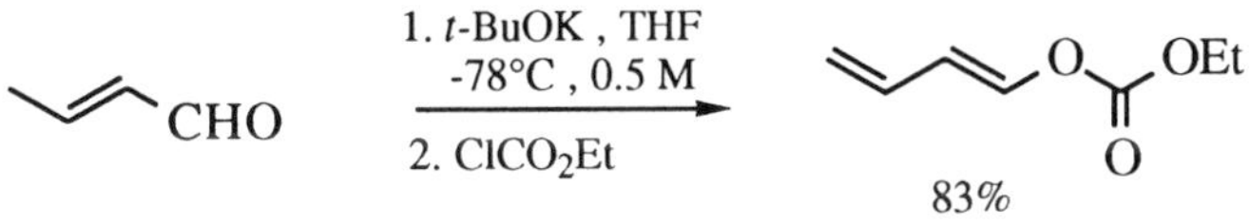

CHO

1. *t*-BuOK , THF
-78°C , 0.5 M
2. $ClCO_2Et$

O OEt O

83%

De Cusati, P.F.; Olofson, R.A. *Tetrahedron Lett.*, ***1990***, *31*, 1405.

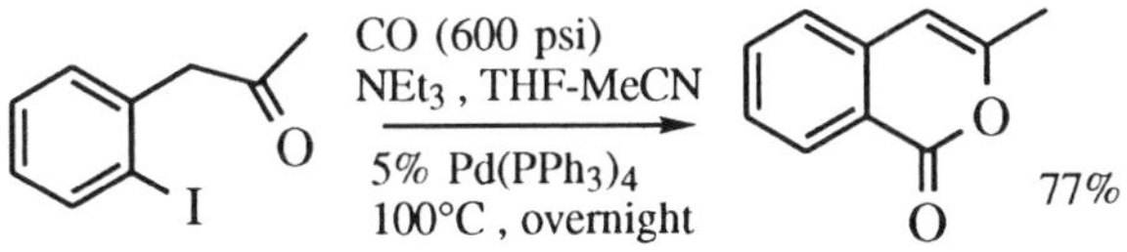

O I

CO (600 psi)
NEt_3 , THF-MeCN
5% $Pd(PPh_3)_4$
100°C , overnight

O O 77%

Shimoyama, I.; Zhang, Y.; Wu, G.; Nigishi, E. *Tetrahedron Lett.*, ***1990***, *31*, 2841.

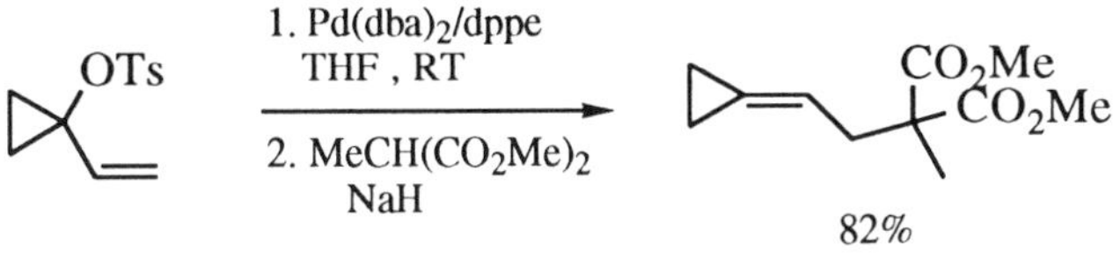

OTs

1. $Pd(dba)_2$/dppe
THF , RT
2. $MeCH(CO_2Me)_2$
NaH

CO_2Me CO_2Me

82%

Stolle, A.; Salaün, J.; de Meijere, A. *Tetrahedron Lett.*, ***1990***, *31*, 4593.
Stolle, A.; Salaün, J.; de Meijere, A. *SynLett*, ***1991***, 321.

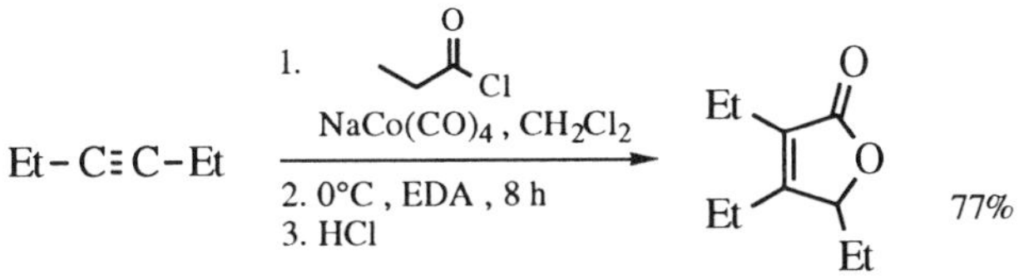

Et-C≡C-Et

1. O Cl
$NaCo(CO)_4$, CH_2Cl_2
2. 0°C , EDA , 8 h
3. HCl

O Et O Et Et 77%

Krafft, M.E.; Pankowski, J. *Tetrahedron Lett.*, ***1990***, *31*, 5139.
Krafft, M.E.; Pankowski, J. *SynLett*, ***1991***, 865.

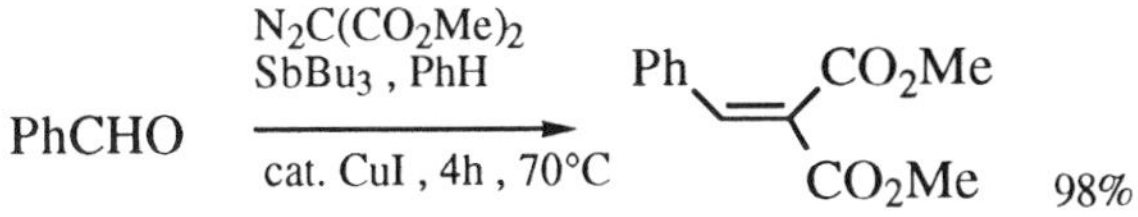

PhCHO

$N_2C(CO_2Me)_2$
$SbBu_3$, PhH
cat. CuI , 4h , 70°C

Ph CO_2Me CO_2Me 98%

also works with ketones

Liao, Y.; Huang, Y-Z. *Tetrahedron Lett.*, ***1990***, *31*, 5897.

CO_2Me O

1. t-BuOK, -30°C, 14 h
2. PhI, Pd(dba)$_2$, 2 h
5% dppe

Ph CO_2Me O 88%

Fournet, G.; Balme, G.; Van Hemelryck, B.; Gore, J. *Tetrahedron Lett.*, ***1990***, *31*, 5147.

=C n-C_6H_{13} n-C_5H_{11} CO_2Me

MeOH, CO, RT
3% Pd(PPh$_3$)$_4$, 1 h

n-C_5H_{11} n-C_6H_{13} CO_2Me

(1:1 E:Z) 97%

Nokami, J.; Maihara, A.; Tsuji, J. *Tetrahedron Lett.*, ***1990***, *31*, 5629.

CO_2Et CO_2Et

, 30% EtOH
bakers yeast, 48 h
pH 7.2, 37°C

CO_2Et CO_2Et 74%

(endo:exo 0:100)

Rama Rao, K.; Srinivasan, T.N.; Bhanumathi, N. *Tetrahedron Lett.*, ***1990***, *31*, 5959.

Me_3Si CO_2Me CO_2Me

prepared from alkenes and
2-2-(silylethyl) dienes

Ph$_3$C$^+$ BF$_4^-$, MeCN
RT, 5 h
MeO_2C CO_2Me

CO_2Me MeO_2C CO_2Me CO_2Me

Hosomi, A.; Masunari, T.; Tominaga, Y.; Yanagi, T.; Hojo, M. *Tetrahedron Lett.*, ***1990***, *31*, 6201.

Me_3Si-C≡C Me OH

1. iBuMgBr
cat. Cp$_2$TiCl$_2$
2. CO$_2$
3. H$_3$O$^+$

O O Me 63%

Ito, T.; Okamoto, S.; Sato, F. *Tetrahedron Lett.*, ***1990***, *31*, 6399.

O Ph Si O Ph CO_2Me

PhMe, 160°C
168 h

O $SiPh_2$ O CO_2Me 62%

HCl
MeOH
20°C

OH O O 54%

Craig, D.; Reader, J.C. *Tetrahedron Lett.*, ***1990***, *31*, 6585.

1. $ClCO_2iBu$
2. CH_2N_2
3. hν, EtOH
4. Ac_2O, Py

n-$C_{10}H_{21}$ epoxide CO_2H → AcO, n-$C_{10}H_{21}$, CO_2Et 24% overall

Thijs, L.; Dommerholt, F.J.; Leemhuis, F.M.C.; Zwanenburg, B. *Tetrahedron Lett.*, ***1990***, *31*, 6589.

H-C≡C-CO_2Me → (Me_3SnH, THF; 2% $Pd(PPh_3)_4$) → CO_2Me, $SnMe_3$ 71%

Cochran, J.C.; Bronk, B.S.; Terrence, K.M.; Phillips, H.K. *Tetrahedron Lett.*, ***1990***, *31*, 6621.

CO_2Me → (1. LDA; 2. MeI; 3. mCPBA; 4. TFA, $CHCl_3$) → O, =O 40%

Cory, R.M.; Ritchie, B.M.; Shrier, A.M. *Tetrahedron Lett.*, ***1990***, *31*, 6789.

Me, I, Br, I → (2 CO_2Me; $Pd(OAc)_2$, NEt_3; MeCN, 100°C) → MeO_2C, Me, Br, CO_2Me 46%

Tao, W.; Nesbitt, S.; Heck, R.S. *J. Org. Chem.*, ***1990***, *55*, 63.

Cl, Cl, CO_2Me, CO_2Me → (Cu°, DMSO, 25°C; 5 h) → CO_2Me, MeO_2C 73% yield 64% conversion

70% E isomer

Tezuka, Y.; Hashimoto, A.; Ushizaka, K.; Imai, K. *J. Org. Chem.*, ***1990***, *55*, 329.

n-$C_5H_{11}CO_2Et$ → (1. $LiCHBr_2$; BuLi, TIPS-Cl; 2. Ph CHO, $TiCl_4$, CH_2Cl_2; 3. MeOH, -78°C→ RT) → CO_2Me, n-C_5H_{11}, Ph 91%

Kowalski, C.J.; Sakdarat, S. *J. Org. Chem.*, ***1990***, *55*, 1977.

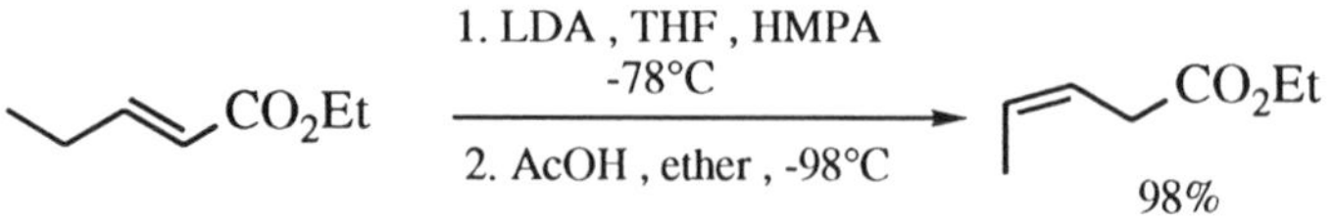

1. LDA , THF , HMPA -78°C
2. AcOH , ether , -98°C

CO_2Et → CO_2Et 98%

Piers, E.; Gavai, A.V. *J. Org. Chem.*, ***1990***, *55*, 2374.

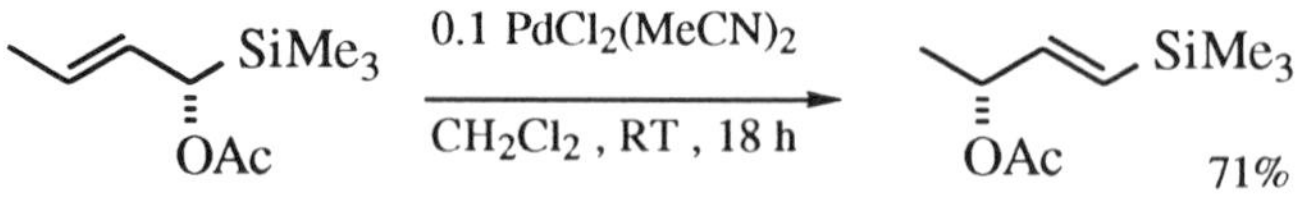

$0.1\ PdCl_2(MeCN)_2$ / CH_2Cl_2 , RT , 18 h

$SiMe_3$, OAc → $SiMe_3$, OAc 71%

Panek, J.S.; Sparks, M.A. *J. Org. Chem.*, ***1990***, *55*, 5564.

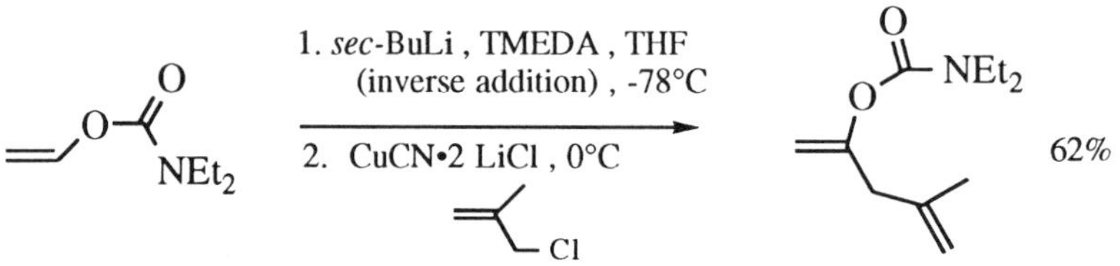

1. *sec*-BuLi , TMEDA , THF (inverse addition) , -78°C
2. CuCN•2 LiCl , 0°C, Cl

O, O, NEt_2 → O, O, NEt_2 62%

Sengupta, S.; Snieckus, V. *J. Org. Chem.*, ***1990***, *55*, 5680.

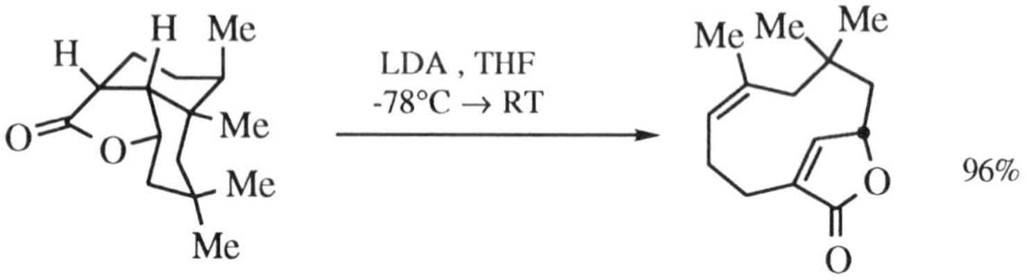

LDA , THF -78°C → RT

H, H, Me, O, O, Me, Me, Me → Me, Me, Me, O, O 96%

Wender, P.A.; Manly, C.J. *J. Am. Chem. Soc.*, ***1990***, *112*, 8579.

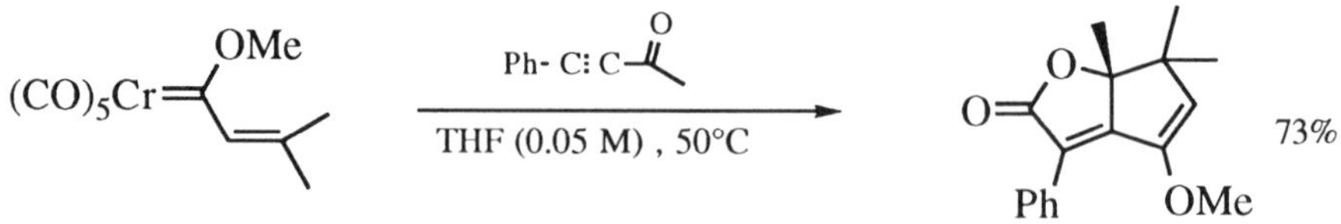

Ph- C≡C–C(O)– / THF (0.05 M) , 50°C

$(CO)_5Cr$, OMe → O, O, Ph, OMe 73%

Brandvold, T.A.; Wulff, W.D.; Rheingold, A.L. *J. Am. Chem. Soc.*, ***1990***, *112*, 1645.

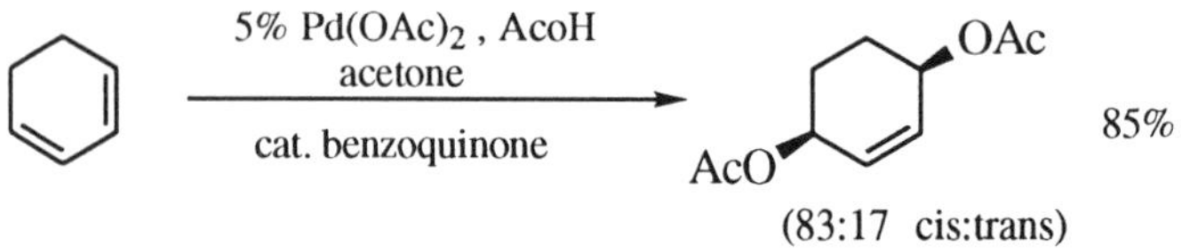

5% $Pd(OAc)_2$, AcoH acetone / cat. benzoquinone

→ OAc, AcO 85%

(83:17 cis:trans)

Bäckvall, J-E.; Grandberg, K.L.; Hopkins, R.B. *Acta Chem. Scand.*, ***1990***, *44*, 492.

1. NaH
2. 5% $Pd_2(dba)_3 \cdot CHCl_3$
15% (bicyclic phosphite), 40°C
DMSO, 5 min

(4 : 1) 67%

Hanzawa, Y.; Ishizawa, S.; Ito, H.; Kobayashi, Y.; Taguchi, T. *J. Chem. Soc., Chem. Commun.*, ***1990***, 394.

Bu_3Sn ... SePh

Bu_3SnH, AIBN, PhH
5 mM, reflux, 48 h

34%

no lactone was formed

Baldwin, J.E.; Adlington, R.M.; Mitchell, M.B.; Robertson, J. *J. Chem. Soc., Chem. Commun.*, ***1990***, 1574.

1. O_3, CH_2Cl_2, -78°C
2. $Ph_3P=CHCO_2Me$

CO_2Me

89%

Hon, Y-S.; Lu, L.; Li, S-Y. *J. Chem. Soc., Chem. Commun.*, ***1990***, 1627.

CO_2Et CO_2Et I

Bu (diene), DMF
5% Pd)OAc)$_2$, Bu_4NCl
Na_2CO_3, PPh_3, 60°C
1 d

CO_2Et CO_2Et 87%

Larock, R.C.; Fried, C.A. *J. Am. Chem. Soc.*, ***1990***, *112*, 5882.

Me HO C≡C-H

t-$BuSiMe_2H$, PhH
CO, DBU, 100°C
$Rh_2(CO)_{12}$

Me O O 79% + Me HO OHC $SiMe_2$*t*-Bu 6%

Matsuda, I.; Ogiso, A.; Sato, S. *J. Am. Chem. Soc.*, ***1990***, *112*, 6120.

C≡C-CO_2Me

$IrH_5(P[iPr]_3$ - PBu_3
PhMe, 80°C, 30 h

CO_2Me

87%

Ma, D.; Lu, X. *Tetrahedron*, ***1990***, *46*, 3189.

Ts, CO_2Et, PhMe; $(BzO)_2$, reflux

MeO_2C CO_2Me → Ts CO_2Et MeO_2C CO_2Me 64%

Chuang, C-P. *SynLett*, ***1990***, 527.

Ph CO_2H — 1. NEt_3, CH_2Cl_2, RT; 2. PBu_3, PhSeCl, THF RT → Ph (O) SePh 65%

Batty, D.; Crich, D. *Synthesis*, ***1990***, 273.

e$^-$ (Pt electrodes), 24 h; $[(bpy)(trpy)RuO]^{+2}$; pH 8 buffer → 42%

trpy = 2,2',6',2"-terpyridine

Campos, J.L.; De Giovani, W.F.; Romero, J.R. *Synthesis*, ***1990***, 597.

CO_2Me — PhSSPh, THF, reflux → CO_2Me 81%

(98:2 E:Z)

Miyata, O.; Shinada, T.; Ninomiya, I.; Naito, T. *Synthesis*, ***1990***, 1123.

$EtO-P(O)(EtO)-CH_2CO_2Et$ — cyclohexanone, LiBr; NEt_3, 25°C → CO_2Et 79%

Rathke, M.W.; Bouhlel, E. *Synth. Commun.*, ***1990***, *20*, 869.

PhCHO — $Br_2C(CO_2Et)_2$, THF; 2 eq. Bu_2Te, 50°C → Ph CO_2Et CO_2Et 93%

Zhou, Z-L.; Shi, L-L.; Huang, Y-Z. *Synth. Commun.*, ***1991***, *21*, 1027.

BCl_2 — N_2CHCO_2Et, ether; -65°C → CO_2Et 65%

Brown, H.C.; Salunkhe, A.M. *SynLett*, ***1991***, 684.

CO_2Me — MeOH, 80°C → MeO_2C, Me

Gruseck, U.; Heuschmann, M. *Chem. Ber.*, ***1990***, *123*, 1911, 1905.

Ph-C≡C- ... CO_2Et — 1. Me_2CuLi, ether, -20°C 2. 2N H_2SO_4, 25°C → Me, Ph, C, CO_2Et, H 79%

Krause, N. *Chem. Ber.*, ***1990***, *123*, 2173.

OH, CO_2Me, Cl — 1. 19% HCl, 25°C, 1 d 2. NEt_3, 25°C, 2 d → O, O (>96% ee) 66%

Tsuboi, S.; Sakamoto, J.; Sakai, T.; Utaka, M. *SynLett*, ***1991***, 867.

Ph- -OTf — 1. $Pd(OAc)_2$, TEBA, K_2CO_3 2. Ag_2CO_3, PhMe, 80°C → Ph- , O, O 48%

Arcadi, A.; Bernocchi, E.; Cacchi, S.; Marinelli, F. *Tetrahedron*, ***1991***, *47*, 1525.

H-C≡C- , MeO_2C, CO_2Me — PhI, *t*-BuOK, DMSO 5% $Pd(dba)_2$, 14 h 30°C → Ph, CO_2Me, CO_2Me 80%

Fournet, G.; Balme, G.; Gore, J. *Tetrahedron*, ***1991***, *47*, 6293.

CO_2Me, CO_2Me — 20% BuSSBu hν (>300 nm) → MeO_2C, MeO_2C 77%

Singleton, D.A.; Huval, C.C.; Church, K.M.; Priestley, E.S. *Tetrahedron Lett.*, ***1991***, *32*, 5765.

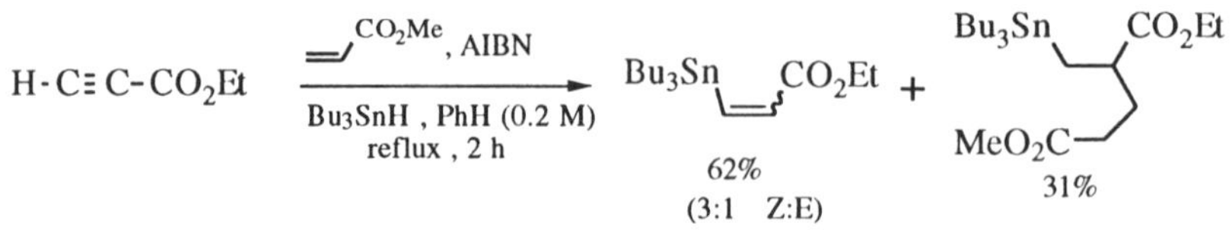

Lee, E.; Hur, C.U. *Tetrahedron Lett.*, ***1991***, *32*, 5101.

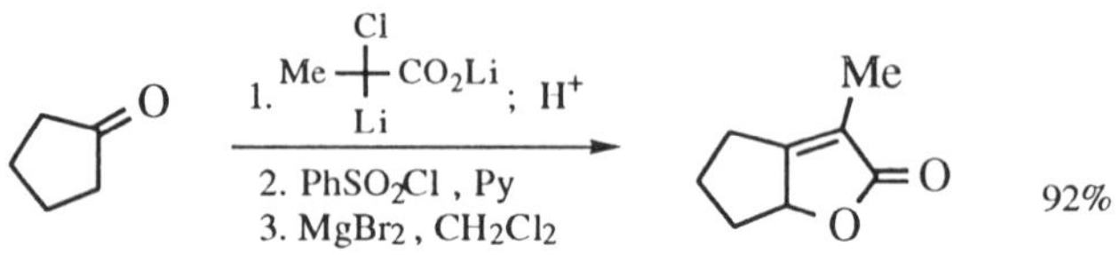

Black, T.H.; McDermott, T.S.; Brown, G.A. *Tetrahedron Lett.*, ***1991***, *32*, 6501.

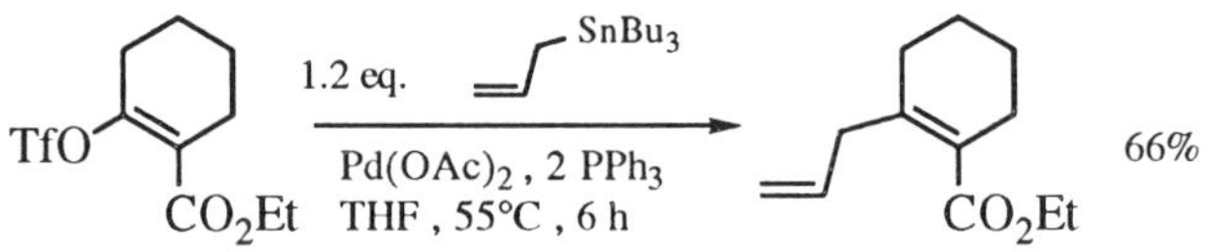

Houpis, I.N. *Tetrahedron Lett.*, ***1991***, *32*, 6675.

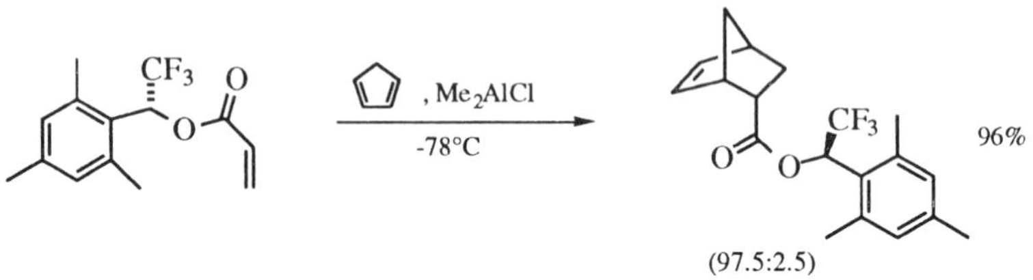

Corey, E.J.; Cheng, X-M.; Cimprich, K.A. *Tetrahedron Lett.*, ***1991***, *32*, 6839.

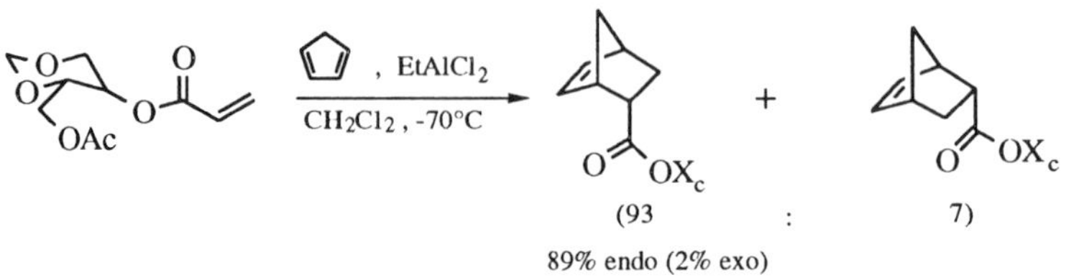

Gras, J-L.; Pellissier, H. *Tetrahedron Lett.*, ***1991***, *32*, 7043.

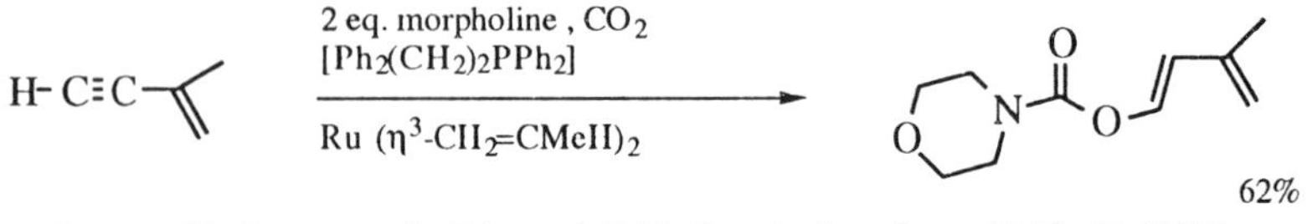

Höfer, J.; Doucet, H.; Bruneau, C.; Dixneuf, P.H. *Tetrahedron Lett.*, ***1991***, *32*, 7409.

$EtO_2P(O)CH_2C(O)SEt$; NaH , THF , 20°C , 19 h — 87%

Liu, H-J.; Rose, P.A.; Sasaki, D.J. *Can. J. Chem.*, ***1991***, *69*, 934.

NEt_3 , CO , MeOH ; RT , 3 h — 23%

NHBn

N H CO_2Me

Shaw, R.; Lathburg, D.; Anderson, M.; Gallagher, T. *J. Chem. Soc., Perkin Trans. I*, ***1991***, 659.

Bu−C≡C−H — HCOOH , 120°C , 15 h ; [Ru(*p*-cymene)(PPh_3)Cl_2] → Bu, OCHO — 78%

Neveux, M.; Bruneau, C.; Dixneuf, P.H. *J. Chem. Soc., Perkin Trans. I*, ***1991***, 1197.
Philippot, K.; Devanne, D.; Dixneuf, P.H. *J. Chem. Soc., Chem. Commun.*, ***1990***, 1199.

1. Cl_2CLiCO_2Li
2. $PhSO_2Cl$, Py
3. $MgBr_2$, ether

Cl — 71%

Black, T.H.; McDermott, T.S. *J. Chem. Soc., Chem. Commun.*, ***1991***, 184.
Black, T.H.; McDermott, T.S. *Synth. Commun.*, ***1990***, *20*, 2959.

Fe(CO)$_4$(acetone) , PhMe ; 145°C , sealed tube , 60 h ; CO (55 psi) — Ph — 87%

Pearson, A.J.; Dubbert, R.A. *J. Chem. Soc., Chem. Commun.*, ***1991***, 202.

Ph−C≡C−Ph — CO_2Me , Py , 80°C ; Ru(COD)(COT) , 20 h → Ph, Ph, CO_2Me — 87%

Mitsudo, T.; Zhang, S-W.; Nagao, M.; Watanabe, Y. *J. Chem. Soc., Chem. Commun.*, ***1991***, 598.

CO_2Et OH Me — cyclohexyl-CHO; Pd-$SnCl_2$, 80°C; DMI-H_2O, 72 h → O, O, Me + O, O, Me

(96 : 4) 38%

Masuyama, Y.; Nimura, Y.; Kurusu, Y. *Tetrahedron Lett.*, ***1991***, *32*, 225.

O S O — 1. AgOTf, Br; CH_2Cl_2, RT; 2. DBU, -78°C → O S O H Me H 63%

(84:6:6:4)

Tahir, S.H.; Olmstead, M.M.; Kurth, M.J. *Tetrahedron Let.*, ***1991***, *32*, 335.

PhCHO — $BrCH_2CO_2Et$, PBu_3; 25% $Pd(PPh_3)_4$, 24 h; 110°C → Ph CO_2Et 85%

(86:14 → 100:0 E:Z for various aryl aldehydes

Shen, Y.; Zhou, Y. *Tetrahedron Lett.*, ***1991***, *32*, 513.

Me Me N OEt P N O O Me Li — iPrCHO, THF; 2266-tetramethyl piperidine → Me CO_2Et + Me CO_2Et

(92 : 8) 93%

Patois, C.; Savignac, P. *Tetrahedron Lett.*, ***1991***, *32*, 1317.

TfO — OAc, NEt_3; $Pd(OAc)_2$, PPh_3; DMF, 60°C, 7 h → AcO 79%

(3.3:1 E:Z)

Ciattini, P.G.; Morera, E.; Ortar, G. *Tetrahedron Lett.*, ***1991***, *32*, 1579.

1. t-BuOK
2. Pd(0), dppe

1 eq.	t-BuOK/0°C/24 h	40	43
	/20°C/9 h	35	45
0.2 eq.	t-BuOK/78°C/3 h	0	75
+18-crown-6	/78°C/18 h	76	0

Monteiro, N.; Balme, G.; Gore, J. *Tetrahedron Lett.*, ***1991***, *32*, 1645.

$CH_2(CO_2Et)_2$, THF
vinyl bromide, 65°C
$Pd(dba)_2$/dppe

(91 : 9) 69%
(35:65 E:Z)

Chaptal, N.; Colovray-Gotteland, V.; Grandjean, C.; Cazes, B.; Goré, J. *Tetrahedron Lett.*, ***1991***, *32*, 1795.

5% $AlMe_3$, ether
reflux, 62 h

48%

Harvey, D.F.; Brown, M.F. *Tetrahedron Lett.*, ***1991***, *32*, 2871.

13 kbar, CH_2Cl_2
7 d

76%
(60:40 diastereomers)

Prapansiri, V.; Thornton, E.R. *Tetrahedron Lett.*, ***1991***, *32*, 3147.

NEt_3, KBr
5% $Pd(OAc)_2$, 10% PPh_3
aq. DMF, 70°C, 2 h

71%

Mandai, T.; Ogawa, M.; Yamaoki, H.; Nakata, T.; Murayama, H.; Kawada, M.; Tsuji, J. *Tetrahedron Lett.*, ***1991***, *32*, 3397.

Bu_3SnH, AIBN
PhMe, 80°C
1 h

58%

C_5-H α:β (90:10)

Iwasa, S.; Yamamoto, M.; Kohmoto, S.; Yamada, K. *J. Org. Chem.*, ***1991***, *56*, 2849.

Ph_3SnH, AIBN

92%

Clive, D.L.J.; Daigneault, S. *J. Org. Chem.*, ***1991***, *56*, 3801.

$Rh_2(OAc)_4$

58%

Davies, H.M.L.; Clark, T.J.; Smith, H.D. *J. Org. Chem.*, ***1991***, *56*, 3817.

PhI, CO (300 psi), PhH
$Pd(OAc)_2$, PPh_3, NEt_3
120°C, 24 h

65%

Huang, Y.; Alper, H. *J. Org. Chem.*, ***1991***, *56*, 4534.

5% $PdCl_2(PhCN)_2$
AcOH, RT

81%

Ma, S.; Lu, X. *J. Org. Chem.*, ***1991***, *56*, 5120.
Ma, S.; Lu, X. *J. Chem. Soc., Chem. Commun.*, ***1990***, 733.

1. TMSOTf, CH_2Cl_2
-78°C, 10 h
2. aq. $NaHCO_3$

85%
(30:1 syn:anti)

Panek, J.S.; Yang, M. *J. Org. Chem.*, ***1991***, *56*, 5755.

1. $TaCl_5$, Zn, DME, PhH
2. PhH, Py
3. EtCHO
4. aq. NaOH

n-$C_{10}H_{25}C\equiv C-CO_2Et$ → n-$C_{10}H_{25}$, CO_2Et, HO + n-$C_{10}H_{25}$, CO_2Et, HO

(95 : 5) 76%

Takai, K.; Tezuka, M.; Utimoto, K. *J. Org. Chem.*, ***1991***, *56*, 5980.

MeO_2C–CH=CH–CO_2Me

Ph, Ph, O, $TiCl_2$, CH_2Cl_2, ether, 20°C

CO_2Me, CO_2Me 79%

(92% ee)

Devine, P.N.; Oh, T. *J. Org. Chem.*, ***1992***, *57*, 396.

Ph–(cyclohexenyl)–OTf + $C\equiv C$-H, CO_2H

$Pd(OAc)_2(PPh_3)_2$, NEt_3, Bu_4NCl, MeCN, 60°C, 45 min

O, O, Ph 74%

Arcadi, A.; Burini, A.; Cacchi, S.; Delmastro, M.; Marinelli, F.; Pietroni, B.R. *J. Org. Chem.*, ***1992***, *57*, 976.

N_2, CO_2Me

1. =/OEt, $Rh_2(Ooctyl)_4$, pentane
2. Et_2AlCl

CO_2Me, EtO 75%

Davies, H.M.L.; Hu, B. *J. Org. Chem.*, ***1992***, *57*, 3186.

OH, OTf

$Pd(PPh_3)_4$, CO

NBu_3, LiCl, 65°C

O, O 83%

Crisp, G.T.; Meyer, A.G. *J. Org. Chem.*, ***1992***, *57*, 6972.

1. Mg*, THF, RT
2. acetone
3. CO_2
4. H_3O^+

O, O 68%

Mg* = activated Mg

Xiong, H.; Rieke, R.D. *J. Org. Chem.*, ***1992***, *57*, 7007.

Ph_3SnH, AIBN
PhH, reflux, 7 h
4 mM

63%

Baldwin, J.E.; Adlington, R.M.; Ramcharitar, S.H. *Tetrahedron,* ***1992****, 48,* 3417.

PhCHO

$BrZnCH_2CN$, THF
-13°C, 16 h

cat.

(93% ee, S)

76%

Kang, S-K.; Kim, S-G.; Lee, J-S. *Tetrahedron Asymmetry,* ***1992****, 3,* 1139.

CO_2Me, 80°C
N-methylpiperidine
4% Ru(COD)(COT)
30 min
also with conjugated amides

(90 : 10) 75%

Mitsudo, T.; Zhang, S-W.; Kondo, T.; Watanabe, Y. *Tetrahedron Lett.,* ***1992****, 33,* 341.

$Ni(CO)_4$, *t*-BuOH

79%

Llebaria, A.; Camps, F.; Moretó, J.M. *Tetrahedron Lett.,* ***1992****, 33,* 3683.

1. 2 eq. BuLi
2. iPrCHO
3. TFAA
4. DBU

68%

Bonete, P.; Nájera, C. *Tetrahedron Lett.,* ***1992****, 33,* 4065.

Ph-C≡C-H

$Pd(PPh_3)_4$, PhH
CO (15 atm), 30 h
100°C

(95 : 4) 88%

Itoh, K.; Miura, M.; Nomura, M. *Tetrahedron Lett.,* ***1992****, 33,* 5369.

PhCHO , CH_2Cl_2 / alumina — 93%

Varma, R.S.; Varma, M. *Tetrahedron Lett.*, ***1992***, *33*, 5937.

hv , EtOH , 0°C — (91 : 9)

Charlton, J.L.; Pham, V.C.; Pete, J-P. *Tetrahedron Lett.*, ***1992***, *33*, 6073.
Piva, O.; Pete, J-P. *Tetrahedron Lett.*, ***1990***, *31*, 5157.

TsNCO , RT / 3 h — (2.3 : 1) 87%

Friesen, R.W.; Kolaczewska, A.E.; Khazanovich, N. *Tetrahedron Lett.*, ***1992***, *33*, 6715.

Me−C≡C−CO_2Bn → cat. $Pd(OAc)_2$, RT / AcOH , LiOAc / 8.5 h → 78%

Lu, X.; Zhu, G.; Ma, S. *Tetrahedron Lett.*, ***1992***, *33*, 7205.

1. $LiN(SiMe_3)_2$, DME
2. $Cp_2Zr(H)Cl$, 0°C → RT
— 65%

Godfrey, A.G.; Ganem, B. *Tetrahedron Lett.*, ***1992***, *33*, 7461.

EtCHO , BzCl , Et_4NCl / Bu_2SnCl_2 — 84%

Yano, K.; Baba, A.; Matsuda, H. *Bull. Chem. Soc. Jpn.*, ***1992***, *65*, 66.

$Pd_2(dba)_3$•$CHCl_3$ / AcOK , PPh_3 , PhMe / reflux — 77%

Trost, B.M.; Brieden, W.; Barringhaus, K.H. *Angew. Chem. Int. Ed. Engl.*, ***1992***, *31*, 1335.

3% $Pd(PPh_3)_4$, RT
dioxane, 3 h, CO

66%

Bando, T.; Tanaka, S.; Fugami, K.; Yoshida, Z.; Tamaru, Y. *Bull. Chem. Soc. Jpn.*, ***1992***, *65*, 97.

C≡C-H
H-N
C≡C-H

$\{(MeCN)_2[E\text{-}EtO_2CCH{=}CHCO_2Et]_2Co\}$

CO_2Et, PhMe, 0°C, 30 min

H-N CO_2Et CO_2Et 86%

Zhou, Z.; Costa, M.; Chiusoli, G.P. *J. Chem. Soc., Perkin Trans. I*, ***1992***, 1399, 1407.

H-C≡C-CO_2Et

1. BuHCu(CN(ZnI
2. CHO, ICH_2ZnI

Bu O O 76%

(80:20 cis:trans)

Sidduri, A.R.; Knochel, P. *J. Am. Chem. Soc.*, ***1992***, *114*, 7579.

, TMSOTf, MeOTs
CH_2Cl_2, -45°C

66%

(98:2 'para:meta')

Hashimoto, Y.; Nagashima, T.; Hasegawa, M.; Saigo, K. *Chem. Lett.*, ***1992***, 1353.

N CO_2Me

Ph, 2 eq. TBACN
EtCN, 5 eq. K_2CO_3
-45°C, 30 min

N CO_2Et CO_2Me

53%

TBACN = tetrabutylammonium cerium (IV) nitrate

Narasaka, K.; Okauchi, T.; Tanaka, K.; Murakami, M. *Chem. Lett.*, ***1992***, 2099.

BuCHO

1. Br CO_2Et
Sn, THF, aq. NH_4Cl
2. H^+

Bu O O

81%

Zhou, J.; Lu, G.; Wu, S. *Synth. Commun.*, ***1992***, *22*, 481.

Related Methods: Section 60A (Protection of Aldehydes). Section 180A (Protection of Ketones). Also via Acetylenic Esters: Section 306 (Alkyne - Ester). Alkenyl Acids: Section 322 (Carboxylic Acid - Alkene). β-Hydroxy-esters: Section 327 (Alcohol - Ester).

SECTION 363: ETHER, EPOXIDE, THIOETHER - ETHER, EPOXIDE, THIOETHER

See Section 60A (Protection of Aldehydes) and Section 180A (Protection of Ketones) for reactions involving formation of Acetals and Ketals.

acetone , MeCN , RT

"near quant."

Bujons, J.; Camps, F.; Messeguer, A. *Tetrahedron Lett.*, ***1990***, *31*, 5235.

PhSeSePh , MeOH

$(NH_4)_2(O_3SO\text{-}OSO_3)$

reflux , 1.5 h

OH

O SePh

91%

Tiecco, M.; Testaferri, L.; Tingoli, M.; Bartoli, D.; Balducci, R. *J. Org. Chem.*, ***1990***, *55*, 429.

1. Cl_2Ce $OCeCl_2$

2. H^+

58%

Mudryk, B.; Shook, C.A.; Cohen, T. *J. Am. Chem. Soc.*, ***1990***, *112*, 6389.

Ph OMe OMe

5% $PbBr_2$, Al , TFA

THF ,1 h

Ph OMe Ph OMe

83%

Dhimane, H.; Tanaka, H.; Torii, S. *Bull. Soc. Chim. Fr.*, ***1990***, *127*, 283.

$n\text{-}C_8F_{17}I$

OEt , 1.1 NEt_3 , RT

2.5% $[CpFe(CO)_2]_2$, MeOH

12 h

$n\text{-}C_8H_{17}$ OMe OEt

95%

Sasaoka, S.; Uno, M.; Joh, T.; Imazaki, H.; Takahashi, S. *J. Chem. Soc., Chem. Commun.*, ***1991***, 86.

e^-, Pt anode, $LiClO_4$
50% MeOH-THF

(5.3 : 1) 73%

Moeller, K.D.; Marzabadi, M.R.; New, D.G.; Chiang, M.Y.; Keith, S. *J. Am. Chem. Soc.*, ***1990***, *112*, 6123.

PhSeCl, CH_2Cl_2
RT, 24 h

74%

(1:1.4 α:β)

Kang, S.H.; Hwang, T.S.; Kim, W.J.; Lim, J.K. *Tetrahedron Lett.*, ***1991***, *32*, 4015.

Me_3SiSPh, -78°C
5% TMSOTf, 0.3 h
CH_2Cl_2

93%

Kusche, A.; Hoffmann, R.; Münster, I.; Keiner, P.; Brückner, R. *Tetrahedron Lett.*, ***1991***, *32*, 467.

1. $(Me_3Si)_3SiH$, PhMe
2. AIBN, 90°C, 1 h

84%

Arya, P.; Lesage, M.; Wayner, D.D.M. *Tetrahedron Lett.*, ***1991***, *32*, 2853.

PhSSPh, PhSeSePh, hν
45°C, 30 h

74%

Ogawa, A.; Tanaka, H.; Yokoyama, H.; Obayashi, R.; Yokoyama, K.; Sonoda, N. *J. Org. Chem.*, ***1992***, *57*, 111.

PhSeSePh, DCN, hν
MeCN

60%

Pandey, G.; Somasekhar, B.B.V. *J. Org. Chem.*, ***1992***, *57*, 4019.

MeSeSeMe , $SnCl_4$
CH_2Cl_2 , RT , 0.2 h

Me Me → MeSe SeMe Me Me 82%

Hermans, B.; Colard, N.; Hevesi, L. *Tetrahedron Lett.*, ***1992***, *33*, 4629.

e^- , Hg cathode , Pt anode
NEt_3 , Et_4NClO_4 , MeCN

O O Br → O O 85%

Bhuvaneswari, N.; Venkatachalam, C.S.; Balasubramanian, K.K. *Tetrahedron Lett.*, ***1992***, *33*, 1499.

Bu_3SnH , AIBN , PhH
80°C , 45 min

O O I O → O O O 81%

Dalla, V.; Pale, P. *Tetrahedron Lett.*, ***1992***, *33*, 7857.

OH

MeO–C6H4–$TeCl_3$
$CHCl_3$

O Cl Te Cl OMe 98%

Comasseto, J.V.; Grazini, M.V.A. *Synth. Commun.*, ***1992***, *22*, 1061.

PhSeTMS , cat. TMSOTf
CH_2Cl_2 , -25°C , 15 min

n-C_5H_{11} OMe OMe → n-C_5H_{11} OMe SePh 91%

Sakakibara, M.; Katsumata, K.; Watanabe, Y.; Toru, T.; Ueno, Y. *SynLett*, ***1992***, 965.

SECTION 364: ETHER, EPOXIDE, THIOETHER - HALIDE, SULFONATE

$XeF_2(SPh)_2$
CH_2Cl_2 , -20°C

→ SePh F 63%

Uneyama, K.; Kanai, M. *Tetrahedron Lett.*, ***1990***, *31*, 3583.

$n\text{-}C_7H_{15}CO_2Me$ → 1. [MeO–C$_6$H$_4$–P(=S)S]$_2$, PhMe, 110°C, 36 h; 2. $Et_2N\text{-}SF_3$, CH_2Cl_2 → $n\text{-}C_7H_{15}CF_2OMe$ 55%

Brunnelle, W.H.; McKinnis, B.R.; Narayanan, B.A. *J. Org. Chem.*, ***1990***, *55*, 768.

PhS–CH$_2$CF$_3$ → e^- (10 F/mol), MeOH, Pt anode, RT, Et_4NOTs → PhS–CH(OMe)CF$_3$ 93%

Fuchigami, T.; Yamamoto, K.; Nakagawa, Y. *J. Org. Chem.*, ***1991***, *56*, 137.

$CH_2{=}CH\text{-}n\text{-}C_4F_9$ → [F_2, H_2O, MeCN], 25°C, via HOF•MeCN → epoxide $n\text{-}C_4F_9$ 70%

Hung, M.H.; Smart, B.E.; Feiring, A.E.; Rozen, S. *J. Org. Chem.*, ***1991***, *56*, 3187.

→ $CF_3CH_2O^-Na^+$, 110°C, 10% CuBr, 5 d → $PhOCH_2CF_3$ 70%

Keegstra, M.A.; Brandsma, L. *Rec. Trav. Chim. Pays-Bas*, ***1991***, *110*, 299.

Cycloheptene → 1. $MeS\text{-}S^{\oplus}Me_2$ $BF_4^{\ominus}$, CH_2Cl_2; 2. NEt_3, 3 HF, 0°C; 3. aq. NH_3 → 1-fluoro-2-(SMe)cycloheptane 70%

Haufe, G.; Alvernhe, G.; Anker, D.; Laurent, A.; Saluzzo, C. *J. Org. Chem.*, ***1992***, *57*, 714.
Saluzzo, C.; Alvernhe, G.; Anker, D. *Tetrahedron Lett.*, ***1990***, *31*, 663.

$CH_3CH_2CH(OMe)_2$ → MnO_2-TMSCl → $CH_3CHClCH(OMe)_2$ 99%

Bellesia, F.; Boni, M.; Ghelfi, F.; Grandi, R.; Pagnoni, U.M.; Pinetti, A. *Tetrahedron*, ***1992***, *48*, 4579.

Alkene (Me, O*t*-Bu) → I_2, MeCN, -20°C → RT → tetrahydrofuran (Me, CH$_2$I) 94%

Marek, I.; Lefrançois, J-M.; Normant, J-F. *Tetrahedron Lett.*, ***1992***, *33*, 1747.

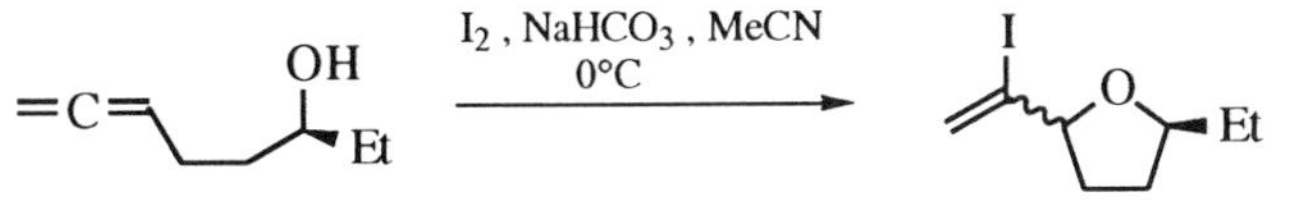

I_2, $NaHCO_3$, MeCN, 0°C — 70% (37:63 cis:trans)

Walkup, R.D.; Guan, L.; Kim, S.W.; Kim, Y.S. *Tetrahedron Lett.*, ***1992***, *33*, 3969.

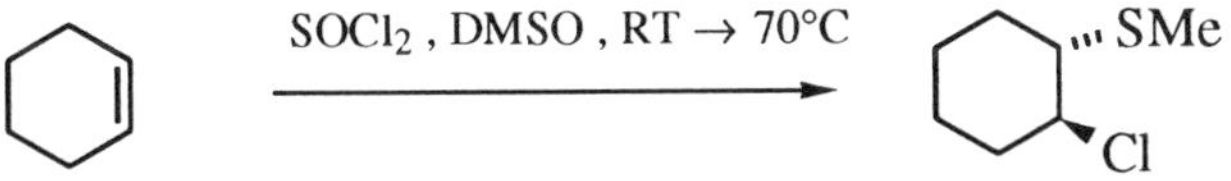

$SOCl_2$, DMSO, RT → 70°C — SMe, Cl — 91%

Bellesia, F.; Boni, M.; Ghelfi, F.; Pagnoni, U.M.; Pinetti, A. *Synth. Commun.*, ***1992***, *22*, 1101.

SECTION 365: ETHER, EPOXIDE, THIOETHER - KETONE

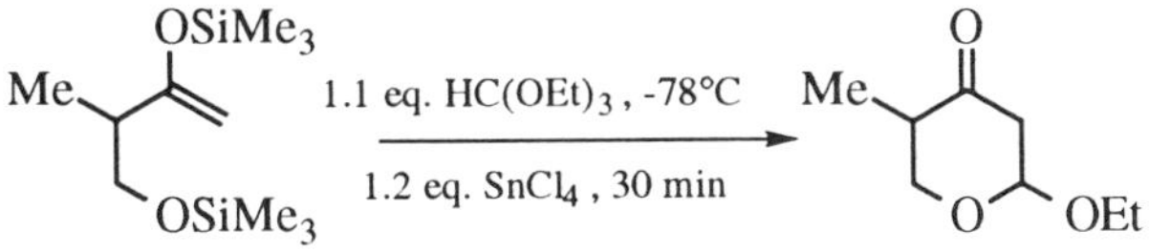

$OSiMe_3$, Me, $OSiMe_3$ — 1.1 eq. $HC(OEt)_3$, -78°C; 1.2 eq. $SnCl_4$, 30 min — O, Me, O, OEt — 88%, 93:7 cis:trans

Martin, V.A.; Perron, F.; Albizati, K.F. *Tetrahedron Lett.*, ***1990***, *31*, 301.

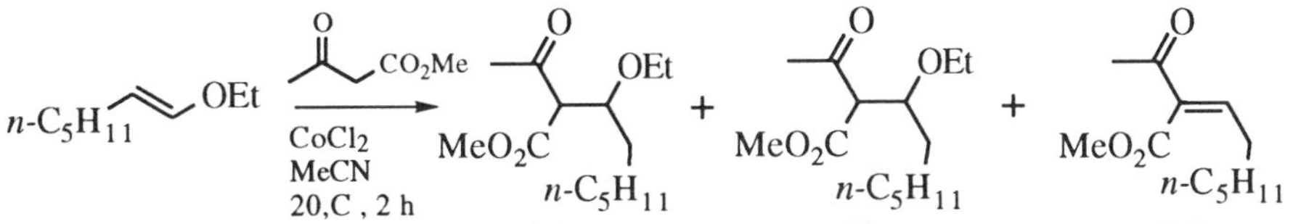

n-C_5H_{11}, OEt — O, CO_2Me; $CoCl_2$, MeCN, 20,C, 2 h — O, OEt, MeO_2C, n-C_5H_{11} + O, OEt, MeO_2C, n-C_5H_{11} + O, MeO_2C, n-C_5H_{11} — (3.1 : 1) 63% 28%

Iqbal, J.; Srivastava, R.R.; Khan, M.A. *Tetrahedron Lett.*, ***1990***, *31*, 1485.

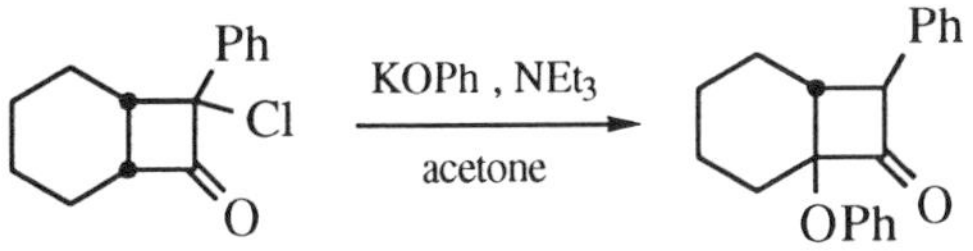

Ph, Cl, O — KOPh, NEt_3, acetone — Ph, OPh, O — 75%

Hassner, A.; Naidorf-Meir, S.; Gottlieb, H.E. *Tetrahedron Lett.*, ***1990***, *31*, 2181.

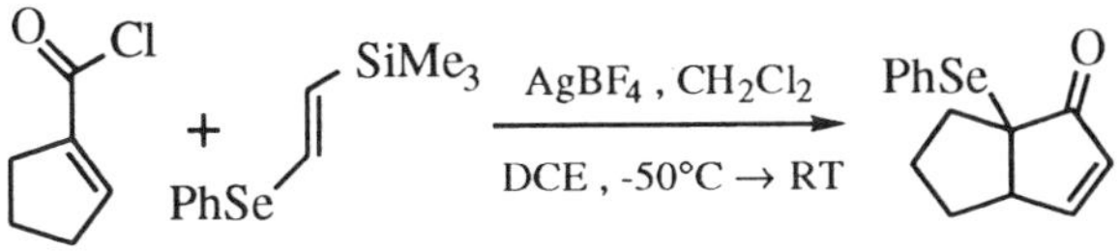

O, Cl + $SiMe_3$, PhSe — $AgBF_4$, CH_2Cl_2, DCE, -50°C → RT — PhSe, O — 47%

Yamazaki, S.; Hama, M.; Yamabe, S. *Tetrahedron Lett.*, ***1990***, *31*, 2917.

1. Bu_2BOTf, iPr_2NEt, CH_2Cl_2, n-PrCHO
2. Me_3SiOTf, iPr_2NEt, CH_2Cl_2, -78°C → 20°C

SPh

Me, n-C_3H_7, O + Me, n-C_3H_7, O

(70 : 30)

25% overall

Paterson, I.; Osborne, S. *Tetrahedron Lett.*, ***1990***, *31*, 2213.

MOMO, MeO, OMOM, Me → hν → MOMO, Me, OH, MeO, O, Me

no yields

Kraus, G.A.; Thomas, P.J.; Schwinden, M.D. *Tetrahedron Lett.*, ***1990***, *31*, 1819.

OBu, PhMe; reflux, 30 min

Me, Me, Me → Me, O, Bu

84%

Coleman, R.S.; Grant, E.B. *Tetrahedron Lett.* ***1990***, *31*, 3677.

MeO → e^-, Pt anode, $LiClO_4$, CH_2Cl_2, Ac_2O, 1-2 mA/cm^2 → MeO, Me, O

56%

(*p:o* = 21:1)

Gatti, N. *Tetrahedron Lett.*, ***1990***, *31*, 3933.

I, OH → PhC≡CH, $NHEt_2$, $PdCl_2$(dppf), 120°C, 20 atm CO, 2 h, autoclave → O, O, Ph

81%

Kalinin, V.N.; Shostakovsky, M.V.; Ponomaryav, A.B. *Tetrahedron Lett.*, ***1990***, *31*, 4073.

O O O → PPh_3, CCl_4, THF, RT, overnight → O, O

78%

Arimoto, H.; Nishiyama, S.; Yamamura, S. *Tetrahedron Lett.*, ***1990***, *31*, 5619.

Bu OH OTBS → 3 NIS, 3 Li_2CO_3, 16 MeOH, PhH, hν (infrared lamp), 35°C → Bu O OMe OTBS 77%

McDonald, C.E.; Holcomb, H.; Leathers, T.; Ampadu-Nyarko, F.; Frommer, J. Jr. *Tetrahedron Lett.*, ***1990***, *31*, 6283.

MeO O CO_2iPr OH → Ph-Pb(OAc)$_3$, TMG, THF, 3 d, cat. Pd(OAc)$_2$, PPh_3 → MeO O Ph O 71%

TMG = N,N,N',N'-tetramethylguanidine

Barton, D.H.R.; Donnelly, D.M.X.; Finet, J-P.; Guiry, P.J.; Kielty, J.M. *Tetrahedron Lett.*, ***1990***, *31*, 6637.

O NO_2 → 2 eq. PhSNa, 70°C, 3 h → SPh O 78%

Tamura, R.; Kusama, Y.; Oda, D. *J. Org. Chem.*, ***1990***, *55*, 595.

O Cl → 1. $MeOCH_2CHN_2$, ether, NEt_3; 2. $Rh_2(OAc)_4$ → O OMe 95%

Hudlicky, T.; Olivo, H.F.; Natchus, M.G.; Umpierrez, E.F.; Pandolfi, E.; Volonterio, C. *J. Org. Chem.*, ***1990***, *55*, 4767.

Et−C≡C−Bu → PCWP, H_2O_2, $CHCl_3$, reflux, 24h → O O 57%

PCWP = peroxotungstophosphate

Ishii, Y.; Sakata, Y. *J. Org. Chem.*, ***1990***, *55*, 5545.

Ph CO_2Me OH → NaOCl, aq. MeCN, RT → Ph O CO_2Me O 92%

Foucaud, A.; le Rouillé, E. *Synthesis*, ***1990***, 787.

O=(lactone) →

MeO
1. Li−C≡C−CH=CH−OMe, THF
2. K_2CO_3, MeOH
3. 30% $HClO_4$, CH_2Cl_2
))))))))

O O O 48%

Crimmins, M.T.; O'Mahony, R. *J. Org. Chem.*, ***1990***, *55*, 5894.

O Ph Ph

PL_5

$+CH-CH_2+_{0.4}$

Ph O H H Ph 94%

(97% ee)

PL = H N H O N H 33

Itsuno, S.; Sakakura, M.; Ito, K. *J. Org. Chem.*, ***1990***,*55*, 6047.

Ph Cl O

1. Ph−CH(OSiMe₃)CN, -78°C → 0°C
2. NH_4Cl, H_2O
3. NEt_3•3 HF, 3 d
4. KOH, ether, 15 h

O Ph O Ph 39%

Hünig, S.; Marschner, C. *Chem. Ber.*, ***1990***, *123*, 107.

O Ph O

PhI(OH)OTs, MeCN

O Ph O 75%

Prakash, O.; Pahuja, S.; Goyal, S.; Sawhney, S.N.; Moriarty, R.M. *SynLett*, ***1990***, 337.

O O C

BF_3•OEt_2, CH_2Cl_2
-78°C

O O O 82%

Ricci, A.; Degl'Innocenti, A.; Capperucci, A.; Faggi, C.; Seconi, G.; Favaretto, L. *SynLett*, ***1990***, 471.

$(CO)_5Cr$=C(OMe)Me

Ph, ether, hv
CO

Me Ph MeO O + MeO Ph Me O

(10 : 1) 44%

Söderberg, B.C.; Hegedus, L.S.; Sierra, M.A. *J. Am. Chem. Soc.*, ***1990***, *112*, 4364.

t-Bu, OSiMe$_3$, 5% TMSOTf; MeCN, -40°C, 30 min

SPh, OMe, OMe → SPh, OMe O (92:8 anti:syn) 92%

Saigo, K.; Kudo, K.; Hashimoto, Y.; Kimoto, H.; Hasegawa, M. *Chem. Lett.*, ***1990***, 941.

O, C≡C-H, O → Bu_3SnH, AIBN, hν; 0.8 h, 20°C → O, $SnBu_3$, O, H 62%

Nishida, A.; Takahashi, H.; Takeda, H.; Takada, N.; Yonemitsu, O. *J. Am. Chem. Soc.*, ***1990***, *112*, 902.

Ph, Ph, Ph, Ph → Co(acac)$_3$, AcOH; reflux, 150 min → Me, O, Me, Ph, Ph, O, O, Ph, Ph, Me, O, Me 70%

Nishino, H.; Yoshida, T.; Kurosawa K. *Bull. Chem. Soc. Jpn.*, ***1991***, *64*, 1108.

Me, OH, Me, OH → 1. TsOH, EtCHO; 2. 2 eq. $SnCl_4$, CH_2Cl_2; -78°C – -23°C → O, Me, Me, O, Et 49%

Hopkins, M.H.; Overman, L.E.; Rishton, G.M. *J. Am. Chem. Soc.*, ***1991***, *113*, 5354.
Brown, M.J.; Harrison, T.; Herrinton, P.M.; Hopkins, M.H.; Hutchinson, K.D.; Mishra, P.; Overman, L.E. *J. Am. Chem. Soc.*, ***1991***, *113*, 5365.

PhS → 2 eq. OSiMe$_3$, Ph, MeCN, RT; 2 eq. CAN → O, Ph, SPh 75%

Narasaka, K.; Okauchi, T. *Chem. Lett.*, ***1991***, 515.

O, Ph → 1. Bu_2BOTf, iPr_2NEt; CH_2Cl_2, -78°C → 0°C; 2. PhSCl, -78°C → O, Ph, SPh 95%

Paterson, I.; Osborne, S. *SynLett*, ***1991***, 145.

acetone

86%

Adam, W.; Hadjiarapoglou, L.; Smerz, A. *Chem. Ber.*, **1991**, *124*, 227.

1. Meldrum's acid/Py
2. Ph, Cl, HO, Ph
3. K_2CO_3, DMF

>90%

Eid, C.N. Jr.; Konopelski, J.P. *Tetrahedron Lett.*, **1991**, *32*, 461.

$VO(OiPr)Cl_2$, N_2
iPrOH-CH_2Cl_2
0°C (3 h); RT (5 h)

(>98 : <2) 67%

Hirao, T.; Mikami, S.; Mori, M.; Ohshiro, Y. *Tetrahedron Lett.*, **1991**, *32*, 1741.

$PhCH(OMe)_2$, Py
TMSOTf, CH_2Cl_2
-78°C → RT

91%

Kim, S.; Kim, Y.G.; Park, J.H. *Tetrahedron Lett.*, **1991**, *32*, 2043.

OTMS OTMS OMe
TMSOTf

56%

Molander, G.A.; Cameron, K.O. *J. Org. Chem.*, **1991**, *56*, 2617.

EtOH, H_2O, RT

90%

also reacts with other nucleophiles

Weichert, A.; Hoffmann, H.M.R. *J. Org. Chem.*, **1991**, *56*, 4098.

1. 3 eq. LDA , TMSCl TEA
2. $TiCl_4$, BnCHO
3. H_2SO_4 , AcOH

OH

O

O

Ph

78%

also reacts with $TiCl_4$/ketones

Kelly, S.E.; Vanderplas, B.C. *J. Org. Chem.*, ***1991***, *56*, 1325.

PhSeSePh , MeOH
$(NH_4)_2S_2O_8$, 1.5 h

MeO OMe

30%

Tiecco, M.; Testaferri, L.; Tingoli, M.; Bartoli, D.; Marini, F. *J. Org. Chem.*, ***1991***, *56*, 5207.
Tiecco, M.; Testaferri, L.; Tingoli, M.; Bartoli, D. *J. Org. Chem.*, ***1990***, *55*, 4523.

N_2

$Rh_2(OAc)_4$

84%

Pirrung, M.C.; Zhang, J.; McPhail, A.T. *J. Org. Chem.*, ***1991***, *56*, 6269.
Pirrung, M.C.; Zhang, J. *Tetrahedron Lett.*, ***1992***, *33*, 5987.

1. [a. MeO—CH2—C(O)NMe$_2$ / Tf_2O; b. collidine , DCE]
2. aq. CCl_4

OMe

71%

Genicot, C.; Gobeaux, B. Ghosez, L *Tetrahedron Lett.*, ***1991***, *32*, 3827.

t-BuOOH , *t*-BuOK
THF/NH_3

Bn_2N Me Me

Me Me NBn_2 + Me Me NBn_2

(95 : 5) 65%

Reetz, M.T.; Lauterbach, E.H. *Tetrahedron Lett.*, ***1991***, *32*, 4477.

BF_3•OEt_2 , MeCN
$Tl(NO_3)_3$, 1 h

92%

Singh, O.V.; Kapil, R.S.; Garg, C.P.; Kapoor, R.P. *Tetrahedron Lett.*, ***1991***, *32*, 5619.

OMe Br → $Cp_2Zr(R)Cl$, MeCN → OMe, O

R = Me 86%
R = iBu 72%

Barr, K.J.; Watson, B.T.; Buchwald, S.L. *Tetrahedron Lett.*, ***1991***, *32*, 5465.

O, O, Me → 0.1 $MeAlCl_2$, 2 h, -65°C → O, O, H, Me 69%

(31:69 dr)

Rogers, C.; Keay, B.A. *Tetrahedron Lett.*, ***1991***, *32*, 6477.

$(CO)_5Cr$=C(Me)(OMe) → OEt, hν; CO (5 bar), 80°C (λ > 400 nm) → Me, MeO, OEt, O + Me, MeO, OEt, O (6 : 1) 87%

Köbbing, S.; Mattay, J. *Tetrahedron Lett.*, ***1992***, *33*, 927.

O + $SiMe_3$, SePh → $SnCl_4$ → PhSe, Me_3Si, O 66%

Yamazaki, S.; Fujitsuka, H.; Yamabe, S.; Tamura, H. *J. Org. Chem.*, ***1992***, *57*, 5610.

Me, O, Cl → PhCHO , THF , 60°C; Bu_3Sn-N(Et)CO_2Me → O, Ph, Me, O 59%

(13:87 cis:trans)

Shibata, I.; Yamasaki, H.; Baba, A.; Matsuda, H. *J. Org. Chem.*, ***1992***, *57*, 6909.
Shibata, I.; Yamasaki, H.; Baba, A.; Matsuda, H. *SynLett*, ***1990***, 490.

N_2, O, OMe, O, Me → $Rh_2(OAc)_4$, CH_2Cl_2 → O, OMe, O, Me + O, OMe, O, Me (24 : 1)

Spero, D.M.; Adams, J. *Tetrahedron Lett.*, ***1992***, *33*, 1143.

1. MsN_3 , NEt_3 , MeCN , 25°C
2. , CH_2Cl_2 , $Rh_2(HCO_3)_2$

CO_2Me O O 54%

McCarthy, N.; McKervey, M.A.; Ye, T.; McCann, M.; Murphy, E.; Doyle, M.P. *Tetrahedron Lett.*, ***1992***, *33*, 5983.

$OSiMe_3$ CH_2Cl_2 , 0°C , 30 min N-Ts Ph- S N · SPh Ts

O SPh 75%

Magnus, P.; Rigollier, P. *Tetrahedron Lett.*, ***1992***, *33*, 6111.

O Te F O hν , PhH , reflux O Te F O 96%

Chen, C.; Crich, D.; Papadatos, A. *J. Am. Chem. Soc.*, ***1992***, *114*, 8313.

O O S Ph 1. TMSOTf , $EtN(iPr)_2$ CH_2Cl_2 2. $MgBr_2$•OEt_2 O SPh 90%

Kosugi, H.; Hoshino, K.; Uda, H. *J. Chem. Soc., Chem. Commun.*, ***1992***, 560.

HO PhS OH TMSOTf , PhMe 0°C PhS O + PhS O

(97 : 3) 62%

Kudo, K.; Saigo, K.; Hashimoto, Y.; Saito, K.; Hasegawa, M. *Chem. Lett.*, ***1992***, 1449.

OAc 1. PhCHO , $SnCl_2$ 2. EtOH 3. NCS , CH_2Cl_2 Ph EtO O 84%

Masuyama, Y.; Kobayashi, Y.; Yanagi, R.; Kurusu, Y. *Chem. Lett.*, ***1992***, 2039.

SECTION 366: ETHER, EPOXIDE, THIOETHER - NITRILE

NHCHO; 1. CH_2Cl_2, dimethyldioxirane, -40°C; 2. $(CF_3SO_2)_2O$, iPr_2NEt, -78°C; N≡C; Br; 41%; 8%

Baldwin, J.E.; O'Neil, I.A. *Tetrahedron Lett.*, ***1990***, *31*, 2047.

1. Ph_3C-C≡N-O , PhH, 25°C; 2. I_2 , CH_2Cl_2; I; O; CN; 85%

(1:4 cis:trans)

Kurth, M.J.; Rodriguez, M.J.; Olmstead, M.M. *J. Org. Chem.*, ***1990***, *55*, 283.

Ph; OEt; OEt; TMS-CN , $CoCl_2$; CH_2Cl_2; Ph; OEt; CN; 98%

Soga, T.; Takenoshita, H.; Yamada, M.; Mukaiyama, T. *Bull. Chem. Soc. Jpn.*, ***1990***, *63*, 3122.

Me; CHO; $ClCH_2CN$, aq. NaOH; Et_3BnNCl , 20°C; Me; O; CN; 77%

Sadanandam, Y.S.; Leelavati, P. *J. Ind. Chem. Soc.*, ***1990***, *67*, 253.

Cl; n-C_9H_{19}; SPh; Me_3SiCN , $SnCl_4$; CH_2Cl_2; CN; n-C_9H_{19}; SPh; 68%

Fortes, C.C.; Okino, E.A. *Synth. Commun.*, ***1990***, *20*, 1943.

OMe; N-OAc; Ph-C≡C-$AlEt_2$, $ZnCl_2$; CH_2Cl_2 , 0°C; OMe; Ph; CN; 91%

Fujioka, H.; Yamanaka, T.; Yakuma, K.; Miyazaki, M.; Kita, Y. *J. Chem. Soc., Chem. Commun.*, ***1991***, 533.

SECTION 367: ETHER, EPOXIDE, THIOETHER - ALKENE

Enol ethers are found in this section as well as alkenyl ethers.

n-C_3H_7–C≡C–CH_2OCO_2Me + MeCO–CH(CO_2Et)–CH(CO_2Et)–COMe $\xrightarrow[\text{dppe, THF, 48h, reflux}]{0.05\ Pd(dba)_2}$ 65%

Geng, L.; Lu, X. *Tetrahedron Lett.*, ***1990***, *31*, 111.

$\xrightarrow[0°C,\ 20\ min]{Bu_3SnLi,\ THF}$ 98%

Fujiwara, S.; Katsumura, S.; Isoe, S. *Tetrahedron Lett.*, ***1990***, *31*, 691.

1. $Me_2PhSiLi$ 2. n-BuLi, THF, 25°C 3. MeI — $OSiMe_2Ph$, Me — 40%

Brook rearrangement

Koreeda, M.; Koo, S. *Tetrahedron Lett.*, ***1990***, *31*, 831.

NPhth, MeO_2C, O + OAc, EtO $\xrightarrow[230\ h]{125°C}$ NPhth, OAc, MeO_2C, O, OEt + NPhth, OAc, MeO_2C, O, OEt (1 : 1.3) 97%

Tietze, L.F.; Hartfiel, U. *Tetrahedron Lett.*, ***1990***, *31*, 1697.

MeO, OMe, $SiMe_3$ $\xrightarrow[TMSOTf]{OSiMe_3}$ OMe 75%

Lee, T.V.; Roden, F.S.; Yeoh, H.T-L. *Tetrahedron Lett.*, ***1990***, *31*, 2063.

Ph–C≡C–H $\xrightarrow[CH_2Cl_2,\ 30\ min]{B(SeMe)_3,\ RT}$ Ph, MeSe (61%) + Ph, SeMe, MeSe (18%)

Kataoka, T.; Yoshimatsu, M.; Shimizu, H.; Hori, M. *Tetrahedron Lett.*, ***1990***, *31*, 5927.

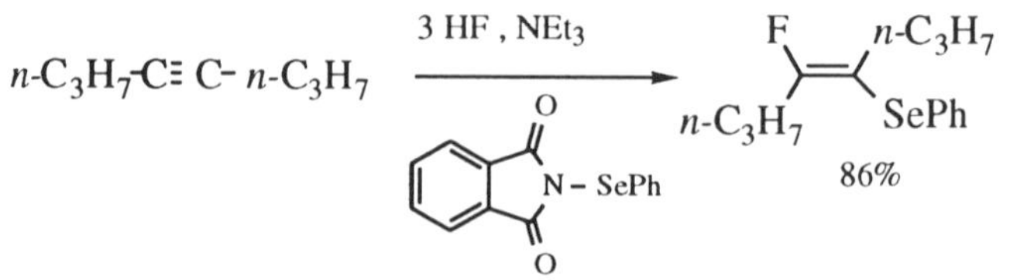
3 HF , NEt3
n-C3H7-C≡C-n-C3H7
N - SePh
F
n-C3H7
n-C3H7
SePh
86%

Saluzzo, C.; Alvernhe, G.; Anker, D. *Tetrahedron Lett.*, ***1990***, *31*, 2127.

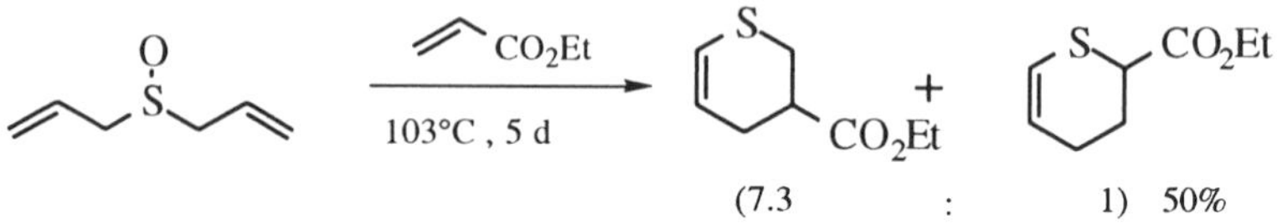
CO2Et
103°C , 5 d
CO2Et
CO2Et
(7.3 : 1) 50%

Block, E.; Zhao, S.H. *Tetrahedron Lett.*, ***1990***, *31*, 5003.

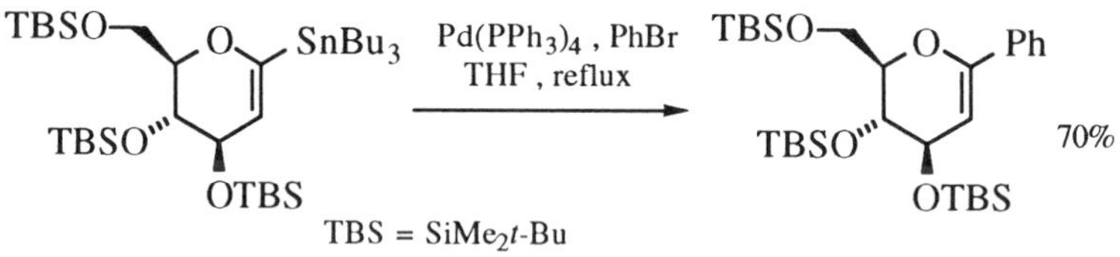
TBSO
SnBu3
TBSO
OTBS
Pd(PPh3)4 , PhBr
THF , reflux
TBS = SiMe2t-Bu
TBSO
Ph
TBSO
OTBS
70%

Friesen, R.W.; Sturino, C.F. *J. Org. Chem.*, ***1990***, *55*, 2572.

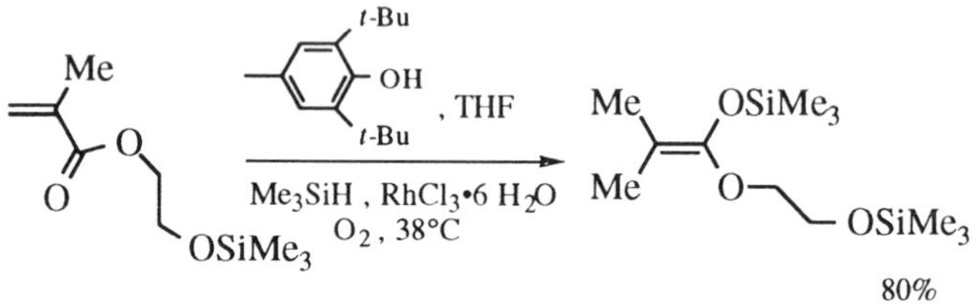
Me
OSiMe3
t-Bu
OH
t-Bu
, THF
Me3SiH , RhCl3•6 H2O
O2 , 38°C
Me
OSiMe3
Me
OSiMe3
80%

Revis, A.; Hilty, T.K. *J. Org. Chem.*, ***1990***, *55*, 2972.

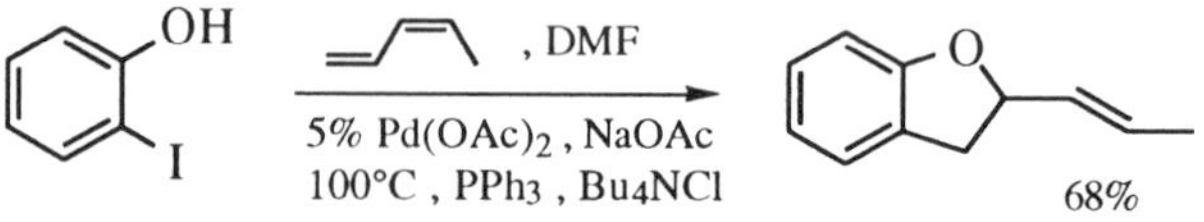
OH
I
, DMF
5% Pd(OAc)2 , NaOAc
100°C , PPh3 , Bu4NCl
68%

Larock, R.C.; Berrios-Peña, N.; Narayanan, K. *J. Org. Chem.*, ***1990***, *55*, 3447.

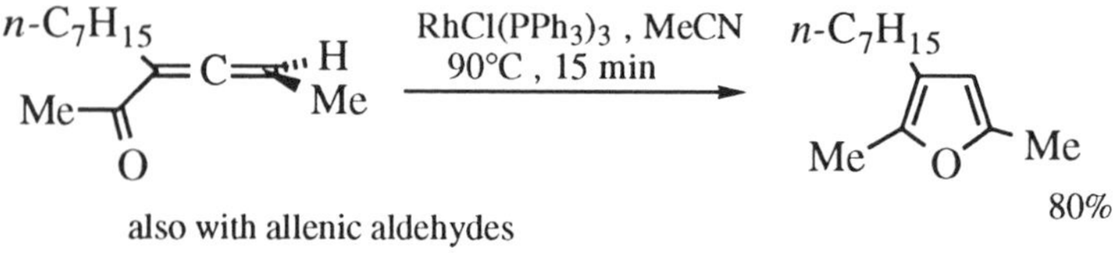
n-C7H15
H
Me
Me
RhCl(PPh3)3 , MeCN
90°C , 15 min
n-C7H15
Me
Me
80%

also with allenic aldehydes

Marshall, J.A.; Robinson, E.D. *J. Org. Chem.*, ***1990***, *55*, 3450.

CCl_4 , 135°C , 12 h

Me_3Si OAc → Me_3Si OAc 94%

Clark, D.L.; Chou, W-N.; White, J.B. *J. Org. Chem.*, ***1990***, *55*, 3975.
Chou, W-N.; White, J.B. *Tetrahedron Lett.*, ***1991***, *32*, 157.

1. LiHMDS , THF-HMPA
2. $PhN(SO_2CF_3)_2$
3. Bu_2CuLi

Bu 60%

Tsushima, K.; Murai, A. *Chem. Lett.*, ***1990***, 761.

$SnBu_3$
$GaCl_3$, CH_2Cl_2
-23°C , 30°C

SEt, Ph, SEt → SEt, Ph, Me 77%

Saigo, K.; Hashimoto, Y.; Kihara, N.; Hara, K.; Hasegawa, M. *Chem. Lett.*, ***1990***, 1097.

$TeCl_4$, CH_2Cl_2 , RT

S, S, Ph, Me → S, S, Ph, Me 58%

Tani, H.; Inamasu, T.; Tamura, R.; Suzuki, H. *Chem. Lett.*, ***1990***, 1323.

, 0°C
$SnCl_4$, CH_2Cl_2 , 3 h

Ph, MeO_2C, O → Ph, MeO_2C, O 63%

Sera, A.; Ohara, M.; Yamada, H.; Egashira, E.; Ueda, N.; Setsune, J. *Chem. Lett.*, ***1990***, 2043.

$TiCl_4$, Zn
dioxane , reflux

Ph, O, O, O, Me → O, Ph, Me 55%

Banerji, A.; Nagak, S.K. *J. Chem. Soc., Chem. Commun.*, ***1990***, 150.

$MeO_2C-C\equiv C-CO_2Me$ —Al_2O_3 , PhOH→ PhO, MeO_2C, CO_2Me 91%

Kodomari, M.; Sakamoto, T.; Yoshitomi, S. *J. Chem. Soc., Chem. Commun.*, ***1990***, 701.

Me, H·C≡C, OSiPh$_2$iPr; Et/Me CHO, CH_2Cl_2; $TiCl_4$, -78°C → -40°C; Cl, Et, Me, O, Me 57%

Arya, P.; Chan, T-H. *J. Chem. Soc., Chem. Commun.*, ***1990***, 967.

PhCHO; Te(iBu)$_2$ Br; KOH$_{(solid)}$, RT, 8 h THF (trace water); Ph O 96% (70:30 cis:trans)

Zhou, Z-L.; Sun, Y-S.; Shi, L-L.; Huang, Y-Z. *J. Chem. Soc., Chem. Commun.*, ***1990***, 1439.

O Me Cl; 2 eq. EtSH, 12 h cyclohexane -5°C → RT; SEt SEt Me 75-88%

Mursakulov, I.G.; Ramazanov, Ê.A.; Keriomov, F.F.; Abbasov, I.M.; Zefirov, N.S. *Zhur. Org. Khim.*, ***1990***, *26*, 1638 (Engl., p. 1416).

O O; 2,6-di-*t*-butylpyridine 2 eq. $SnCl_4$, -10°C 300 min; O 25% + Cl O 13%

Blumenkopf, T.A.; Bratz, M.; Castañeda, A.; Look, G.C.; Overman, L.E.; Rodriguez, D.; Thompson, A.S. *J. Am. Chem. Soc.*, ***1990***, *112*, 4386.

O O Ph; Cp_2TiMe_2, PhMe, 65°C; O Ph 41%

Petasis, N.A.; Bzowej, E.I. *J. Am. Chem. Soc.*, ***1990***, *112*, 6392.

2 Ph_3P=CHPh; 1. $(Se)_n$, 90°C 2.; Se Ph 34%

Okuma, K.; Komiya, Y.; Kaneko, I. ; Tachibana, Y.; Iwata, E.; Ohta, H. *Bull. Chem. Soc. Jpn.*, ***1990***, *63*, 1653.

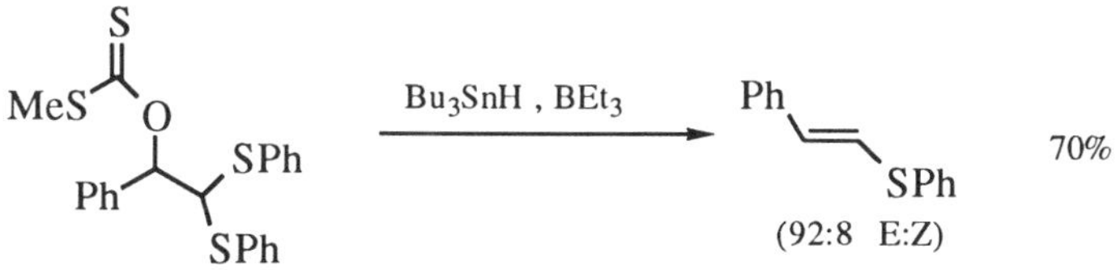

Bu_3SnH , BEt_3

SPh, SPh, Ph, MeS, O, S → Ph, SPh (92:8 E:Z) 70%

Nozaki, K.; Oshima, K.; Utimoto, K. *Bull. Chem. Soc. Jpn.*, ***1990***, *63*, 2578.

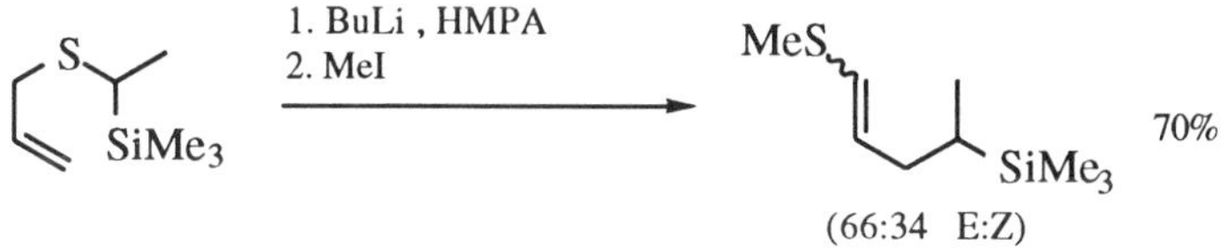

1. BuLi , HMPA
2. MeI

S, $SiMe_3$ → MeS, $SiMe_3$ (66:34 E:Z) 70%

Miura, K.; Oshima, K.; Utimoto, K. *Bull. Chem. Soc. Jpn.*, ***1990***, *63*, 2584.

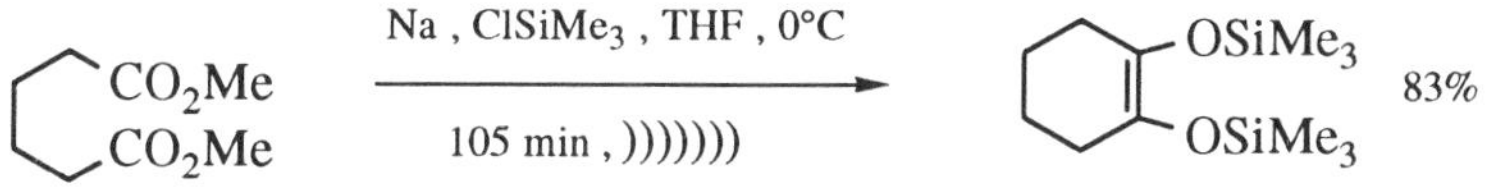

Na , $ClSiMe_3$, THF , 0°C
105 min ,)))))))

CO_2Me, CO_2Me → $OSiMe_3$, $OSiMe_3$ 83%

Fadel, A.; Canet, J-L.; Salaün, J. *SynLett*, ***1990***, 89.

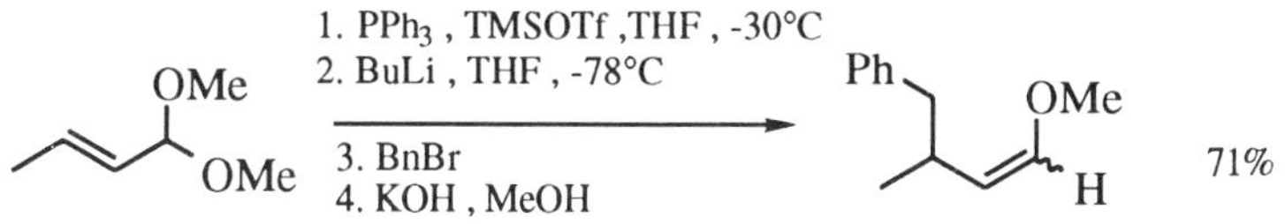

1. PPh_3 , TMSOTf ,THF , -30°C
2. BuLi , THF , -78°C
3. BnBr
4. KOH , MeOH

OMe, OMe → Ph, OMe, H 71%

Kim, S.; Kim, Y.C. *SynLett*, ***1990***, 115.

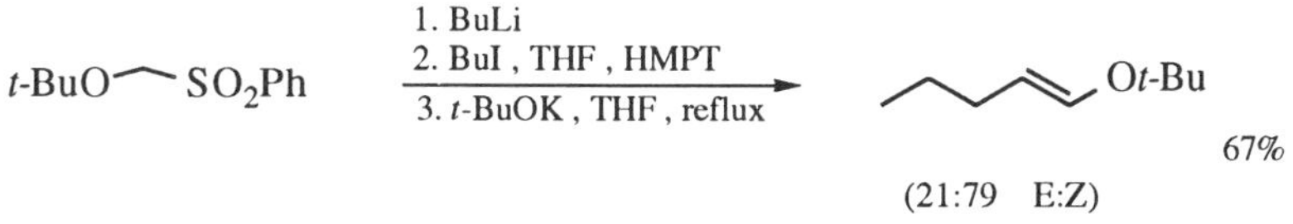

1. BuLi
2. BuI , THF , HMPT
3. *t*-BuOK , THF , reflux

t-BuO SO_2Ph → O*t*-Bu (21:79 E:Z) 67%

Julia, M.; Uguen, D.; Verpeaux, J-N.; Zhang, D. *SynLett*, ***1991***, 505.

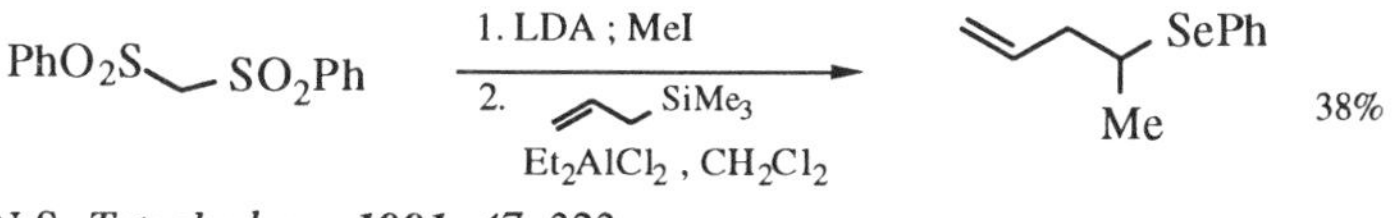

1. LDA ; MeI
2. $SiMe_3$, Et_2AlCl_2 , CH_2Cl_2

PhO_2S SO_2Ph → SePh, Me 38%

Simpkins, N.S. *Tetrahedron*, ***1991***, *47*, 323.

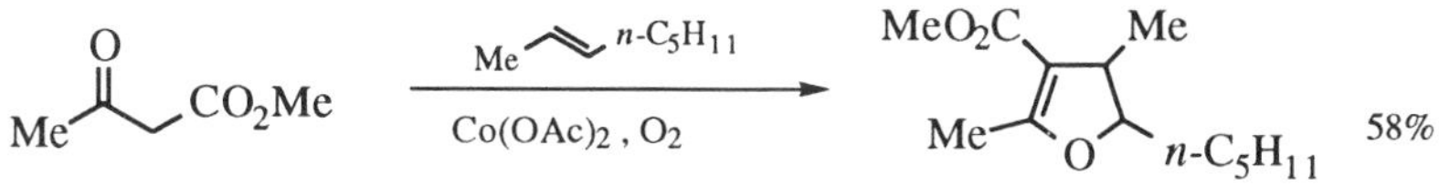

Me n-C_5H_{11}
$Co(OAc)_2$, O_2

O, Me CO_2Me → MeO_2C, Me, Me, O, n-C_5H_{11} 58%

Iqbal, J.; Bhatia, B.; Nayyar, N.K. *Tetrahedron*, ***1991***, *47*, 6457.

$Bu_2Sn(OTf)_2$, DCE
$Bu_2Sn(SPh)_2$, 35°C
12 h

OMe OMe PhS SPh 71%

(72:28 E:Z)

Sato, T.; Otera, J.; Nozaki, H. *SynLett*, ***1991***, 903.

1. PhCHO , PhSH , TsOH
2. BuLi
3. $MeCH_2X$
4. BF_3

N N N H Me Ph SPh 92%

Katritzky, A.R.; Afridi, A.S.; Kuzmierkiewicz, W. *Helv. Chim. Acta*, ***1991***, *74*, 1931.

Bu−C≡C−SMe

1. EtCu(CN)ZnEt•2 LiCl
0°C , 1 h
2. Br

Et Bu SMe 92%

Rao, S.A.; Knochel, P. *J. Am. Chem. Soc.*, ***1991***, *113*, 5735.

OMe SiMe3

e⁻ (RVC anode)
$LiClO_4$, CH_2Cl_2
20% CD_3OD

OMe OCD_3 + OMe OCD_3 $SiMe_3$ OMe

60% 10%

Hudson, C.M.; Marzabadi, M.R.; Moeller, K.D.; New, D.G. *J. Am. Chem. Soc.*, ***1991***, *113*, 7372.

S O Tol

MgBr
-78°C → RT

O S- Tol + S-Tol

0% 85%

Iwata, C.; Maezaki, N.; Kurumada, T.; Fukuyama, H.; Sugiyama, K.; Imanishi, T. *J. Chem. Soc., Chem. Commun.*, ***1991***, 1408.

Me_3Si O

10% Et_3Al , CH_2Cl_2
15°C , 20 h

$OSiMe_3$ 82%

Fujiwara, T.; Suda, A.; Takeda, T. *Chem. Lett.*, ***1991***, 1619.

CS_2 , 2 eq. MeI , MeCN
KF , Al_2O_3 , RT

SMe
SMe
58%

Villemin, D.; Ben Alloum, A. *Synthesis*, ***1991***, 301.

Ph OCO_2Me OCO_2Me

10% $Pd_2(dba)_3$•$CHCl_3$
80% $P(OiPr)_3$, THF
RT

Ph OCO_2Me
56%

Trost, B.M.; Tometzki, G.B. *Synthesis*, ***1991***, 1235.

O

$SnBu_3$, CH_2Cl_2
t-$BuMe_2SiOTf$

Me $OSiMe_2$*t*-Bu
82%

Kim, S.; Lee, J.M. *Synth. Commun.*, ***1991***, *21*, 25.

MeO
MeO Br

1. Me_3Si SPh
$TiCl_4$
2. $(Bu_3Sn)_2$, PhH
hν , 10°C

MeO
33%

95% for step 1; 35% for step 2

Ward, D.E.; Kaller, B.F. *Tetrahedron Lett.*, ***1991***, *32*, 843.

OMe

10% O Cl Ti O Cl , -30°C
CH_2Cl_2 , 48 h
MeO_2C-CHO

OMe O CO_2Me + OMe O CO_2Me
(78 : 22) 77%
(94% ee) (>90% ee)

Terada, M.; Mikami, K.; Nakai, T. *Tetrahedron Lett.*, ***1991***, *32*, 935.

=C= OMe H

$HSnPh_3$
$Pd(PPh_3)_4$

Ph_3Sn OMe + Ph_3Sn OMe
(9 : 91) 72%

Koerber, K.; Gore, J.; Vatele, J-M. *Tetrahedron Lett.*, ***1991***, *32*, 1187.

2% Aliquat 366, 0°C
THF, t-BuOK, 2h

68%

Bailey, W.J.; Zhou, L-L. *Tetrahedron Lett.*, ***1991***, *32*, 1539.

PhSCl

72%

(90% de)

Nativi, C.; Palio, G.; Taddei, M. *Tetrahedron Lett.*, ***1991***, *32*, 1583.

DME/PhH, $ZnCl_2$
microwaves (600 W)
5 min

with no microwaves, at 140°C, 0% yield

82%

Stambouli, A.; Chastrette, M.; Soufiaoui, M. *Tetrahedron Lett.*, ***1991***, *32*, 1723.

NaH, THF

95%

Pegram, J.J.; Anderson, C.B. *Tetrahedron Lett.*, ***1991***, *32*, 2197.

0.2 eq. $AgNO_3$, H_2O
$CaCO_3$, acetone, RT

95%

Marshall, J.A.; Wang, X. *J. Org. Chem.*, ***1991***, *56*, 960.

Li_2CO_3, acetone-AcOH
2 eq. *p*-benzoquinone
5% $Pd(OAc)_2$, 20°C

86%

(>98% trans)

Bäckvall, J-E.; Andersson, P.G. *J. Org. Chem.*, ***1991***, *56*, 2274.

Ph-C≡C-Me → ($AgOSO_2Tol$, MeCN, PhSeCl, RT) → $TolO_2SO$(Ph)C=C(Me)SePh 75%

Back, T.G.; Muralidharan, K.R. *J. Org. Chem.*, ***1991***, *56*, 2781.
Back, T.G.; Muralidharan, K.R. *Tetrahedron Lett.*, ***1990***, *31*, 1957.

PhCHO → (1. $IZn(CN)Cu(CH_2)_4$-ZnI, THF, -60°C → -25°C; 2. I_2, -70°C → RT) → PhCH(OH)$(CH_2)_4$I 59%

Hudlicky, T.; Barbieri, G. *J. Org. Chem.*, ***1991***, *56*, 4598.

$CH_2=C(Me)CH=CHOSiMe_3$ → (1. MeLi; 2. $PhMe_2SiCl$) → $CH_2=C(Me)CH=CHOSiMe_2Ph$ 93%

Duhamel, L.; Guillemont, J.; Poirier, J-M.; Chabardes, P. *Tetrahedron Lett.*, ***1991***, *32*, 4495, 4499.

$PrCH=C(SPh)SnBu_3$ → (1. $SnCl_4$; 2. Ph-OMe, MS 4Å) → $PrCH=C(SPh)(C_6H_4$-4-OMe) 80%

Takeda, T.; Kanamori, F.; Masuda, M.; Fujiwara, T. *Tetrahedron Lett.*, ***1991***, *32*, 5567.
Takeda, T.; Kanamori, F.; Matsusita, H.; Fujiwara, T. *Tetrahedron Lett.*, ***1991***, *32*, 6563 [with NBS; $AlCl_3$].

MeC(=O)CH(Me)OH → (t-BuO_2C-C≡C-CO_2t-Bu; K_2CO_3, acetone, 80°C, 20 h) → 2,3-dihydrofuran: 3-OH, 3-Me, 2-Me, 4-CO_2t-Bu, 5-CO_2t-Bu 88%

Jauch, J.; Schurig, V. *Tetrahedron Lett.*, ***1991***, *32*, 4687.

PhS-C≡C-Me → ($Pd(PPh_3)_4$, PhH, Bu_3SnH) → PhS(Bu_3Sn)C=CHMe 79%

Magriotis, P.A.; Brown, J.T.; Scott, M.E. *Tetrahedron Lett.*, ***1991***, *32*, 5047.

t-BuO I

1. 2 eq. t-BuLi, ether
2. allylmagnesium bromide
3. $ZnBr_2$
4. aq. HCl

Me

Ot-Bu 81%

(95:5 dr)

Marek, I.; Lefrançois, J-M.; Normant, J.F. *Tetrahedron Lett.*, ***1991***, *32*, 5969.
Marek, I.; Normant, J.F. *Tetrahedron Lett.*, ***1991***, *32*, 5973.
Marek, I.; Alexakis, A.; Normant, J-F. *Tetrahedron Lett.*, ***1991***, *32*, 6337. (95% yield)

H-C≡C–CH(OEt)(OEt)

1. $Bu_3SnCu(Bu)CNLi_2$
THF, -78°C, 2 h
2. H_2O

Bu_3Sn ... OEt OEt 84%

Beaudet, I.; Parrain, J-L.; Quintard, J-P. *Tetrahedron Lett.*, ***1991***, *32*, 6333.

1. $Hg(O_2CCF_3)_2$
CH_2Cl_2
2. $NaBH_4$, OH^-

(92 : 8) 96%

p-TsOH, PhH
RT

Balasubramanian, T.; Balasubramanian, K.K. *Tetrahedron Lett.*, ***1991***, *32*, 6633.

Ph Cl Cl

1. $(Bu_3Sn)_2Se$
2. TBAF•3 H_2O, THF
3.

Se Ph 79%

(59:41 endo:exo)

Segi, M.; Kato, M.; Nakajima, T. *Tetrahedron Lett.*, ***1991***, *32*, 7427.

Ph-C≡C-H

PhSeSePh, neat
hν (sunlight/pyrex)

PhSe Ph SePh 83%

(82:18 E:Z)

Ogawa, A.; Yokoyama, H.; Yokoyama, K.; Masawaki, T.; Kambe, N.; Sonoda, N. *J. Org. Chem.*, ***1991***, *56*, 5721.
Ogawa, A.; Obayashi, R.; Sekiguchi, M.; Masawaki, T.; Kambe, N.; Sonoda, N. *Tetrahedron Lett.*, ***1992***, *33*, 1329.

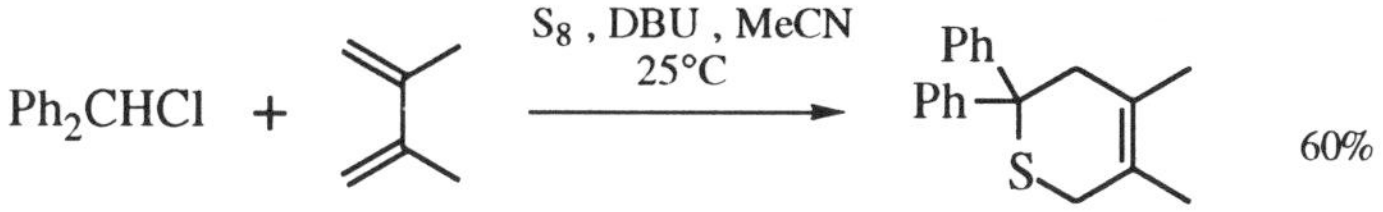

Abelman, M.M. *Tetrahedron Lett.*, ***1991***, *32*, 7390.

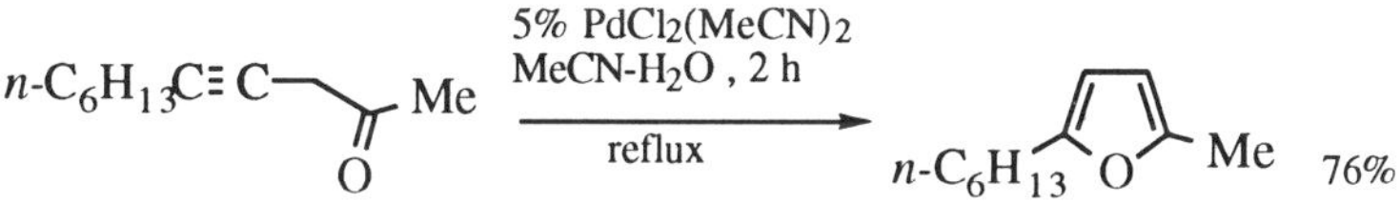

Fukuda, Y.; Shiragami, H.; Utimoto, K.; Nozaki, H. *J. Org. Chem.*, ***1991***, *56*, 5816.
Kataoka, Y.; Tezuka, M.; Takai, K.; Utimoto, K. *Tetrahedron*, ***1992***, *48*, 3495.

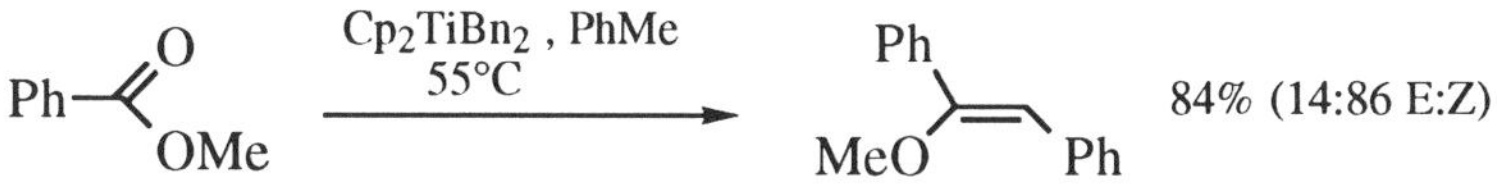

Petasis, N.A.; Bzowej, E.I. *J. Org. Chem.*, ***1992***, *57*, 1327.

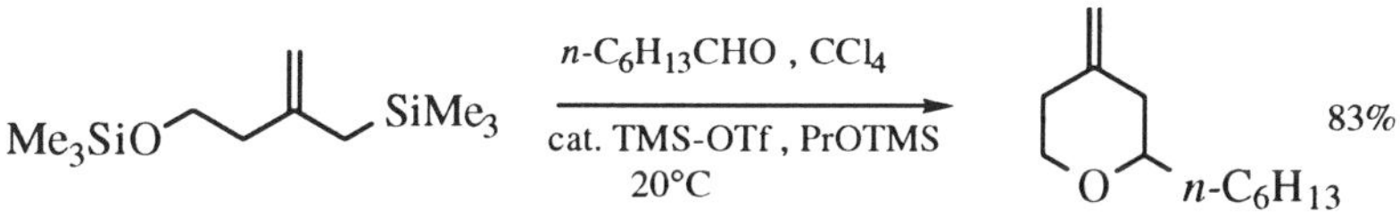

Markó, I.E.; Mekhalfia, A.; Bayston, D.J.; Adams, H. *J. Org. Chem.*, ***1992***, *57*, 2211.
Mekhalfia, A.; Markó, I.E.; Adams, H. *Tetrahedron Lett.*, ***1991***, *32*, 4783.

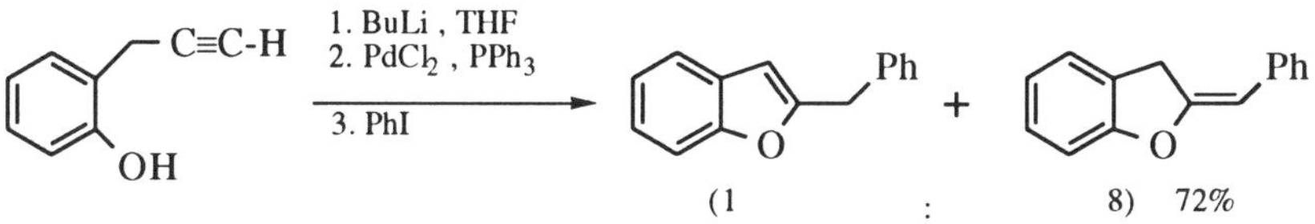

Luo, F-T.; Schreuder, I.; Wang, R-T. *J. Org. Chem.*, ***1992***, *57*, 2213.

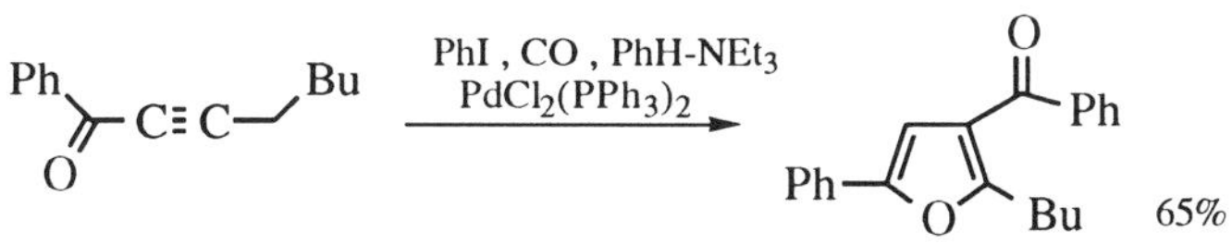

Okuro, K.; Furuune, M.; Miura, M.; Nomura, M. *J. Org. Chem.*, ***1992***, *57*, 4754.

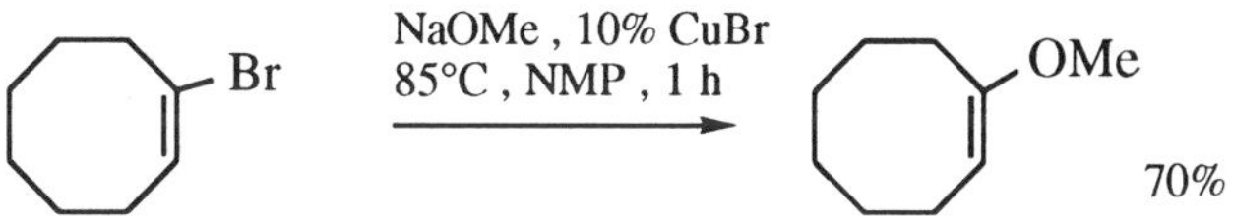

Keegstra, M.A. *Tetrahedron*, ***1992***, *48*, 2681.

Ph
O
O
O
Ph

Ti-graphite , THF
reflux , 8 h

Ph
Ph
O

88%

Fürstner, A.; Jumbam, D.N. *Tetrahedron,* ***1992***, *48*, 5991.
Fürstner, A.; Jumbam, D.N.; Weidmann, H. *Tetrahedron Lett.,* ***1991***, *32*, 6695.

O
Br O

t-BuOK

O
O

Aben, R.W.M.; Scheeren, J.W. *Rec. Trav. Chim. Pays-Bas,* ***1992***, *109*, 399.

Me
$OSiMe_2t$-Bu

$Fe(acac)_3$/2,2'-bipy
$AlEt_3$

Me Me
$OSiMe_2t$-Bu
(E + Z)

"good yield"

Takacs, J.M.; Myoung, Y.C. *Tetrahedron Lett.,* ***1992***, *33*, 317.

Ph
S
O
$SnBu_3$

AcCl , CH_2Cl_2

PhS OAc

50%

Beddoes, R.L.; MacLeod, D.; Moorcroft, D.; Quale, P.; Zhao, Y. *Tetrahedron Lett.,* ***1992***, *33*, 417.

O

1. $(CF_3SO_3)_2Hg{\bullet}NMe_2Ph$
$MeNO_2$, -20°C
2. NaCl
3. $NaBH_4$, NaOH

O
H

23%

Gopalan, A.S.; Prieto, R.; Mueller, B.; Peters, D. *Tetrahedron Lett.,* ***1992***, *33*, 1679.

I
MeO O
Me

1. Co(I) salophen , THF
2. hv (sunlamp)

H
MeO
O
Me

43%

Ali, A.; Harrowven, D.C.; Pattenden, G. *Tetrahedron Lett.,* ***1992***, *33*, 2851.

n-C_5H_{11}-C≡C-O ⟶ ($LiAlH_4$, THF) n-C_5H_{11} ... O ... Ph

Solà, L.; Castro, J.; Moyano, A.; Pericàs, M.A.; Riera, A. *Tetrahedron Lett.*, ***1992***, *33*, 2863.

Br-C≡C ... OAc ⟶ (K_2CO_3, MeOH, 10 min) 99%

Grandjean, D.; Pale, P.; Chuche, J. *Tetrahedron Lett.*, ***1992***, *33*, 4905.

F_3C-CO-$SiPh_3$ ⟶ (BuLi, THF) F_2C=C($OSiPh_3$)Bn 95%

Jin, F.; Jiang, B.; Xu, Y. *Tetrahedron Lett.*, ***1992***, *33*, 1221.

SMe, OSi(iPr)$_3$ ⟶ (n-C_6H_{13}CHO, 1.2 eq. Me_2AlCl) OH, SMe, n-C_6H_{13}, OSi(iPr)$_3$ 94%

(85:15 threo:erythro, 83% ee)

Tanino, K.; Shoda, H.; Nakamura, T.; Kuwajima, I. *Tetrahedron Lett.*, ***1992***, *33*, 1337.

⟶ ($Pd(OAc)_2$, (R)-BINAP, PhH, PhOTf, 9 d, Me_2N NMe_2) Ph + Ph

(71 : 29) quant.

(>96% ee) (17% ee)

Ozawa, F.; Kubo, A.; Hayashi, T. *Tetrahedron Lett.*, ***1992***, *33*, 1485.

(EtO)$_2$P(O)CH_3 ⟶ (1. 3 eq. LDA, THF, -78°C; 2. 2 eq. PhSeBr, 60°C; 3. PhCHO, RT) Ph-CH=C(SePh)SePh 97%

Shin, W.S.; Lee, K.; Oh, D.Y. *Tetrahedron Lett.*, ***1992***, *33*, 5375.

$SiMe_3$ ⟶ ($BuCH(OMe)_2$, CH_2Cl_2, TMSOTf, -78°C) MeO, Bu 85%

Hojo, M.; Ohsumi, K.; Hosomi, A. *Tetrahedron Lett.*, ***1992***, *33*, 5981.

1. PhSeBr, THF, -78°C
2. H_3O^+

71%

Chung, H.S.; Oh, D.Y. *Tetrahedron Lett.*, ***1992***, *33*, 5097.

, EtOH

$NEt_3 \cdot CaCl_2$

65%

Pinto, I.L.; Buckle, D.R.; Rami, H.K.; Smith, D.G. *Tetrahedron Lett.*, ***1992***, *33*, 7597.

1. BuLi, THF, -78°C; *t*-BuOK; BuLi
2. Me-furan-CHO
3. SiO_2/H_2O, microwaves 5 min
4. TBSCl, imidazole, DMF

(1 : 1) 75%

Wang, W-B.; Roskamp, E.J. *Tetrahedron Lett.*, ***1992***, *33*, 7631.

>180°C

(3 : 1) quant.

Detert, H.; Anthony-Mayer, C.; Meier, H. *Angew. Chem. Int. Ed. Engl.*, ***1992***, *31*, 791.

1. NaOMe, MeOH
2. Cl_3CCO_2H, Ac_2O, PhMe, reflux
3. mCPBA, CH_2Cl_2

38%

Chan, W.H.; Lee, A.W.M.; Chan, E.T.T. *J. Chem. Soc., Perkin Trans I*, ***1992***, 945.

t-BuOK, *t*-BuOH

18-crown-6

70%

Marshall, J.A.; BuBay, W.J. *J. Am. Chem. Soc.*, ***1992***, *114*, 1450.

$\xrightarrow[\text{RT, 48 h}]{Ph_2C=CH_2\text{, }ZnI_2\text{, PhH}}$ 87%

Dvorak, D.; David, S.; Arnold, Z.; Ivana, C.; Petricek, V. *Coll. Czech. Chem. Commun.*, ***1992***, *57*, 2337.

1. F^- 2. allyl-$SiMe_3$, $TfOH_2^+$ $B(OTf)_4^-$, -5°C, 30 min

(23 : 1) 85%

Davis, A.P.; Hegarty, S.C. *J. Am. Chem. Soc.*, ***1992***, *114*, 2745.

PhSH, $Pd(OAc)_2$ / THF, 67°C, 16 h — 86%

Kuniyasu, H.; Ogawa, A.; Sato, K-I.; Ryu, I.; Kambe, N.; Sonoda, N. *J. Am. Chem. Soc.*, ***1992***, *114*, 5902.
Kuniyasu, H.; Ogawa, A.; Sato, K-I.; Ryu, I.; Sonoda, N. *Tetrahedron Lett.*, ***1992***, *33*, 5525.

Bu-C≡C-SMe, PhMe, $TiCl_2(OiPr)_2$, 0°C / pet ether — 92% (>98% ee)

Narasaka, K.; Hayashi, Y.; Shimadzu, H.; Niihata, S. *J. Am. Chem. Soc.*, ***1992***, *114*, 8869.

PhC≡CH, $PdCl_2(PPh_3)_2$, CuI, NEt_3, DMF / 60°C, 16 h — 77%

Kundu, N.G.; Pal, M.; Mahanty, J.S.; Dasgupta, S.K. *J. Chem. Soc., Chem. Commun.*, ***1992***, 41.

$Bu_2Sn(CH_2CH=CH_2)_2$ / Et_4NCl, DCE, 60°C, 1 h — 76%

Yano, K.; Hatta, Y.; Baba, A.; Matsuda, H. *Synthesis*, ***1992***, 693.

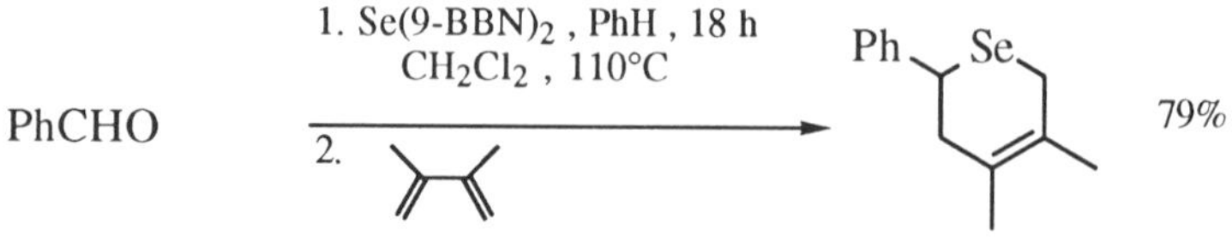

1. $Se(9\text{-}BBN)_2$, PhH , 18 h
CH_2Cl_2 , 110°C

PhCHO → 2. [diene] → Ph, Se — 79%

Shimada, K.; Jin, N.; Fujimura, M.; Nagano, Y.; Kudoh, E.; Takikawa, Y. *Chem. Lett.*, ***1992***, 1843.

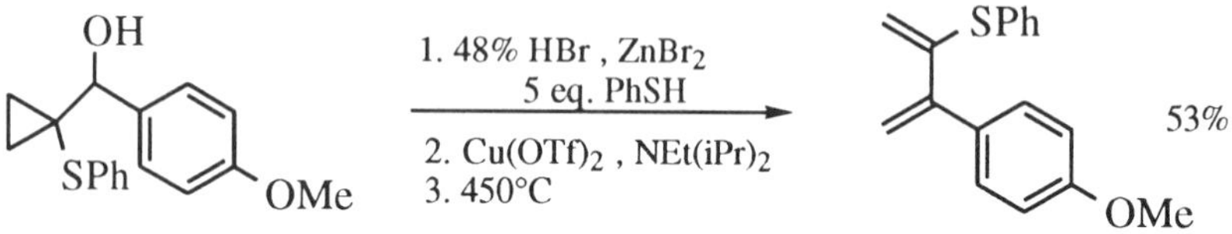

OH, SPh, OMe

1. 48% HBr , $ZnBr_2$
5 eq. PhSH
2. $Cu(OTf)_2$, $NEt(iPr)_2$
3. 450°C

SPh, OMe — 53%

Kwon, T.W.; Smith, M.B. *Synth. Commun.*, ***1992***, *22*, 2273.

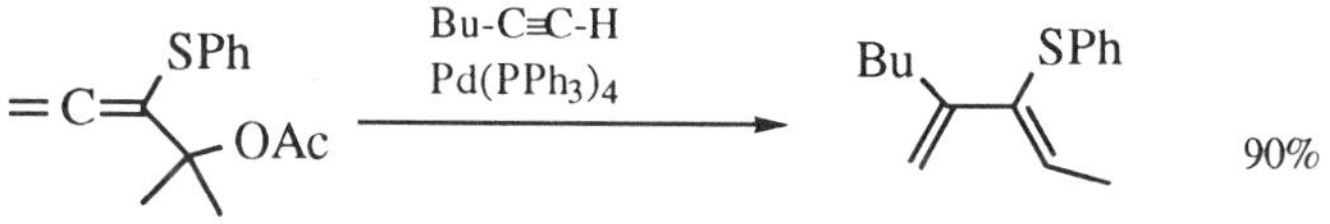

=C= SPh, OAc

Bu-C≡C-H
$Pd(PPh_3)_4$

Bu, SPh — 90%

Ni, Z.; Padwa, A. *SynLett*, ***1992***, 869.

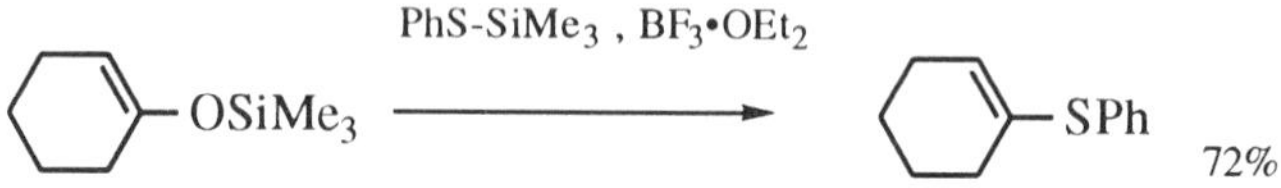

$OSiMe_3$ → PhS-$SiMe_3$, BF_3•OEt_2 → SPh — 72%

Degl'Innocenti, A.; Ulivi, P.; Capperucci, A.; Mordini, A.; Reginato, G.; Ricci, A. *SynLett*, ***1992***, 499.

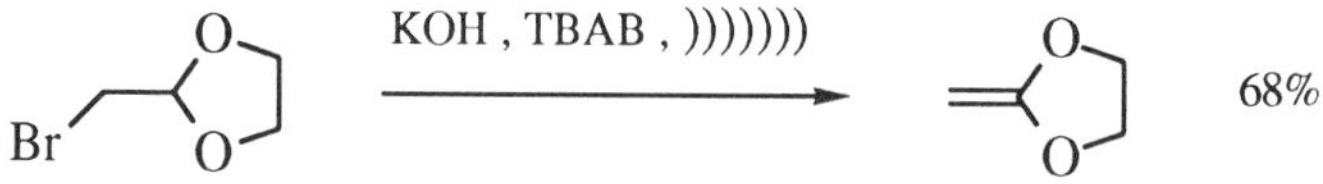

Br, O, O → KOH , TBAB ,))))))) → O, O — 68%

Díez-Barra, E.; de la Hoz, A.; Díaz-Ortiz, A.; Prieto, P. *SynLett*, ***1992***, 893.

Related Methods: Section 180A (Protection of Ketones)

SECTION 368: HALIDE, SULFONATE - HALIDE, SULFONATE

Halocyclopropanations are found in Section 74F (Alkyls from Alkenes).

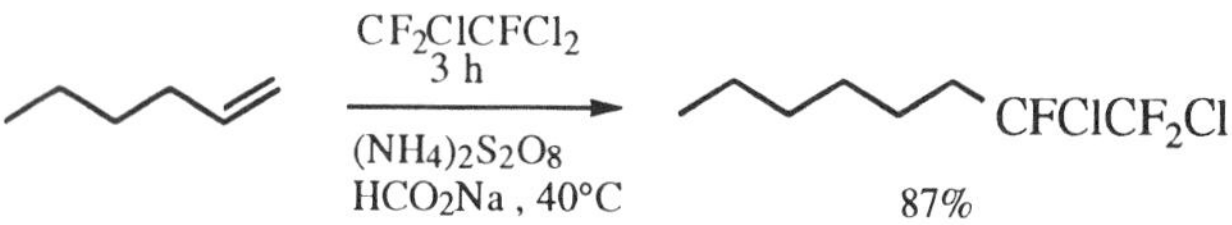

$CF_2ClCFCl_2$
3 h
$(NH_4)_2S_2O_8$
HCO_2Na , 40°C

→ $CFClCF_2Cl$ — 87%

Hu, C-M.; Qing, F-L. *Tetrahedron Lett.*, ***1990***, *31*, 1307.

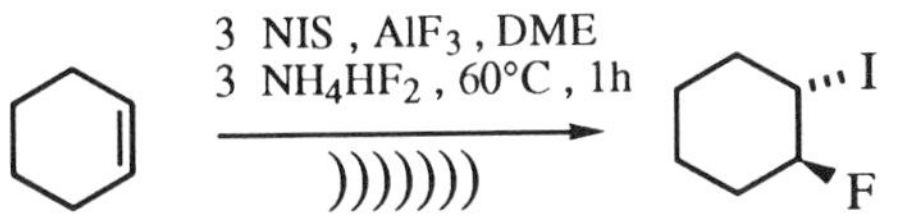

3 NIS , AlF_3 , DME
3 NH_4HF_2 , 60°C , 1h
))))))

I
F

Ichihara, J.; Funabiki, K.; Hanafusa, T. *Tetrahedron Lett.*, ***1990***, *31*, 3167.

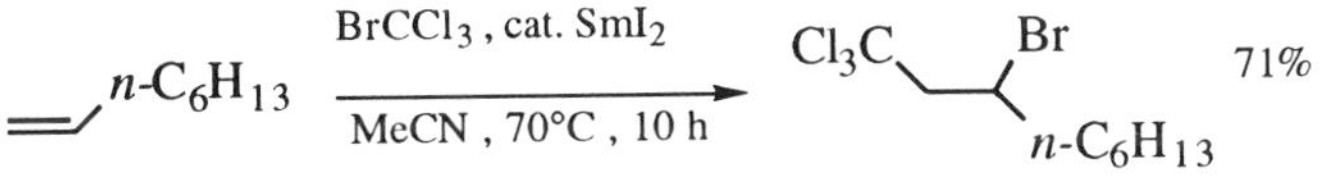

n-C_6H_{13}

$BrCCl_3$, cat. SmI_2
MeCN , 70°C , 10 h

Cl_3C Br n-C_6H_{13} 71%

Ma, S.; Lu, X. *J. Chem. Soc., Perkin Trans. I*, ***1990***, 2031.

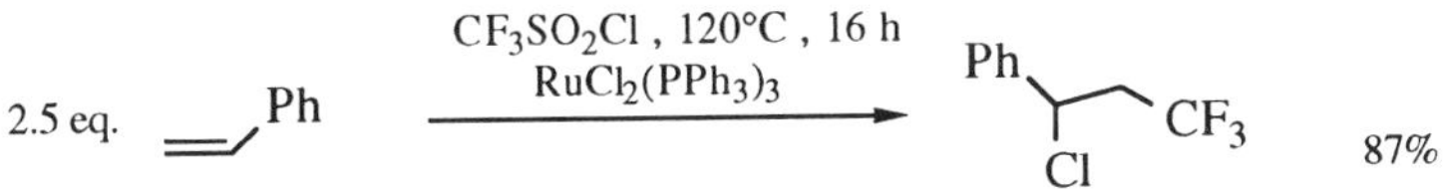

2.5 eq. Ph

CF_3SO_2Cl , 120°C , 16 h
$RuCl_2(PPh_3)_3$

Ph CF_3 Cl 87%

Kamigata, N.; Fukushima, T.; Terakawa, Y.; Yoshida, M.; Sawada, H. *J. Chem. Soc., Perkin Trans. I*, ***1991***, 627.

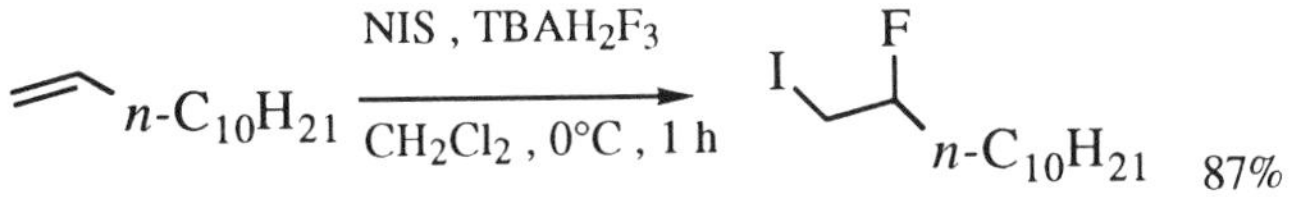

n-$C_{10}H_{21}$

NIS , $TBAH_2F_3$
CH_2Cl_2 , 0°C , 1 h

F I n-$C_{10}H_{21}$ 87%

$TBAH_2F_3$ = tetrabutylammonium dihydrogen trifluoride

Kuroboshi, M.; Hiyama, T. *Tetrahedron Lett.*, ***1991***, *32*, 1215.
Kuroboshi, M.; Hiyama, T. *SynLett.*, ***1991***, 185.

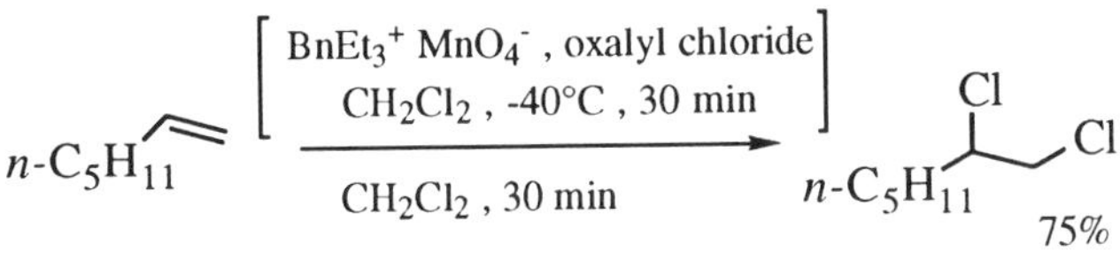

n-C_5H_{11}

[$BnEt_3^+$ MnO_4^- , oxalyl chloride
CH_2Cl_2 , -40°C , 30 min]
CH_2Cl_2 , 30 min

Cl Cl n-C_5H_{11} 75%

Markó, I.E.; Richardson, P.F. *Tetrahedron Lett.*, ***1991***, *32*, 1831.

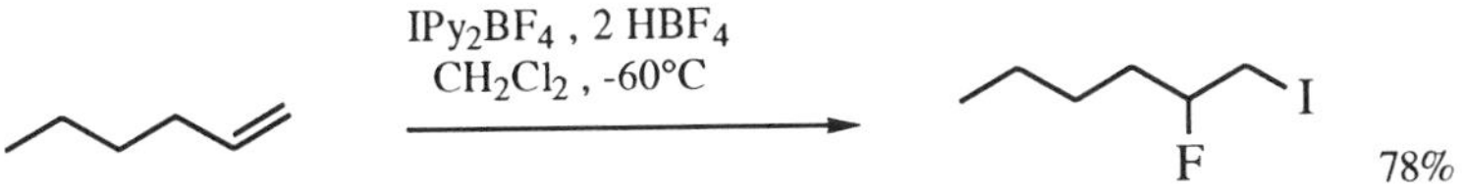

IPy_2BF_4 , 2 HBF_4
CH_2Cl_2 , -60°C

I F 78%

Barluenga, J.; Campos, P.J.; González, J.M.; Suárez, J.L.; Asensio, G. *J. Org. Chem.*, ***1991***, *56*, 2234.

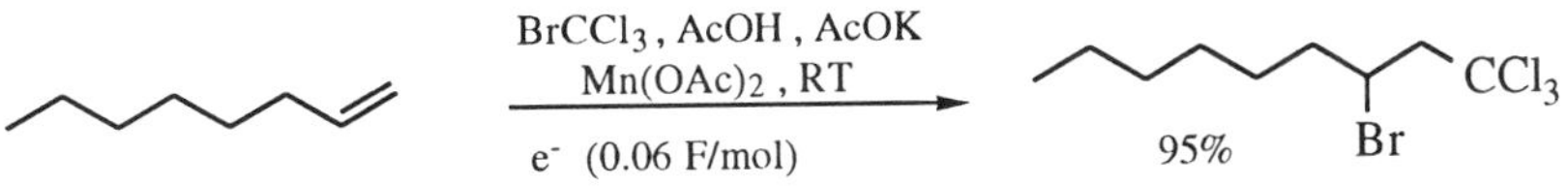

$BrCCl_3$, AcOH , AcOK
$Mn(OAc)_2$, RT
e^- (0.06 F/mol)

CCl_3 95% Br

Nohair, K.; Lachaise, I.; Paugram, J-P.; Nédétec, J-Y. *Tetrahedron Lett.*, ***1992***, *33*, 213.

SECTION 369: HALIDE, SULFONATE - KETONE

Me, O, SiMe$_3$, Me, SO$_2$Ph → MgBr$_2$•OEt$_2$, ether; 20°C, 4 h → Me, O, SiMe$_3$, Me, Br — 49%

Hewkin, C.T.; Jackson, R.F.W. *Tetrahedron Lett.*, ***1990***, *31*, 1877.

Ph, SPh, O → e$^-$, MeCN, Et$_3$N•3 HF; Pt electrodes, RT; 1.10 V, 1.6 F mol^{-1} → F, Ph, SPh, O — 84%

Brigaud, T.; Laurent, E. *Tetrahedron Lett.*, ***1990***, *31*, 2287.

O, Me, Na$^{\oplus}$, O → S$^{\oplus}$, CF$_3$, $^{\ominus}$OTf; DMF, -65°C → RT, overnight → O, Me, CF$_3$, O — 83%

Umemoto, T.; Ishihara, S. *Tetrahedron Lett.*, ***1990***, *31*, 3579.

O, CO$_2$Me → (O$_2$N–C$_6$H$_4$–SO$_2$–O)$_2$; CH$_2$Cl$_2$, 0°C → O, CO$_2$Me, O–SO$_2$–C$_6$H$_4$–NO$_2$ — 65%

Hoffman, R.V.; Wilson, A.L.; Kim, H-O. *J. Org. Chem.*, ***1990***, *55*, 1267.

CCl$_3$, O → 2% RuCl$_2$(PPh$_3$)$_3$; PhH, 155°C, 24 h → O, Cl, Cl, Cl — 77% + O, Cl, Cl, Cl — 8%

Lee, G.M.; Weinreb, S.M. *J. Org. Chem.*, ***1990***, *55*, 1281.

OAc → [CH$_3$CCl$_2$CO$_2$F]; aq. KI → O, F — 40-50%

Rozen, S.; Hebel, D. *J. Org. Chem.*, ***1990***, *55*, 2621.

1. $MeOTf$, $TiCl_4$, ether
2. 1,8-bis-(dimethylamino)-naphthalene, $MeCN$
3. H_3O^+

35%

Welch, J.T.; De Corte, B.; De Kimpe, N. *J. Org. Chem.*, ***1990***, *55*, 4981.

CO_2Me

1. $SOCl_2$, CH_2Cl_2, RT, 1 h
2. 50% aq. H_2SO_4, 6 h reflux

Cl 81%

De Kimpe, N.; Brunet, P. *Synthesis*, ***1990***, 595.

Me CO_2Et Br

1. $ClCH_2Cl$, THF $LiN(c\text{-}C_6H_{11})_2$
2. aq. HCl

Me Cl Br Cl 70%

Barluenga, J.; Llavona, L.; Yus, M.; Concellón, J.M. *Synthesis*, ***1990***, 1003.

OH Ph Me

mCPBA, HCl, DMF
25°C, 6 h

Ph Cl 80%

Kim, H.J.; Kim, H.R.; Ryu, E.K. *Synth. Commun.*, ***1990***, *20*, 1625.

OH Me OMe O

$CuCl_2$-DMSO, 1 h
dioxane, reflux

OMe Me Cl O 80%

Sahasrabhuddhe, A.S.; Ghiya, B.J. *Ind. J. Chem.*, ***1990***, *29B*, 61.

OH Ph

$TsSO_2NCl_2$, MeCN
40°C, 1.5 h

Ph Cl 80%

Kim, Y.H.; Lee, I.S.; Lim, S.C. *Chem. Lett.*, ***1990***, 1125.

Br CO_2Et

1. LiBr, $ClCH_2I$, -78°C
2. 1.6 eq. MeLi, -78°C
3. HCl - ether
4. HCl - H_2O

Br Cl O 64%

Barluenga, J.; Llavona, L.; Concellón, J.M. *J. Chem. Soc., Perkin Trans I*, ***1990***, 417.

1. 2 eq. $BrCH_2Cl$
Br_2CH_2
2. 2 eq. LDA , -78°C
3. Me_2CuLi
4. aq. HCl , 0°C

CO_2Et → O, Cl 83%

Barluenga, J.; Llavona, L.; Yus, M.; Concellón, J.M. *J. Chem. Soc., Perkin Trans. I*, ***1991***, 2890.

Ph, O, O, Ph — $PhICl_2$, CH_2Cl_2 → O, O, Ph, Ph, Cl 54%

Moskovkina, T.V.; Vysotskii, V.I. *Zhur. Org. Khim.*, ***1991***, *27*, 717.

1. LDA
2. $F\text{-}N(SO_2Ph)_2$, THF

O, Ph → O, Ph, F 85%

also for fluorination of Ar-H

Differding, E.; Ofner, H. *SynLett*, ***1991***, 187.

HO, Me — $PhI(OAc)_2$, I_2 cyclohexane, hν (sunlamp) → O, I, Me 78%

Ellwood, C.W.; Pattenden, G. *Tetrahedron Lett.*, ***1991***, *32*, 1583.

1. $NaN(SiMe_3)_2$, ether -78°C
2. -78°C → 20°C

O_2, S, N F, S, O_2

O → O, F 80%

Davis, F.A.; Han, W. *Tetrahedron Lett.*, ***1991***, *32*, 1631.

Br — 1. BuLi , CuBr•DMS 2. TFAA → O, CF_3 84%

Kerdesky, F.A.J.; Basha, A. *Tetrahedron Lett.*, ***1991***, *32*, 2003.

NBS , cat. $AgBF_4$

70-80%

Angara, G.J.; McNelis, E. *Tetrahedron Lett.*, ***1991***, *32*, 2099.

$Cu(NO_3)_2$, I_2

MeCN , RT

quant.

Cort, A.D. *J. Org. Chem.*, ***1991***, *56*, 6708.

Tl $(SO_2Tol)_3$, MeCN

92%

Khanna, M.S.; Garg, C.P.; Kapoor, R.P. *Tetrahedron Lett.*, ***1992***, *33*, 1495.
Khanna, M.S.; Garg, C.P.; Kapoor, R.P. *SynLett*, ***1992***, 393.

, DMSO

MeOH-NEt_3 , CH_2Cl_2

94%

Raina, S.; Bhuniya, D.; Singh, V.K. *Tetrahedron Lett.*, ***1992***, *33*, 6021.

PhI(OH)OTs
MeCN
)))))))

91%

Tuncay, A.; Dustman, J.A.; Fisher, G.; Tuncay, C.I.; Suslick, K.S. *Tetrahedron Lett.*, ***1992***, *33*, 7647.

KI-KIO_3-H_2SO_4 , 45 h

86%

Okamoto, T.; Kakinami, T.; Nishimura, T.; Hermawan, T.; Kajigaeshi, S. *Bull. Chem. Soc. Jpn.*, ***1992***, *65*, 1731.

1. [AgF/PhSeBr/)))))))]

2. 30% aq. H_2O_2

52%

Usuki, Y.; Iwaoka, M.; Tomoda, S. *J. Chem. Soc., Chem. Commun.*, ***1992***, 1148.

Ph-C≡C-Me → oxone, HCl, DMF; 25°C, 2 h → PhC(O)CCl$_2$Me 94%

Kim, K.K.; Kim, J.N.; Kim, K.M.; Kim, H.R.; Ryu, E.K. *Chem. Lett.*, ***1992***, 603.

SECTION 370: HALIDE, SULFONATE - NITRILE

Pd(PPh$_3$)$_4$, HMPA; proton sponge, 65°C

NC, I, N, CO_2Ph → NC, I, N, CO_2Ph 67%

Mori, M.; Kubo, Y.; Ban, Y. *Heterocycles*, ***1990***, *31*, 433.

N, Cl, Me, PO_2Cl_2; 30% H_2O_2, CH_2Cl_2; -20°C

O → Cl, O 67%

Rodriguez, J.; Dulcère, J-P. *SynLett*, ***1991***, 477.

SECTION 371: HALIDE, SULFONATE - ALKENE

1. PhSeCl, MeCN; AgF, RT, 18 h
2. O_3, CCl_4, -2°C
3. iPr$_2$NH, reflux; 16 h

MeO → MeO, F 44%

McCarthy, J.R.; Matthews, D.P.; Barney, C.L. *Tetrahedron Lett.*, ***1990***, *31*, 973.

PhS-C≡C-CH$_2$OH →
1. Red-Al, ether, 0°C
2. I_2, -78°C → 23°C, 1h
3. HO, OH, NaO_2C, CO_2K

→ PhS, I, OH 80%

Magriotis, P.A.; Doyle, T.J.; Kim, K.D. *Tetrahedron Lett.*, ***1990***, *31*, 2541.

H-C≡C-CH(Me)OTs → CF_3Cu, DMF → $F_3CHC=C=CHMe$ 65%

Burton, D.J.; Hartgraves, G.A.; Hsu, J. *Tetrahedron Lett.*, ***1990***, *31*, 3699.

$H-C\equiv C-CH_2Br$ $\xrightarrow[20°C]{C_6F_{13}Cu\ ,\ DMSO}$ $C_6F_{13}HC=C=CH_2$ 75%

Hung, H-H. *Tetrahedron Lett.*, ***1990***, *31*, 3703.

Br, Br; HO, B(OH)$_2$; Pd(PPh$_3$)$_4$, TlOH , THF; 23°C; HO; Br; 61%

Roush, W.R.; Moriarty, K.J.; Brown, B.B. *Tetrahedron Lett.*, ***1990***, *31*, 6509.

$Ph-C\equiv C-H$ — IPy$_2$ BF$_4$, AcOH; CH$_2$Cl$_2$, HBF$_4$; 12 h → Ph, I, AcO 53%

Barluenga, J.; Rodríguez, M.A.; Campos, P.J. *J. Org. Chem.*, ***1990***, *55*, 3104.
Barluenga, J.; Rodríquez, M.A.; Campos, P.J. *Tetrahedron Lett.*, ***1990***, *31*, 2751.
Barluenga, J.; Campos, P.J.; López, F.; Llorente, I.; Rodríguez, M.A. *Tetrahedron Lett.*, ***1990***, *31*, 7375.
Barluenga, J.; Rodríguez, M.A.; González, J.M.; Campos, P.J. *Tetrahedron Lett.*, ***1990***, *31*, 4207.

O; F$_3$C, OEt — 1. Dibal , THF; 2. (EtO)$_2$POCFCO$_2$CO$_2$Et Li; -78°C - +25°C → F$_3$C, F, CO$_2$Et (83:17 E:Z) 63%

Thenappan, A.; Burton, D.J. *J. Org. Chem.*, ***1990***, *55*, 4639.

F, CO$_2$TMP, SiMe$_3$ — 1. LDA; 2. PhCHO → Ph, F, CO$_2$TMP 49% (1:29 E:Z)

TMP = 2,4,6-trimethylphenyl

Welch, J.T.; Herbert, R.W. *J. Org. Chem.*, ***1990***, *55*, 4782.

CH$_2$I$_2$ — 1. KHMDS , -100°C; 2. PhO$_2$S, Li, *n*-C$_6$H$_{13}$ → *n*-C$_6$H$_{13}$, I 66% (84:16 E:Z)

Charreau, P.; Julia, M.; Verpeaux, J.N. *Bull. Soc. Chim. Fr.*, ***1990***, *127*, 275.

Br, Br — morpholine , DMSO , 85°C; KOH , EtOH → Br 80%

Bandodakar, B.S.; Nagendrappa, G. *Synthesis*, ***1990***, 843.

$CBrCl_3$, 2 eq. VCl_2
DMF, 80°C, 20 h

CCl_3

63%

Hirao, T.; Ohshiro, Y. *SynLett, **1990***, 217.

Me_3SiCl, NaI, MeCN
0.5 H_2O, RT, 1 h

Bu-C≡C-Et

Bu Et

I

>98%

Kamiya, N.; Chikami, Y.; Ishii, Y. *SynLett, **1990***, 675.

1. $AlCl_3$, 150°C, 3 d
2. 2 eq. Zn, DMF
3. triglyme, 25°C, 48 h

Cl

$CF_3CFBrCF_2Br$

CF_3 CF_2 + CF_3 CF_2

Morken, P.A.; Lu, H.; Nakamura, A.; Burton, D.J. *Tetrahedron Lett., **1991**, 32*, 4271.

1. *sec*-BuLi
2. n-$C_6H_{13}CHO$
3. DAST, CH_2Cl_2

$F_2C=CH_2$

n-C_6H_{13}

CF_3 50-60%

DAST = diethylaminosulfur trifluoride

Tellier, F.; Sauvêtre, R. *Tetrahedron Lett., **1991**, 32*, 5963.

OMe
OMe

$SiMe_3$, CH_2Cl_2
$SnBr_2$-AcBr, RT, 1 h

Br

66%

Oriyama, T.; Iwanami, K.; Tsukamoto, K.; Ichimura, Y.; Koga, G. *Bull. Chem. Soc. Jpn., **1991**, 64*, 1410.

O
Ph

4 eq. $MoCl_5$, 24 h

Cl
Ph Cl
Cl

34%

Hirao, T.; Nagata, S.; Yamaguchi, M.; Oshiro, Y. *Bull. Chem. Soc. Jpn., **1991**, 64*, 1717.

CHO

$(CF_3CO)_2O$, Zn-Cu, MS
$CF_3CCl_2CO_2Me$, THF
60°C

CO_2Me
CF_3 88%

(65:35 Z:E)

Allmendinger, T.; Lang, R.W. *Tetrahedron Lett., **1991**, 32*, 339.

0.33 PhC≡CAlCl , 0°C
benzene - CH_2Cl_2

44%

Bhaskar Ready, G.; Hanamoto, T.; Hiyama, T. *Tetrahedron Lett.*, ***1991***, *32*, 521.

Me_3SiCl , NaI
MeCN

50%

Amouroux, R.; Ejjiyar, S. *Tetrahedron Lett.*, ***1991***, *32*, 3059.

• TMEDA
cat. $Pd(PPh_3)_4$, THF

96%

Jiang, B.; Xu, Y. *J. Org. Chem.*, ***1991***, *56*, 7336.
Jiang, B.; Xu, Y. *Tetrahedron Lett.*, ***1992***, *33*, 511[coupling of two vinyl halides].

H- C≡C- CO_2Et → LiI , AcOH , MeCN; reflux , 21 h → I, CO_2Et 87%

Ma, S.; Lu, X.; Li, Z. *J. Org. Chem.*, ***1992***, *57*, 709.
Ma, S.; Lu, X. *Tetrahedron Lett.*, ***1990***, *31*, 7653.
Ma, S.; Lu, X. *J. Chem. Soc., Chem. Commun.*, ***1990***, 1643 [with LiBr].

$TiCl_4$, CH_2Cl_2

54%

Mayr, H.; Bäuml, E.; Cibura, G.; Koschinsky, R. *J. Org. Chem.*, ***1992***, *57*, 768.

4.2 eq. I_2 , RT , 24 h

57%

Johnson, C.R.; Adams, J.P.; Braun, M.P.; Senanayake, C.B.W.; Wovkulich, P.M.; Uskoković, M.R. *Tetrahedron Lett.*, ***1992***, *33*, 917.

CHO → (1. Tf_2O, CH_2Cl_2, $N(iBu)_2$ base, 0°C, 12 h; 2. MgI_2; 3. DBU) → vinyl iodide (96:4 E:Z) 26-47%

Martínez, A.G.; Alvarez, R.M.; González, S.M.; Subramanian, L.R.; Conrad, M. *Tetrahedron Lett.*, ***1992***, *33*, 2043.

CF_3CH_2OTs → (1. 2 eq. BuLi, THF; 2. BBu_3, LiF; 3. CuI, THF-HMPA; 4. 2% $Pd_2(dba)_3$-PPh_3, PhI, RT, 1 h) → F_2C=C(Bu)(Ph) 90%

Ichikawa, J.; Minami, T.; Sonoda, T.; Kobayashi, H. *Tetrahedron Lett.*, ***1992***, *33*, 3779.

t-Bu cyclohexanone (O) → (1. LDA; 2. Cl-pyridyl-NTf_2, -78°C, 2h) → *t*-Bu cyclohexenyl OTf 87%

Comins, D.L.; Dehghani, A. *Tetrahedron Lett.*, ***1992***, *33*, 6299.

Br, SPh → ($Bu_3SnSnBu_3$, hν (sunlamp); 1-iodo-1-methylcyclohexane (I), PhH, 30 min) → Br

Curran, D.P.; Yoo, B. *Tetrahedron Lett.*, ***1992***, *33*, 6931.

O, N, Cbz → ($POCl_3$, DMF, RT; $ClCHCCl_2$, 3 d) → Cl, N, Cbz 96%

Al-awar, R.S.; Joseph, S.P.; Comins, D.L. *Tetrahedron Lett.*, ***1992***, *33*, 7635.

n-C_4H_9-C≡C-H → (1. thexylBH_2, THF; 2. $(Me_2N)_3P$=O, $CuCl_2$, H_2O/THF; 3. 20°C → 70°C) → Cl-CH=CH-Bu 72%

Masuda, Y.; Hoshi, M.; Arase, A. *J. Chem. Soc., Perkin Trans. I*, ***1992***, 2725.

BF_3 , O_2 (cat.)
PhMe , 30°C
2 h

C≡C-H
C≡C-H

I

C≡C-H

49%
(1:9 E:Z)

5%

Albrecht, U.; Warthow, R.; Hoffmann, H.M.R. *Angew. Chem. Int Ed. Engl.*, ***1992***, *31*, 910.

SPh
Cl
Cl

$AgBF_4$, $AlCl_3$, 10 min
microwaves (650 W)

SPh
Cl

75%

Villemin, D.; Labiad, B. *Synth. Commun.*, ***1992***, *22*, 2043.

Ph— —$SnMe_3$ CFS , MeOH → Ph— —F 48%

CFS = cesium fluoroxysulfate

Hodson, H.F.; Madge, D.J.; Widdowson, D.A. *SynLett*, ***1992***, 831.

SECTION 372: KETONE - KETONE

hν (Hg lamp) , 2.5 h
Ni(II)/Co(II) filter
25°C , air

95%

Macias, F.A.; Molinillo, J.M.G.; Collado, I.G.; Massanet, G.M.; Rodriquez-Luis, F. *Tetrahedron Lett.*, ***1990***, *31*, 3063.

, $FeCl_3$, 20°C
claynick , 24 h

90%

claynick = [K-10 montrmorillonite/NiBr/MeCN/280°C]

Laszlo, P.; Montaufier, M-T.; Randriamahefa, S.L. *Tetrahedron Lett.*, ***1990***, *31*, 4867.

SePh

$MeC(O)CH_2SnBu_3$
hν (100 W Hg lamp)
PhH , RT , 3 h

Me

82%

Watanabe, Y.; Yoneda, T.; Ueno, Y.; Toru, T. *Tetrahedron Lett.*, ***1990***, *31*, 6669.

LDA , THF , 25°C , 6 h; I–CH$_2$C(=CH$_2$)–O–CH$_2$–OMe — 53%

Gu, X-P.; Kirito, Y.; Ikeda, I.; Okahara, M. *J. Org. Chem.*, ***1990***, *55*, 3390.

$EtCHO$, $SnCl_2$, CH_2Cl_2 , RT — 88%

Padwa, A.; Hornbuckle, S.F.; Zhang, Z.; Zhi, L. *J. Org. Chem.*, ***1990***, *55*, 5297.

CO_2t-Bu; 1. Me, CO_2Me, Cs_2CO_3 , $CHCl_3$; 2. PTSA — CO_2Me, Me, H, Me — 73%

Ruel, R.; Hogan, K.T.; Deslongchamps, P. *SynLett*, ***1990***, 516.

Br, Ph, O; $OSnBu_3$, Me, , HMPT; RT , 1 h — Me, Ph, O 73% + Me, O, Ph, O 12%

Baba, A.; Yasuda, M.; Yano, K.; Shibata, I.; Matsuda, H. *J. Chem. Soc., Perkin Trans. I*, ***1990***, 3205.

Me, NO_2; 1. O^- Li^+, Bu , THF , -78°C; 2. 10% HCl , RT , 5 h — Bu, O, Me, O — 65%

Miyashita, M.; Awen, B.Z.E.; Yoshikoshi, A. *Synthesis*, ***1990***, 563.

$OSiMe_3$; $PhIF_2 \cdot BF_3$; -78°C → -10°C — 74%

Zefirov, N.S.; Samoniya, N.Sh.; Jutateladze, T.G.; Zhdankin, V.V. *Zhur. Org. Khim.*, ***1991***, *27*, 220 (Engl., p. 194).

$PdCl_2(MeCN)_2$, H_2O, THF, RT

n-C_6H_{13} 63%

Mahrwald, R.; Schick, H. *Angew. Chem. Int. Ed. Engl.*, ***1991***, *30*, 593.

$OSiMe_2t$-Bu, Ph, DMF

2.4 eq. $Mn(pic)_3$, 0°C

Ph, OH, Ph, Ph 80%

$Mn(pic)_3$ = manganese (III) 2-pyridinecarboxylate

Iwasawa, N.; Hayakawa, S.; Isobe, K.; Narasaka, K. *Chem. Lett.*, ***1991***, 1193.

I_2, DMSO, 155°C, 10 h

Ph, Ph, Ph, Ph 85%

Yusybov, M.S.; Filimonov, V.D. *Synthesis*, ***1991***, 131.

2 CO_2Et

1. Na, ether
2. $SOCl_2$, -18°C → RT

75%

Olah, G.A.; Wu, A. *Synthesis*, ***1991***, 1177.

1. mCPBA
2. $LiClO_4$, PhMe toluene

quant.

French, L.G.; Fenlon, E.E.; Charlton, T.P. *Tetrahedron Lett.*, ***1991***, *32*, 851.

1. (cyclopentenone), $CsHCO_3$, MeOH, 20°C, 2 h
2. 4N H_2SO_4, 60°C, 4 h
3. 3 MsCl, 5 NEt_3, cat. DMAP, CH_2Cl_2, 20°C, 30 min

24% overall

d'Angelo, J.; Gomez-Pardo, D. *Tetrahedron Lett.*, ***1991***, *32*, 3063.

2 $PhCO_2Me$ → 1. e^-, $SmCl_3$, Mg anode; 2. H_3O^+ → PhC(O)C(O)Ph 68%

Hébri, H.; Duñach, E.; Heintz, M.; Troupel, M.; Périchon, J. *SynLett*, ***1991***, 901.

PhC(O)Cl → 1. [Ph–CH(OTMS)–P(O)(OEt)$_2$ /LDA]; 2. H_3O^+ → PhC(O)C(O)Ph 83%

Olah, G.A.; Wu, A. *J. Org. Chem.*, ***1991***, *56*, 902.

$CH_3CH_2CH_2CO_2Me$ → 2 eq. BuLi, -110°C → BuC(O)C(O)CH$_2$CH$_2$CH$_3$ 66%

Seyferth, D.; Weinstein, R.M.; Hui, R.C.; Wang, W-L.; Archer, C.M. *J. Org. Chem.*, ***1991***, *56*, 5768.

HO–CH(CH$_3$)–C≡C–CH(CH$_3$)–OH → $IrH_5(PiPr_3)_2$, PhMe, reflux, 40 h → $CH_3C(O)CH_2CH_2C(O)CH_3$ 70%

Lu, X.; Ji, J.; Ma, D.; Shen, W. *J. Org. Chem.*, ***1991***, *56*, 5774.

→ 2 eq. $Mn(OAc)_3$•2 H_2O, EtOH, 24 h → 51%

Snider, B.B.; Buckman, B.O. *J. Org. Chem.*, ***1992***, *57*, 322.

OTf → 1. methyl vinyl ketone, CH_2Cl_2, 5% $Bu_2Sn(OTf)_2$, -78°C, 5 h; 2. 1N HCl, THF, RT → 70%

Sato, T.; Wakahara, Y.; Otera, J.; Nozaki, H. *Tetrahedron*, ***1991***, *47*, 9773.
Sato, T.; Wakahara, Y.; Otera, J.; Nozaki, H. *Tetrahedron Lett.*, ***1990***, *31*, 1581.

1. $iPrNH_2$, $TiCl_4$ - ether
16 h , 0°C → RT
2. LDA , THF
3. EtI
4. oxalic acid , H_2O/CH_2Cl_2
reflux , 1 h

58%

De Kimpe, N.; D'Hondt, L.; Stanoeva, E. *Tetrahedron Lett.*, ***1991***, *32*, 3879.

1. BuLi , THF , -78°C
2. PhCHO
3. NBS , aq. acetone
-5°C

Ph Ph 66%

Page, P.C.B.; Graham, A.E.; Park, B.K. *Tetrahedron*, ***1992***, *48*, 7265.

n-C_9H_{19} Br

1. $Na_2Fe(CO)_4$, CO , THF
2. CuCl , 25°C , 1.5 h
3. aq. CAN

n-$C_{10}H_{21}$ n-$C_{10}H_{21}$ 80%

Devasagayara, A.; Periasamy, M. *Tetrahedron Lett.*, ***1992***, *33*, 1227.

$OSiMe_3$

3 eq. $VO(OEt)Cl_2$
-75°C → -40°C

71%

Fujii, T.; Hirao, T.; Ohshiro, Y. *Tetrahedron Lett.*, ***1992***, *33*, 5823.

Me

n-$C_6H_{13}Br$
Me_2NAc

n-C_6H_{13} Me 70%

Yamashita, M.; Tashika, H.; Uchida, M. *Bull. Chem. Soc. Jpn.*, ***1992***, *65*, 1257.

1. BuLi , ether-hexane , -50°C
2. H_3PO_4 , H_2O , 20°C
12 h

Me Bu 58%

Wedler, C.; Schick, H. *Synthesis*, ***1992***, 543.

SECTION 373: KETONE - NITRILE

BnCN , 20°C , 3 h
2 eq. $HRh(CO)(PPh_3)_3$

Ph CN O 85%

also with conjugated esters, but in lower yield

Paganelli, S.; Schionato, A.; Botteghi, C. *Tetrahedron Lett.*, ***1991***, *32*, 2807.

$n-C_6H_{13}-I$ — CN, PhH, 80°C; Bu_3SnH, AIBN; CO (80 atm), 2 h → $n-C_6H_{13}$ O CN 74%

Ryu, I.; Kusano, K.; Yamazaki, H.; Sonoda, N. *J. Org. Chem.*, ***1991***, *56*, 5003.

$PhCO_2Me$ — e^-, (Mg anode, Ni/Cd cathode); $Bu_4N\ BF_4$, PhBr, MeCN → Ph CN O 83%

Barhadi, R.; Gal, J.; Heintz, M.; Troupel, M. *J. Chem. Soc., Chem. Commun.*, ***1992***, 50.

$PhCO_2Me$ — 1. e^-, 10% $SmCl_3$, MeCN, *t*-BuOH; 2. H_3O^+ → Ph CN O 75%

Hébri, H.; Duñach, E.; Périchon, J. *SynLett*, ***1992***, 293.

REVIEWS:

"Stereocontrolled Construction of Complex Cyclic Ketones via oxy-Cope Rearrangement" Paquette, L.A. *Angew. Chem. Int. Ed. Engl.*, ***1990***, *29*, 609.

SECTION 374: KETONE - ALKENE

For the oxidation of allylic alcohols to alkene ketones, see Section 168 (Ketones from Alcohols and Phenols)

For the oxidation of allylic methylene groups ($C=C-CH_2 \rightarrow C=C-C=O$), see Section 170 (Ketones from Alkyls and Methylenes).

For the alkylation of alkene ketones, also see Section 177 (Ketones from Ketones) and for conjugate alkylations see Section 74E (Alkyls from Alkenes).

$(PhIO)_n$, RT , 15h
n-Bu_4NF
THF/CH_2Cl_2

OSiMe$_3$

O

+ PhI

72%

Moriarty, R.M.; Vaid, R.K.; Hopkins, E.; Vaid, B.K.; Prakash, O. *Tetrahedron Lett.*, ***1990***, *31*, 197.

O
CO_2Et
S*t*-Bu
$Mn(OAc)_3$, $Cu(OAc)_2$
PbO_2 , Py , AcOH
2d , RT
O
CO_2Et
91%

Breuilles, P.; Ugen, D. *Tetrahedron Lett.*, ***1990***, *31*, 357.

O
1. PhCHO , CH_2Cl_2
$TiCl_4$, $Ti(OiPr)_4$, PPh_3
-50°C → RT , 8h
2. 10% K_2CO_3
O
Ph
83%

Takanami, T.; Suda, K.; Ohmori, H. *Tetrahedron Lett.*, ***1990***, *31*, 677.

EtO_2C
EtO_2C
+
N
1. 5% $Pd(OAc)_2$
2% PPh_3 , MeCN
70°C , 7 h
2. H_3O^+
EtO_2C
EtO_2C
H
H
O
64% (8:1 trans:cis)

Takacs, J.M.; Zhu, J. *Tetrahedron Lett.*, ***1990***, *31*, 1117.

Me
PivO
O
OTBDMS
N
N
MeCN , reflux
Me
OTBDMS
PivO
H
H
O
80-90%

McLoughlin, J.I.; Brahma, R.; Campopiano, O.; Little, R.D. *Tetrahedron Lett.*, ***1990***, *31*, 1377.

OCO_2Et
O
cat. $TiCl_4$, RT , 5 d
OCO_2Et
O
+
O
79%
20%

De Cusati, P.F.; Olofson, R.A. *Tetrahedron Lett.*, ***1990***, *31*, 1409.

1. $Tl(OAc)_3$, AcOH reflux, 3 h
2. ice, H_2O

96%

Singh, O.V.; Kapoor, R.P. *Tetrahedron Lett.*, ***1990***, *31*, 1459.

HO OH OH OH

1. DMP, MeOH $HClO_4$
2. PCC, PhH
3. $(MeO)_2POCH_2Li$

42%

Ali, S.M.; Ramesh, K.; Borchardt, R.T. *Tetrahedron Lett.*, ***1990***, *31*, 1509.

N S O n-$C_{15}H_{31}$ O

5 eq. , 5°C
CH_2Cl_2
hν
(150 W lamp)

n-$C_{15}H_{31}$ S N O 79% + n-$C_{15}H_{31}$ O 85%

Barton, D.H.R.; Sarma, J.C. *Tetrahedron Lett.*, ***1990***, *31*, 1965.

Ph_3P=Me

1. AcCl
2. BuLi, HCHO
3. 150-170°C 30 min

Me O Me 66%

LeRoux, J.; Le Corre, M. *Tetrahedron Lett.*, ***1990***, *31*, 2591.

Ph_3P= O CF_3

1. BuLi, THF -78°C
2. PhCHO

Ph CF_3 Bu O 63%

Shen, Y.; Wang, T. *Tetrahedron Lett.*, ***1990***, *31*, 3161.

$OSiMe_2t$-hexyl OCO_2Me

BF_3•OEt_2, CH_2Cl_2
0°C, 15 min

O 43% ee good yield

Ollivier, J.; Legros, J-Y.; Fiaud, J-C.; de Meijere, A.; Salaün, J. *Tetrahedron Lett.*, ***1990***, *31*, 4135.

PhI(OAc)$_2$, MeOH; 67%

Callinan, A.; Chen, Y.; Morrow, G.W.; Swenton, J.S. *Tetrahedron Lett.*, ***1990***, *31*, 4551.

Co(CO)$_6$, NMO; CH$_2$Cl$_2$, RT; 85%

Shambayati, S.; Crowe, W.E.; Schreiber, S.L. *Tetrahedron Lett.*, ***1990***, *31*, 5289.

Bu$_3$SnH, AIBN; PhH, reflux; 1.5 h; 90%

Lange, G.L.; Gottardo, C. *Tetrahedron Lett.*, ***1990***, *31*, 5985.

EtAlCl$_2$, PhH; 50 min; 70%

Fujiwara, T.; Takeda, T. *Tetrahedron Lett.*, ***1990***, *31*, 6027.

OSiMe$_3$, Me, Me, SiMe$_2$*t*-Bu; 1. Ph, Ph$_3$C$^+$ ClO$_4^-$; 2. aq. HBF$_4$, THF, 20°C; Me, Me, Me, Ph; (>95% ee); 48%

Lohray, B.B.; Zimbiniski, R. *Tetrahedron Lett.*, ***1990***, *31*, 7273.

SePh, Me; SnBu$_3$, hν, 1h; Me; 76%

Toru, T.; Okumura, T.; Ueno, Y. *J. Org. Chem.*, ***1990***, *55*, 1277.

1. $ClHC=CCl_2$, KH, THF
2. BuLi
3. HMPA I
4. $Co_2(CO)_8$, iso-octane reflux, 16 h

Ph OH → Ph O O

38% (3.2:1 diastereomeric ratio)

Poch, M.; Valentí, E.;' Moyano, A.; Pericàs, M.A.; Castro, J.; DeNicola, A.; Greene, A.E. *Tetrahedron Lett.*, ***1990***, *31*, 7505.

MMPP, H_2O
15 min

O → O O 99%

MMPP = Mg monoperoxy phthalate

Domínguez, C.; Csáky, A.G.; Plumet, J. *Tetrahedron Lett.*, ***1990***, *31*, 7669.

$Cr(CO)_5$ O^- Me_4N^+

1. AcCl
2. Me-C≡C OH
3. PhMe, reflux

O O 34%

(65:35 trans:cis)

Herndon, J.W.; Matasi, J.J. *J. Org. Chem.*, ***1990***, *55*, 786.

O O → $Me_2Cu(CN)Li_2$ → O Me 91%

(2.5:1 Z:E)

Cheng, M.; Hulce, M. *J. Org. Chem.*, ***1990***, *55*, 964.

O Me CO_2Et Me

2 $Mn(OAc)_3$•2 H_2O
$Cu(OAc)_2$•H_2O
4 d, 25°C

O Me CO_2Et + O Me CO_2Et Me OAc 70%

39% (2.5:1 β:α)

Kates, S.A.; Dombroski, M.A.; Snider, B.B. *J. Org. Chem.*, ***1990***, *55*, 2427.

n-C_3H_7-C≡C-n-C_3H_7 → iPrCHO, THF, 20 h; P(n-C_8H_{17})$_3$, 100°C; Ni(COD)$_2$ → 93% (93:7 E:Z)

Tsuda, T.; Kiyoi, T.; Saegusa, T. *J. Org. Chem.*, ***1990***, *55*, 2554.

OTf → 1. CH$_2$=C(OEt)SnMe$_3$, 3 eq. LiCl; 2% Pd(PPh$_3$)$_4$, THF; reflux, 16 h; 2. 2 NH_4Cl, THF, RT; 3 h → 75%

Kwon, H.B.; McKee, B.H.; Stille, J.K. *J. Org. Chem.*, ***1990***, *55*, 3114.

Ph → 1. disiamyl borane; 2. NaOH, H_2O_2; 3. oxalyl chloride/DMSO; NEt_3, -5°C; 4. 1N HCl/MeOH, reflux; 15 min → O, Ph 79%

Collins, S.; Hong, Y.; Kataoka, M.; Nguyen, T. *J. Org. Chem.*, ***1990***, *55*, 3395.

O → 1. LiC(SPh)$_3$; 2. I, SiMe$_3$; 3. AgOTf → O, H, O, H, PhS, SPh 32%

Posner, G.H.; Asirvatham, E.; Hamill, T.G.; Webb, K.S. *J. Org. Chem.*, ***1990***, *55*, 2132.

Me, Cl, CO_2Et → $PhSCH_2Li$/TMEDA; THF, -70°C → O, PhS, Me 54%

Mathew, J. *J. Org. Chem.*, ***1990***, *55*, 5294.

O, Me → 1. NBS, CCl_4, reflux; 2. aniline, CCl_4, RT; 15 h → O, Me 48%

Shimazaki, M.; Huang, Z-H.; Goto, M.; Suzuki, N.; Ohta, A. *Synthesis*, ***1990***, 677.

PhI , 5% $BnClPd(PPh_3)_2$
10% CuI , DMF , RT

Bu_3Sn O O O Ph O O 99%

Liebeskind, L.S.; Fengl, R.W. *J. Org. Chem.*, ***1990***, *55*, 5359.

1. $ZnCl_2$, ether
)))))) , 25°C
2. $CuBr \bullet SMe_2$, HMPA
MOMO
C≡C-CO_2Et
n-C_5H_{11}

$OSiMe_3$ OEt O CO_2Et n-C_5H_{11} OMOM

Crimmins, M.T.; Nantermet, P.G. *J. Org. Chem.*, ***1990***, *55*,4235.

Bu_3SnH , AIBN
80°C , PhH

O Me Br Ph C C O Me CHPh 80%

(44:56 cis:trans ring juncture)

Boger, D.L.; Mathvink, R.J. *J. Org. Chem.*, ***1990***, *55*, 5442.

1.[t-BuC(O)F / Me_2S / BF_3]
CH_2Cl_2
2. aq. $NaHCO_3$

Me O Me t-Bu 72%

(1:2 E:Z)

Verteletskii, P.V.; Balenkova, E.S. *Zhur. Org. Khim.*, ***1990***, *26*, 2446 (Engl., p. 2113).

$OSiMe_3$, $SnCl_2$
AcCl , CH_2Cl_2 , RT , 2h

Ph OMe OMe O Ph quant.

Oriyama, T.; Iwanami, K.; Miyauchi, Y.; Koga, G. *Bull. Chem. Soc. Jpn.*, ***1990***, *63*, 3716.

, DME
4% $Pd_2(dba)_3 \bullet CHCl_3$ / SiO_2
dppe

$OSiMe_3$ O 75%

Baba, T.; Nakano, K.; Nishiyama, S.; Tsurya, S. *J. Chem. Soc., Chem. Commun.*, ***1990***, 348.

-20°C , 24 h

99%

(88% ee , S)

Maruoka, K.; Banno, H.; Yamamoto, H. *J. Am. Chem. Soc.*, ***1990***, *112*, 7791.

PDC , CH_2Cl_2 , RT 32 h

95%

Mukhopadhyay, A.; Suryawanshi, S.N.; Bhakuni, D.S. *Ind. J. Chem.*, ***1990***, *29B*, 1060.

1. *t*-BuLi , THF , -78°C
2. PhCHO
3. $(CO_2H)_2$, MeOH , RT 15 h

36%

Giles, M.; Hadley, M.S.; Gallagher, T. *J. Chem. Soc., Chem. Commun.*, ***1990***, 831.

, $EtAlCl_2$
CH_2Cl_2 , 0°C

77%

Hayashi, Y.; Niihata, S.; Narasaka, K. *Chem. Lett.*, ***1990***, 2091.

B-Br , PhCHO
PhH , RT , 1 d

62%

Shimizu, H.; Hara, S.; Suzuki, A. *Synth. Commun.*, ***1990***, *20*, 549.

$PhSO_2Me$

1. BuLi , BF_3•OEt_2
2. *n*-C_5H_{11} epoxide
3. Jones oxidation
4. DBU

60%

Marczak, S.; Wicha, J. *Synth. Commun.*, ***1990***, *20*, 1511.

1. TMSCl , NEt_3
2. $Pb(OAc)_2$, MeCN
benzoquinone

52%

Kim, M.; Applegate, L.A.; Park, O-S.; Vasudevan, S.; Watt, D.S. *Synth. Commun.*, ***1990***, *20*, 985.

N_2 Ph

H-C≡C-CH_2OMe
CH_2Cl_2 , 4 h
$Rf_2(pfb)_4$
syring pump addition

MeO Ph + Ph OMe

(82 : 18) 80%

pfb = perfluorobutyrate

Dole, M.P.; Bagheri, V.; Claxton, E.E. *J. Chem. Soc., Chem. Commun.*, ***1990***, 46.

CHO Me

O_2N NO_2 NH_2
30% H_2O_2 , 3N HCl
MeCN , 70°C , 3 h

Me

72%

Gupta, R.B.; Franck, R.W. *SynLett*, ***1990***, 355.

Me_3SiO $OSiMe_3$ Me

PhCHO , $TiCl_4$, -78°C
CH_2Cl_2 , 4 h

Ph O Me

66%

Peterson, J.R.; Kirchhoff, E.W. *SynLett*, ***1990***, 394.

Me_3SiO H Me_3SiO H

1. TMSOTf
2. $TiCl_4$
3. $Pb(OAc)_4$

MeO H H

48%

Lee, T.V.; Porter, J.R.; Roden, F.S. *Tetrahedron*, ***1991***, *47*, 139.

Bu_3SnH , AIBN
PhH , reflux

77%

Baldwin, J.E.; Adlington, R.M.; Robertson, J. *Tetrahedron*, ***1991***, *47*, 6795.

MeI , CO (1 atm) , 40°C
$Mn_2(CO)_{10}$, CTAB
CH_2Cl_2 , NaOH

n-C_5H_{11} 40% + n-C_5H_{11} 7%

CTAB = cetyltrimethylammonium bromide

Satyanarayana, N.; Alper, H. *J. Chem. Soc., Chem. Commun.*, ***1991***, 8.

MMP , MeCN , 20°C , 1 h
cat. Fe^{III} (TF_4PS_4P)
pH 3 (tartrate buffer)

85%

MMP = magnesium monoperoxyphthalate
Fe (TF_4PS_4P) = iron-meso-tetrakis(2,3,5,6-tetrafluorophenyl) tetrasulfonatoporphyrin

Artaud, I.; Aziza, K.B.; Chopard, C.; Mansuy, D. *J. Chem. Soc., Chem. Commun.*, ***1991***, 31.

1. BzCl-$AlCl_3$, 0°C
CH_2Cl_2
2. $C_7H_{15}COCl$-$AlCl_3$
CH_2Cl_2 , 0°C → RT

55%

Babudri, F.; Fiandanese, V.; Marchese, G.; Naso, F. *J. Chem. Soc., Chem. Commun.*, ***1991***, 237.

$BnMe_3N\ Br_3$, NaOAc
CH_2Cl_2 , RT

quant.

Kajigaeshi, S.; Morikawa, Y.; Fujisaki, S.; Kakinami, T.; Nishihara, K. *Bull. Chem. Soc. Jpn.*, ***1991***, *64*, 336.

PhCHO , InI , THF

74%

Araki, S.; Butsugan, Y. *Bull. Chem. Soc. Jpn.*, ***1991***, *64*, 727.

Bu, I → B-n-C_8H_{17}; CO, $Pd(PPh_3)_4$, 25°C; dioxane, K_3PO_4 → O, n-C_8H_{17}, Bu 98%

Ishiyama, T.; Miyaura, N.; Suzuki, A. *Bull. Chem. Soc. Jpn.*, ***1991***, *64*, 1999.

n-C_5H_{11}, C≡C, Bu, OMe → 10% H_2O in MeOH; 5% $NaAuClO_4$; reflux, 2 h → O, n-C_5H_{11}, Bu 79%

Fukuda, Y.; Utimoto, K. *Bull. Chem. Soc. Jpn.*, ***1991***, *64*, 2013.

O, Me → PhCHO; TMSOTf → Ph, O, Me 67%

Sato, T.; Hayase, K. *Bull. Chem. Soc. Jpn.*, ***1991***, *64*, 3384.

Me, CHN_2, Me, O → $Rh_2(OAc)_4$ → H, Me, Me, H, O 40%

Honda, T.; Ishige, H.; Tsubuki, M.; Naito, K.; Suzuki, Y. *J. Chem. Soc., Perkin Trans. I*, ***1991***, 954.

n-C_6H_{13} → Bu, C≡C, H, $Co_2(CO)_6$ → n-C_6H_{13}, O, Bu + O, Bu, n-C_6H_{13} (1 : 1) 41%

Krafft, M.E.; Juliano, C.A.; Scott, I.L.; Wright, C.; McEachin, M.D. *J. Am. Chem. Soc.*, ***1991***, *113*, 1693.

$OSiMe_3$ → Pd(2-furyl)$_3$Cl$_2$, xylene; reflux, 36 h → O 84%

Hettrick, C.M.; Scott, W.J. *J. Am. Chem. Soc.*, ***1991***, *113*, 4903.

Ph, I-Ph (+) BF_4^- → (methyl vinyl ketone: O, Me), DMF; cat. $Pd(OAc)_2$, RT; $NaHCO_3$ → Ph, O, Me 73%

Moriarty, R.M.; Epa, W.R.; Awasthi, A.K. *J. Am. Chem. Soc.*, ***1991***, *113*, 6315.

1. *t*-$BuCO_2SO_2F$, CH_2Cl_2 -70°C; 2. NEt_3 → O, *t*-Bu 57%

Gavrishova, T.I.; Shastin, A.V.; Balenkova, E.S. *Zhur. Org. Khim.*, ***1991***, *27*, 673 (Engl., p. 580).

O, EtO-P-Me, OEt → 1. BuLi, THF; 2. PhCN, -78°C → -5°C; 3. MeO-C_6H_4-CHO, -5°C → RT; 4. 5N H_2SO_4, RT, 1 h → O, Ph, OMe 86%

Lee, K.; Oh, D.Y. *Synthesis*, ***1991***, 213.

Me_3Sn, $SnMe_3$ → 1. Cl, O, Cl, O; THF, reflux, 2 h; 2. $PdBnCl(PPh_3)_2$ 0°C → RT, 2h → O, O 65%

Degl'Innocenti, A.; Dembech, P.; Mordini, A.; Ricci, A.; Seconi, G. *Synthesis*, ***1991***, 267.

O, O, Cu. C≡C, Me, Me, OMe → *n*-Bu-I, THF, -78°C → -35°C → O, O, *n*-C_4H_9 61%

Majetich, G.; Leigh, A.J.; Condon, S. *Tetrahedron Lett.*, ***1991***, *32*, 605.
Majetich, G.; Leigh, A.J. *Tetrahedron Lett.*, ***1991***, *32*, 609.

O → 0.016 M, 1M $LiBF_4$, 72 h, 25°C → O quant.

Smith, D.A.; Houk, K.N. *Tetrahedron Lett.*, ***1991***, *32*, 1549.

EtCHO , $CoCl_2$, Ac_2O , MeCN, 25°C , 8 h — 85%

Iqbal, J.; Srivastava, R.R. *Tetrahedron Lett.*, ***1991***, *32*, 1663.

CHO; LDA , TMSCl, -78°C → RT — 98%

Herscovici, J.; Boumaïza, L.; Antonakis, K. *Tetrahedron Lett.*, ***1991***, *32*, 1791.

Ph, O, OEt; CO_2Et , PhCN; 6 P, 3, SO_3Na, , 24 h; $Pd(dba)_2$ — Ph, CO_2Et, 97%

Safi, M.; Sinou, D. *Tetrahedron Lett.*, ***1991***, *32*, 2025.

OH, Ph; 10% $CpRuCl(PPh_3)_2$, dioxane , reflux , 1 h — Ph, 74%

Trost, B.M.; Kulawiec, R.J. *Tetrahedron Lett.*, ***1991***, *32*, 3039.

Ph; O_2N—C6H4—OH; 5% $Pd(PPh)_3)_4$, THF, reflux , 1 h — Ph, 83%

Kim, S.; Uh, K.H.; Lee, S.; Park, J.H. *Tetrahedron Lett.*, ***1991***, *32*, 3395.

$SiMe_3$, H, Me; 1. $EtAlCl_2$, PhMe; 2. aq. NH_4Cl — H, Me 44% + H, Me 23%

Majetich, G.; Hull, K.; Casares, A.M.; Khetani, V. *J. Org. Chem.*, ***1991***, *56*, 3958.

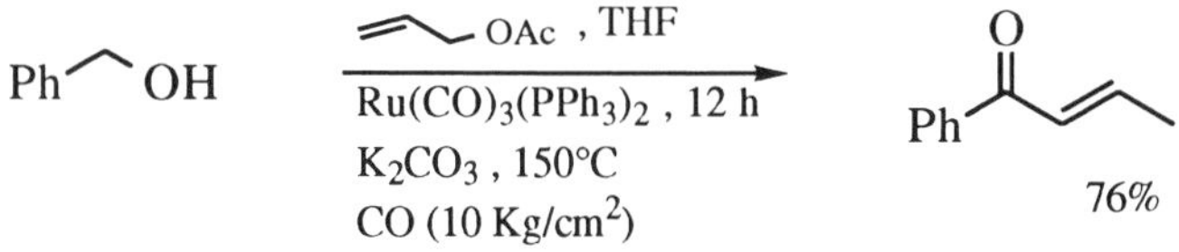

Ph OH → (OAc, THF; $Ru(CO)_3(PPh_3)_2$, 12 h; K_2CO_3, 150°C; CO (10 Kg/cm^2)) → 76%

Kondo, T.; Mukai, T.; Watanabe, Y. *J. Org. Chem.*, ***1991***, *56*, 487.

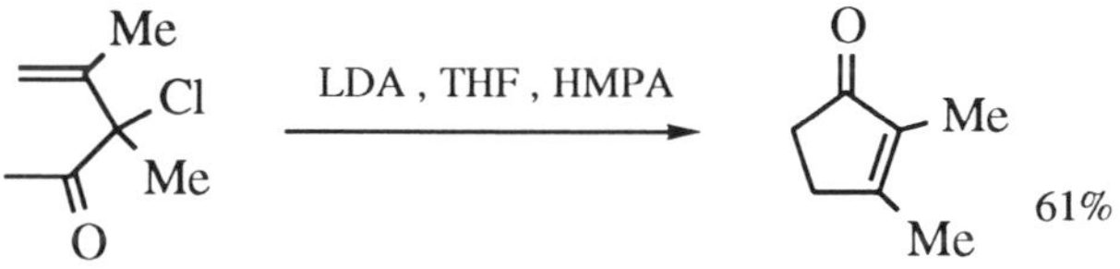

Me Cl Me O → (LDA, THF, HMPA) → O Me Me 61%

Mathew, J. *J. Org. Chem.*, ***1991***, *56*, 713.

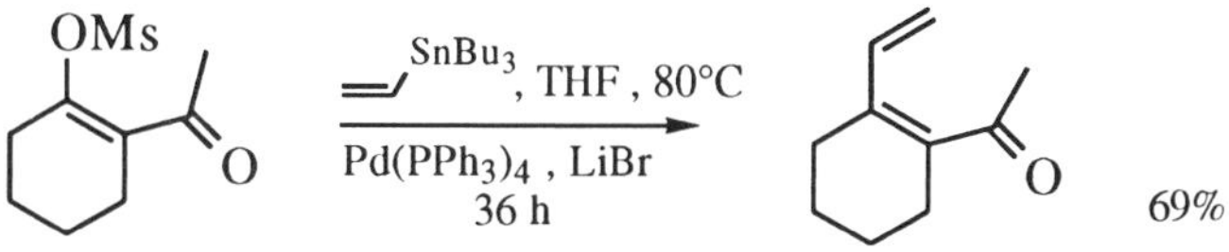

OMs O → ($SnBu_3$, THF, 80°C; $Pd(PPh_3)_4$, LiBr, 36 h) → O 69%

Hettrick, C.M.; Kling, J.K.; Scott, W.J. *J. Org. Chem.*, ***1991***, *56*, 1489.

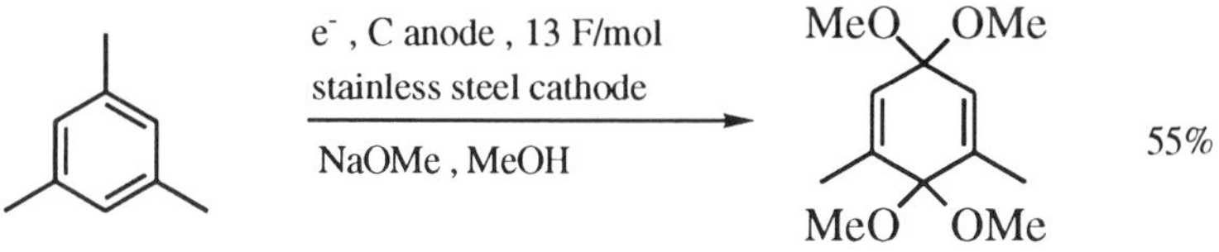

→ (e$^-$, C anode, 13 F/mol, stainless steel cathode; NaOMe, MeOH) → MeO OMe MeO OMe 55%

Barba, I.; Chinchilla, R.; Gómez, C. *J. Org. Chem.*, ***1991***, *56*, 3673.

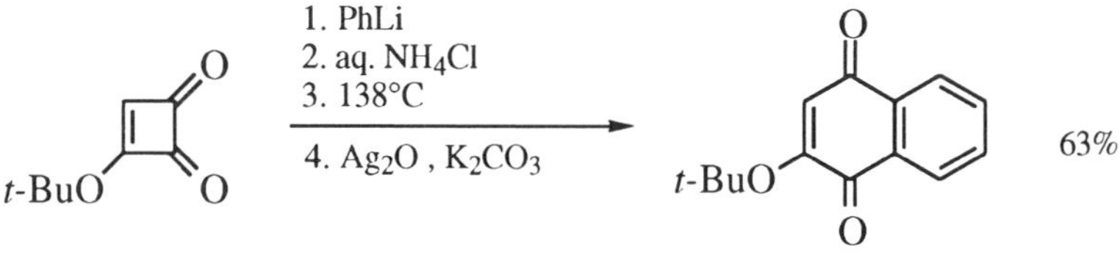

t-BuO O O → (1. PhLi 2. aq. NH_4Cl 3. 138°C; 4. Ag_2O, K_2CO_3) → O *t*-BuO O 63%

Heerding, J.M.; Moore, H.W. *J. Org. Chem.*, ***1991***, *56*, 4048.

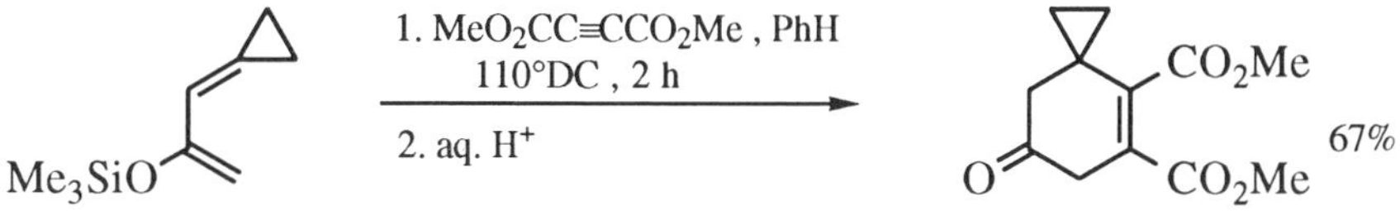

Me_3SiO → (1. $MeO_2CC{\equiv}CCO_2Me$, PhH 110°DC, 2 h; 2. aq. H^+) → CO_2Me O CO_2Me 67%

Thiemann, T.; Kohlstruk, S.; Schwär, G.; de Meijere, A. *Tetrahedron Lett.*, ***1991***, *32*, 3483.

1. $PdCl_2(PhCN)_2$
2 eq. NEt_3
$NaCH(CO_2Me)_2$
2. CO
3. vinyl-$SnMe_3$

88%

Masters, J.J.; Hegedus, L.S.; Tamariz, J. *J. Org. Chem.*, ***1991***, *56*, 5666.

2 eq. $Mn(OAc)_3$
$Cu(OAc)_2$, AcOH

50% + 18%

Snider, B.B.; Merritt, J.E. *Tetrahedron*, ***1991***, *47*, 8663.

hv, CH_2Cl_2, EtI

48%

Sato, T.; Takezoe, K. *Tetrahedron Lett.*, ***1991***, *32*, 4003.

t-Bu, $OSiMe_3$
10% Me_3SiOTf, CH_2Cl_2
-78°C → RT

(94 : 6) 68%

Kudo, K.; Saigo, K.; Hashimoto, Y.; Houchigai, H.; Hasegawa, M. *Tetrahedron Lett.*, ***1991***, *32*, 4311.

1. Mg, THF
2. Me_3SnCl
3. AcCl, $CHCl_3$, 12 h
2% $BnPd(PPh_3)_2Cl$
65°C

81%

Lang, K-T.; Kim, S.S.; Lee, J.C. *Tetrahedron Lett.*, ***1991***, *32*, 4341.

$(BuCH=CH)_2Zn$, THF-DMF
$PdCl_2(PPh_3)_2$, 2 BuLi

87%

Negishi, E.; Owczarczyk, Z.R.; Swanson, D.R. *Tetrahedron Lett.*, ***1991***, *32*, 4453.

1. PhC(O)CH$_2$Br, t-BuOK , THF , RT
2. 2M HCl/CH_2Cl_2, RT

OEt, Br, CO_2Et → O, CO_2Et, Ph 82%

Hatanaka, M.; Himeda, Y.; Ueda, I. *Tetrahedron Lett.*, ***1991***, *32*, 4521.

1. hν
2. MeOH

O, O, O, O → O, CO_2Me 57% + O, O 17%

Liao, C-C.; Wei, C-P. *Tetrahedron Lett.*, ***1991***, *32*, 4553.

10% $PdCl_2(PPh_3)_2$
20% CuI , NMP
air , benzoquinone

Me, Me, Me, $SnBu_3$, O, O → Me, Me, Me, Me, Me, Me, O, O, O, O 80%

Liebeskind, L.S.; Riesinger, S.W. *Tetrahedron Lett.*, ***1991***, *32*, 5681.

1. 2 eq. Mg , THF
2. CuBr•SMe_2
3. 3-chlorocyclopentenone, -30°C

Br, Br → O 65°C

Canonne, P.; Boulanger, R.; Angers, P. *Tetrahedron Lett.*, ***1991***, *32*, 5861.

20 eq. CH_2=CHC(O)n-C_5H_{11}
1.5 eq., $Pd(OAc)_2$
2 eq. K_2CO_3, 0.2 NaI
CH_2Cl_2

$OSiMe_3$, t-$BuMe_2SiO$ → O, n-C_5H_{11}, O, t-$BuMe_2SiO$ 62%

Larock, R.C.; Lee, N.H. *Tetrahedron Lett.*, ***1991***, *32*, 5911.

PhO_2S — 1. cyclohexadiene, 35 h; 2. NaOEt, EtOH, 0°C, 25 min — 91%

Leon, F.M.; Carretero, J.C. *Tetrahedron Lett.*, ***1991***, *32*, 5405.

PhSSPh, AIBN, hν — 82%

Kim, S.; Lee, S. *Tetrahedron Lett.*, ***1991***, *32*, 6575.

$OSiMe_3$, C≡C-H, OMe, MeO — $SnCl_4$, CH_2Cl_2, -70°C → -23°C — OMe, H (6:1 dr) — 68%

Johnson, T.O.; Overman, L.E. *Tetrahedron Lett.*, ***1991***, *32*, 7361.

1% $IrH_5(iPr_3)_2$, 4% PBu_3, PhH, reflux, 31 h — 73%

Guo, C.; Lu, X. *Tetrahedron Lett.*, ***1991***, *32*, 7549.

CO_2Et — TMS-OTf, CH_2Cl_2, RT — CO_2Et — 31%

Andrews, J.F.P.; Regan, A.C. *Tetrahedron Lett.*, ***1991***, *32*, 7731.

n-C_6H_{13}, $OSiMe_3$ — 1. $SnCl_4$, CH_2Cl_2, 0°C; 2. TMEDA, 20°C, 15 min — n-C_6H_{13} — 90%

Nakahira, H.; Ryu, I.; Ikebe, M.; Oku, Y.; Ogawa, A.; Kambe, N.; Sonoda, N.; Murai, S. *J. Org. Chem.*, ***1992***, *57*, 17.
Nakahira, H.; Ryu, I.; Han, L.; Kambe, N.; Sonoda, N. *Tetrahedron Lett.*, ***1991***, *32*, 229 [with $TeCl_4$].

OTMS, $LiClO_4$, ether; RT, 3 h — OH — O — O — O — 85%

Pearson, W.H.; Schkeryantz, J.M. *J. Org. Chem.*, ***1992***, *57*, 2986.

Me−C≡C— O O Me — BF_3•OEt_2, CH_2Cl_2; 24 h — O — Me — O — 58%

Sisko, J.; Balog, A.; Curran, D.P. *J. Org. Chem.*, ***1992***, *57*, 4341.

Me O iPrO OMe Ph — 1. BuLi 2. TFAA 3. 10% aq. NH_4Cl; 4. 140°C, xylene 5. CAN, aq. MeCN — O Me Bu O — 57%

Liebeskind, L.S.; Granberg, K.L.; Zhang, J. *J. Org. Chem.*, ***1992***, *57*, 4345.

Bu_3Sn $SnBu_3$ — $Pd(PPh_3)_4$, dioxane reflux; Cl O — O O — 63%

Pérez, M.; Castaño, A.M.; Escavarren, A.M. *J. Org. Chem.*, ***1992***, *57*, 5047.

O C≡C−Ph — 1. 2 eq. EtMgBr, THF Cp_2TiCl_2, -78°C; 2. CO, $CHCl_3$ — Ph O= O — 49%

Grossman, R.B.; Buchwald, S.L. *J. Org. Chem.*, ***1992***, *57*, 5841.

H n-C_5H_{11} H-C≡C $SiMe_2t$-Bu — O; $TiCl_4$ — n-C_5H_{11} O $SiMe_2t$-Bu — 51%

Danheiser, R.L.; Dixon, B.R.; Gleason, R.W. *J. Org. Chem.*, ***1992***, *57*, 6094.

Bu_3Sn-CH=CH_2
5% $Pd(OAc)_2$, 3 LiCl
dppp , DMF , 90°C
24 h

O
O_2
S
O
F

O

86%

Badone, D.; Cecchi, R.; Guzzi, U. *J. Org. Chem.*, ***1992***, *57*, 6321.

PhMe , 100°C , CO (3 atm)
$PdCl_2(PPh_3)_2$

n-C_6H_{13}
O
O
$SnBu_3$
O
Cl

n-C_6H_{13}
O
O
O

58%

Baldwin, J.E.; Adlington, R.M.; Ramcharitar, S.H. *Tetrahedron*, ***1992***, *48*, 2957.

Cl

Me-C≡C-CO_2Me
$Ni(CO)_4$, MeOH , RT

O H CO_2Me
Me
MeO_2C H

72%

Camps, F.; Llebaria, A.; Moretó, J.M.; Pagès, L. *Tetrahedron Lett.*, ***1992***, *33*, 113.

SO_2Ph

1. $MeNO_2$, DBU , RT
2. t-BuOK , THF

O

>95%

Bäckvall, J-E.; Ericsson, A.M.; Plobeck, N.A.; Juntunen, S.K. *Tetrahedron Lett.*, ***1992***, *33*, 131.

CF_3CH_2OTs

1. 2 eq. BuLi , THF
2. Bu_3B
3. CuI
4. BzCl , THF/HMPA
RT

O
Ph
F_2C
Bu

78%

Ichikawa, J.; Hamada, S.; Sonoda, T.; Kobayashi, H. *Tetrahedron Lett.*, ***1992***, *33*, 337.

O
CO_2t-Bu

1. 2 eq. LDA
2. $Ph_3P^{\oplus}$ $Br^{\ominus}$
3. Me_2CHCHO

CO_2t-Bu
O

62%

(16:1 E,Z:E,E)

White, J.D.; Jensen, M.S. *Tetrahedron Lett.*, ***1992***, *33*, 577.

$\diagup\!\!\diagup SnBu_3$, NMP

5% $Pd(PhCN)_2Cl_2$
10% CuI , 90°C , 20 min

Me Me 81%

Johnson, C.R.; Adams, J.P.; Braun, M.P.; Senanayake, C.B.W. *Tetrahedron Lett.*, ***1992***, *33*, 919.

OH
n-C_3H_7
OMe

1. $Tl(NO_3)_2$, MeOH
K_2CO_3
2. hν (0.02 M)
λ (310 nm/U glass)

O
n-C_3H_7 CO_2Me 50%

Hong, F-T.; Lee, K-S.; Liao, C-C. *Tetrahedron Lett.*, ***1992***, *33*, 2155.

$OSi(iPr)_3$

1. PhIO , TMS-N_3 , CH_2Cl_2
-15°C
2. TBAF , THF , 0°C

68%

Magnus, P.; Evans, A.; Lacour, J. *Tetrahedron Lett.*, ***1992***, *33*, 2933.

$SiMe_3$

$FeCl_3$, 0.02 M
CH_2Cl_2 , -3°C
2 h

49%

Kang, K-T.; Kim, S.S.; Lee, J.C.; Sun, U.J. *Tetrahedron Lett.*, ***1992***, *33*, 3495.

MeO n-C_5H_{11}

1. 2 eq. BrClCHLi , ether-THF
-100°C
2. PhCHO , Et_2AlCl , Zn
cat. CuBr , THF , -10°C

Ph Cl
n-C_5H_{11}
59%

Narita, S.; Takahashi, A.; Sato, H.; Aoki, T.; Yamada, S.; Shibasaki, M. *Tetrahedron Lett.*, ***1992***, *33*, 4041.

1. LHMDS
2. PhCH(Me)CHO
3. Me_3SiCl
4. 500°C , <1 sec

O $OSiMe_3$ Ph + O $OSiMe_3$ Ph

(>98 : 2) 72%

Ahmar, M.; Bloch, R.; Mandville, G.; Romain, I. *Tetrahedron Lett.*, ***1992***, *33*, 2501.

Ph, Ph, C, Me, Me — mCPBA , CH_2Cl_2 → Ph, Ph, Me, Me, O 50%

Sugita, H.; Mizuno, K.; Saito, T.; Isagawa, K.; Otsuji, Y. *Tetrahedron Lett.*, ***1992***, *33*, 2539.

$OSiMe_3$, Ph — 1. $Hg(O_2CCF_3)_2$ CH_2Cl_2 2. aq. Na_2CO_3 → O, Ph 70%

Kim, S.; Uh, K.H. *Tetrahedron Let.*, ***1992***, *33*, 4325.

O, CO_2Me, Me — [$Pd(OAc)_2$, PPh_3 PhMe , RT] K_2CO_3 , 16 h → Me, O, MeO_2C 76%

Gravel, D.; Benoît, S.; Kumanovic, S.; Sivavamakrishnan, H. *Tetrahedron Lett.*, ***1992***, *33*, 1407, 1403.

Ph, SPh, SPh — $OSiMe_3$ $SnCl_4$ → O, SPh, Ph 71%

Hunter, R.; Michael, J.P.; Walter, D.S. *Tetrahedron Lett.*, ***1992***, *33*, 5413.

O, H, Br — SmI_2 , THF/MeOH 25°C , 2 in → O, H + O, H

(98 : 2) 85%

with NaHTe , EtOH (<1 : >99) 95%

Beerli, R.; Brunner, E.J.; Borschberg, H-J. *Tetrahedron Lett.*, ***1992***, *33*, 6449.

$Fe(CO)_3$ — 8 eq. $AlCl_3$, CO (100 atm) 4 eq. $Fe(CO)_5$, CH_2Cl_2 100°C , 12 h → O 70%

Franck-Neumann, M.; Michelotti, E.L.; Simler, R.; Vernier, J-M. *Tetrahedron Lett.*, ***1992***, *33*, 7361.

Bu_3SnH, AIBN → 54%

Yamamoto, M.; Furusawa, A.; Iwasa, S.; Kohmoto, S.; Yamada, K. *Bull. Chem. Soc. Jpn.*, ***1992***, *65*, 1550.

$AgBF_4$, PhH, 80°C, 1 h → 93%

Shigemasa, Y.; Oikawa, H.; Ohrai, S.; Sashiwa, H.; Saimoto, H. *Bull. Chem. Soc. Jpn.,*. ***1992***, *65*, 2594.

1. $OSiMe_3$, CP_2Et, $BF_3 \cdot OEt_2$, CH_2Cl_2, -78°C → 0°C 2. siO_2, PhH, reflux → 88%

Sugimura, H.; Yoshida, K. *Bull. Chem. Soc. Jpn.*, ***1992***, *65*, 3209.

Ph-C≡C-Ph, reflux, 1% aq. dioxane, 6 h, syringe pump addition → 78% (85:15 trans:cis)

Tumer, S.U.; Herndon, J.W.; McMullen, L.A. *J. Am. Chem. Soc.*, ***1992***, *114*, 8394.

1. Cl_3COCl, Zn, ether))))))) 2. Bu_3SnH, AIBN, 80°C 3. Me_3SiI, ZnI_2 4. DBU → 48%

Dowd, P.; Zhang, W. *J. Am. Chem. Soc.*, ***1992***, *114*, 10084.

1. 2 eq. $Co_2(CO)_8$, DMF 2. reflux → 71%

Iwasawa, N. *Chem. Lett.*, ***1992***, 473.

1. $(CO)_5Cr$=C(OEt)Me
2. $[FeCl_4]$ $[Fe(DMF)_3Cl_2]$

53%

Mori, M.; Watanuki, S. *J. Chem. Soc., Chem. Commun.*, ***1992***, 1082.

n-C_7H_{15} / SO_2Ph

1. LDA, -78°C
2. NO_2
3. O_3, CH_2Cl_2
4. Me_2S, 0°C

O, n-C_7H_{15} 84%

Awen, B.Z.; Miyashita, M.; Shiratani, T.; Yoshikoshi, A.; Irie, H. *Chem. Lett.*, ***1992***, 767.

CO_2Et

1. CoTDCPP, O_2, Et_3SiH
iPrOH - CH_2Cl_2
2. Ac_2O - DMAP
CH_2Cl_2

O, CO_2Et 82%

CoTDCPP = [5,10,15,20-tetra(2,6-dichlorophenyl)porphinato] cobalt (II)

Matsushita, Y.; Sugamoto, K.; Matsui, T. *Chem. Lett.*, ***1992***, 2165.

OH, OMe

$PhI(OAc)_2$, EtOH, RT

O, EtO OMe 78%

Fleck, A.E.; Hobart, J.A.; Morrow, G.W. *Synth. Commun.*, ***1992***, *22*, 179.

O

1. $OSiMe_3$, Ph, $TiCl_4$, 1 h
2. TFAA, 15 min
3. NEt_3
4. H_2O

Ph, O 57%

Bouchlel, E.; Ben Hassine, B. *Synth. Commun.*, ***1992***, *22*, 2183.

OH, n-C_3H_7, n-C_3H_7

$Pd(OAc)_2$, PPh_3
PhMe, 110°C, 30 h

O 69%

Guo, C.; Lu, X. *SynLett*, ***1992***, 405.

REVIEWS:

"Diels-Alder Reactions of Cycloalkenones in Organic Synthesis"
Fringuelli, F.; Taticchi, A.; Wenkert, E. *Org. Prep. Proceed. Int.*, ***1990***, *22*, 131.

"Regiospecific Alkylation of Cyclohexenones"
Podraza, K.F. *Org. Prep. Proceed. Int.*, ***1991***, *23*, 219.

"Symposia-in-Print - Organotitanium Reagents in Organic Chemistry"
Reetz, M.T. (Ed.) *Tetrahedron*, ***1992***, *48*, 5557-5754.

"Recent Studies on the Peterson Olefination Reaction"
Barrett, A.G.M.; Hill, J.M.; Wallace, E.M.; Flygare, J.A. *SynLett*, ***1991***, 764.

SECTION 375: NITRILE - NITRILE

1. cyclopentene, 80°C; 2. Bu_3SnH, 0.1 M

(89 : 19) 61%

Curran, D.P.; Seong, C.M. *J. Am. Chem. Soc.*, ***1990***, *112*, 9401.

SECTION 376: NITRILE - ALKENE

PhCHO → Ph–CH=CH–C≡N 95%

Cl–CH(C≡N)SiMe$_3$; Zn*, THF, reflux, 30 min

35:65 E:Z

Palomo, C.; Aizpurua, J.M.; Aurrekoetxea, N. *Tetrahedron Lett.*, ***1990***, *31*, 2209.

OMe, N-OH → OMe, CN 61%

allyl-SiMe$_3$; 10% TMSOTf, CH_2Cl_2, 0°C, 6 h

Fujioka, H.; Miyazaki, M.; Yamanaka, T.; Yamamoto, H.; Kita, Y. *Tetrahedron Lett.*, ***1990***, *31*, 5951.

NO_2 → CN 37%

1. K_2CO_3, H_2O, CS_2, TBAB, CH_2Cl_2; 2. aq. NaOH, CS_2, RT

Albanese, D.; Landini, D.; Penso, M.; Pozzi, G. *Synth. Commun.*, ***1990***, *20*, 965.

Me_3Si–CH_2–CN → 1. LDA, THF, -78°C; 2. MOM-Cl, -78°C; 3. NH_4Cl, H_2O → Cl–CH_2–C(Me)=CH–CN 76% (58:42 Z:E)

Mauzé, B.; Miginiac, L. *Synth. Commun.*, ***1990***, *20*, 2251.

PhCHO → $BrCH_2CN$, PBu_3, Zn; 100°C, 16 h → PhCH=CHCN 68% (75:25 E:Z)

Zheng, J.; Yu, Y.; Shen, Y. *Synth. Commun.*, ***1990***, *20*, 3277.

Bu–C≡C–H → 1. siamyl$_2$BH, THF, -15°C; 2. CuCN, THF; 3. $(Me_2N)_3PO$; 4. $Cu(OAc)_2$•H_2O, THF → Bu–CH=CH–CN 65%

Masuda, Y.; Hoshi, M.; Arase, A. *J. Chem. Soc., Chem. Commun.*, ***1991***, 748.

$(EtO)_2P(O)$–C(CN)=CH–Et → 1. LDA, THF; 2. iPrCHO, -78°C → RT → iPr–CH=CH–C(CN)=CH–Et 55%

Janecki, T. *Synthesis*, ***1991***, 167.

PhCHO → BnCN, K_2CO_3, MeOH; 65°C, 15 min → PhCH=C(Ph)CN 95%

Ladhar, F.; El Gharbi, R. *Synth. Commun.*, ***1991***, *21*, 413.

TsN=S(O)(Ph)–CH=CH–CH_2CH_2Ph → LiCN, DMF, RT → CH_2=C(CN)–CH_2CH_2Ph 81%

Bailey, P.L.; Jackson, R.F.W. *Tetrahedron Lett.*, ***1991***, *32*, 3119.

Ph–CH=CH–CH_2–N_3 → 1. PPh_3; 2. Ph_2C=C=O, RT → Ph_2C(CN)–CH(Ph)–CH=CH_2 51%

Molina, P.; Alajarin, M.; Lopez-Leonardo, C. *Tetrahedron Lett.*, ***1991***, *32*, 4041.

PhCHO + $CH_2(CN)_2$ —ZnCl$_2$, 100°C, 10 min→ PhCH=C(CN)$_2$ 91%

Rao, P.S.; Venkataraatnam, R.V. *Tetrahedron Lett.*, ***1991***, *32*, 5821.

1. PhH, 80°C
2. hν
3. Bu_3SnH

n-C_3H_7 n-C_3H_7 + H C C I NC CN → NC CN n-C_3H_7 n-C_3H_7 82%

Curran, D.P.; Seong, C.M. *Tetrahedron*, ***1992***, *48*, 2157, 275.

O Me C≡C $SiMe_3$ —Amberlyst 15, 2-methoxy-1,3-dioxolane (O, O, OMe)→ O Me C 80%

Schinzer, D.; Kabbara, J.; Ringe, K. *Tetrahedron Lett.*, ***1992***, *33*, 8017.

SECTION 377: ALKENE - ALKENE

O O N N Sn(n-Bu)$_3$ + I (E/Z, 89:1) → O O N N 63% (90:10, E:Z)

5% PdCl$_2$(MeCN)$_2$, DMF, 20°C, 24h

Kende, A.S.; DeVita, R.J. *Tetrahedron Lett.*, ***1990***, *31*, 307.

1. BuLi, THF, 0°C
2. Me_3SiCl
3. BuLi, -70°C

Ph_3P Br$^-$ → Ph_3P $SiMe_3$

n-C_6H_{13}CHO
2h, 0°C → 20°C

n-C_6H_{13} H $SiMe_3$ 82%

1E,3E : 1E,3Z 80:20

Shen, Y.; Wang, T. *Tetrahedron Lett.*, ***1990***, *31*, 543.

Ph Me + Ph N Ph —Na, DMF, 90°C, 2 h→ Ph Ph 54%

Paventi, M.; Hay, A.S. *Synthesis*, ***1990***, 878.

$[Me_3SiCl/NaI]$

MeCN , 15 min

OH OH

(EZ/EE = 0.23) 35%

Hill, R.K.; Pendalwar, S.L.; Kielbasinski, K.; Baevsky, M.F.; Nugara, P.N. *Synth. Commun.*, ***1990***, *20*, 1877.

C_6H_{13} H AcO

$Ni(COD)_2$, dppb

THF , 20°C

15 h

a metallo-Ene reaction

C_6H_{13} + C_6H_{13}

(>99 : <1) 79%

Oppolzer, W.; Keller, T.H.; Kuo, D.L.; Pachinger, W. *Tetrahedron Lett.*, ***1990***, *31*, 1265.

Bu−C≡C−B(n-C_6H_{13})$_3$

1. $O\cdot CO_2Et$

$Pd(PPh_3)_4$, THF

-78°C → 40°C , 24 h

2. AcOH , 25°C , 6h

(87:13 E:Z) 70%

Chen, Y.; Li, N-S.; Deng, M-Z. *Tetrahedron Lett.*, ***1990***, *31*, 2405.

TfO

TBAF , THF

5% $Pd(PPh_3)_4$, 50°C

30 min

n-C_6H_{13} $SiMeF_2$

n-C_6H_{13}

99%

Hatanaka, Y.; Hiyama, T. *Tetrahedron Lett.*, ***1990***, *31*, 2719.

O S Ph

1. 4 eq. t-BuLi , -78°C

2. MeOH , -78°C

72%

Theobald, P.G.; Okamura, W.H. *J. Org. Chem.*, ***1990***, *55*, 741.

OTf Ph

$(\quad)_2Cu(CN)Li_2$

THF , -78°C

Ph

82%

Lipshutz, B.H.; Elworthy, T.R. *J. Org. Chem.*, ***1990***, *55*, 1695.

n-C_6H_{13} MgBr ZnBr + CO_2Et CO_2Et EtO_2C THF , -78°C → -20°C CO_2Et n-C_6H_{13} 83%

(82:18 Z:E)

Tucker, C.E.; Rao, S.A.; Knochel, P. *J. Org. Chem.*, ***1990***, *55*, 5446.

Et_3SiO O $TiCl_3$ / C_8K / DME $OSiEt_3$ 82%

Clive, D.L.J.; Murthy, K.S.K.; Zhang, C.; Hayward, W.D.; Daigneault, S. *J. Chem. Soc., Chem. Commun.*, ***1990***, 509.

CO_2Et PPh_3 ⊕ ⊖ Br O Br , 12 h $NaHCO_3$, CH_2Cl_2 RT EtO_2C 96%

Hatanaka, M.; Himeda, Y.; Ueda, I. *J. Chem. Soc., Chem. Commun.*, ***1990***, 526.

Ph - C≡C - H

1. BuLi
2. $AlMe_2Cl$
3. $Cp_2Ti(Me)Cl$
4. CH_2=CH_2 , HMPA/THF
5. $HCl_{(gas)}$, -78°C
6. aq. $NaHCO_3$

Me Ph + Ph Me

(4 : 1)

Dennehy, R.D.; Whitby, R.J. *J. Chem. Soc., Chem. Commun.*, ***1990***, 1060.

$SiMe_2Ph$ OH KH , THF , RT 79%

(99.6:0.4 trans,trans:trans,cis)

Fleming, I.; Morgan, I.T.; Sarkar, A.K. *J. Chem. Soc., Chem. Commun*, ***1990***, 1575.

Ph I

1. $SiMe_3$, $Pd(OAc)_2$ $AgNO_3$, NEt_3 , 50°C , 3 h
2. HI

Ph $SiMe_3$ 25%

Karabelas, K.; Hallberg, A. *Acta Chem. Scand.*, ***1990***, *44*, 257.

MeMgI , 5% $NiCl_2(PPh_3)_2$; PhH , 20 h — 61%

Ni, Z-J.; Yang, P-F.; Ng, D.K.P.; Tzeng, Y-L.; Luh, T-Y. *J. Am. Chem. Soc.*, ***1990***, *112*, 9356.

0.5 eq. $Mo(CO)_6$, H_2O ; 20 h — (1:1 E:Z) 50%

Masuyama, Y.; Maekawa, K.; Kurihara, T.; Kurusu, Y. *Bull. Chem. Soc. Jpn.*, ***1991***, *64*, 2311.

PhCHO — 1. Me_3SiCH_2MgCl 2. CrO_3•Py 3. CH_2=CHMgBr 4. AcOH , NaOAc — 56%

Brown, P.A.; Bonnert, R.V.; Jenkins, P.R.; Lawrence, N.J.; Selim, M.R. *J. Chem. Soc., Perkin Trans. I*, ***1991***, 1893.

2.5% $Pd_2(dba)_3$•$CHCl_3$; P(*o*-Tol)$_3$, RT , PhMe — 85%

Trost, B.M.; Shi, Y. *J. Am. Chem. Soc.*, ***1991***, *113*, 701.
Trost, B.M.; Lautens, M.; Chan, C.; Jebaratnam, D.J.; Mueller, T. *J. Am. Chem. Soc.*, ***1991***, *113*, 636.

5% P(*o*-Tol)$_3$; DMAD , DCE — (2.8 : 1) 60%

Trost, B.M.; Trost, M.K. *J. Am. Chem. Soc.*, ***1991***, *113*, 1850.

1. (sulfoxide) 2. Py , reflux — 46%

new allene equivalent

Williams, R.V.; Chauhan, K. *J. Chem. Soc., Chem. Commun.*, ***1991***, 1672.

1. $CH_2=C(CH_3)CH_2MgCl$, 4 BF_3•OEt_2, CH_2Cl_2-THF , 0°C
2. PPh_3
3. 2 eq. NaH , THF

Se, O, N, S, Ph — Ph — 98%

Shibaa, K.; Shiono, H.; Mitsunobu, O. *Chem. Lett., 1991*, 661.

t-Bu, O — Me_3S=O^+ I^- , NaH / DMSO , 130°C , 30 min — t-Bu — 89%

Yurchenko, A.G.; Kyrij, A.B.; Likhotvorik, I.R.; Melnik, N.N.; Zaharzh, P.; Bzhezovski, V.V.; Kushko, A.D. *Synthesis, 1991*, 393.

OEt — 1. Br_2 , CH_2Cl_2 , -78°C; 2. $ZnCl_2$, $OSiMe_3$; 3. Zn , EtOH — 87%

Ohm, S.; Bäuml, E.; Mayr, H. *Chem. Ber., 1991, 124*, 2785.

O — t-BuOK , DMSO , 140°C — 59%

Arenz, T.; Vostell, M.; Frauenrath, H. *SynLett, 1991*, 23.

Cl, OAc — MgCl , THF / CuCN•2 LiCl , 0°C, 0.3 h — OAc 88% + OAc 8%

Yanagisawa, A.; Noritake, Y.; Nomura, N.; Yamamoto, H. *SynLett, 1991*, 251.

O, H, $SnMe_3$, O, O — 1. LDA ; $PhN(SO_2CF_3)_2$, -48°C → RT; 2. 5% $Pd(PPh_3)_4$, THF , RT — H, O, O — 60%

Piers, E.; Friesen, R.W.; Keay, B.A. *Tetrahedron, 1991,47*, 4555.

1. t-BuLi , TMSCl
2. Na , DMAN
$NHEt_2$

SO_2Ph → $SiMe_3$ 67%

(E/Z > 10)

Chan, T.H.; Labrecque, D. *Tetrahedron Lett.*, ***1991***, *32*, 1149.

Cp_2Ti acetone , PhMe , 60°C >98%

Doxsee, K.M.; Mouser, J.K.M. *Tetrahedron Lett.*, ***1991***, *32*, 1687.

Cl C≡C Cl

2 eq. n-C_8H_{17} - Zn Br
CuCN•2 LiBr , THF

n-C_8H_{17} n-C_8H_{17} 93%

Zhu, L.; Rieke, R.D. *Tetrahedron Lett.*, ***1991***, *32*, 2865.

Bu SO_2Ph

t-BuOH , t-BuOK
80°C

Bu 78%

Sellén, M.; Bäckvall, J-E.; Helquist, P. *J. Org. Chem.*, ***1991***, *56*, 835.

O

1. Me_3Si Me B
THF , 9 h
2. NaOH , H_2O_2 , 3 h

Me 67%

Wang, K.K.; Liu, C.; Gu, Y.G.; Burnett, F.N.; Sattsangi, P.D. *J. Org. Chem.*, ***1991***, *56*, 1914.

Cl n-C_5H_{11}

$PdCl(crotyl)_2$
n-C_5H_{11} $ZrCp_2Cl$
THF , -78°C – RT

n-C_5H_{11} n-C_5H_{11} + n-C_5H_{11} n-C_5H_{11}

(22 : 78) 87%

Hutzinger, M.W.; Oehlschlager, A.C. *J. Org. Chem.*, ***1991***, *56*, 2918.

OH

1. Ph-S(O) $C(OEt)_3$, H^+
2. 100°C

EtO_2C 75%

Posner, G.H.; Crouch, R.D.; Kinter, C.M.; Carry, J-C. *J. Org. Chem.*, ***1991***, *56*, 6981.

1.5 eq. BuMgBr, 0.3 eq. CuI, THF, 2 h, -78°C → -30°C

OAc; n-C_7H_{15} → n-C_7H_{15} ... Bu (90:9 EE:EZ) 83%

Nakanishi, N.; Matsubara, S.; Utimoto, K.; Kozima, S.; Yamaguchi, R. *J. Org. Chem.*, ***1991***, *56*, 3278.

1. BuLi, ether, -60°C; 2. MeOH, RT

PhS ... OMe → PhS ... Bu (60:40 1E,3E:1Z:3E) 50%

Gaonac'h, O.; Maddaluno, J.; Chauvin, J.; Duhamel, L. *J. Org. Chem.*, ***1991***, *56*, 4045.

1. Bu ... Li; 2. $EtCO_2H$

Bu, Br, B, O, O → Bu, Bu 70%

Brown, H.C.; Bhat, N.G.; Iyer, R.R. *Tetrahedron Lett.*, ***1991***, *32*, 3655.

$Pd(PPh_3)_4$, AcOH, reflux, 8 h

Me → Me 80%

Zair, T.; Santelli-Rouvier, C.; Santelli, M. *Tetrahedron Lett.*, ***1991***, *32*, 4501.

Cl, THF, -30°C; $Pd(dba)_2$, PPh_3, 40 min

82%

Nyström, J-E.; Vågberg, J.O.; Söderberg, B.C. *Tetrahedron Lett.*, ***1991***, *32*, 5247.

$POCl_3$, Py

MeO, OH, OMe → MeO, OMe 90%

Runk, C.S.; Haverty, S.M.; Klein, D.A.; McCrea, W.R.; Strobel, E.D.; Pinnick, H.W. *Tetrahedron Lett.*, ***1992***, *33*, 2665.

DMSO, 160°C; 5 h

Ph Ph

85%

Patro, B.; Ila, H.; Junjappa, H. *Tetrahedron Lett.*, ***1992***, *33*, 809.

PhI, cat. $Pd(OAc)_2$, DMF
cat. PPh_3, $Tl(OAc)_3$

n-C_6H_{13} → Ph n-C_6H_{13}

81%

Jeffery, T. *Tetrahedron Lett.*, ***1992***, *33*, 1989.

H-C≡C–OH

1. In_2Br_3; DMF, 100°C-140°C; RT
2. H_2O

OH + OH

(65 : 35) 91%

Araki, S.; Imai, A.; Shimizu, K.; Butsugan, Y. *Tetrahedron Lett.*, ***1992***, *33*, 2581.

⊕ I-Ph ⊖ BF_4

Ph SnBu₃; $PdCl_2(MeCN)_2$, 1 h; DMF

Ph

63%

Moriarty, R.M.; Epa, W.R. *Tetrahedron Lett.*, ***1992***, *33*, 4095.

OH; Et; C≡C-n-C_6H_{13}

1. Bu_3SnH, AIBN, 90°C
2. MsCl, NEt_3, CH_2Cl_2; 0°C → RT

Et; =C=; n-C_6H_{13}

71%

Konoike, T.; Araki, Y. *Tetrahedron Lett.*, ***1992***, *33*, 5093.

$SnBu_3$; Ph

$SiMe_3$, 1 h; $SnCl_4$, -78°C

Ph + Ph

(76 24) 76%

Takeda, T.; Takagi, Y.; Takano, H.; Fujiwara, T. *Tetrahedron Lett.*, ***1992***, *33*, 5381.

H-C≡C–$SiMe_3$

1. $Cp_2Zr(H)Cl$
2. n-$C_{10}H_{21}CHO$, CH_2Cl_2; cat. $AgClO_4$, RT, 1 h
3. reflux, 1 h

n-$C_{10}H_{21}$

(97:3 E:Z) 82%

Maeta, H.; Suzuki, K. *Tetrahedron Lett.*, ***1992***, *33*, 5969.

$Ph\text{-}C{\equiv}C\text{-}SnMe_3$ → 1. $(Mes)_2BH$ 2. Mes-Li 3. PhCHO 4. TFAA → PhCH=C=CHPh 61%

Mes = mesityl

Pelter, A.; Smith, K.; Jones, K.D. *J. Chem. Soc., Perkin Trans. I*, ***1992***, 747.

MeO_2C / MeO_2C (Me) → 5% $Pd(OAc)_2$, THF; 2 eq. PPh_3, 65°C → MeO_2C / MeO_2C (Me) 95%

Takacs, J.M.; Zhu, J.; Chandramouli, S. *J. Am. Chem. Soc.*, ***1992***, *114*, 773.

$t\text{-}BuMe_2SiO$ ... $C{\equiv}C\cdot H$ → (Br), PhMe; 5% $Pd(OAc)_2$, 15% PPh_3, NEt_3 → $t\text{-}BuMe_2SiO$ ... 62%

Trost, B.M.; Pfrengle, W.; Urabe, H.; Dumas, J. *J. Am. Chem. Soc.*, ***1992***, *114*, 1923.

Me_3Si ... O → 1. PhLi, ether 2. NaH; MeI, TAS 3. $ZnBr_2$, CH_2Cl_2 RT, 1 h → Ph 84% (92:8 E:Z)

Fujiwara, T.; Suda, A.; Takeda, T. *Chem. Lett.*, ***1992***, 1631.

PhCHO → 1. Me_3Si ... Br, $CrCl_2$, cat. $NiCl_2$, DMF 2. aq. HCl, THF → Ph ... 74% (>10:1 trans:cis)

Angell, R.; Parsons, P.J.; Naylor, A.; Tyrrell, E. *SynLett*, ***1992***, 599.

REVIEWS:

"Synthetic and Mechanistic aspects of 1,3-Diene Photooxidation"
Clennan, E.L. *Tetrahedron*, ***1991***, *47*, 1343.

"Nickel (0)-Mediated Intramolecular Cyclizations of Enynes, Dienynes, *bis*-Dienes, and Diynes"
Tamao, K.; Kobayashi, K.; Ito, Y. *SynLett*, ***1992***, 539.

SECTION 378: OXIDES - ALKYNES

t-Bu–C≡C–$\overset{\oplus}{I}$-Ph $\ominus$OTs —[P(OMe)$_3$, RT]→ t-Bu–C≡C–P(O)(OMe)(MeO) 90%

Lodaya, J.S.; Koser, G.F. *J. Org. Chem.*, ***1990***, *55*, 1513.

Et–C(=O)–CH$_2$–SO$_2$Ph —[iPr$_2$NH , Tf$_2$O , 0°C; CH$_2$Cl$_2$]→ Et–C≡C–SO$_2$Ph 81%

Clasby, M.C.; Craig, D. *SynLett*, ***1992***, 825.

SECTION 379: OXIDES - ACID DERIVATIVES

NO ADDITIONAL EXAMPLES

SECTION 380: OXIDES - ALCOHOLS, THIOLS

1,3-dithiane —[NaIO$_4$, MeOH]→ 1,3-dithiane 1,3-dioxide —[1. pyridine, THF; 2. n-BuLi , -45°C; 3. n-BuCHO, -78°C , 10 min; 4. 2N HCl]→ 49% + 34%

Aggarwal, V.K.; Davies, I.W.; Maddock, J.; Mahon, M.F.; Molloy, K.C. *Tetrahedron Lett.*, ***1990***, *31*, 135.

Ph–S(O)–C(=CH$_2$)–SiMe$_3$ —[PhCHO , TBAF; THF , -70°C]→ Ph–S(O)–C(=CH$_2$)–CH(OH)Ph 75%

Cheng, H-C.; Yan, T. *Tetrahedron Lett.*, ***1990***, *31*, 673.

(EtO)$_2$P(O)–CH$_2$CH$_2$–Br —[1. Zn , THF , 30°C; 2. CuCN , 2 LiCl; 3. PhCHO]→ (EtO)$_2$P(O)–CH$_2$CH$_2$–CH(OH)Ph 96%

also with ketones, acid chlorides, conjugate addition

Retherford, C.; Chou, T-S.; Schelkun, R.M.; Knochel, P. *Tetrahedron Lett.*, ***1990***, *31*, 1833.

PhO_2S SOTol → (MeCN, piperidine (N-H); 0°C, 5 h; isobutyraldehyde CHO) → PhO_2S (E)-allylic alcohol (OH) 86%

Dominguez, E.; Carretero, J.C. *Tetrahedron Lett.*, ***1990***, *31*, 2487.

Ph-S(O)-CH₂CH₂OH → (1. 2.2 eq. LDA, THF, -7 - -20°C; 2. MeI, -78°C → RT) → (7 : 1) 85%

Ohta, H.; Matsumoto, S.; Sugai, T. *Tetrahedron Lett.*, ***1990***, *31*, 2895.

PhO_2S, Me_3Si, Me, OMOM → (1. BuLi, THF, -78°C; 2. KF, MeOH) → PhO_2S, Bu, Me, OMOM + PhO_2S, Bu, Me, OMOM (89 : 11) 85%

Alcaraz, C.; Carretero, J.C.; Domínguez, E. *Tetrahedron Lett.*, ***1991***, *32*, 1385.

→ (2 PhO_2S SO_2Ph; $Pd(PPh_3)_4$, dppp, THF, RT) → PhO_2S, PhO_2S, OH + PhO_2S, PhO_2S, OH (9.7 : 1) 77%

Trost, B.M.; Granja, J.R. *Tetrahedron Lett.*, ***1991***, *32*, 2193.

n-$C_5H_{11}CHO$ → ($EtNO_2$, TBAF•3 H_2O, THF, NEt_3, 0°C; t-$BuMe_2SiCl$) → n-C_5H_{11}, OH, NO_2 92% (67:33 erythro:threo)

Fernández, R.; Gasch, C.; Gómez-Sánchez, A.; Vílchez, J.E. *Tetrahedron Lett.*, ***1991***, *32*, 3225.

Bu CHO → (PhO_2S S-C₆H₄-Cl; MeCN, piperidine, RT) → Bu, OH, SO_2Ph 79%

Trost, B.M.; Grese, T.A. *J. Org. Chem.*, ***1991***, *56*, 3189.

cyclohexene oxide (O) → (montmorillonite, 15 h; $TolSO_2Na$, H_2O/acetone; PhH, RT) → OH, SO_2Tol 75%

Biswas, G.K.; Bhattacharyya, P. *Synth. Commun.*, ***1991***, *21*, 569.

O O
Tol S S Tol

1. BuLi , THF
2. PhCHO

O S Tol
Ph S O
OH Tol

70%

Solladié, G.; Colobert, F.; Ruiz, P.; Hamdouch, C.; Carreño, M.C.; García Ruano, J.L. *Tetrahedron Lett.*, ***1991***, *32*, 3695.

OH
PhO_2S

OAc , iPr_2O , 5.5 h
lipase-PS

OH
PhO_2S
46%
(>98% ee ,S)

+ OAc
PhO_2S
47%
(>95% ee , R)

Domínguez, E.; Carretero, J.C.; Fernández-Mayoralas, A.; Conde, S. *Tetrahedron Lett.*, ***1991***, *32*, 5159.

O S S O

1. NaHDMS , 0°C
2. PhCHO , 0°C
3. HCl , EtOH

O S S O Ph OH + O S S O Ph OH

(96 : 4) 87%

Aggarwal, V.K.; Franklin, R.J.; Rice, M.J. *Tetrahedron Lett.*, ***1991***, *32*, 7743.

O
S
Ph

1. LDA , THF , -78°C
2. iPrCHO , -78°C , 5 min
3. H_3O^+

O OH
S
Bn
+
O OH
S
Bn

(4 : 1) 90%

Caset, M.; Muxherjee, I.; Trabsa, H. *Tetrahedron Lett.*, ***1992***, *33*, 127.

O
NO_2

$Zn(BH_4)_2$, DME
5 min

OH
NO_2
73%

(40:60)

Ranu, B.C.; Das, A.R. *Tetrahedron Lett.*, ***1992***, *33*, 2363.

O
Ph

1. $EtO-\overset{O}{P}-CH_3$ / BuLi , THF
OEt -78°C
2. $BF_3{\bullet}OEt_2$, -70°C
2 h

O OH
EtO-P Ph
OEt

quant.

Racha, S.; Li, Z.; El-Subbagh, H.; Abushanab, E. *Tetrahedron Lett.*, ***1992***, *33*, 5491.

SECTION 381: OXIDES - ALDEHYDES

1. dimethyldioxirane, acetone/CH_2Cl_2, -10°C → 20°C, 2 h
2. RT, 12 h

99%

Adam, W.; Hadjiarapoglou, L.; Klicic, J. *Tetrahedron Lett.*, ***1990***, *31*, 6517.

1. CH_2=N-NMe_2 CH_2Cl_2, RT
2. O_3
3. DMS

74%

Lassaletta, J-M.; Fernández, R. *Tetrahedron Lett.*, ***1992***, *33*, 3691.

SECTION 382: OXIDES - AMIDES

1. MeCN, 30°C, pyrrolidine
2. Hg^{+2}, aq. MeCN 30°C

>70%

treatment with Na_2S led to thioamides

Manjunatha, S.G.; Reddy, K.V.; Rajappa, S. *Tetrahedron Lett.*, ***1990***, *31*, 1327.

1. $Pb(OAc)_4$, DMF
2. $P(OMe)_3$, $TiCl_4$

76%

Corcoran, R.C.; Green, J.M *Tetrahedron Lett.*, ***1990***, *31*, 6827.

1. $BnNH_2$, 30°C, THF, 18 h
2. NEt_3, THF, 19 h Ph-C≡C-C(O)Cl
3. *t*-BuOH, THF, 30°C cat. Triton B, 48 h

49%

Patra, R.; Maiti, S.B.; Chatterjee, A.; Chakravarty, A.K. *Tetrahedron Lett.*, ***1991***, *32*, 1363.

NO, 3 eq. Et_3SiH, DCE
5% Co (salen) derivative
RT

77%

Kato, K.; Mukaiyama, T. *Chem. Lett.*, ***1990***, 1395.

1. LDA,
2. LDA, CH_2=$CHCO_2Me$

60%

Hua, D.H.; Bharathi, S.N.; Panangadan, J.A.K.; Tsujimoto, A. *J. Org. Chem.*, ***1991***, *56*, 6998.

BF_3•OEt_2
THF, -78°C

76%

Green, D.L.C.; Thompson, C.M. *Tetrahedron Lett.*, ***1991***, *32*, 5051.

SECTION 383: OXIDES - AMINES

1. $(CH_2O)_n$
NaOMe
2. $P(OMe)_3$
$TiCl_4$

48%

Ha, H-J.; Nam, G-S.; Park, K.P. *Tetrahedron Lett.*, ***1990***, *31*, 1567.

DBU, *o*-xylene
reflux, 18 h

79%

Chucholowski, A.W.; Uhlendorf, S. *Tetrahedron Lett.*, ***1990***, *31*, 1949.

1. $BnNH_2$, CH_2Cl_2/MeOH
2. NaOMe

quant.

Padwa, A.; Norman, B.H. *J. Org. Chem.*, ***1990***, *55*, 4801.

1. BuLi , THF
2. EtCH=NPh

(94 : 6) 68%

Pyne, S.G.; Dikic, B. *Tetrahedron Lett.*, ***1990***, *31*, 5231.

$TolSO_2CH_2Cl$
Bu_3SnH , AIBN
PhH , reflux

54%

(84:16 cis:trans)

Renaud, P.; Schubert, S. *Angew. Chem. Int. Ed. Engl*, ***1990***, *29*, 433.

L-Selectride , CH_2Cl_2
-20°C , 6 h

57%

(97:3 RR:RS)

Ogura, K.; Tomori, H.; Fujita, M. *Chem. Lett.*, ***1991***, 1407.

1. LDA , THF , -78°C
2. BnBr

(34 66) 96%

Bell, S.I.; Parvez, M.; Weinreb, S.M. *J. Org. Chem.*, ***1991***, *56*, 373.

, 25°C

90%

Padwa, A.; Gareau, Y.; Harrison, B.; Norman, B.H. *J. Org. Chem.*, ***1991***, *56*, 2713.

NO_2 , O_3 , <0°C
$CHCl_3$, 2.5 h

(4.4 : 1) 98%

Suzuki, H.; Ishibashi, T.; Murashima, T.; Tsukamoto, K. *Tetrahedron Lett.*, ***1991***, *32*, 6591.

n-$C_3H_7C{\equiv}N$-O

)))))))

4 : 1 86%

a 79%

a 95%

a = $NiCl_2$•6 H_2O, $NaBH_4$, MeOH

Armstrong, S.K.; Warren, S.; Collington, E.W.; Naylor, A. *Tetrahedron Lett.*, ***1991***, *32*, 4171.

cat. OsO_4

NMO

60%

Priebe, W.; Grynkiewicz, G. *Tetrahedron Lett.*, ***1991***, *32*, 7353.

1. $PhCH_2NH_2$
2. $ZnBr_2$, Dibal

3 h

(>97 : <3) 80%

Ruano, J.L.G.; Lorente, A.; Rodríguez, J.H. *Tetrahedron Lett.*, ***1992***, *33*, 5637.

1. t-BuLi ; BnBr
2. HCl-H_2O , pH 2-3
3. H_2 , $Pd(OH)_2/C$ MeOH

(76%ee , R) 26% overall

Maury, C.; Royer, J.; Husson, H-P. *Tetrahedron Lett.*, ***1992***, *33*, 6127.

95 mol % $AgOSO_2CF_3$

CH_2Cl_2

(major) 81% (95% dr)

Davies, I.W.; Gallagher, T.; Lamont, R.B.; Scopes, D.I.C. *J. Chem. Soc., Chem. Commun.*, ***1992***, 335.

O, EtO-P, EtO, OH — HN_3, PhH, DEAD, PPh_3 / CH_2Cl_2, -10°C → RT — O, EtO-P, EtO, N_3 88%

Gajda, T.; Matusiak, M. *Synthesis*, ***1992***, 367.

Me, NMe_2 — 1. *t*-BuLi, pentane, RT; 2. EtSSEt, THF, 0°C; 3. $NaBO_3 \cdot 4H_2O$, AcOH, RT, 15 h — Me, NMe_2, S-Et, O 13% + Me, NMe_2, S-Et, O 76%

Shimazaki, M.; Takahashi, M.; Komatsu, H.; Ohta, A.; Kajii, K.; Kodama. Y. *Synthesis*, ***1992***, 555.

SECTION 384: OXIDES - ESTERS

CO_2Me — 1. LDA, THF, -78°C; 2. $TolSO_2F$, THF, RT, 3 h; 3. aq. NH_4Cl — CO_2Me, SO_2Tol 87%

Kende, A.S.; Mendoza, J.S. *J. Org. Chem.*, ***1990***, *55*, 1125.

CO_2Me — $EtNO_2$, K_2CO_3, 2 h, Aliquat 336,)))))) — CO_2Me, NO_2 90%

Jouglet, B.; Blanco, L.; Rousseau, G. *SynLett*, ***1991***, 907.

O, Tol, S, CO_2t-Bu, CO_2H — CH_2Cl_2, -20°C — O, S, Tol, CO_2t-Bu, CO_2H + O, S, Tol, CO_2H, CO_2t-Bu 75-85%

(89 : 6)

Alonso, I.; Carretero, J.C.; Ruano, J.L.G. *Tetrahedron Lett.*, ***1991***, *32*, 947.

PhO_2S, Me, OH — 1. $CH_2(OMe)_2$, P_2O_5, $CHCl_3$; 2. BuLi, THF, -78°C; 3. Amberlite IR, MeOH, CF_3SO_3H — PhO_2S, O, O, Me 42%

Carretero, J.C.; Rojo, J. *Tetrahedron Lett.*, ***1992***, *33*, 7407.

Amberlyst A21, 1 h; MeOH, reflux

NO_2 → CO_2Me / NO_2 96%

Ballini, R.; Petrini, M.; Polzonetti, V. *Synthesis*, ***1992***, 355.

SECTION 385: OXIDES - ETHERS, EPOXIDES, THIOETHERS

1. PhSeBr, THF, -78°C
2. $HgCl_2$, $AgNO_2$ MeCN
3. H_2O_2, MeOH NaOH, 5 min

NO_2 O 34%

Nájera, C.; Yus, M.; Karlsson, U.; Gogoll, A.; Bäckvall, J-E. *Tetrahedron Lett.* ***1990***, *31*, 4199.

$TolSO_2CH_2Cl$ → (OEt; Bu_3SnH, AIBN, PhH) → $TolO_2S$ OEt 40%

Renaud, P. *Tetrahedron Lett.*, ***1990***, *31*, 4601.

PhSH, MeCN, RT; 0.1 eq. NEt_3

NO_2 SPh (88:12 trans:cis, 88%)

NO_2

1. PhS^-, THF, RT
2. AcOH, -78°C

NO_2 SPh (0:100 trans:cis, 88%)

Kamimura, A.; Sasatani, H.; Hashimoto, T.; Kawai, T.; Hori, K.; Ono, N. *J. Org. Chem.*, ***1990***, *55*, 2437.

O O → (Br; PhO_2S Br; NaH, *t*-BuOK, THF) → O O SO_2Ph 80%

Padwa, A.; Murphree, S.S.; Yeske, P.E. *J. Org. Chem.*, ***1990***, *55*, 4241.

Me Ph $SiMe_3$ SO_2Tol → (1. MeLi 2. SO_2) → Ph Me S SO_2Tol 40%

Braverman, S.; van Asten, P.F.T.M.; van der Linden, J.B.; Zwanenburg, B. *Tetrahedron Lett.*, ***1991***, *32*, 3867.

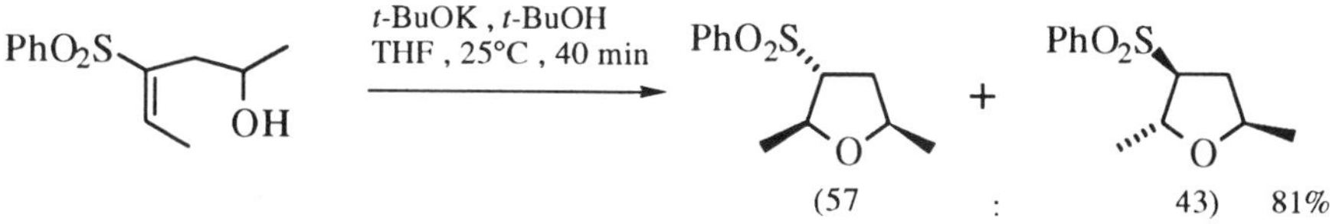

t-BuOK , t-BuOH
THF , 25°C , 40 min

(57 : 43) 81%

Craig, D.; Smith, A.M. *Tetrahedron Lett.*, ***1992***, *33*, 695.

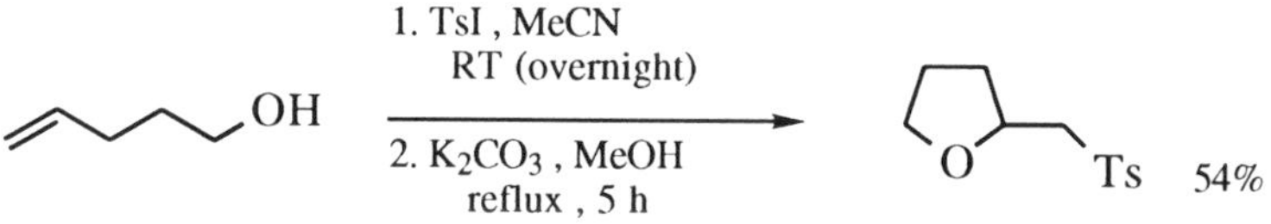

1. TsI , MeCN
RT (overnight)
2. K_2CO_3 , MeOH
reflux , 5 h

54%

Edwards, G.L.; Walker, K.A. *Tetrahedron Lett.*, ***1992***, *33*, 1779.

SECTION 386: OXIDES - HALIDES, SULFONATES

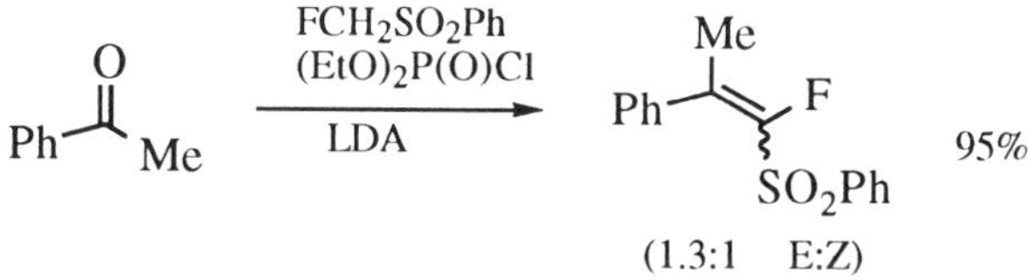

FCH_2SO_2Ph
$(EtO)_2P(O)Cl$
LDA

95%

(1.3:1 E:Z)

McCarthy, J.R.; Matthews, D.P.; Edwards, M.L.; Stemerick, D.M.; Jarvi, E.T. *Tetrahedron Lett.*, ***1990***, *31*, 5449.

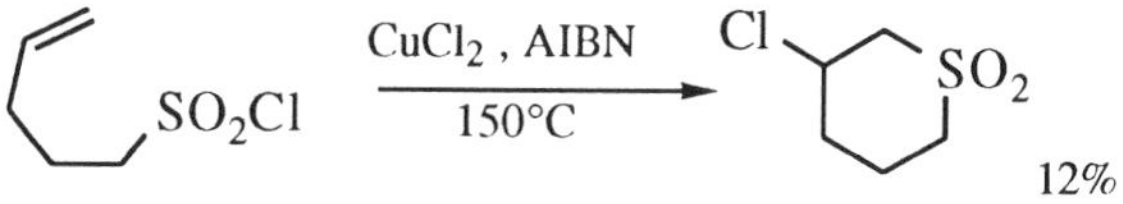

$CuCl_2$, AIBN
150°C

12%

Culshaw, P.N.; Walton, J.C. *Tetrahedron Lett.*, ***1990***, *31*, 6433.

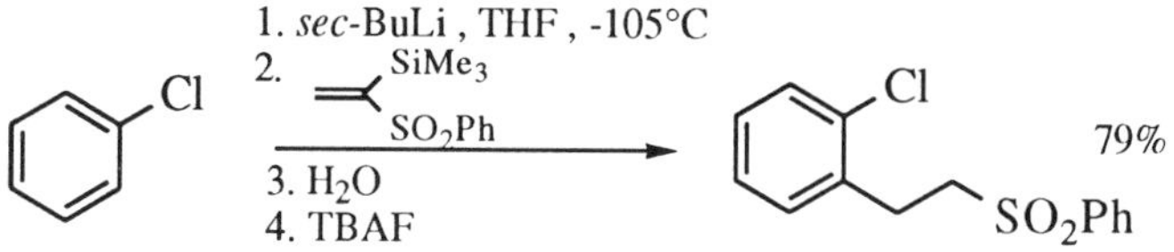

1. *sec*-BuLi , THF , -105°C
2. $SiMe_3$ / SO_2Ph
3. H_2O
4. TBAF

79%

Iwao, M. *J. Org. Chem.*, ***1990***, *55*, 3622.

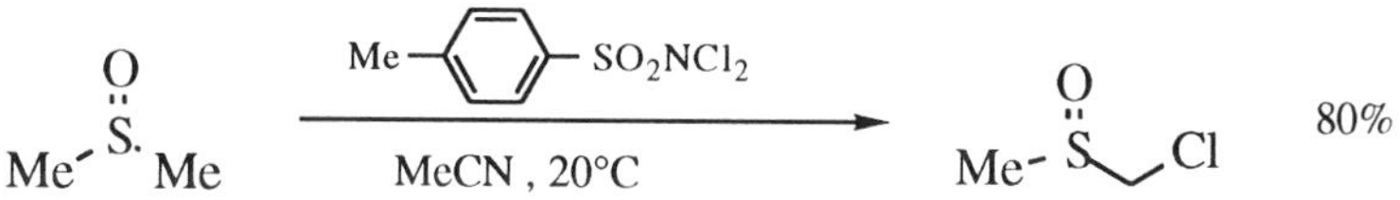

Me-C_6H_4-SO_2NCl_2
MeCN , 20°C

80%

Kim, Y.H.; Lim, S.C.; Kim, H.R.; Yoon, D.C. *Chem. Lett.*, ***1990***, 79.

F, I, F (C) – P(O)(OEt)(OEt) → (1-pentene, $Pd(PPh_3)_4$, RT) → I, F, F, P(O)(OEt)(OEt) product 85%

Yang, Z-Y.; Burton, D.J. *Tetrahedron Lett.*, ***1991***, *32*, 1019.

Bz_2O_2, $PhSO_2CH_2I$
100°C (sealed tube)
PhH, 5 h

SO_2Ph, I 66%

Masnyk, M. *Tetrahedron Lett.*, ***1991***, *32*, 3259.

HO, N → 1. $NaHCO_3$, aq. dioxane RT, 48 h; trichloroisocyanuric acid (Cl-N, O, Cl); 2. aq. NaOH → Cl, NO_2 78%

Walters, T.R.; Zajac, W.W. Jr.; Woods, J.M. *J. Org. Chem.*, ***1991***, *56*, 316.

EtO-P(O)(EtO)-CHF_2
1. LDA, THF
2. $C_6H_{13}CHO$, THF, -78°C
3. PhOC(O)Cl
4. Bu_3SnH, AIBN reflux

EtO-P(O)(OEt)-CF_2-CH_2-C_6H_{13} 73%

Martin, S.F.; Dean, D.W.; Wagman, A.S. *Tetrahedron Lett.*, ***1992***, *33*, 1839.

SECTION 387: OXIDES - KETONES

Ph, O, S, Me, *t*-Bu, *t*-Bu
1. LDA, THF -78°C
2. cyclopentenone
3. aq. NH_4Cl

Ph, O, S, Me, H, O, *t*-Bu, *t*-Bu 54%

(85:15 diastereomeric ratio)

Swindell, C.S.; Blase, F.R.; Eggleston, D.S.; Krause, J. *Tetrahedron Lett.*, ***1990***, *31*, 5409.

1. Ph-C≡CH , heat
2. heat ,

96%

Pfeifer, K-P.; Himbert, G. *Tetrahedron Lett.*, ***1990***, *31*, 5725.

PhO_2S Br Br

excess NaOMe
MeOH , 25°C

SO_2Ph CO_2Me 52%

Murphree, S.S.; Muller, C.L.; Padwa, A. *Tetrahedron Lett.*, ***1990***, *31*, 6145.

1. LDA , THF , -78°C
2. Ph_2PCl , THF , -78°C
3. O_2 , THF , RT

88%

Mikolajczyk, M.; Kiełbasiński, P.; Wieczorek, M.W.; Blaszczyk, J.; Kolbe, A. *J. Org. Chem.*, ***1990***, *55*, 1198.

1. LDA ; n-$C_9H_{19}CHO$
2. LDA

n-C_9H_{19} 93%

Satoh, T.; Fujii, T.; Yamakawa, K. *Bull. Chem. Soc. Jpn.*, ***1990***, *63*, 1266.
Satoh, T.; Hayashi, Y.; Mizu, Y.; Yamakawa, K. *Tetrahedron Lett.*, ***1992***, *33*, 7181.

1. CH_2Cl_2 , cat. $(\ldots CO_2)_4 Rh_2$ (N-SO_2Ph)
2. aq. NaOH
3. aq. NH_4Cl

SO_2Ph Me >90% (30% ee)

Kennedy, M.; McKervey, M.A.; Maguire, A.R.; Roos, G.H.P. *J. Chem. Soc., Chem. Commun.*, ***1990***, 361.

$MeNO_2$, t-BuOK , DMSO
RT

NO_2 55%

Jäger, V.; Seidel, B.; Guntrum, E. *Synthesis*, ***1991***, 629.

$OSiMe_2t$-Bu
Ph
DMF, RT, 1 d
2.5 eq. $Mn(pic)_3$
O_2N Ph O 74%
+ Ph O 6%
$OSiMe_2t$-Bu

$Mn(pic)_3$ = manganese (III) 2-pyridinecarboxylate

Narasaka, K.; Iwakura, K.; Okauchi, T. *Chem. Lett.*, ***1991***, 423.

2 eq. $Mn(OAc)_3 \cdot 2\ H_2O$
$Cu(OAc)_2 \cdot H_2O$, 25°C
AcOH, 14 h
40%

Snider, B.B.; Wan, B.Y-F.; Buckman, B.O.; Foxman, B.M. *J. Org. Chem.*, ***1991***, *56*, 328.

KH, DMSO, 20°C
45 min
74%

carbanion accelerated Claisen rearrangement

Denmark, S.E.; Marlin, J.E. *J. Org. Chem.*, ***1991***, *56*, 1003.

60% aq. AcOH, 100°C
$PhSO_2Na$, 12 h
63%

Padwa, A.; Bullock, W.H.; Dyszlewski, A.D.; McCombie, S.W.; Shankar, B.B.; Ganguly, A.K. *J. Org. Chem.*, ***1991***, *56*, 3556.
Padwa, A.; Bullock, W.H.; Dyszlewski, A.D. *J. Org. Chem.*, ***1990***, *55*, 955 [with vinyl sulfides].

1. LDA, ether
2. $(EtO)_2POCl$
3. air
79%

Lee, K.; Wiemer, D.F. *J. Org. Chem.*, ***1991***, *56*, 5556.

1. $P(OEt_3)$
2. NBS, CH_2Cl_2, -5°C
78%

McClure, C.K.; Grote, C.W. *Tetrahedron Lett.*, ***1991***, *32*, 5313.

O, O, S, Tol → $SnMe_3$, $TiCl_4$ / CH_2Cl_2, -78°C → O, H, O, S, Tol, H, Me + O, H, O, S, Tol, H, Me

(40 : 1) erythro+threo

Pan, L-R.; Tokoroyama, T. *Tetrahedron Lett.*, ***1992***, *33*, 1469.

O, NO_2 → 1. 3 eq. LDA, THF-HMPA -35°c; 2.EtI, -35°C → RT, 3 h → Et, O, NO_2 + Et, O, NO_2

(84 : 16) 65%

Ballini, R.; Bartoli, G.; Castagnani, R.; Marcantoni, E.; Petrini, M. *SynLett*, ***1992***, 64.

REVIEWS:

"Recent Progress in the Synthesis and Reactivity of Nitoketones"
Rosini, G.; Ballini, R.; Petrini, M.; Marotta, E.; Righi, P. *Org. Prep. Proceed. Int.*, ***1990***, *22*, 707.

SECTION 388: OXIDES - NITRILES

Ph- CO_2Et → 1. R(+)- Tol- S(O)- Me, LDA, THF; 2. EtAlCN, PhMe → Ph, HO, CN, S, Tol, O

90%

García Ruano, J.C.; Martín Castro, A.M.; Rodriguez, J.H. *Tetrahedron Lett.*, ***1991***, *32*, 3195.

SECTION 389: OXIDES - ALKENES

Me, SO_2Ph, S, O_2 → 1. *n*-BuLi; 2. Br; 3. Na(Hg), RT, H_3PO_4, 4 h → Me, S, O_2

63%

Chou, S-S.P.; Sun, C-M. *Tetrahedron Lett.*, ***1990***, *31*, 1035.

SO_2Ph, Br → 1. LDA, THF, -78°C; 2. Me, CO_2Et; 3. aq. HCl → PhO_2S, Me, CO_2Et

75%

Ghera, E.; Yechezkel, T.; Hassner, A. *Tetrahedron Lett.*, ***1990***, *31*, 3653.

SMe, SO_2Tol, Et — SiO_2 / $CDCl_3$, RT, 100 min → Et, SMe, $TolO_2S$ + MeS, Et, $TolO_2S$

(95 : 5) 72%

Ogura, K.; Yahata, N.; Fujimori, T.; Fujita, M. *Tetrahedron Lett.*, ***1990***, *31*, 4621.

O, P, O, Ph — 1. 2 LDA, THF, -78°C / 2. 2 TIPS OTf, -78°C → Ph, P-OH, O 81%

De, B.; Corey, E.J. *Tetrahedron Lett.*, ***1990***, *31*, 4831.

SO_2Ph — 3 eq. OiBu / BuSSBu, hν → OiBu, PhO_2S 81% (67:33 cis:trans)

with MeSSMe, 31:69 cis:trans, 66%

Singleton, D.A.; Church, K.M.; Lucero, M.J. *Tetrahedron Lett.*, ***1990***, *31*, 5551.

Ph, O, S, Me, TBDMSN — 1. 2.2 eq. LiTMP, THF / 2. 5 eq. DMF / 3. TsCl / 4. $Bu_2Cu(CN)Li_2$, ether, -40°C → 0°C → Ph, O, S, TBDMSN, Bu 48%

Paley, R.S.; Snow, S.R. *Tetrahedron Lett.*, ***1990***, *31*, 5853.

NO_2 — HC≡CCO_2Me, KF / DMSO, Bu_4NCl → NO_2, CO_2Me 90%

Anderson, D.A.; Hwu, J.R. *J. Org. Chem.*, ***1990***, *55*, 511.

Ph, O, S, O, O — $Pd_2(dba)_3$, $P(OEt)_3$ / PhH, dioxane, RT / 18 h → Ph, O, O, S, O, O

Tamaru, Y.; Nagao, K.; Bando, T.; Yoshida, Z. *J. Org. Chem.*, ***1990***, *55*, 1823.

$TolSO_2SePh$, $CHCl_3$
AIBN , reflux

$C\equiv C-H$ → SO_2Tol, SePh

92% (8:1 E:Z)

Back, T.E.; Lai, E.K.Y.; Muralidharan, K.R. *J. Org. Chem.*, ***1990***, *55*, 4595.

hν , BuSSBu
neat

SO_2Ph → PhO_2S

Singleton, D.A.; Church, K.M. *J. Org. Chem.*, ***1990***, *55*, 4780.

1. $(EtO)_2PCl$, NEt_3
ether
2. 80°C , 3 h

Ph, CO_2Me, OH → EtO-P(O)(EtO), Ph, CO_2Me 68%

Janecki, T.; Bodalski, R. *Synthesis*, ***1990***, 799.

1. EtO·P(O)(OEt)·Cl , -78°C
2. PhCHO

PhO_2S–C(Li)(Li)–Cl → Ph, Cl, SO_2Ph 82%

(93:7 E:Z)

Lee, J.W.; Oh, D.Y. *Synth. Commun.*, ***1990***, *20*, 273.

1. Ph-ethylene, $PhSO_2I$, hν
CCl_4 , 8 h
2. NEt_3

CO_2Me, SO_2Ph → PhO_2S, Ph, CO_2Me 58%

Harvey, I.W.; Phillips, E.D.; Whitham, G.H. *J. Chem. Soc., Chem. Commun.*, ***1990***, 481.

5% $Pd(OAc)_2$
30% $P(OiPr)_3$
dioxane , BSA
100°C

O, $SiMe_3$, OAc, SO_2Ph → SO_2Ph + Me, H, SO_2Ph

(1 : 7.2) 72%

Trost, B.M.; Grese, T.A. *J. Am. Chem. Soc.*, ***1991***, *113*, 7363.

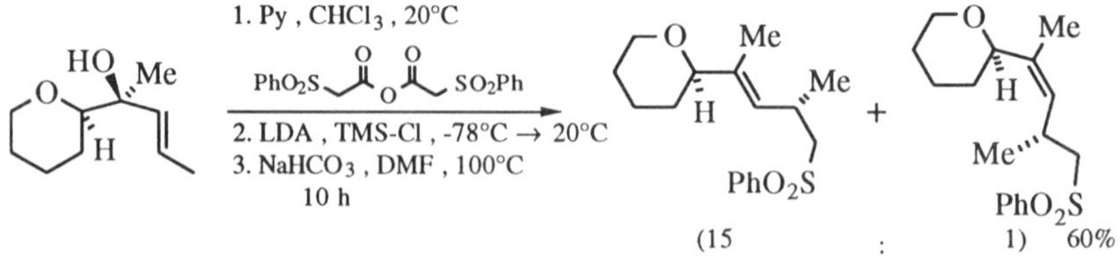

HO Me O H

1. Py, $CHCl_3$, 20°C

PhO_2S–CH_2–C(O)–O–C(O)–CH_2–SO_2Ph

2. LDA, TMS-Cl, -78°C → 20°C

3. $NaHCO_3$, DMF, 100°C

10 h

O Me H Me PhO_2S + O Me H Me PhO_2S

(15 : 1) 60%

Davidson, A.H.; Eggleton, N.; Wallace, I.H. *J. Chem. Soc., Chem. Commun.*, ***1991***, 378.

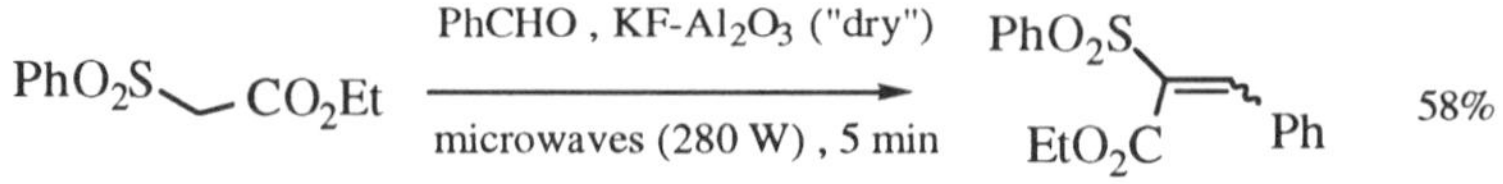

PhO_2S–CH_2–CO_2Et → PhCHO, $KF-Al_2O_3$ ("dry"), microwaves (280 W), 5 min → PhO_2S, EtO_2C, Ph 58%

Villemin, D.; Ben Alloum, A. *Synth. Commun.*, ***1991***, *21*, 63.

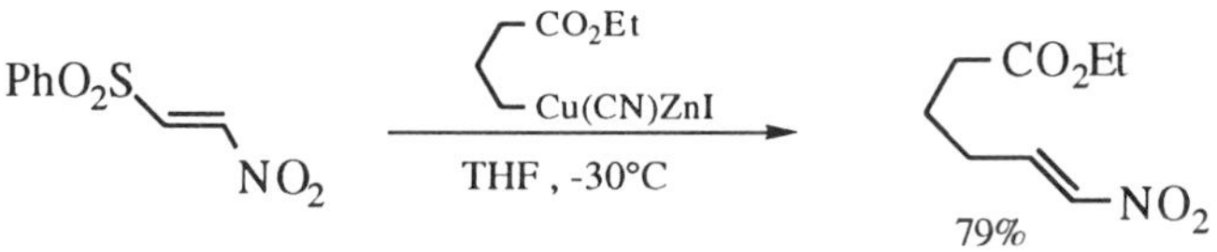

PhO_2S–CH=CH–NO_2 → CO_2Et, Cu(CN)ZnI, THF, -30°C → CO_2Et, NO_2 79%

Retherford, C.; Knochel, P. *Tetrahedron Lett.*, ***1991***, *32*, 441.

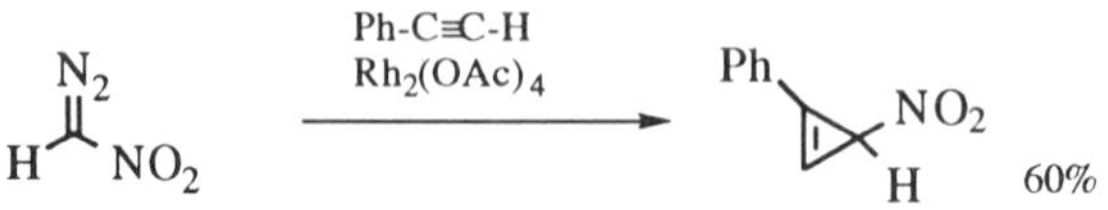

N_2, H, NO_2 → Ph-C≡C-H, $Rh_2(OAc)_4$ → Ph, NO_2, H 60%

O'Bannon, P.E.; Dailey, W.P. *J. Org. Chem.*, ***1991***,*56*, 2258.

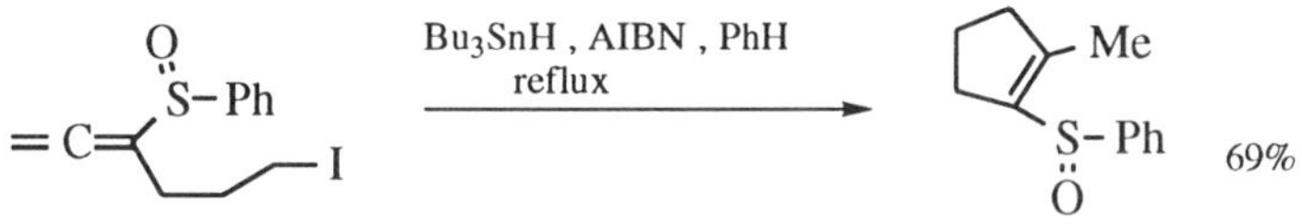

=C=, O, S–Ph, I → Bu_3SnH, AIBN, PhH, reflux → Me, S–Ph, O 69%

Crandall, J.K.; Ayers, T.A. *Tetrahedron Lett.*, ***1991***, *32*, 3659.

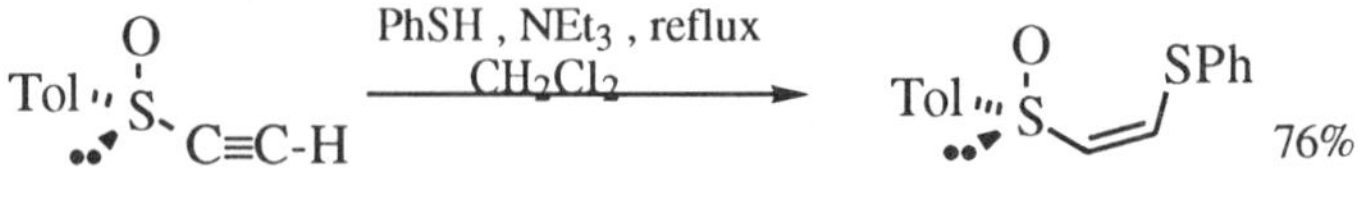

Tol, O, S, C≡C-H → PhSH, NEt_3, reflux, CH_2Cl_2 → Tol, O, S, SPh 76%

Guillot, C.; Maignan, C. *Tetrahedron Lett.*, ***1991***, *32*, 4907.

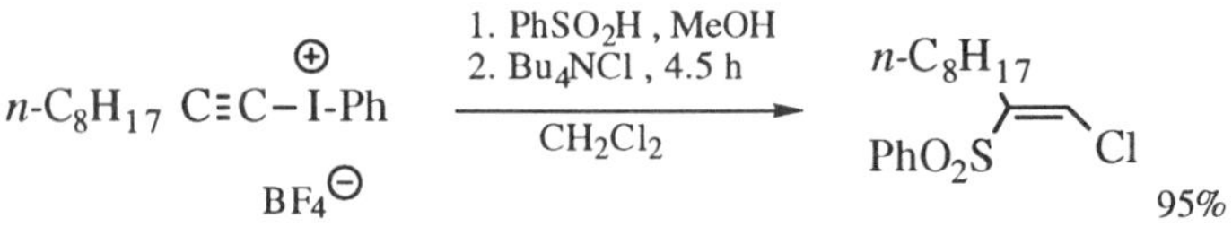

n-C_8H_{17} C≡C–$I^{\oplus}$-Ph $BF_4^{\ominus}$ → 1. $PhSO_2H$, MeOH; 2. Bu_4NCl, 4.5 h; CH_2Cl_2 → n-C_8H_{17}, PhO_2S, Cl 95%

Ochiai, M.; Oshima, K.; Masaki, Y. *Tetrahedron Lett.*, ***1991***, *32*, 7711.

1. $TolSO_2I$, CH_2Cl_2 , RT
2. 1N NaOH

$TolO_2S$ ... n-C_6H_{13} 75%

n-C_6H_{13}

(70:30 Z:E)

Vaultier, M.; El Louzi, A.; Titouani, S.L.; Soufiaoui, M. *SynLett,* ***1991*** 267.

Bu_2CuLi , THF , -5°C

Ph S Br → Ph S Bu 65%

(80:20 E:Z)

Cardellicchio, C.; Fiandnese, V.; Naso, F. *J. Org. Chem.,* ***1992***, *57*, 1718.

alumina , CH_2Cl_2

40°C , 7 h

OH NO_2 → NO_2 77%

Ballini, R.; Castagnani, R.; Petrini, M. *J. Org. Chem.,* ***1992***, *57*, 2160.

$Pd(dppe)_2$, BSA
dioxane

PhO_2S SO_2Ph OAc → PhO_2S SO_2Ph + PhO_2S SO_2Ph

(20 : 80) 44%

Trost, B.M.; Vos, B.A.; Brzezowski, C.M.; Martina, D.P. *Tetrahedron Lett.,* ***1992***, *33*, 717.

1. EtBr
2. NaOEt , THF

Ph_3P O Me → Ph–P(O) Ph OEt Me 86%

Ruder, S.M.; Norwood, B.K. *Tetrahedron Lett.,* ***1992***, *33*, 861.

1. LDA , THF , -78°C
2. t-$BuMe_2SiOTf$, -78°C

O O S Tol → t-$BuMe_2SiO$ O S Tol 85%

Solladié, G.; Maugein, N.; Morreno, I.; Almario, A.; Carreño, M.C.; Garcia-Ruano, J.L. *Tetrahedron Lett.,* ***1992***, *33*, 4561.

Solladié, G.; Ruiz, P.; Colobert, F.; Carreño, M.C.; Garcia-Ruano, J.L. *Synthesis,* ***1991***, 1011.

Me Me; SO_2Ph SO_2Ph, 3% dppp, THF; 1% $[(\pi\text{-}C_3H_7)PdCl]_2$, 3% NaOMe, 100°C; Me Me Me SO_2Ph SO_2Ph 95%

Trost, B.M.; Zhi, L. *Tetrahedron Lett.*, ***1992***, *33*, 1831.

Tol O S O; 1. NEt_3, DMF, 20°C 2. Me_3SiCl; Tol O S Me_3SiO 51%

Bonfand, E.; Gosselin, P.; Maignan, C. *Tetrahedron Lett.*, ***1992***, *33*, 2347.

H-C≡C-S O Tol; ZnI_2, PhH, RT, 15 h; I S O Tol 87%

de la Pradilla, R.F.; Morente, M.; Paley, R.S. *Tetrahedron Lett.*, ***1992***, *33*, 6101.

OMe; 20 eq. TsNa, 4 eq. $Cu(OAc)_2$; OMe Ts 56%

Chuang, C-P. *Tetrahedron Lett.*, ***1992***, *33*, 6311.

C≡C-H; 1. O_2, PhSh 2. oxone 3. PhSCl, NEt_3, CH_2Cl_2 4. NEt_3, CH_2Cl_2 5. H_2O_2, AcOH; SO_2Ph SO_2Ph <70%

Ni, Z.; Wang, X.; Rodriguez, A.; Padwa, A. *Tetrahedron Lett.*, ***1992***, *33*, 7303.

REVIEWS:

"The Chemistry of Vinyl Sulphones"
Simpkins, N.S. *Tetrahedron*, ***1990***, *46*, 6951.

Symposia-in-Print: "Nitroalkanes and Nitroalkenes in Synthesis"
Barrett, A.G.M. *Tetrahedron*, ***1990***, *46*, 7313-5798.

"1,2-bis(Arylsulfonyl)alkenes"
Cossu, S.; DeLucchi, O.; Fabbri, D.; Licini, G.; Pasquato, L. *Org. Prep. Proceed Int.*, ***1991***, *23*, 573.

SECTION 390: OXIDES - OXIDES

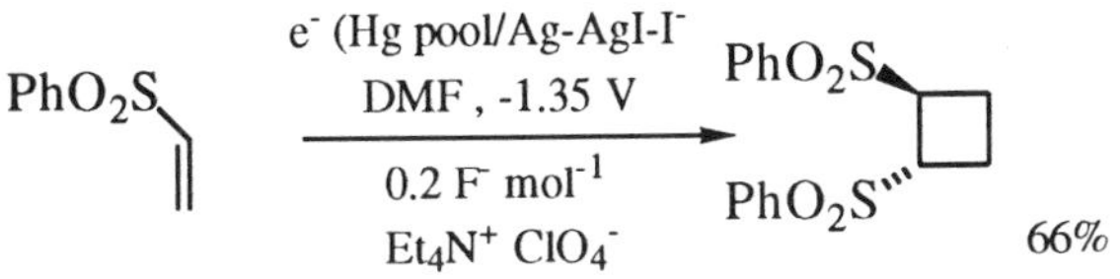
PhO2S
e- (Hg pool/Ag-AgI-I-
DMF , -1.35 V
0.2 F- mol-1
Et4N+ ClO4-
PhO2S
PhO2S
66%

Delaunay, J.; Mabon, G.; Orliac, A.; Simonet, J. *Tetrahedron Lett.*, ***1990***, *31*, 667.

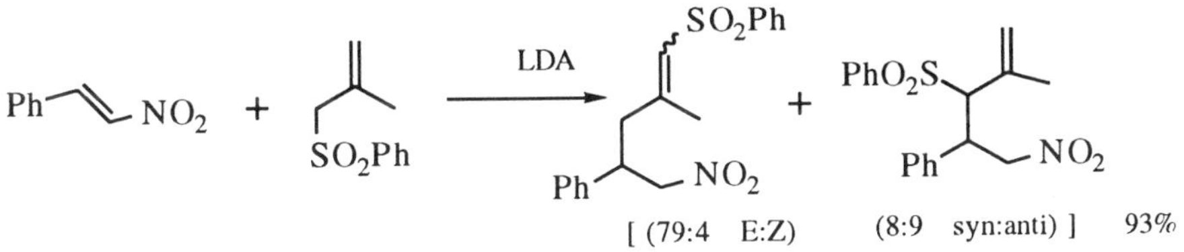
Ph
NO2
+
SO2Ph
LDA
SO2Ph
Ph
NO2
+
PhO2S
Ph
NO2
[(79:4 E:Z)
(8:9 syn:anti)]
93%

Ghera, E.; Ben-Yaakov, E.; Yechezkel, T.; Hassner, A. *Tetrahedron Lett.*, ***1992***, *33*, 2741.

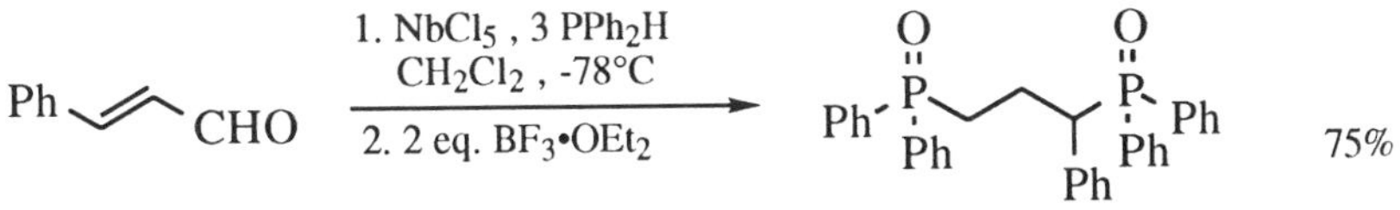
Ph
CHO
1. NbCl5 , 3 PPh2H
CH2Cl2 , -78°C
2. 2 eq. BF3•OEt2
O
O
Ph
P
Ph
P
Ph
Ph
Ph
75%

Hashimoto, T.; Maeta, H.; Matsumoto, T.; Morooka, M.; Ohba, S.; Suzuki, K. *SynLett*, ***1992***, 340.